T0309779

NATURAL PRODUCTS
IN CHEMICAL BIOLOGY

NATURAL PRODUCTS IN CHEMICAL BIOLOGY

Edited by

NATANYA CIVJAN

A JOHN WILEY & SONS, INC., PUBLICATION

Library of Congress Cataloging-in-Publication Data:

Natural products in chemical biology / edited by Natanya Civjan.
    p. cm.
  Includes bibliographical references and index.
  ISBN 978-1-118-10117-9 (hardback)
  1. Biological products. 2. Natural products. 3. Biochemistry. I. Civjan, Natanya.
  QH345.N3466 2012
  572–dc23

                                                                    2011049643

Printed in the United States of America

10 9 8 7 6 5 4 3 2 1

CONTENTS

PREFACE

Natural products have long been of interest to the fields of chemistry, biology, and medicine, because of their immense molecular diversity, biological activities, and medicinal properties. Interdisciplinary studies have yielded further insights into the ecological significance, biological roles, and therapeutic potential of these important molecules.

Containing carefully selected reprints from the award-winning *Wiley Encyclopedia of Chemical Biology* as well as new contributions, this book features a collection of chapters on topics in natural products research. The chapters cover the diversity of natural products in plants, marine invertebrates, and microbes, the biosynthetic pathways of some major classes of natural products, and the role of natural products in drug discovery and medicine.

Written by prominent scholars, this book will be of interest to advanced students and researchers, both in academia and industry, entering the field of chemical biology or related fields such as biochemistry, medicine, and pharmaceutical sciences.

NATANYA CIVJAN

CONTRIBUTORS

Toshiyuki Akiyama, Natural Products Research Laboratories, School of Pharmacy, University of North Carolina, Chapel Hill, North Carolina

Mariana Ávalos, Instituto de Investigaciones Biomédicas, Universidad Nacional Autónoma de México, México

Adelbert Bacher, Institute of Biochemistry, Department Chemie, Technische Universität München, München, Germany

Jörg Bohlmann, Michael Smith Laboratories and Departments of Botany and Forest Sciences, University of British Columbia, British Columbia, Canada

Tonia J. Buchholz, Life Sciences Institute, University of Michigan, Ann Arbor, Michigan

Russell J. Cox, School of Chemistry, University of Bristol, Bristol, United Kingdom

Wolfgang Eisenreich, Institute of Biochemistry, Department Chemie, Technische Universität München, München, Germany

Markus Fischer, Institute of Food Chemistry, University of Hamburg, Hamburg, Germany

Nobuhiro Fusetani, Hokkaido University, Hakodate, Japan

A. Leslie Gunatilaka, Southwest Center for Natural Products Research, University of Arizona, Tucson, Arizona

Silvia Guzmán-Trampe, Instituto de Investigaciones Biomédicas, Universidad Nacional Autónoma de México, México

Ilka Haase, Institute of Food Chemistry, University of Hamburg, Hamburg, Germany

Alison M. Hill, School of Biosciences, University of Exeter, United Kingdom

Hideji Itokawa, Natural Products Research Laboratories, School of Pharmacy, University of North Carolina, Chapel Hill, North Carolina

Manuel Jiménez-Estrada, Instituto de Química, Universidad Nacional Autónoma de, México

Christopher I. Keeling, Michael Smith Laboratories, University of British Columbia, British Columbia, Canada

David G. I. Kingston, Department of Chemistry, Virginia Polytechnic Institute and State University, Blacksburg, Virginia

Jeffrey D. Kittendorf, University of Michigan, Ann Arbor, Michigan

Kuo-Hsiung Lee, Natural Products Research Laboratories, School of Pharmacy, University of North Carolina, Chapel Hill, North Carolina

Mohamed A. Marahiel, Philipps-University of Marburg, FB Chemie/Biochemie, Marburg, Germany

Susan L. Morris-Natschke, Natural Products Research Laboratories, School of Pharmacy, University of North Carolina, Chapel Hill, North Carolina

David J. Newman, Natural Products Branch, Developmental Therapeutics Program, Division of Cancer Treatment and Diagnosis, National Cancer Institute, Frederick, Maryland

Sarah E. O'Connor, John Innes Centre, Norwich Research Park, Colney, Norwich

Romina Rodríguez-Sanoja, Instituto de Investigaciones Biomédicas, Universidad Nacional Autónoma de México, México

Felix Rohdich, Institute of Biochemistry, Department Chemie, Technische Universität München, München, Germany

Beatriz Ruiz, Instituto de Investigaciones Biomédicas, Universidad Nacional Autónoma de México, México

Georg Schoenafinger, Philipps-University of Marburg, FB Chemie/Biochemie, Marburg, Germany

Sergio Sanchez, Instituto de Investigaciones Biomédicas, Universidad Nacional Autónoma de México, México

David H. Sherman, University of Michigan, Ann Arbor, Michigan

Thomas J. Simpson, School of Chemistry, University of Bristol, Bristol, United Kingdom

Sheo B. Singh, Merck Research Laboratories, Rahway, New Jersey

Sergey B. Zotchev, Department of Biotechnology, Norwegian University of Science and Technology, Norway

PART I

CHEMICAL DIVERSITY OF NATURAL PRODUCTS

1

PLANT NATURAL PRODUCTS

A. LESLIE GUNATILAKA

Southwest Center for Natural Products Research, University of Arizona, Tucson, Arizona

Plants contain numerous natural products (secondary metabolites) that may not participate directly in their growth and development but play an important role in ecological
interactions with other organisms. Despite immense chemical diversity, which originates from simple carbohydrates produced because of photosynthesis, plant natural
products are formed from only a few biosynthetic building blocks that consist of
acetate, mevalonate, and shikimate. These basic building blocks undergo a variety
of biosynthetic transformations and combinations that lead to numerous classes of
plant natural products including, but not limited to, carbohydrates, fatty acids and
their esters, aromatic polyketides (phenols and quinones), terpenoids and steroids,
phenyl propanoids (lignans and lignin, coumarins, flavonoids, and isoflavonoids), and
alkaloids. Summarized in this chapter are representative members of these important
classes of plant natural products with special emphasis on their chemical diversity.
The chapter concludes with a brief discussion on recent methods for the maximization
of chemical diversity and the production of natural products from plants.

The number of different plant species on the surface of the earth has been
estimated to be over 250,000 (1, 2), and only a fraction of these have been
investigated for their constituent natural products (3). Plants are known to produce over 100,000 natural products (4). However, according to Verpoorte (5)
"extrapolations of the number of species studied and the number of compounds
known suggests that, from all plant species, at least a million different compounds
could be isolated". The vast majority of these compounds, commonly referred
to as secondary metabolites, does not seem to participate directly in the growth
and development of plants (6). In their natural environments, plants coexist and

Natural Products in Chemical Biology, First Edition. Edited by Natanya Civjan.
© 2012 John Wiley & Sons, Inc. Published 2012 by John Wiley & Sons, Inc.

interact with other organisms in a variety of ecosystems (7), and the possible roles that these natural products play in plants, especially in the context of ecological interactions, are being speculated about, appreciated, and debated (8, 9). Although the functional role that secondary metabolites play in the producing organism is a matter of controversy (10, 11), the chemical diversity of plant natural products is well recognized, and it has been suggested that the chemical diversity of plant natural products is far greater than their functional diversity (9, 10, 12, 13).

Two models exist to explain the abundant chemical diversity of plant natural products. In the first model, secondary metabolites produced by plants are believed to be involved in physiological responses during the interactions with their biotic and abiotic environments, especially as elements of their defense arsenals. In this model, the diversity of compounds produced by plants is explained by considering the great diversity of plant life strategies and the vast number of accompanying defense strategies (14–16). The second model is an evolutionary model that makes the assumption that potent biological activity is a rare property for any natural product to possess and therefore is of no value to the producer organism (8). This model, based on the Screening Hypothesis of Jones and Firn (10, 17, 18), suggests that organisms that make and "screen" many chemicals will have an increased likelihood of enhanced fitness simply because the greater the chemical diversity the greater the chances of producing the rare metabolites with useful and potent biological activities.

1.1   ORIGIN OF NATURAL PRODUCTS IN PLANTS

It is intriguing that despite the immense chemical diversity exhibited by them, plant natural products are derived from only a few building blocks: acetate (which contains two carbon atoms), mevalonate (which contains five carbon atoms), and shikimate (which contains nine carbon atoms). These building blocks are derived in turn from simple carbohydrates produced because of the light-catalyzed reduction of atmospheric carbon dioxide by higher (green) plants during photosynthesis. The products formed by the condensation of the above building blocks (small biosynthetic units) are additionally elaborated ("tailored" or "decorated") by numerous enzyme-catalyzed reactions such as cyclization, elimination, rearrangement, reduction, oxidation, methylation, and so forth. Chemical diversity that results from these "decoration" reactions will be considered under each biosynthetic class of plant natural products.

1.2   CARBOHYDRATES

Although not as structurally diverse as other classes of natural products, carbohydrates are among the most abundant chemical constituents of plants. All animals and most microorganisms depend on plant-derived carbohydrates for their nourishment and survival. Simple carbohydrates (aldoses and ketoses), the first formed products of photosynthesis, are used by plants to make their food reserves, as starter units for the synthesis of plant secondary metabolites, and to make sugar derivatives (glycosides) of products of secondary metabolism. Plant carbohydrates consist of monosaccharides (pentoses and hexoses), disaccharides, oligosaccharides, and polysaccharides. Monosaccharides of plant origin include the stereoisomeric forms of hexose sugars β-D-glucose (**A1**), β-D-galactose (**A2**), β-D-mannose (**A3**), α-L-rhamnose (**A4**), β-D-xylose (**A5**), α-L-arabinose (**A6**), β-D-ribose (**A7**), and β-D-fructose (**A8**) (Fig. 1.1). L-ascorbic acid (**A9**), commonly known as vitamin C and occurring in most fresh fruits and vegetables, is a monosaccharide derived from D-glucose. As the name implies, disaccharides are formed from two monosaccharide units and contain a C−O−C link. Among the common plant dissacharides, maltose (**A10**) and lactose (**A11**) contain respectively C(1α) −O−C(4) and C(1β) −O−C(4) links formed between two D-glucose (**A1**) units, whereas sucrose (**A12**) contains the C(1α) −O−C(2β) link formed between D-glucose (**A1**) and D-fructose (**A8**) units (Fig. 1.1). Polysaccharides that are polymeric monosaccharides perform two

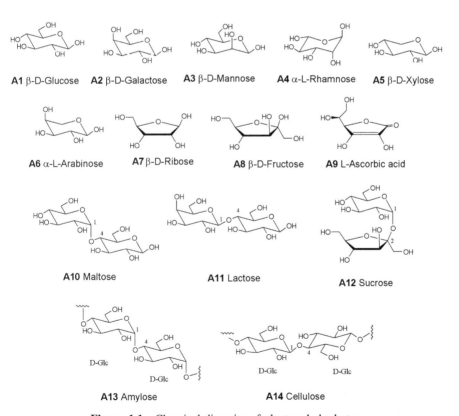

Figure 1.1 Chemical diversity of plant carbohydrates.

major functional roles, namely, the food reserves and structural elements in plants. Amylose (**A13**), an example of a storage polysaccharide, is a linear polymer that contains 1000–2000 C(1α)–C(4) linked glucopyranose units. Cellulose is an example of a structural polysaccharide composed of a linear chain of ca. 8000 residues of C(1β)–C(4) linked glucopyranose units. It is noteworthy that cellulose, the main constituent in plant cell walls, is the most abundant organic material on earth.

1.3   PRODUCTS OF ACETATE PATHWAY

The two-carbon precursor, acetyl coenzyme A (acetyl–CoA), is the initial substrate for synthesis of the carbon backbone of plant polyketides. As the name implies, polyketides are naturally occurring polymers of ketene (CH_2CO) and contain alternating carbonyl and methylene groups derived from the acetate pathway. Polyketides and their derivatives are ubiquitous and are found in all organisms known to produce secondary metabolites. Because of their immense structural diversity, a unified classification of polyketides has yet to emerge (19). Plant polyketides are represented by two major classes of metabolites: fatty acids and aromatic compounds.

1.3.1   Fatty Acids and Their Esters

Plants contain both saturated and unsaturated fatty acids mostly as esters of the trihydroxy alcohol, glycerol. As they are derived from the linear combination of acetate (C_2) units, common fatty acids possess an even number of carbon atoms and contain a straight chain. Thus, fatty acids that contain an odd number of carbons are rare in nature. Over 300 fatty acids belonging to 18 structural classes occur in plants (20). Among these, the more common are saturated fatty acids that contain 16 or 18 carbon atoms such as lauric acid, myristic acid, palmitic acid, and stearic acid, and the unsaturated analogs of stearic acid, namely, oleic acid and linoleic acid. Oils produced by many plants constitute glycerol esters of both saturated and unsaturated fatty acids. The other classes of fatty acids are defined by the number and arrangement of double or triple bonds and various other functional groups.

1.3.2   Aromatic Polyketides in Plants

Aromatic natural products of polyketide origin are less prevalent in plants compared with microorganisms. The majority of the plant constituents that contain aromatic structures are known to arise from the shikimate pathway. Unlike those derived from the shikimate pathway, aromatic products of the polyketide pathway invariably contain a *meta* oxygenation pattern because of their origin from the cyclization of polyketides. Phenolic compounds such as chrysophanol-anthrone (**B1**), and emodin-anthrone (**B2**), and the anthraquinones, aloe-emodin (**B3**) and

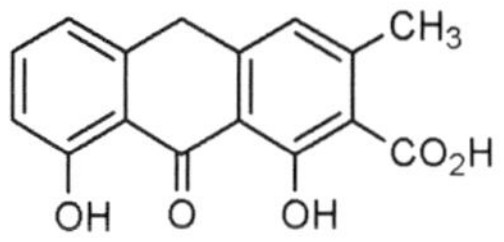

B1 Chrysophanol-anthrone 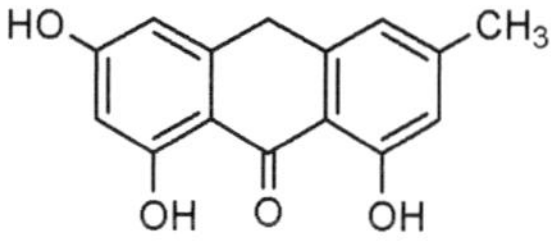   **B2** Emodin-anthrone 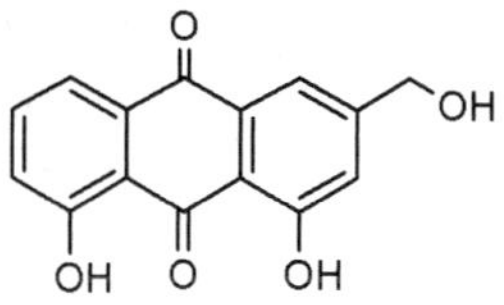   **B3** Aloe-emodin

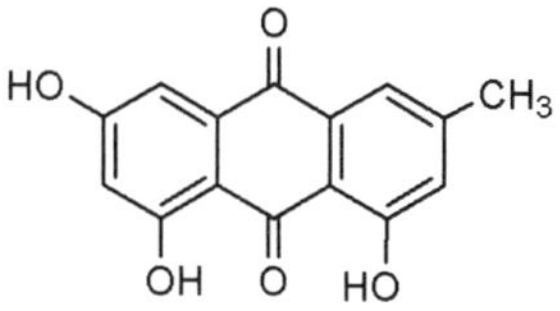

B4 Emodin 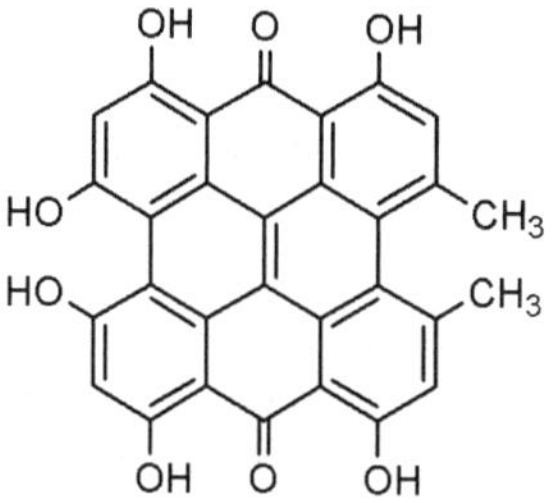   **B5** Hypericin

Figure 1.2 Examples of some aromatic polyketides in plants.

emodin (**B4**) (Fig. 1.2), are products of the polyketide pathway and are found to occur in some plants of the genera *Cassia* (Leguminosae) (21), *Rhamnus* (Rhamnaceae) (22), and *Aloe* (Liliaceae) (23). The dimer of emodin-anthrone (**B2**), namely hypericin, (**B5**) is a constituent of the antidepressant herbal supplement, St. John's wort (*Hypericum perforatum*, Hypericaceae) (24).

1.4 PRODUCTS OF MEVALONATE PATHWAY

Mevalonic acid, a six-carbon building block, is made up from three molecules of the most basic two-carbon precursor, acetyl–CoA. The mevalonate pathway, which involves the intermediary of mevalonic acid, directs acetate into a series of natural products different from those derived directly from the acetate pathway and includes terpenoids and steroids. Terpenoids constitute the most chemically diverse and one of the largest groups of plant natural products, and therefore a detailed discussion on this group of natural products is warranted.

Most terpenoids are derived from mevalonic acid (MVA) through the universal precursor isopentenyl diphosphate (IPP) and its allylic isomer dimethylallyl diphosphate (DMAPP). Thus, the vast majority of terpenoids contain the basic structural residue 2-methylbutane, often less precisely referred to as isoprene units. These C_5 hemiterpene units combine with each other in a variety of ways leading to mono- (C_{10}), sesqui- (C_{15}), di- (C_{20}), sester- (C_{25}), tri- (C_{30}), tetra- (C_{40}), and poly- (C_{5n} ($n = >8$)) terpenes. The primary products of condensation undergo more elaboration (reduction, oxidation, derivatization, etc.) and "decorations" that lead to terpenoid hydrocarbons, alcohols and their glycosides, ethers, aldehydes, ketones, and carboxylic acids and their esters, which makes terpenoids the most diverse class of plant natural products. It is noteworthy that over 40,000 different terpenoids have been isolated and characterized from natural sources including plants (25, 26). Terpenoids also represent a functionally diverse class of natural products. Although all the biological, ecological, and pharmacological functions of terpenoids are yet to be fully understood, they are known to have a variety of functions in the plant kingdom and in human health and nutrition. Many plants are known to produce volatile terpenes for the purpose of attracting specific insects for pollination or to keep away herbivorous animals; some plants produce toxic or bitter-tasting terpenes known as antifeedants to protect them from being eaten by animals. Most importantly, terpenoids also play functional roles in plants as growth regulators (phytohormones) and signaling compounds (sociohormones). Some important pharmacologically active terpenoids include the sesquiterpenoid artemisinin with antimalarial activity (27), the anticancer diterpenoid paclitaxel (Taxol®; Mead Johnson, Princeton, NJ) (28), and the terpenoid indole alkaloids vincristine and vinblastine with anticancer activity (29, 30).

1.4.1 Hemiterpenoids

Compared with other terpenoids, only a few true hemiterpenes are found in nature, and over 90 of these occur as glycosides (31). The most noteworthy example is isoprene (2-methylbut-2-ene), a volatile hemiterpenoid released by many trees. Other natural hemiterpenes include prenol (3-methyl-2-buten-1-ol) in flowers of *Cananga odorata* (Annonaceae) and hops (*Humulus lupulus*, Cannabaceae). Its isomer, (*S*)-(−)-3-methyl-3-buten-2-ol is a constituent of the essential oils of grapefruit, hops, and oranges. Another hemiterpenoid, 4-methoxy-2-methyl-2-butanthiol, is responsible for the characteristic flavor of blackcurrant (*Ribes nigrum*, Saxifragaceae). Numerous plant natural products with ester moieties contain hemiterpenoid-derived carboxylic acid components such as 3-methyl-2-butenoic acid and its isomers, angelic and tiglic acids, as well as its saturated analog, isovaleric acid. The immediate biosynthetic precursors of hemiterpenoids, IPP and DMAPP, are often used by plants as alkylating agents during the formation of some natural products (meroterpenoids) of mixed biosynthetic origin.

1.4.2 Monoterpenoids

Monoterpenoids are responsible for fragrances and flavors of many plants and thus their products are used in perfumery and as spices. To date over 1,500 monoterpenoids are known, and these constitute acyclic, monocyclic, and bicyclic monoterpenoids (32), which occur in nature as hydrocarbons, alcohols, aldehydes, and carboxylic acids and their esters. Several acyclic monoterpenoid hydrocarbons are known, and these include trienes such as β-myrcene (**C1**), α-myrcene **(C2)**, (*Z*)-α-ocimene **(C3)**, (*E*)-α-ocimene **(C4)**, (*Z*)-β-ocimene **(C5)**, and (*E*)-β-ocimene **(C6)**. β-Myrcene and β-ocimene are constituents of basil (*Ocimum basilicum*, Labiatae) and bay (*Pimenta acris*, Myrtaceae), pettitgrain (*Citrus vulgaris*, Rutaceae) leaves, strobiles of hops (*Humulus lupulus*, Cannabaceae), and several other essential oils. Unsaturated acyclic monoterpene alcohol constituents of plants and their derived aldehydes play a significant role in the perfume industry. Some common acyclic monoterpene alcohols and aldehydes include geraniol **(C7)**, linalool **(C8)** (a constituent of coriander oil), (*R*)-3,7-dimethyloctanol **(C9)** (of geranium oil), citranellol **(C10)** (of rose oil), geranial **(C11)** (of lemon oil), and citranellal **(C12)** (of citronella oil) (Fig. 1.3).

Monoterpene precursors undergo a variety of cyclization and rearrangement reactions leading to diverse monocyclic and bicyclic monoterpenoids that consist of hydrocarbons, alcohols, and ketones. Compared with cyclopentane and cyclohexane analogs, cyclopropane and cyclobutane monoterpenoids that contain irregular monoterpene carbon skeletons are rare in nature. (+)-*Trans*-chrysanthemic acid **(D1)** and (+)-*trans*-pyrethric acid **(D2)** esters, which are known to occur in flower heads of *Chrysanthemum cinerariaefolium* (Compositae), are two important examples of cyclopropane monoterpenoids. Noteworthy examples of cyclobutane monoterpenoids are (1*S*,2*S*)-fragranol **(D3)**, which occurs in the roots of *Artemisia fragrans* (Asteraceae), and junionone **(D4)**, which occurs in the fruits of the juniper tree (*Juniperus communis*, Cupressaceae) (Fig. 1.4).

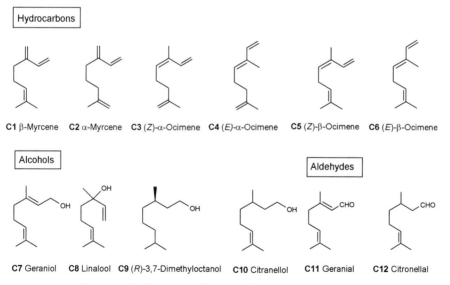

Figure 1.3 Structural diversity of acyclic monoterpenes.

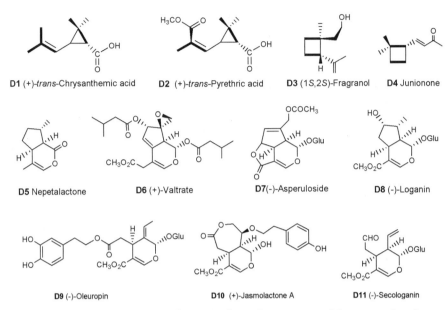

Figure 1.4 Chemical diversity of monocyclic cyclopropane, cyclobutane, and cyclopentane monoterpenoids in plants.

To date about 200 cyclopentane monoterpenoids are known (32), and the majority of these in plants occur as iridoids and *seco*-iridoids that contain the iridane carbon skeleton fused to a six-membered oxygen heterocycle. The simplest iridoid, (+)-nepetalactone (**D5**), is a constituent of the volatile oil of *Nepeta cataria* (Labiatae), which is known to be a powerful cat attractant and stimulant. Other well-known plant-derived iridoids consist of a diverse array of valepotriates known to occur in the popular herbal supplement valerian (*Valeriana officinalis*, Valerianaceae). Most of these valepotriates, including (+)-valtrate (**D6**), the constituent responsible for the tranquilizing properties of the valerian, contain several hydroxyl groups esterified with the C_5 hemiterpene isovaleric acid. Glucosides of iridoids also occur as plant constituents. Important examples are (−)-asperuloside (**D7**) with insect antifeedant activity in *Asperula odorata* (*Galium odoratum*, Rubiaceae) and many other plants and (−)-loganin (**D8**) from the fruits of *Strychnos nux vomica* (Loganiaceae). Although not as prevalent as iridoids, the *seco*-iridoids, (−)-oleuropin (**D9**), (+)-jasmolactone A (**D10**), and (−)-secologanin (**D11**) (Fig. 1.4), have been isolated from many parts of the olive tree (*Olea europaea*, Oleaceae), *Jasminium multiflorum* (Oleaceae) and *Strychnos nux vomica* (Loganiaceae), respectively.

Cyclohexane monoterpenes are a chemically diverse group of monoterpenoids that occur in the plant kingdom mainly as hydrocarbons, alcohols, ketones, aromatic hydrocarbons, and phenols (Fig. 1.5). The saturated hydrocarbon *trans*-*p*-menthane (**E1**) is a constituent of the oil of turpentine and the resin of pine (Pinaceae) trees. Its unsaturated analogs, namely (*R*)-(+)-limonene

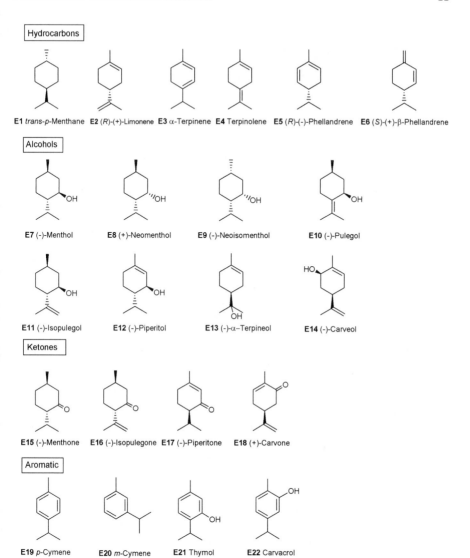

Figure 1.5 Chemical diversity of monocyclic cyclohexane and aromatic monoterpenes.

(**E2**) [present in oil of orange (*Citrus aurantium*) and mandarin (*Citrus reticulata*, Rutaceae) peel oil]; α-terpinene (**E3**) and terpinolene (**E4**) in some *Citrus*, *Juniperus*, *Mentha* and *Pinus* species; (R)-(−)-α-phellandrene (**E5**) in *Eucalyptus phellandra* (Myrtaceae); and (S)-(+)-β-phellandrene (**E6**) in water fennel (*Phellandrium aquaticum*, Umbelliferae), are components of many plant volatile oils. The rich chemical diversity of cyclohexane monoterpene alcohols is apparent from the natural occurrence of all four pairs of *p*-menthan-3-ol enantiomers, for example, (−)-menthol (**E7**) [a major component of peppermint

(*Mentha piperita*, Labiatae) oil], (+)-neomenthol (**E8**) [a constituent of Japanese peppermint (*Mentha arvensis*) oil], and (−)-neoisomenthol (**E9**) in geranium (*Pelargonium roseum*, Geraniaceae) oil. The unsaturated versions of *p*-menthol, namely *p*-menthenol, exhibit extensive regioisomerism and are represented by (−)-pulegol (**E10**) (a constituent of several peppermint (*Mentha gentilis* and *M. spirata* (Labiatae) oils), (−)-isopulegol (**E11**) (in *Mentha rotundifolia* (Labiatae)), (−)-piperitol (**E12**) (in several *Mentha* and *Eucalyptus* species), (−)-α-terpineol (**E13**) (in *Artemesia*, *Eucalyptus*, *Juniperus*, and *Mentha* species), and (−)-carveol (**E14**). Oxidation products of both saturated and unsaturated cyclohexane monoterpene alcohols also occur in nature. Of these, the most abundant in the plant kingdom are (−)-menthone (**E15**) (in peppermint (*Mentha* x *piperita*) oil), (−)-isopulegone (**E16**) (in oil of *Mentha pulegium*), (−)-piperitone (**E17**) (in *Eucalyptus* oil), and (+)-carvone (**E18**) (in ripe fruits of dill (*Anethum graveolens*, Umbelliferae) and caraway (*Carum carvi*, Umbelliferae)).

Aromatic versions of cyclohexane monoterpenes (benzenoid menthanes or cymenes) are also found in nature and are constituents of some plants frequently used as spices. The hydrocarbon *p*-cymene (**E19**) has been found to occur in the oils of cinnamon (*Cinnamonum zeylanicum*), cypress, eucalyptus, thyme, and turpentine, whereas *m*-cymene (**E20**) is a constituent of the oil of blackcurrant (*Ribes nigrum*, Saxifragaceae). The corresponding phenols, thymol (*p*-cymen-3-ol) (**E21**) and carvacrol (*p*-cymen-2-ol) (**E22**), have been found to occur in many plants. Thymol (**E21**) is a constituent of thyme (*Thymus vulgaris*, Labiatae) and *Orthodon angustifolium* (Labiatae). Carvacrol (**E22**) has been found to occur in oils of thyme, marjoram, origanum, and summer savoy.

Additional chemical diversity of monoterpenes is apparent from the natural occurrence of their bicyclic analogs that bear cyclopropane (carane and thujane types), cyclobutane (pinane type), and cyclopentane (camphene/bornane, isocamphane and fenchone types) rings (Figs. 1.6 and 1.7). The carane type of bicyclic monoterpenoids in plants is represented by (+)-3-carene (**F1**) that occurs in *Pinus longifolia* (Pinaceae) and the related carboxylic acid, (+)-chaminic acid (**F2**), in *Chamaecyparis nootkatensis* (Cupressaceae). Compared with caranes, the thujane type of monoterpenoids is more abundant in plants. The hydrocarbon analog (−)-3-thujene (**F3**) has been found to occur in the oils of coriander (*Coriandrum sativum*), *Eucalyptus*, and *Thuja occidentalis* (Cupressaceae). Its regioisomer, (+)-sabinene (**F4**), occurs in *Juniperus sabina* (Cupressaceae). The hydration product of (−)-3-thujene, namely (−)-thujol (**F5**), occurs in plants belonging to the genera *Thuja*, *Artemesia*, and *Juniperus*, whereas the corresponding ketone (+)-3-thujanone (**F6**) is found in the oils of several plants of the families Asteraceae, Labiatae, and Pinaceae.

Pinane-type bicyclic monoterpenoids (Fig. 1.6) occur in the wood of several species of *Pinus*. The most abundant are α- and β-pinenes (**F7** and **F8**, respectively). Allylic hydroxylation products of pinenes, (+)-verbenol (**F9**), (+)-myrtenol (**F10**), and (−)-pinocarveol (**F11**) also occur in nature together with their products of oxidation: (+)-verbenone (**F12**), (+)-myrtenal (**F13**), and

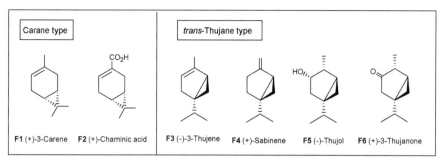

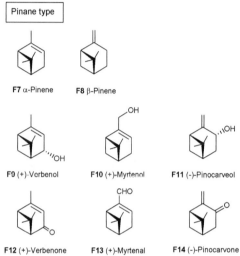

Figure 1.6 Bicyclic cyclopropane and cyclobutane monoterpenoids.

(−)-pinocarvone (**F14**). (+)-Verbenol is a constituent of the oil of turpentine. Its regioisomers, (+)-myrtenol and (−)-pinocarveol, occur in oils of orange (*Citrus sinensis*, Rutaceae) and eucalyptus (*Eucalyptus globulus*, Myrtaceae), respectively.

Cyclopentane bicyclic monoterpenoids that occur in the plant kingdom belong to three major skeletal types: camphane, isocamphane, and fenchane (Fig. 1.7). Camphane-type terpenoid alcohols, (+)-borneol (**G1**) and (−)-isoborneol (**G2**), have been isolated from *Cinnamomum camphora* (Lauraceae) and *Achillea fil-ipendulina* (Asteraceae). A ketone derived from these, (+)-camphor (**G3**), is found in the camphor tree (*Cinnamomum camphora*) and in the leaves of rosemary (*Rosmarinus officinalis*) and sage (*Salvia officinalis*, Labiatae). Camphene (**G4**) and its enantiomer with the isocamphane carbon skeleton are known to occur in the oils of citronella and turpentine. Fenchane-type bicyclic cyclopentane monoterpenoids are commonly found in plants as their ketone derivatives. (−)-Fenchone (**G5**) occurs in the tree of life (*Thuja occidentalis*, Cupressaceae).

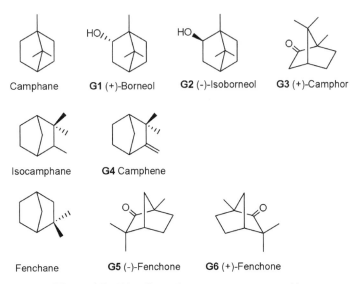

Camphane   G1 (+)-Borneol   G2 (-)-Isoborneol   G3 (+)-Camphor

Isocamphane   G4 Camphene

Fenchane   G5 (-)-Fenchone   G6 (+)-Fenchone

Figure 1.7   Bicyclic cyclopentane monoterpenoids.

Its enantiomer, (+)-fenchone (**G6**), has been isolated from the oil of fennel (*Foeniculum vulgare*, Umbelliferae).

1.4.3   Sesquiterpenoids

The C_{15} terpenoids known as sesquiterpenoids are the most chemically diverse group of terpenoids known in nature. Like monoterpenoids, many sesquiterpenoids contribute to the flavor and fragrances of a variety of plant products. To date about 10,000 sesquiterpenoids are known (32), and in the plant kingdom they commonly occur as hydrocarbons, alcohols, aldehydes, ketones, carboxylic acids, lactones, and oxiranes. The acyclic sesquiterpene hydrocarbons α- and β-farnesenes (**H1** and **H2**, respectively) (Fig. 1.8) are constituents of the oils of orange (*Citrus sinensis*, Rutaceae) and mandarin (*Citrus aurantium* and *C. reticulata*, Rutaceae). The parent alcohol (*E, E*)-farnesol (**H3**) occurs in *Acacia farnensiana* (Mimosaceae), and its regiosiomer (*S*)-(+)-nerolidol (**H4**) is a constituent of the oil of neroli obtained from orange flowers. Some farnesane derivatives that contain furan rings also occur in nature, and these include dendrolasin (**H5**) from sweet potato (*Ipomoea batatas*, Convolvulaceae) and longifolin (**H6**) from the leaves of *Actinodaphe longifolia* (Lauraceae).

The vast chemical diversity of cyclic sesquiterpenoids compared with monoterpenoids results from the number of possible cyclization modes that is enhanced because of increased chain length and the presence of additional double bonds in the acyclic precursor farnesyl diphosphate (FPP). As depicted in Fig. 1.9, these cyclization modes lead to sesquiterpenoids with mono-, bi-, and tricyclic structures. The cyclofarnesane, (*S*)-(+)-abscisic acid (**I1**)

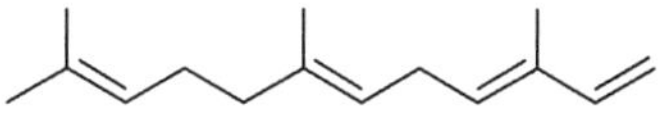

H1 α–Farnesene 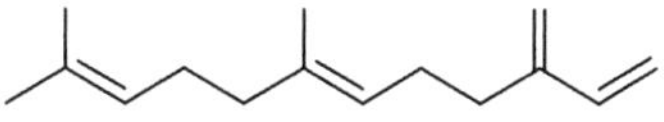 **H2** β–Farnesene

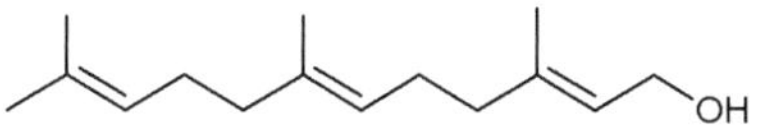

H3 (*E,E*)-Farnesol 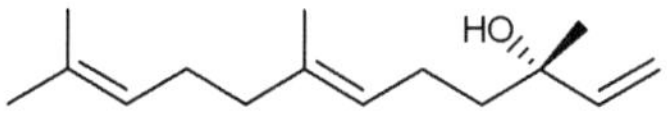 **H4** (*S*)-(+)-Nerolidol

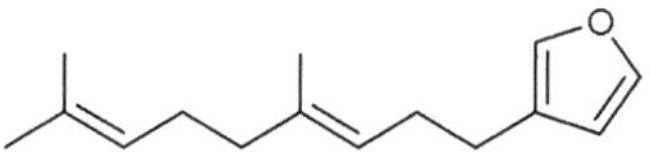

H5 Dendrolasin 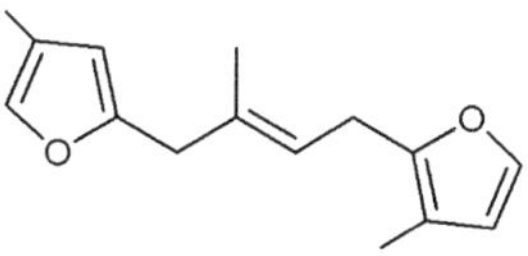 **H6** Longifolin

Figure 1.8 Acyclic plant sesquiterpenoids and their derivatives.

(Fig. 1.10), an antagonist of plant growth hormones essential for plants in controlling flowering, shedding of leaves, and falling of fruits, is a monocyclic sesquiterpene formed because of bond formation between C-6 and C-11. Other common ring closures occur because of bond formation between C-1 and C-6, C-1 and C-10, and C-1 and C-11, which gives rise to bisabolane, germacrane, and humulane types of monocyclic sesquiterpenes (Fig. 1.9). Over 100 bisabolanes, 300 germacranes, and 30 humulanes are known to occur in nature (32). The elemane type of monocyclic sesquiterpenes is structurally related to germacranes as they can arise by a COPE rearrangement involving bond formation between C-1 and C-6 followed by the cleavage of the bond between C-8 and C-9. To date about 50 elemane-type monocyclic sesquiterpenes are known (32). Among the monocyclic sesquiterpenoids, bisabolanes represent one of the important classes known to occur in plants. Common examples of bisabolane in plants include (+)-β-bisabolene (**I2**), β-sesquiphellandrene (**I3**), and (−)-zingiberene (**I4**) in the rhizome of ginger (*Zingiber officinalis*, Zingiberaceae) and sesquisabinene (**I5**) from pepper (*Piper nigrum*, Piperaceae). Germacrane-type sesquiterpenoids contain a 10-membered macrocyclic ring; many of these sesquiterpenoids are constituents of essential oils derived from plants. The germacrane hydrocarbons, germacrenes B (**I6**) and D (**I7**), are found in *Citrus junos* and *C. bergamia* (Rutaceae), respectively. Some elemane-type sesquiterpenoids present in plants are represented by (−)-bicycloelemene (**I8**) from peppermint (*Mentha piperita* and *M. arvensis*) and β-elemenone (**I9**) from *Commiphora abyssinica*. The humulane type of monocyclic sesquiterpenes, which contain an 11-membered macrocyclic ring, is also found in plants and includes regioisomeric α- and β-humulenes (**I10** and **I11**, respectively) from *Lindera strychnifolia* (Lauraceae), and (−)-humulol (**I12**), all of which are important constituents of the essential oils from cloves (*Caryophylli flos*, Caryophyllaceae), hops (*Humulus lupulus*, Cannabaceae), and ginger (*Zingiber*

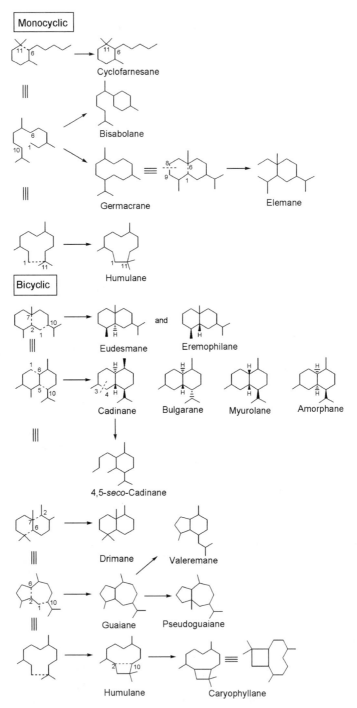

Figure 1.9 Cyclization modes of farnesane skeleton that lead to diverse monocyclic and bicyclic sesquiterpenoids.

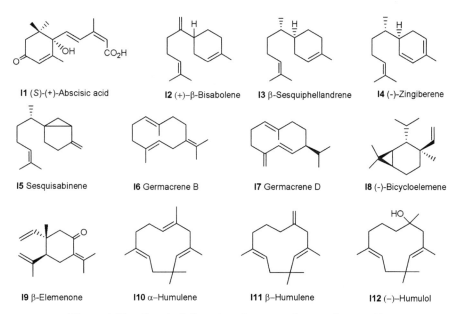

I1 (S)-(+)-Abscisic acid   **I2** (+)-β-Bisabolene   **I3** β-Sesquiphellandrene   **I4** (-)-Zingiberene

I5 Sesquisabinene   **I6** Germacrene B   **I7** Germacrene D   **I8** (-)-Bicycloelemene

I9 β-Elemenone   **I10** α-Humulene   **I11** β-Humulene   **I12** (-)-Humulol

Figure 1.10 Chemical diversity of monocyclic sesquiterpenoids.

zerumbeticum, Zingiberaceae). Bicyclic sesquiterpenoids are formed because of two carbon–carbon bonds, each linking two carbon atoms of the farnesane skeleton together.

1.4.4 Diterpenoids

The diterpenoids, which contain 20 carbon atoms, are represented by acyclic, monocyclic, bicyclic, tricyclic, and tetracyclic structures. Over 5,000 naturally occurring diterpenoids, many of which frequently occur in plant families Araliaceae, Asteraceae, Cistaceae, Cupressaceae, Euphorbiaceae, Leguminosae, Labiatae, and Pinaceae, are known (32). The acyclic diterpenoid alcohol phytol (**J1**) (Fig. 1.11) is a part of the structure of chlorophyll. A group of monocyclic diterpenoids with a 14-carbon macrocyclic ring called cembranes [e.g., cembrene A (**J2**)] also occurs in plants and is represented by over 100 members (32). Among the other cyclic diterpenoids, the most abundant in plants are the bicyclic labdanes [e.g., (−)-forskolin (**J3**)], tricyclic abietanes [e.g., abietic acid (**J4**)], and tetracyclic kauranes [e.g., (−)-kaurane (**J5**)], represented respectively by 500, 200, and 100 members (32). Baccatin III (**J6**), which is the diterpenoid part of the well-known anticancer drug paclitaxel (Taxol®; Mead Johnson, Princeton, NJ), contains a tricyclic skeleton derived from cembrane.

1.4.5 Triterpenoids and Steroids

Triterpenoids and steroids are groups of natural products that contain about 30 carbon atoms. They have a common origin, and their structures can be

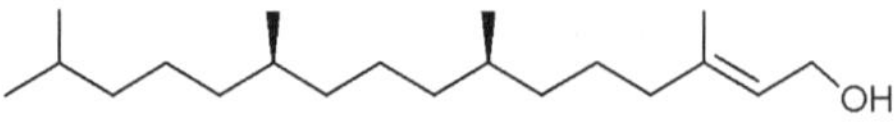

J1 Phytol

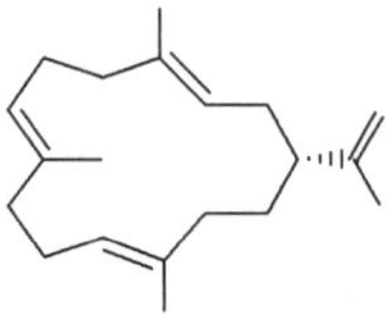

J2 Cembrene A 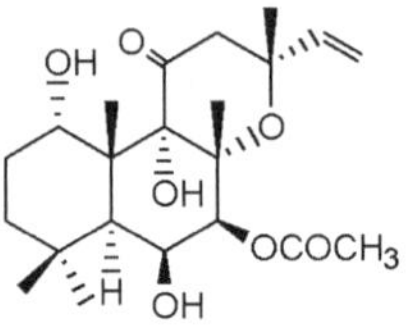   **J3** (-)-Forskolin 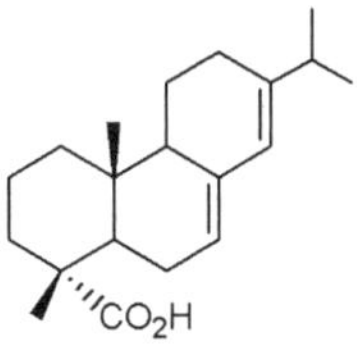   **J4** Abietic acid

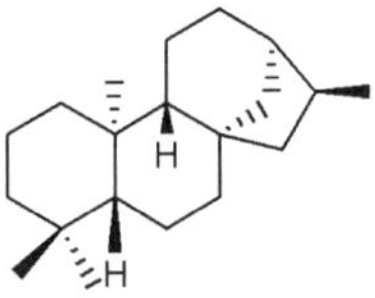

J5 (-)-Kaurane 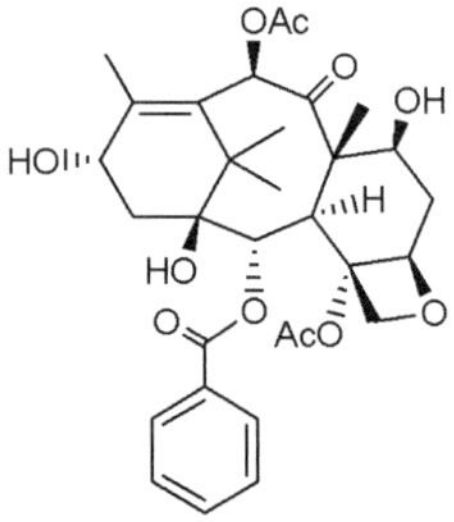   **J6** Baccatin III

Figure 1.11   Some representative examples of diterpenoids in plants.

considered as being derived from that of squalene. Triterpenoids are found mostly in the plant kingdom, whereas steroids occur in plants, animals, and microorganisms. The chemical diversity of plant triterpenoids results from the ability of the C_{30} precursor, squalene, to undergo various modes of cyclization and subsequent "decoration" reactions. The plant triterpenoids belong to two main groups, the tetracyclic and pentacyclic. The tetracyclic triterpenoids, which consist of dammarane (**K1**) and tirucallane (**K2**) among others, are regarded by some authors as methylated steroids. The group of pentacyclic triterpenoids is by far the most diverse and is divided into five main groups: friedelane (**K3**), lupane (**K4**), ursane (**K5**), oleanane (**K6**), and hopane (**K7**) (Fig. 1.12). The steroids are modified triterpenoids that contain the tetracyclic ring system present in lanosterol. Chemical diversity represented by steroids depends mainly on the nature of the side chain attached to the steroid nucleus. Most prevalent in the plant kingdom are stigmastane (**K8**) and cycloartane (**K9**) classes of steroids (Fig. 1.12).

Triterpenoids and steroids frequently occur in many plant species as their glycosides called saponins (33). The chemical diversity of saponins is dependant therefore on both the nature of the 30-carbon moiety and the carbohydrate residue. Some saponins contain carbohydrate residues attached to several different positions of the aglycone (triterpenoid or the steroid) skeleton. Saponins are classified into 11 main structural classes based on the carbon skeletons of their

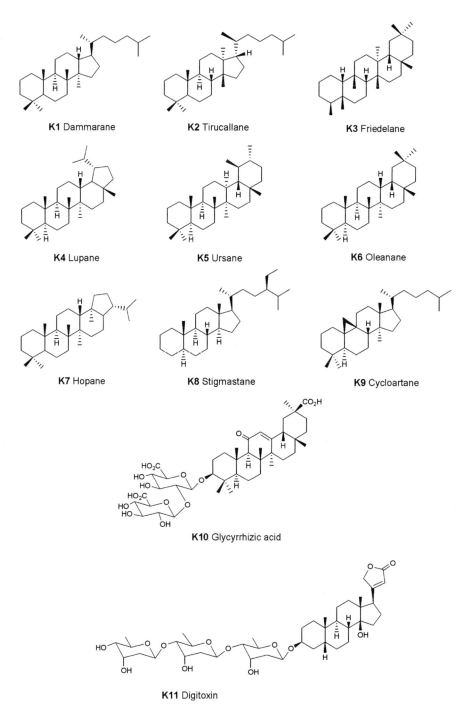

Figure 1.12 Triterpenoid and steroid skeletons common in plants and structures of some saponins.

aglycone moiety (33). In addition to their soap-like behavior in aqueous solution because of this combination of polar (carbohydrate) and nonpolar (aglycone) structural elements, saponins exhibit a diverse range of pharmacological and medicinal properties. Within the plant kingdom saponins are present in two major taxonomic classes, Magneliopsida (dicot) and Liliopsida (monocot) (33). Some important examples of saponins include glycyrrhizic acid (**K10**) from licorice and digitoxin (**K11**) from foxglove (*Digitalis purpurea*).

1.5   PRODUCTS OF SHIKIMATE PATHWAY

The shikimate pathway links the metabolism of carbohydrates to the biosynthesis of aromatic natural products via aromatic amino acids. This pathway, which is found only in plants and microorganisms, provides a major route to aromatic and phenolic natural products in plants. To date, over 8,000 phenolic natural products are known, which accounts for about 40% of organic carbon circulating in the biosphere. Although the bulk of plant phenolics are components of cell wall structures, many phenolic natural products are known to play functional roles that are essential for the survival of plants.

It has been noted that the chemical diversity of plant phenolics is as vast as the plant diversity itself. Most plant phenolics are derived directly from the shikimic acid (simple benzoic acids), shikimate (phenylpropanoid) pathway, or a combination of shikimate and acetate (phenylpropanoid-acetate) pathways. Products of each of these pathways undergo additional structural elaborations that result in a vast array of plant phenolics such as simple benzoic acid and cinnamic acid derivatives, monolignols, lignans and lignin, phenylpropenes, coumarins, stilbenes, flavonoids, anthocyanidins, and isoflavonoids.

1.5.1   Benzoic Acid Derivatives

As apparent from their structures, many benzoic acid derivatives are directly formed from shikimic acid by dehydration, dehydrogenation, and enolization reactions. Gallic acid is a component of gallotannins common in some plants that are used in the tanning of animal hides to make leather. Astringency of some foods and beverages, especially coffee, tea, and wines, is because of their constituent tannins. Other benzoic acid derivatives that occur in plants include protocatechuic acid, 4-hydroxybenzoic acid, and salicylic acid.

1.5.2   Cinnamic Acid Derivatives

Cinnamic acid and its derivatives found in plants originate from the aromatic amino acids L-phenylalanine and L-tyrosine by the elimination of ammonia. Some common natural cinnamic acid derivatives include *p*-coumaric acid, caffeic acid, ferulic acid, and sinapic acid.

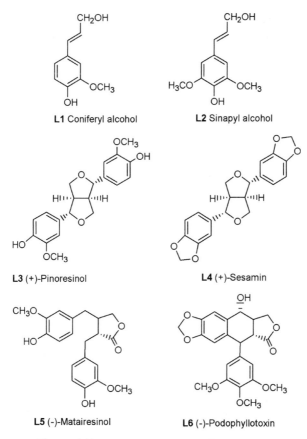

L1 Coniferyl alcohol

L2 Sinapyl alcohol

L3 (+)-Pinoresinol

L4 (+)-Sesamin

L5 (-)-Matairesinol

L6 (-)-Podophyllotoxin

Figure 1.13 Phenyl propanoids of plant origin.

1.5.3 Monolignols, Lignans, and Lignin

The alcohols formed from some cinnamic acid derivatives, namely *p*-coumaryl alcohol, coniferyl alcohol (**L1**), and sinapyl alcohol (**L2**), commonly known as monolignols, undergo dimerization reactions that yield lignans such as (+)-pinoresinol (**L3**), (+)-sesamin (**L4**), (−)-matairesinol (**L5**), and podophyllotoxin (**L6**) (Fig. 1.13). Several thousand lignans are found to occur in nature. Lignins, the structural components of plant cell walls, are polymers of monolignols and/or lignans.

1.5.4 Phenylpropenes

Phenylpropenes are derived from cinnamic acid and its derivatives by a series of reductions and other transformations. Cinnamaldehyde, the first product of the reduction of cinnamic acid, occurs in the bark of cinnamon (*Cinnamomum zeylanicum*, Lauraceae). Several hydrocarbon analogs are also known to occur in plants. Anethole is the main constituent of oils from aniseed (*Pimpinella anisum*, Umbelliferae), fennel (*Foeniculum vulgare*, Umbelliferae), and star anise (*Illicium varum*, Illiciaceae). Eugenol is a major constituent of cinnamon leaf, whereas myristicin occurs in nutmeg (*Myristica fragrans*, Myristicaceae).

1.5.5 Coumarins

Coumarins derive their name from their precursor, *o*-coumaric acid. They occur widely in plants both in the free form and as glycosides and are commonly found in families such as the Umbelliferae and Rutaceae. The parent compound coumarin (**M1**) is found in sweet clover (*Melilotus alba*, Leguminosae) and its hydroxyl derivative umbelliferone (**M2**) has been isolated from several *Ferula* spp. (Umbelliferae). Coumarins with complex structures also occur in plants and are formed by incorporating additional carbons derived from the mevalonate pathway. Alkylation of umbelliferone (**M2**) with dimethylallyl diphosphate (DMAPP) leads to demethylsuberosin (**M3**), which undergoes cyclization yielding marmesin (**M4**), the precursor of naturally occurring furanocoumarins, psorolen (**M5**), and bergapten (**M6**) (Fig. 1.14).

1.5.6 Stilbenes, Flavonoids, Anthocyanidins, and Isoflavonoids

In contrast to other plant phenolics, the basic carbon skeleton of stilbenes, flavonoids, anthocyanidins, and isoflavonoids incorporates elements of both shikimate (phenylpropanoid) and acetate pathways. Plant phenolics derived from this mixed biosynthetic pathway include the well-known cancer chemo-preventative stilbene, resveratrol (**M7**), present in wine; naringenin (**M8**), a flavonone from *Heliotropium* and *Nonea* spp. (Boraginaceae); apigenin (**M9**), a flavone from German chamomile (*Matricaria recuitita*, Asteraceae); rutin (**M10**), a flavonol glycoside from several plants including hawthorn (*Crataegus* spp., Rosaceae); pelargonidin (**M11**), an anthocyanidin responsible for brilliant colors of many flowers; and genistein (**M12**) (Fig. 1.14), an isoflavonoid with oestrogenic activity and thus referred to as a phyto-oestrogen.

1.6 ALKALOIDS

Alkaloids constitute nitrogen-containing natural product bases that occur mainly in plants. About 20% of the flowering plant species are known to produce alkaloids (6). To date, over 12,000 plant-derived alkaloids have been reported, and they are grouped into various classes based on their origin and the nature of the nitrogen-containing moiety. Alkaloids commonly originate from the amino acids, L-ornithine, lysine, nicotinic acid, tyrosine, phenylalanine, tryptophan, anthranilic acid, and histidine, and thus contain pyrrolidine, pyrrolizidine, piperidine, quino-lizidine, indolizidine, pyridine, quinoline, isoquinoline, indole, and imidazole ring systems. Alkaloids are also known to originate from mixed biosynthetic pathways,

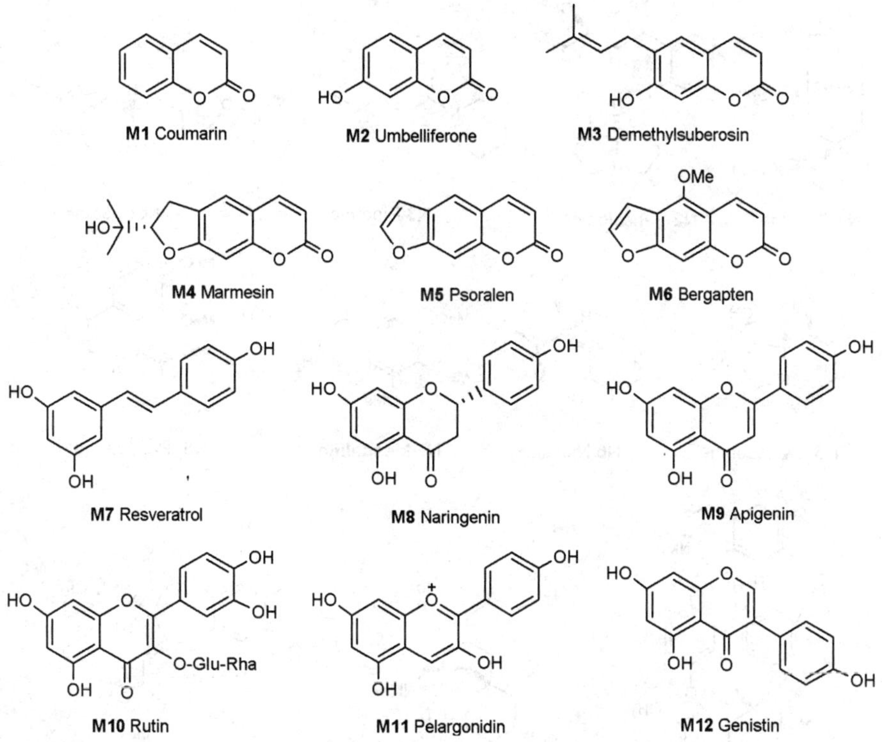

Figure 1.14 Coumarins, stilbenes, flavonoids, anthocyanidins, and isoflavonoids of plant origin.

the most important of which include terpenoid and steroidal alkaloids. A limited number of alkaloids that contain a purine ring (e.g., caffeine) also occur in plants. Of the large number and variety of plant alkaloids, only a few are considered here for the purpose of illustration of their chemical diversity.

1.6.1 Alkaloids Derived from Aliphatic Amino Acids and Nicotinic Acid

Alkaloids that contain pyrrolidine and pyrrolizidine ring systems are derived from the nonprotein amino acid, L-ornithine. Cocaine (**N1**) and (−)-hyoscyamine, the two important pyrrolidine alkaloids that contain a tropane ring system, have been found to occur in coca (*Erythroxylon coca*, Erythroxylaceae) leaves and the whole plant of the deadly nightshade (*Atropa belladonna*, Solanaceae). The hepatotoxic alkaloid senecionine (**N2**) contains a bicyclic pyrrolizidine skeleton derived from two molecules of L-ornithine.

Piperidine alkaloids, for example, piperine (**N3**), and pseudopelletierine, are known to be derived from the amino acid, L-lysine. Piperine is responsible for the pungency of black pepper (*Piper nigrum*, Piperaceae), whereas pseudopelletier-ine is a constituent of the bark of pomegranate (*Punica granatum*, Punicaceae).

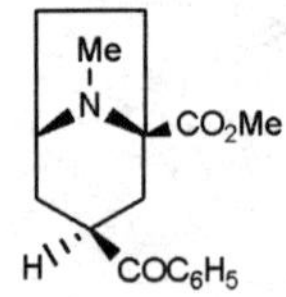

N1 Cocaine 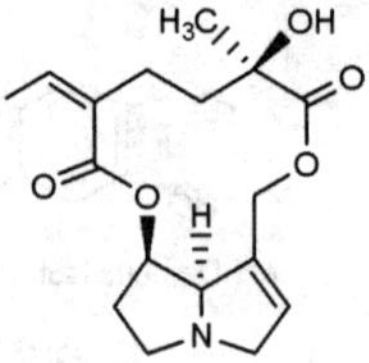   **N2** Senecionine 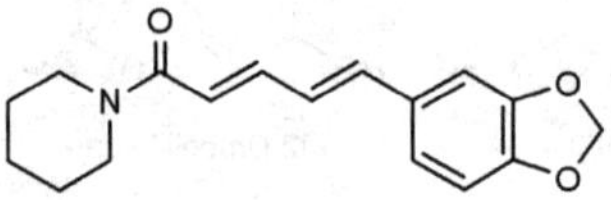   **N3** Piperine 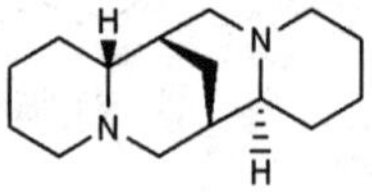   **N4** (-)-Sparteine

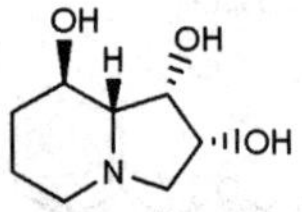

N5 Swainsonine 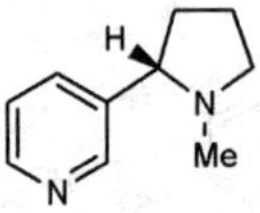   **N6** Nicotine 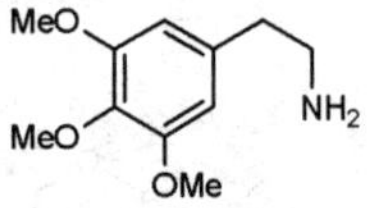   **N7 Mescaline** 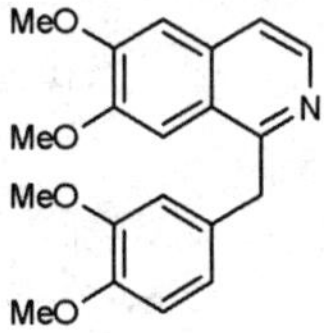   **N8** Papavarine

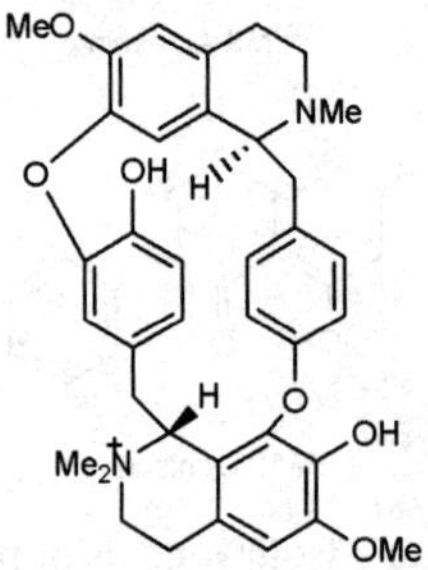

N9 (+)-Tubocurarine 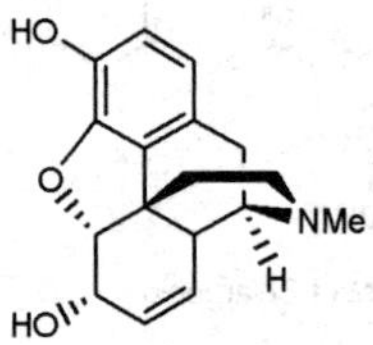   **N10** Morphine 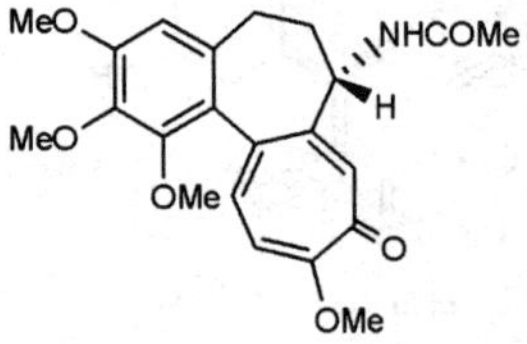   **N11** Colchicine

Figure 1.15   Chemical diversity of plant alkaloids.

The bicyclic ring system in quinolizidine alkaloids such as (−)-sparteine (**N4**) in the broom plant (*Cytisus scoparius*, Leguminosae) is derived from two molecules of L-lysine in a manner similar to the L-ornithine-derived pyrrolizidine ring system. Indolizidine alkaloids derived from L-lysine via the cyclic amino acid, L-pipecolic acid, contain fused six- and five-membered rings with a nitrogen atom at the ring fusion. An important example of an indolizidine alkaloid is swainsonine (**N5**), which occurs in the leguminous plant *Swainsonia canescens*. Alkaloids that contain a pyridine ring also occur in the plant kingdom. Two common plant-derived pyridine alkaloids, nicotine (**N6**) and anabasine, both of which are found in tobacco (*Nicotiana tabacum*, Solanaceae), contain a pyridine and a pyrrolidine or a piperidine ring, respectively (Fig. 1.15).

1.6.2   Alkaloids Derived from Aromatic Amino Acids

Aromatic amino acids that originate from the shikimate pathway also act as precursors to many alkaloids. Alkaloids that contain a phenylethylamine moiety are derived from L-tyrosine or its oxidation product L-dihydroxyphenylalanine (L-DOPA). Mescaline (**N7**) originating from the latter amino acid is known to occur in several cacti and is responsible for the hallucinogenic activity of peyote (*Lophophora williamsii*, Cactaceae). Lophocerine is a tetrahydroisoquinoline alkaloid derived from L-dopamine and found to occur in a different *Lophophora* species, *L. schotti*.

Condensation of two phenylethyl units derived independently from the same or different aromatic amino acid(s) leads to a variety of benzyl-tetrahydroisoquinolines, which, with additional structural modifications, produce a diverse range of alkaloids. (*S*)-Reticuline occurring in several plant species of Annonaceae is an important benzyl-tetrahydroisoquinoline alkaloid that acts as a precursor to several pharmacologically active alkaloids such as papaverine (**N8**), (+)-tubocurarine (**N9**), and morphine (**N10**). Papaverine and morphine are known to occur in opium (*Papaver somniferum*, Papaveraceae) and are responsible for its narcotic activity, whereas (+)-tubocurarine (**N9**) is a muscle relaxant obtained from the arrow poison of the South American Indians, curare (*Chondrodendron tomentosum*, Menispermaceae). Phenethylisoquinoline alkaloids are similar structurally to benzylisoquinolines but as the name implies contain a phenylethyl moiety instead of a benzyl moiety as the pendant aromatic ring. Both (*S*)-autumnaline and the cyclized analog colchicine (**N11**) belonging to this class have been found to occur in the seeds of autumn crocus (*Colchicium autumnale*, Liliaceae).

1.7 MAXIMIZATION OF CHEMICAL DIVERSITY AND PRODUCTION OF NATURAL PRODUCTS IN PLANTS

As is apparent from the foregoing discussion, plants produce a huge array of natural products, many of which are specialized secondary metabolites associated with particular plant species and/or having to play important ecological roles. It is likely that for diversification and survival of the plant kingdom, individual plants had to develop the ability to perform *in vivo* combinatorial chemistry by mixing and matching and evolving the genes required for different secondary metabolite biosynthetic pathways (34, 35). With the elucidation of several secondary metabolic pathways in plants together with the advent of techniques for the introduction of genes into plants and the availability of an increasing number of genes, it has become possible to modulate and diversify secondary metabolite production in transgenic plants and plant cell cultures.

Two general approaches for the production of long-chain polyunsaturated fatty acids usually found in fish oil have been employed, both of which used 18 carbon fatty acids endogenous to plants as the starting substrates (36). Soybean and canola, the oilseed plants rich in omega-6 fatty acids, have been engineered to produce omega-3 polyunsaturated fatty acids such as eicosapentaenoic acid (EPA) and docosohexaenoic acid (DHA) (37, 38).

Chalcone synthase (CHS), the first plant natural product polyketide synthase (PKS) to be characterized at the molecular level (39), catalyzes the condensation of 4-coumaroyl-CoA with three molecules of malonyl-CoA to afford naringenin chalcone, a precursor of the major classes of plant flavonoids. The cloning of a novel type III pentaketide chromone synthase (PCS) from aloe (*Aloe arborescens*, Liliaceae) rich in aromatic polyketides, especially quinones such as aloe-emodin and emodin, resulted in PCS-catalyzed condensation of five molecules of malonyl-CoA to produce 5,7-dihydroxy-2-methyl chromone new to this plant (40). Another novel *Aloe arborescens* type III PKS that produces two hitherto unknown aromatic octaketides, SEK4 and SEK4b, has recently been reported (41). The application of plant cell cultures for the production of the polyketide hypericin from St. John's wort (*Hypericum performatum*, Hypericaceae) has been investigated (42).

To date over 30 plant terpenoid synthases have been cloned as cDNAs, and many of these were found to encode enzymes of secondary metabolism (43). Isolation and analysis of six genomic clones encoding monoterpene ((−)-pinene and (−)-limonene), sesquiterpene ((*E*)-α-bisabolene and δ-selinene) and diterpene (abietadiene) synthases from *Abies grandis*, and a diterpene (taxadiene) synthase from *Taxus brevifolia* have been reported (44). Overexpression of a cotton farnesyl diphosphate synthase (FPPS) in transgenic *Artemesia annua* has resulted in 3- to 4-fold increase in the yield of the sesquiterpenoid antimalarial drug, artemisinin, in hairy roots (45).

Plant cell culture, an environmentally friendly and renewable alternative for the production of plant natural products, has also been investigated to obtain taxane diterpenoids from *Taxus* sp. (46–48) and terpene indole alkaloids from the Madagascar periwinkle (*Catharanthus roseus*) (49). A recent study has provided evidence for the production of novel terpene indole alkaloids using both differentiated *Catharanthus roseus* (seedlings) and hairy root culture (50).

REFERENCES

1. Pimm SL, Russell GJ, Gittleman JL, Brooks TM. The future of biodiversity. Science 1995;269:347–350.
2. Lewis NG, Davin LB. Evolution of lignan and neolignan biochemical pathways. In: Isoprenoids and other natural products: Evolution and function. Nes WD, ed. 1994. ACS Symposium Series, Washington, D.C.
3. Kinghorn AD. Pharmacognosy in the 21st Century. J. Pharm. Pharmacol. 2001;53:135–148.
4. Dixon RA. Natural products and plant disease resistance. Nature 2001;411:843–847.
5. Verpoorte R. Exploration of nature's chemodiversity: the role of secondary metabolites as leads in drug development. Drug Discov. Today 1998;3:232–238.
6. Croteau R, Kutchan TM, Lewis NG. Natural Products (Secondary Metabolites). In: Biochemistry and Molecular Biology of Plants, Fifth Impression. Buchanan BB, Gruissem W, Jones RL, eds. 2005. American Society for Plant Physiologists, Rockville, Maryland.

7. Pietra F. Biodiversity and natural product diversity. 1st edition. 2002. Elsevier Science, Netherlands.

8. Romeo JT, Saunders JA, Barbosa P, eds. Recent Advances in Phytochemistry, Vol 30. Phytochemical Diversity and Redundancy in Ecological Interactions. 1996. Plenum Press, New York.

9. Hartmann, T. Plant-derived secondary metabolites as defensive chemicals in herbivorous insects: a case study in chemical ecology. Planta 2004;219:1–4.

10. Firn RD, Jones CG. Natural products – a simple model to explain chemical diversity. Nat. Prod. Rep. 2003;20:382–391.

11. Tulp M, Bohlin L. Functional versus chemical diversity: is biodiversity important for drug discovery? Trends Pharmacol. Sci. 2002;23:225–231.

12. Tulp M, Bohlin L. Chemical diversity: independent of functional diversity. Trends Pharmacol. Sci. 2002;23:405.

13. McKey D. The distribution of secondary compounds within plants. In: Herbivores: Their Interaction with Plant Secondary Metabolites. Rosenthal G, Janzen D, eds. 1979. Academic Press, New York.

14. Feeny P. The evolution of chemical ecology: contributions from the study of herbivorous insects. In: Herbivores: Their Interactions with Secondary Plant Metabolites. Rosenthal A, Berenbaum MR, eds. 1992. Academic Press, San Diego.

15. Futuyama DJ, Keese Mc. Evolution and coevolution of plants and phytophagous arthropods. In: Herbivores: Their Interactions with Secondary Plant Metabolites Rosenthal A, Berenbaum MR, eds. 1992. Academic Press, San Diego.

16. Harborne JB. Twenty-five years of chemical ecology. Nat. Prod. Rep. 2001;18: 361–379.

17. Jones CG, Firn RD. On the evolution of secondary plant chemical diversity. Philos. Trans. R. Soc. 1991;333:273–280.

18. Firn RD, Jones CG. Do we need a new hypothesis to explain plant VOC emissions? Trends Plant Sci. 2006;11:112–113.

19. Weissman KJ, Leadlay PF. Combinatorial biosynthesis of reduced polyketides. Nat. Rev. Microbiol. 2005;3:925–936.

20. Aitzetmuller K, Matthaus B, Friedrich H. A new database for seed oil fatty acids – the database SOFA. Eur. J. Lipid Sci. Technol. 2003;105:92–103.

21. Viegas C Jr, de Rezende A, Silva DHS, Castro-Gamboa I, Bolzani V da S, Barreiro EJ, Palhares de Miranda AL, Alexandre-Moreira MS, Young MCM. Ethnopharmacological, biological and chemical aspects of the *Cassia* genus. Quim. Nova 2006;29:1279–1286.

22. Fairbairn JW. The analysis and standardization of anthraquinone drugs. Planta Med. 1964;12:260–264.

23. Reynolds T. Aloe chemistry. Curr. Topics Phytochem. 2002;5:235–254.

24. Miskovsky P. Hypericin – a new antiviral and antitumor photosensitizer: mechanism of action and interaction with biological macromolecules. Curr. Drug Targets 2002;3:55–84.

25. Rohdich F, Bacher A, Eisenreich W. Isoprenoid biosynthetic pathways as anti-infective drug targets. Biochem. Soc. Trans. 2005;33:785–791.

26. Withers ST, Keasling JD. Biosynthesis and engineering of isoprenoids small molecules. Appl. Microbiol. Biotechnol. 2007;73:980–990.

27. Lee S. Artemisinin, promising lead natural product for various drug developments. Mini Rev. Med. Chem. 2007;7:411–422.

28. Kingston DGI. Taxol and Its Analogs. In: Anticancer Agents from Natural Products. Cragg GM, Kingston DGI, Newman DJ, eds. 2005. Taylor & Francis, Boca Raton, FL.

29. Wilms K. Chemistry and mechanism of Vinca alkaloids. Planta Med. 1972;22: 324–333.

30. Guéritte F, Fahy J. The Vinca Alkaloids. In: Anticancer Agents from Natural Products. Cragg GM, Kingston DGI, Newman DJ, eds. 2005. Taylor & Francis, Boca Raton, FL.

31. Dembitsky VM. Astonishing diversity of natural surfactants: 7. Biologically active hemi- and monoterpenoid glycosides. Lipids 2006;41:1–27.

32. Connolly JD, Hill RA. Dictionary of Terpenoids, Vols. 1–3. 1991. Chapman and Hall, London.

33. Vincken JP, Heng L, de Groot A, Gruppen H. Saponins, classification and occurrence in the plant kingdom. Phytochemistry 2007;68:275–297.

34. Osboure AE. Metabolic diversity in plants. In Proc. BCPC International Congress: Crop Science and Technology. Brit. Crop Protection Council. 2005. pp. 979–984.

35. Bakht S, Field B, Inagaki Y, Jenner H, Melton R, Mylona P, Qi X, Qin B, Townsend B, Wegel E, Osbourne A. Metabolic diversity in plants. Biol. Plant Microbe Interact. 2006;5:107–112.

36. Graham IA, Larson T, Napier JA. Rational metabolic engineering of transgenic plants for biosynthesis of omega-3 polyunsaturates. Curr. Opin. Biotechnol. 2007;18:142–147.

37. Damude HG, Kinney AJ. Engineering oilseed plants for a sustainable, land-based source of long chain polyunsaturated fatty acids. Lipids 2007;42:179–185.

38. Napier JA. The production of unusual fatty acids in transgenic plants. Annu. Rev. Plant Biol. 2007;58:295–319.

39. Kreuzaler F, Ragg H, Fautz E, Kuhn DN, Hahlbrook K. UV-induction of chalcone synthase mRNA in cell suspension cultures of Petroselinum hortense. Proc. Natl. Acad. Sci. USA. 1983;80:2591–2593.

40. Abe I, Utsumi Y, Oguro S, Morita H, Sano Y, Noguchi H. A plant type III polyketide sythase that produces pentaketide chromone. J. Am. Chem. Soc. 2005;127: 1362–1363.

41. Abe I, Oguro S, Utsumi Y, Sano Y, Noguchi H. Engineered biosynthesis of plant polyketides: chain length control in octaketide-producing plant type III polyketide synthase. J. Am. Chem. Soc. 2005;127:12709–12716.

42. Kirakosyan A, Sirvent TM, Gibson DM, Kaufman PB. The production of hypericins and hyperforin by in vitro cultures of St. John's wort (Hypericum perforatum). Biotechnol. Appl. Biochem. 2004;39:71–81.

43. Bohlmann J, Meyer-Gaven G, Croteau R. Plant terpenoid synthases: molecular biology and phylogenetic analysis. Proc. Natl. Acad. Sci. USA. 1998;95:4126–4133.

44. Trapp SC, Croteau RB. Genomic organization of plant terpene synthases and molecular evolutionary implications. Genetics 2001;158:811–832.

45. Liu Y, Wang H, Ye HC, Li GF. Advances in the plant isoprenoid biosynthetic pathway and its metabolic engineering. J. Integr. Plant Biol. 2005;47:769–782.

46. Frense D. Taxanes: perspectives for biotechnological production. Appl. Microbiol. Biotechnol. 2007;73:1233–1240.

47. Tabata H. Paclitaxel production by plant cell culture technology. Adv. Biochem. Eng. Biotechnol. 2004;87:1–23.

48. Tabata H. Production of paclitaxel and related taxanes by cell suspension cultures of *Taxus* species. Curr. Drug Targets 2006;7:453–461.

49. Pasquali G, Porto DD, Fett-Neto AG. Metabolic engineering of cell cultures versus whole plant complexity in production of bioactive monoterpene indole alkaloids: recent progress related to an old dilemma. J. Biosci. Bioeng. 2006;101:287–296.

50. McCoy E, O'Conner SE. Directed biosynthesis of alkaloid analogs in the medicinal plant *Catharanthus roseus*. J. Am. Chem. Soc. 2006;128:14276–14277.

FURTHER READING

Breitmaier E. Terpenes flavors, fragrances, pharmaca, pheromones. 2006. Wiley-VCH Verlag GmbH & Co. KgaA, Weinheim. pp. 10–116.

Dewick PM. Medicinal natural products. A biosynthetic approach. 1997. John Wiley & Sons Inc., New York. pp. 32–440.

Hartmann T, Dierich B. Chemical diversity and variation of pyrrolizidine alkaloids of the senecionine type: biological need or coincidence? Planta 1998;206:443–451.

Roberts SC. Production and engineering of terpenoids in plant cell culture. Nat. Chem. Biol. 2007;3:387–395.

2

MARINE NATURAL PRODUCTS

Nobuhiro Fusetani

Hokkaido University, Hakodate, Japan

Oceans provide enormous and diverse habitats for marine life. The distinct feature of marine life is the domination of invertebrates, which account for more than 95% of marine animals. Most marine invertebrates are sessile and soft-bodied and lack obvious physical defenses. Instead, they have evolved to defend by chemical means against predation and overgrowth by other fouling organisms. In fact, their secondary metabolites have unusual structural features and potent biologic activities, many of which are not found in terrestrial natural products. This review focuses on bioactive metabolites isolated mainly from marine invertebrates with a special emphasis on the uniqueness of marine natural products.

The world's oceans cover more than 70% of the earth's surface and represent greater than 95% of the biosphere. Species ranging from 3 to 100 million are estimated to inhabit in the oceans. All but 1 of the 35 principal phyla in the animal kingdom are represented in aquatic environments; 8 phyla are exclusively aquatic. Most of them are sessile and soft-bodied organisms, most of which have evolved by chemical means to defend against predators and overgrowth by competing species. As expected, a variety of bioactive metabolites were found in marine invertebrates (see a series of reviews on marine natural products published in *Natural Product Reports* since 1984).

Bergmann's revolutionary discovery of arabinose-containing nucleosides [e.g., spongouridine (1)] from the Caribbean marine sponge *Cryptothera crypta* was a driving force for the birth of a new research field, Marine Natural Products Chemistry, in the early 1970s (1, 2). Nearly 18,000 new compounds, including polyketides, peptides, alkaloids, terpenoids, shikimic acid derivatives, sugars, and

Natural Products in Chemical Biology, First Edition. Edited by Natanya Civjan.
© 2012 John Wiley & Sons, Inc. Published 2012 by John Wiley & Sons, Inc.

a multitude of mixed biogenesis metabolites, have been discovered during the last 40 years according to MarinLit (University of Canterbury, New Zealand). Many marine natural products have structural features previously unreported from terrestrial sources (3). Halogenated, especially brominated, and sulfated compounds are often encountered as marine natural products. Arsenic compounds, in particular arsenobetaines and arsenosugars, are distributed widely in marine algae and invertebrates. Recently, the first polyarsenic organic compound from Nature, arsenicin A (2) was reported from the sponge *Echinochalina bargibanti* (4). Also, several highly bioactive compounds with interesting modes of action have been discovered from marine invertebrates. A considerable percentage of these compounds was suggested or found to be derived from microorganisms. Indeed, certain bivalves and ciguateric fishes become poisonous by feeding toxic dinoflagellates or herbivorous ciguateric fish (5), whereas some cytotoxic metabolites of sponges and tunicates are produced by symbiotic microorganisms (6).

This review describes bioactive compounds isolated from marine algae and invertebrates with an emphasis on their uniqueness. Because of limited space, metabolites from bacteria, cyanobacteria, and fungi cannot be included, although some from cyanobacterial and endosymbiotic bacterial origins are described (some reviews on metabolites of marine bacteria, cyanobacteria, and fungi are provided in the "Further Reading" section). Structures and bioactivities are described for compounds that represent natural product classes, but steroids and carotenoids are not included.

2.1 POLYKETIDES AND FATTY ACID DERIVATIVES

A diverse array of polyketide metabolites are found in marine organisms, which range from simple oxylipins to highly complex polyethers and macrolides. Particularly intriguing are "ladder-shaped" polyethers of dinofragellate origin and sponge macrolides.

2.1.1 Fatty Acid Derivatives

Oxylipins, which are a major metabolite of fatty acids, are observed regularly in seaweeds and opisthobrachs that feed on seaweeds. Aplydilactone (3), which is an unusual oxylipin isolated from the sea hare *Aplysisa kurodai*, activates phospholipase A$_2$ (7). More than 50 halogenated C$_{15}$ acetogenins, which are derived from fatty acids, have been isolated from red algae of the genus *Laurencia*; laurencin (**4**) was first isolated from *L. nipponica* (8). These compounds were reported to be antifeeding and insecticidal. Certain species of soft corals are known to contain prostanoids, of which the most intriguing are punaglandins [punaglandin 1 (**5**)], the first halogenated prostaglandins isolated from the soft coral *Telesto riisei* (3). They are antiviral and cytotoxic.

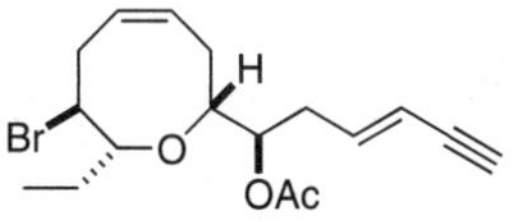

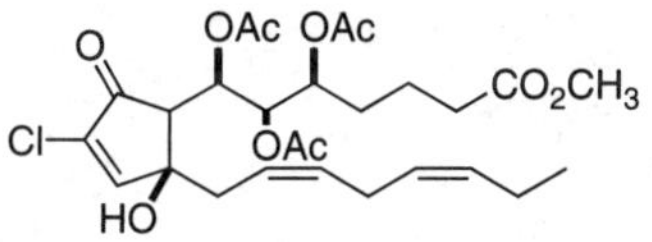

Fatty acid-derived cyclic peroxides are often found in marine sponges of the genus *Plakortis*; the first example was chondrillin (**6**). 1,2-Dioxane and 1,2-dioxolane carboxylates were also discovered from *Plakortis* sponges (9). These cyclic oxides show a range of biologic activities, for example, antimicrobial, cytotoxic, and antimalarial activities.

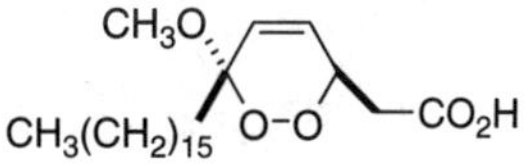

In addition, a variety of bioactive fatty acid derivatives have been isolated from marine organisms, including glycoceramides (10). Perhaps the most unusual example is a highly chlorinated sulfolipid **7** isolated as a cytotoxic principle from the mussel *Mytilus galloprovincialis* (10).

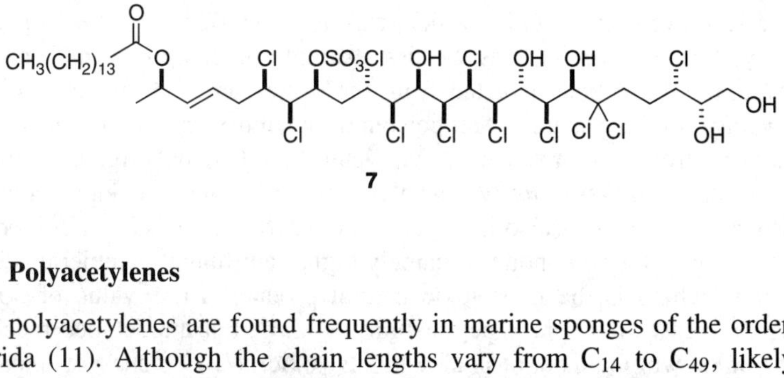

2.1.2 Polyacetylenes

Linear polyacetylenes are found frequently in marine sponges of the order Haplosclerida (11). Although the chain lengths vary from C$_{14}$ to C$_{49}$, likely they are to be derived from C$_{16}$ fatty acids. Polyacetylenes show a wide range of bioactivities, including antimicrobial, cytotoxic, antiviral, and enzyme inhibitory. Corticatic acid A (**8**), which is a C$_{31}$ polyacetylene carboxylic acid isolated from *Petrosia corticata*, is not only antimicrobial, but also it inhibits geranylgeranyltransferase I (12).

2.1.3 Polypropionates

The most prominent source of marine polypropionates are mollusks, in particular opisthobranchs (13). Among a variety of polypropionates, the simplest

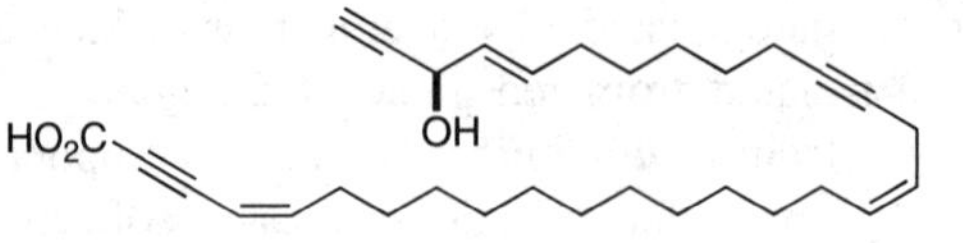

8

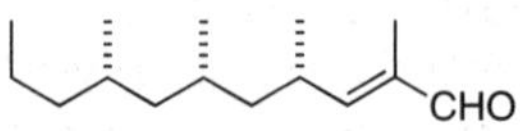

9

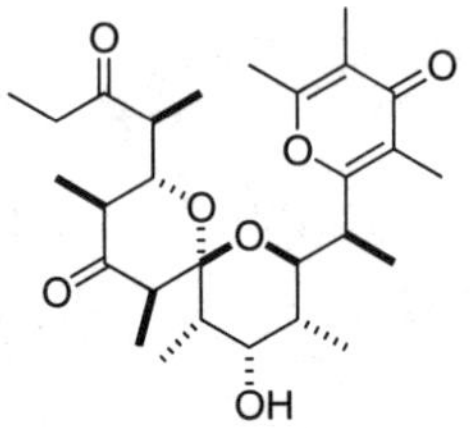

10

one is possibly siphonarienal (**9**) isolated from *Siphonaria grisea*, whereas such unusual pyrone-containing metabolites as siphonarin A (**10**) were isolated from *S. zelandica*. Marine polypropionates not only play defensive roles in mollusks, but also they show antimicrobial, antivial, and cytotoxic activities.

2.1.4 Polyethers

Dinoflagellates, which are unique, aquatic photosynthesizing eukaryotes, are classified in the kingdom Protoctista. They produce a variety of unusual polyketides with potent bioactivities (14). The most unique metabolites are "ladder-shaped" polyethers; brevetoxin B (**11**), which was isolated from *Gymnodinium breve* (currently, *Karenia brevis*), was the first example of this group, and more than 15 brevetoxins have been isolated from the same source. Ciguatoxin (**12**) is a causative agent of ciguatera, a fish poisoning in subtropical and tropical regions, first isolated from a moray eel. It was later found to be originated from the dinoflagellate *Gambierdiscus toxicus* that contains a more complex polyether, maitotoxin (**13**), which is also involved in ciguatera (5, 14). *G. toxicus* produces a variety of polyether compounds, namely highly antifungal gambieric acid (**14**). Maitotoxin, which is the most toxic natural product (LD_{50} value of 50 ng/kg in mice), increases membrane permeability to Ca^{2+}, but the detailed mechanism remains unknown. Ciguatoxin is also highly toxic (1/10 potency of maitotoxin) and activates voltage-gated Na^+ channels, whereas brevetoxins are much less toxic, although their mode of action is similar to that of ciguatoxin.

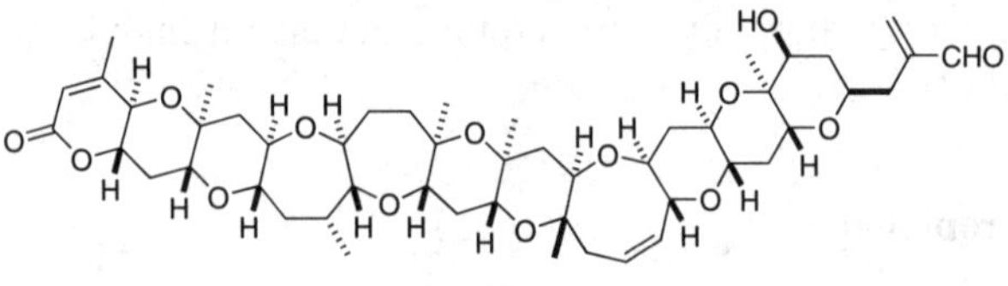

11

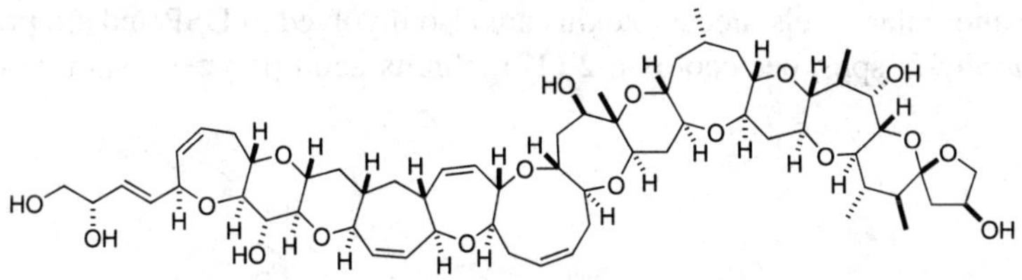

12

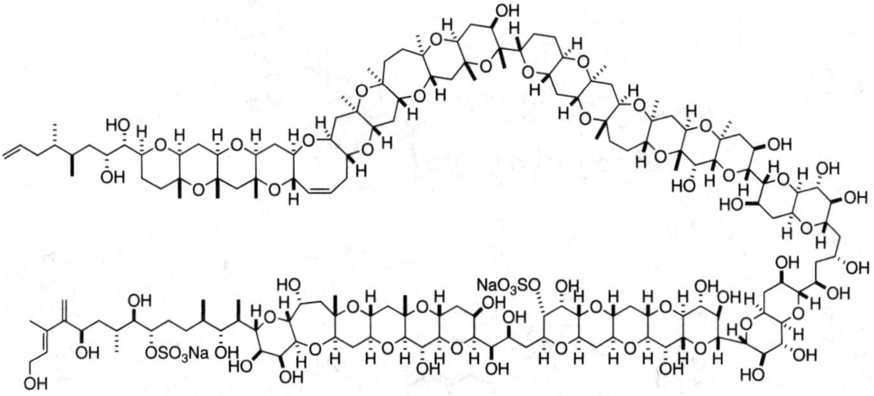

13

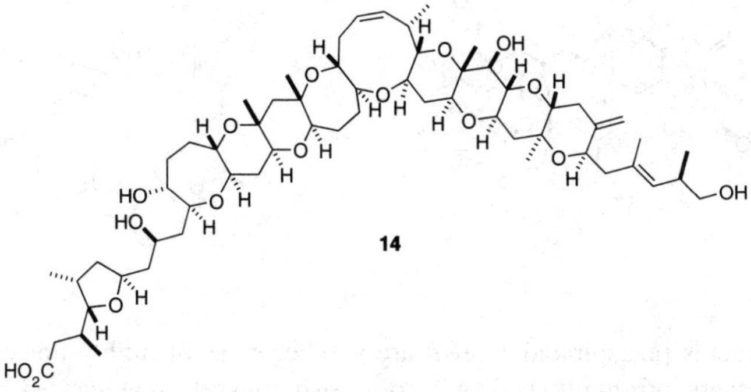

14

A different class of "ladder-shaped" polyethers, yessotoxins [e.g., yessotoxin 1 (**15**)], was isolated from toxic scallops; again its producer is the dinoflagellate *Protoceratium reticulatum* (5). The mode of action seems to be different from that of brevetoxins and ciguatoxins.

The sponge *Halichondria okadai* contains a polyether metabolite named okadaic acid (**16**) that was also isolated as a causative agent of diarrhetic shellfish poisoning (DSP) from mussels and other bivalves (14). However, the real producers are dinoflagellates of the genus *Dinophysis*. It is a potent cancer promoter that was found to be caused by inhibition of protein phosphatases 1 and

2A at nanomolar levels. Pectenotoxins are also involved in DSP and are produced by *Dinophysis* spp.; pectenotoxin 2 (**17**) inhibits actin polymerization (14).

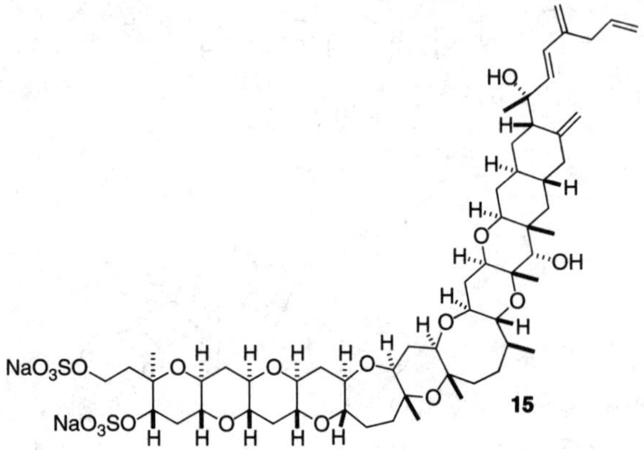

Azaspiracids [azaspiracid-1 (**18**)] are another class of highly unusual polyketide polyethers originally isolated from Irish mussels that caused azaspiracid shellfish poisoning (5). They are produced by the dinoflagellate *Protoperidinium crassipes*. A similar class of polyether toxins named pinnatoxins [pinnatoxin A (**19**)] were reported from the bivalve *Pinna pectinata*; a closely related species *P. attenuata* is known to cause food poisoning in China. Pinnatoxins are likely of dinoflagellate origin and activate Ca^{2+} channels (15).

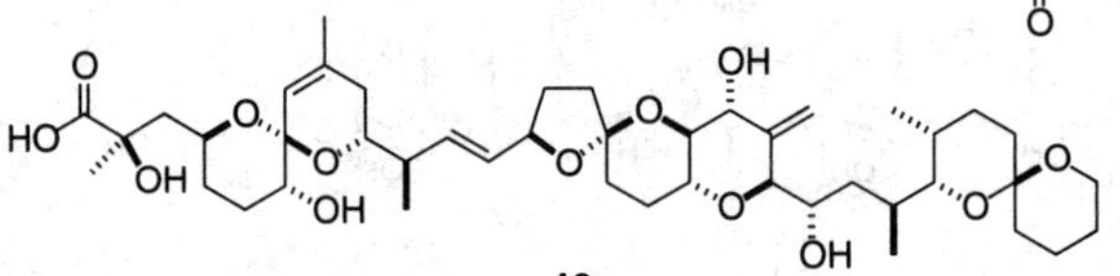

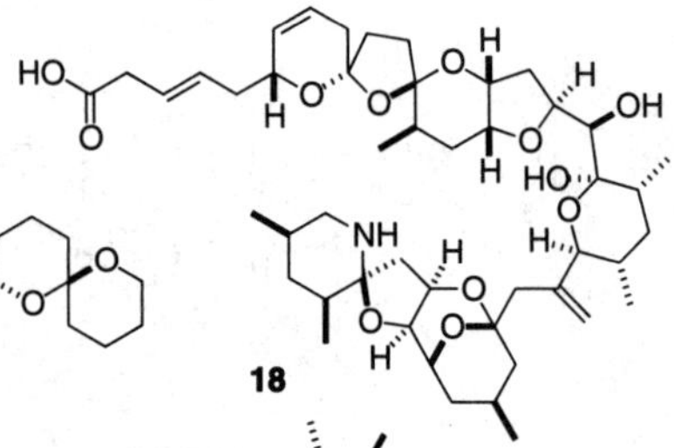

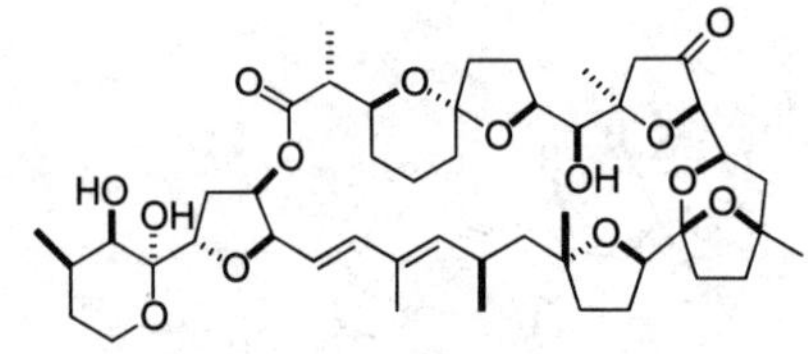

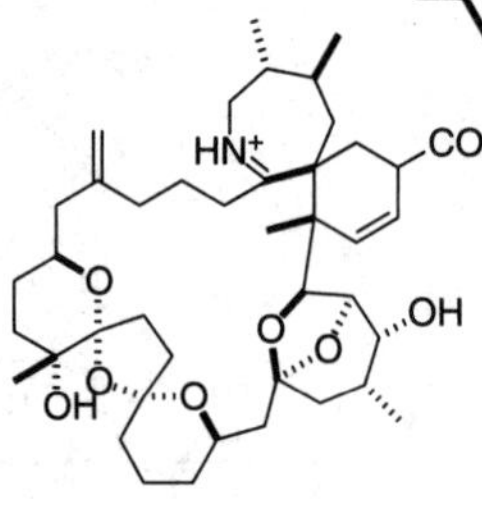

2.1.5 Open-Chain Polyketides

Dinoflagellates of the genus *Amphidinium* produce highly oxygenated, long-chain polyketides named amphidinols [amphidinol 3 (**20**)] that are highly antifungal and hemolytic (5). Palytoxin (**21**), which is found in zoanthids of the genus *Palythoa*, is an unusual polyketide as toxic as ciguatoxin (15). Its congeners were reported from the dinoflagellate *Ostreopsis siamensis*. Palytoxin is involved in several seafood poisonings and increases membrane permeability to Na^+ by acting on Na^+, K^+-ATPase.

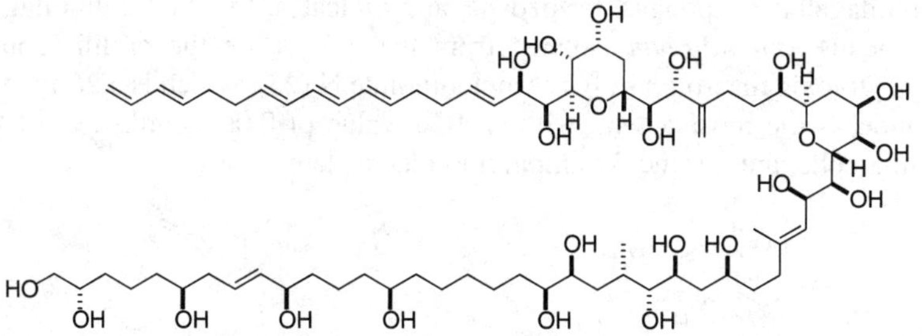

20

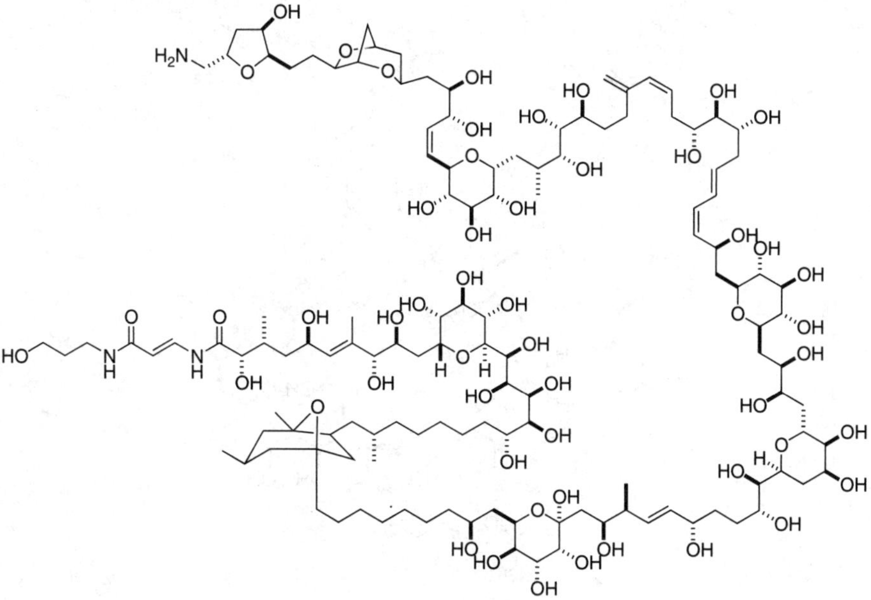

21

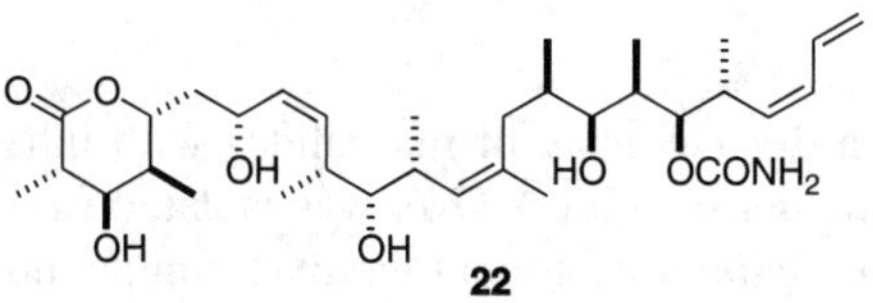

22

Discodermolide (**22**) is a polypropionate-based, unique compound isolated from the Caribbean deep-sea sponge *Discodermia dissoluta*. It is immunosuppressive as well as highly cytotoxic. More significantly, it stabilizes microtubules more potently than taxol (2).

2.1.6 Macrolides

A wide range of macrolides have been isolated from marine organisms, including dinoflagellates, sponges, bryozoans, and tunicates. Symbiotic dinoflagellates of the genus *Amphidinium* isolated from flat worms are the prolific source of highly cytotoxic macrolides (16). Amphidinolide N (**23**), which is a 26-membered macrolide, is the most potent, with an IC_{50} value of 0.05 ng/mL against L1210 leukemia cells, among the 34 amphidinolides isolated.

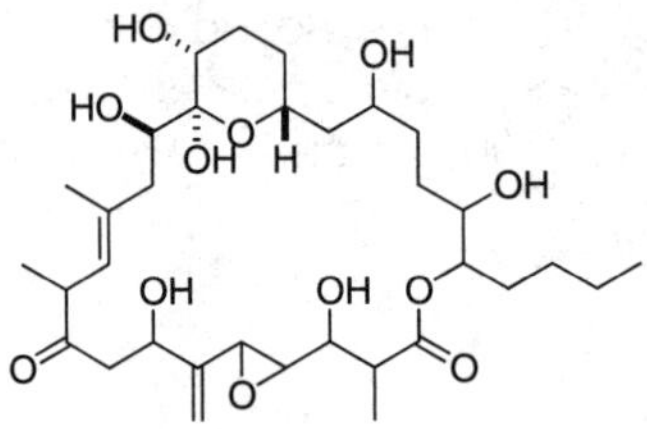

23

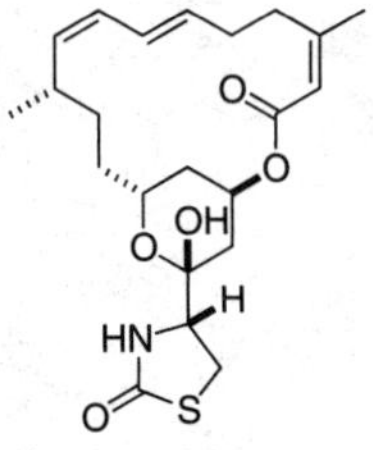

24

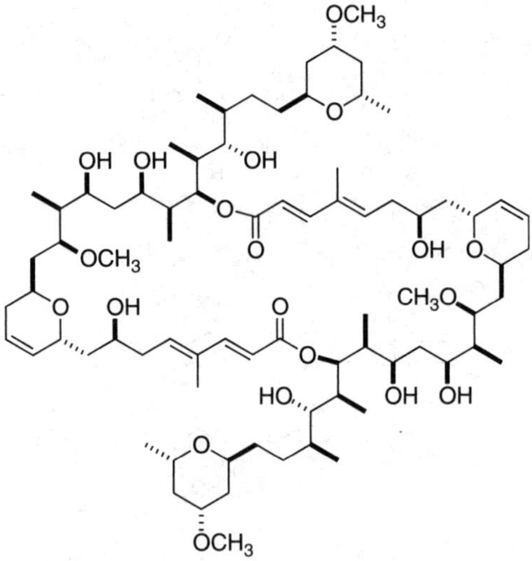

25

Sponges contain a diverse array of macrolides with intriguing activities. The first sponge macrolide, latrunculin A (**24**), was isolated from the Red Sea collection of *Latrunculia magnifica* as an ichthyotoxic compound and later was found to inhibit polymerization of G-actin allosterically. Swinholide A (**25**) is a macrodiolide originally isolated from the Red Sea sponge *Theonella swinhoei*. It is highly antifungal and cytotoxic, but its primary target is G-actin. Recently, it was discovered from cyanobacteria, although its eubacterial origin was predicted (6). Another class of macrolides that inhibit actin polymerization is the tris-oxazole-containing macrolides, namely kabiramides and mycaolides (1). Kabiramide C (**26**) was isolated from eggmasses of a nudibranch of the genus *Hexabranchus*, whereas mycalolides are of sponge origin, which indicates kabiramides were sequestered by nudibranchs from sponges. These macrolides are potently anti-fungal and cytotoxic. They bind G-actin in a molar ratio of 1:1, which thereby inhibits actin polymerization. Similar macrolides named aplyronines were isolated from the sea hare *Aplysis kurodai*; aplyronine A (**27**) shows remarkable antitumor activity (T/C > 500% at 0.08 mg/kg against P388 leukemia cells) as well as a similar mode of action to that of kabiramides (7). Presumably, aplyronines are of cyanobacterial origin.

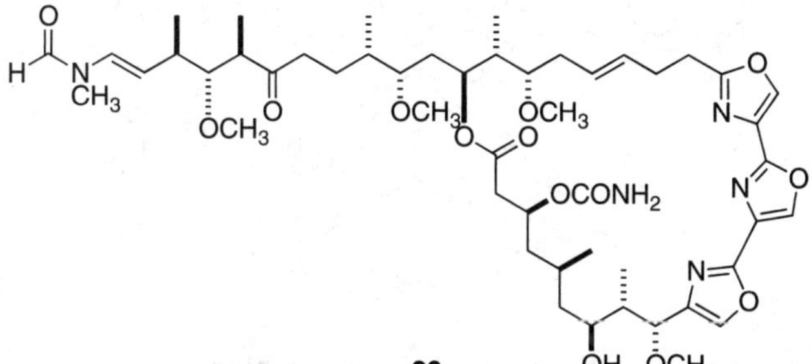

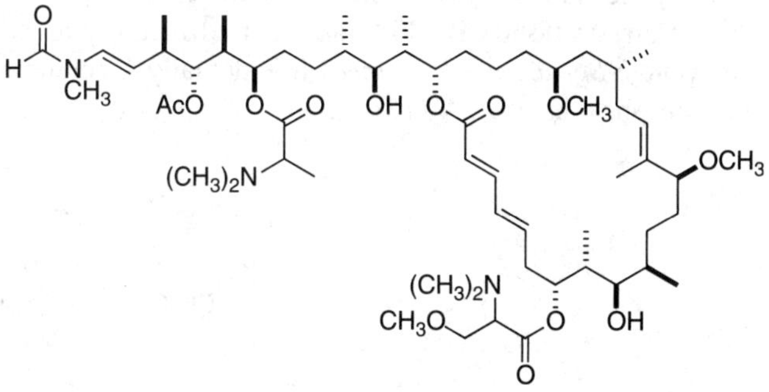

Halichondrin B (**28**), which is a polyether macrolide isolated from *H. okadai*, shows promising antitumor activity and has entered phase I clinical trials (2, 15). It inhibits polymerization of tubulin by binding to the colchicine domain. The macrocyclic portion seems to be essential for the activity (2). Halichondrins have been found in several species of sponges, which indicates their microbial origin (6).

Spongistatins/hyrtiostatins/cinachyrolide A [spongistatin-1 (**29**)] are highly unusual, 35-membered macrolides isolated from several sponges (1). They

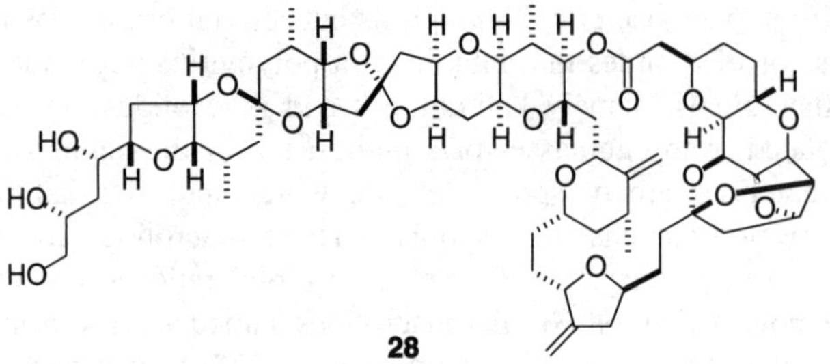

28

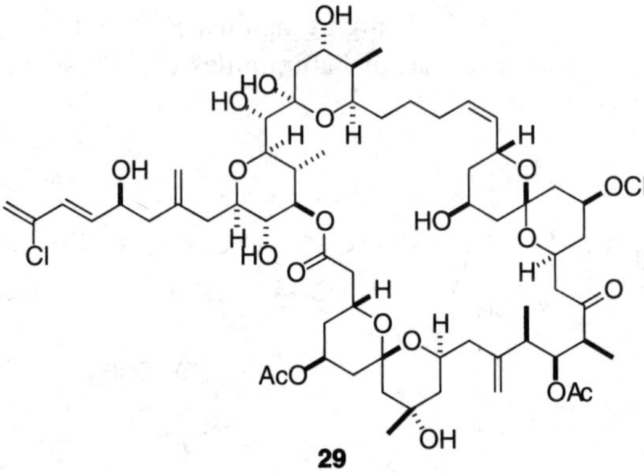

29

inhibit growth of tumor cells at sub-nM levels by binding to the vinca domain of tubulin. Their low yields and occurrence in several different sponges suggest their microbial origin. Spirastrellolide B (**30**) was reported recently as an antimitotic agent from the sponge *Spirastrella coccinea*. It is actually a potent and selective inhibitor of protein phosphatase 2A (12).

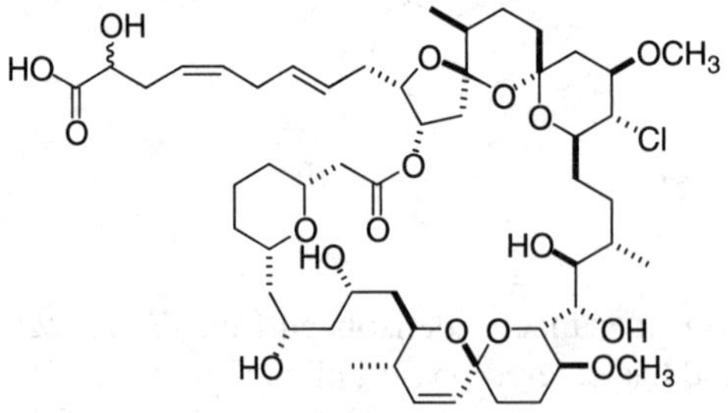

30

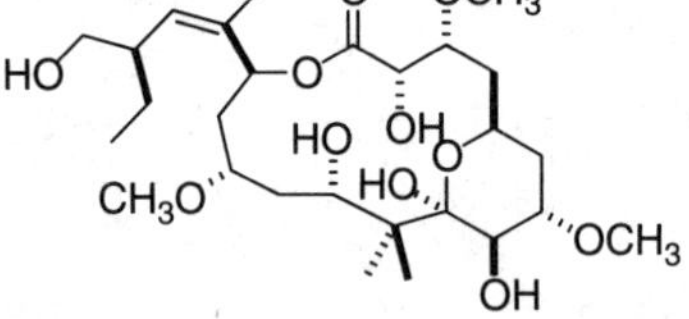

31

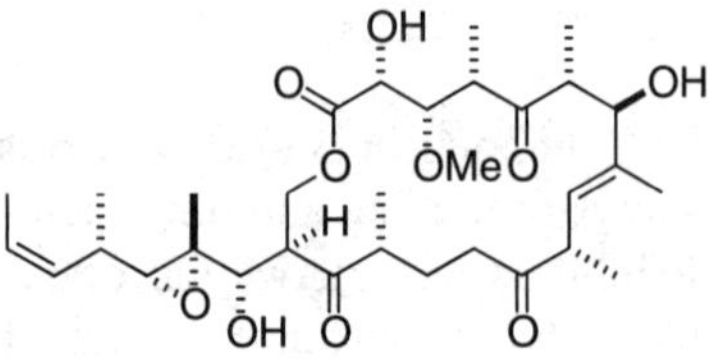

32

Peloruside A (**31**), which is a 16-membered, highly oxidized macrolide from the sponge *Mycale hentscheli*, induces tubulin polymerization (2). 13-Deoxytedanolide (**32**) isolated from Japanese sponges of the genus *Mycale* shows promising antitumor activity. It inhibits protein synthesis by binding to a 70 S large subunit of eukaryotic ribosome (17).

Bryostatins are medicinally important macrolides discovered from the cosmopolitan bryozoan *Bugula neritina* (2). Twenty bryostatins, which all possess a 20-membered ring, are known to date. Bryostatin 1 (**33**) showed good antitumor activity; it selectively modulates protein kinase C (12).

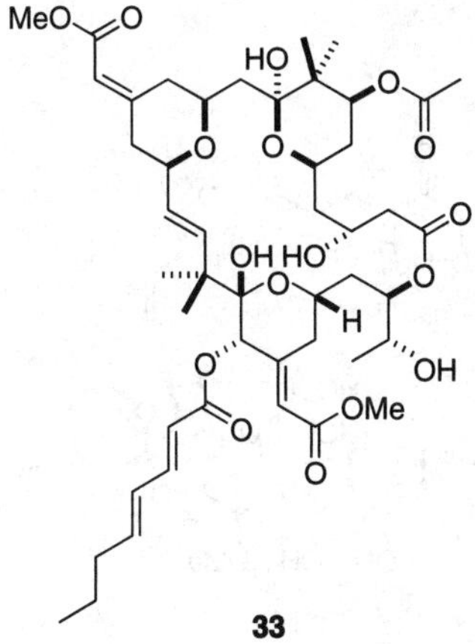

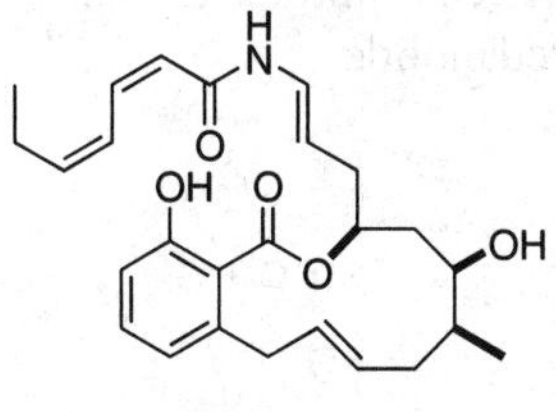

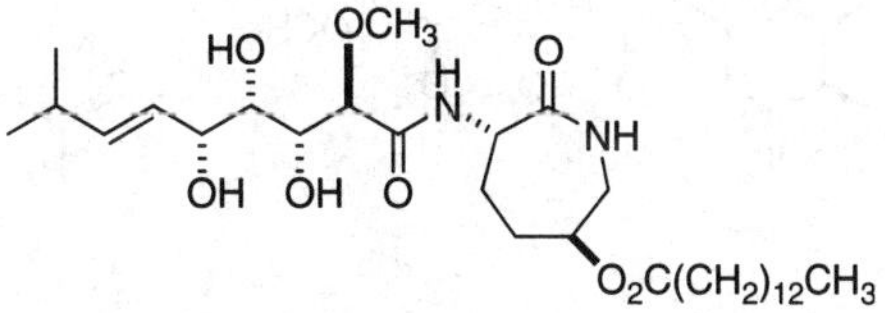

Phorboxazole A (**34**) is an unusual oxazole-containing macrolide isolated from a sponge *Phorbas* sp. (18). It is highly antifungal and cytostatic.

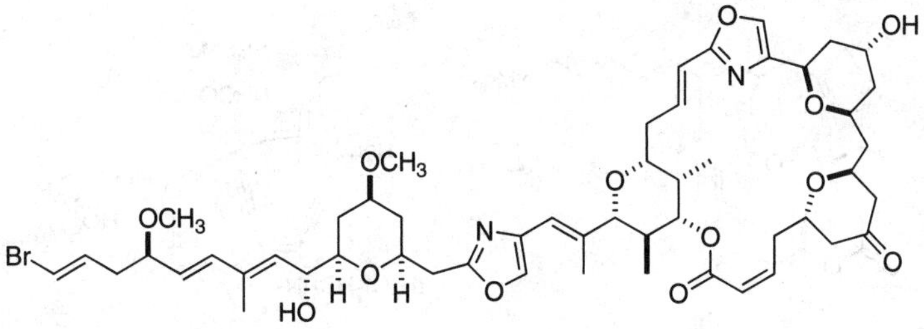

Salicylihalamide A (**35**) is a salicylic acid-containing macrolide enamide isolated from a sponge *Haliclona* sp., and it inhibits V-ATPases at a low nM concentration. Members of this family have been isolated from sponges, tunicates, and bacteria (12).

2.1.7   PKS/NRPS Metabolites

Bengamides were isolated originally as anthelmintic agents from a sponge *Jaspis* sp. Later, bengamide A (**36**) was found to inhibit growth of tumor cells as well as methionine aminopeptidases (2).

Marine natural products of the pederin class (mycalamides, onnamides, and theopederins) isolated from sponges are mixed biogenesis metabolites of polyketide synthase and nonribosomal synthase (1). In fact, biosynthesis gene clusters of this class have been cloned recently using metagenomic techniques from the marine sponge *Theonella swinhoei* (6). These compounds are potently cytotoxic; theopederin A (**37**) inhibits protein synthesis in a similar mode of action to that of 13-deoxytedanolide.

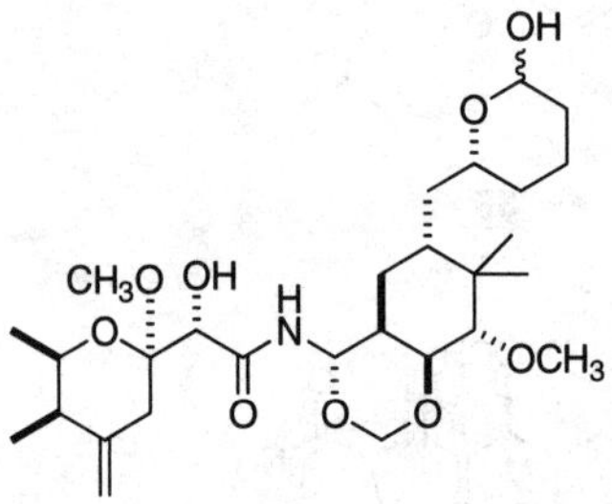

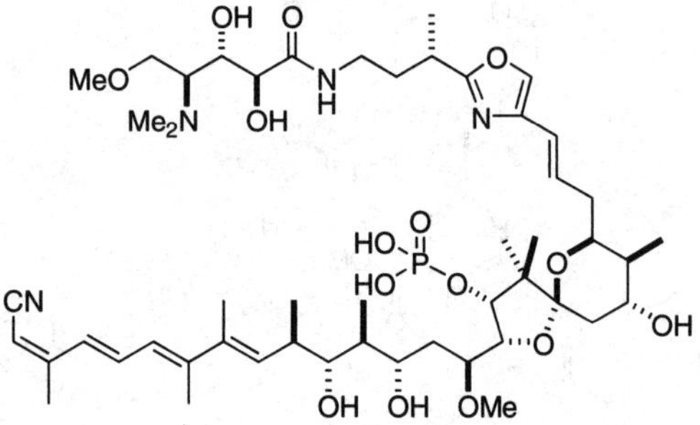

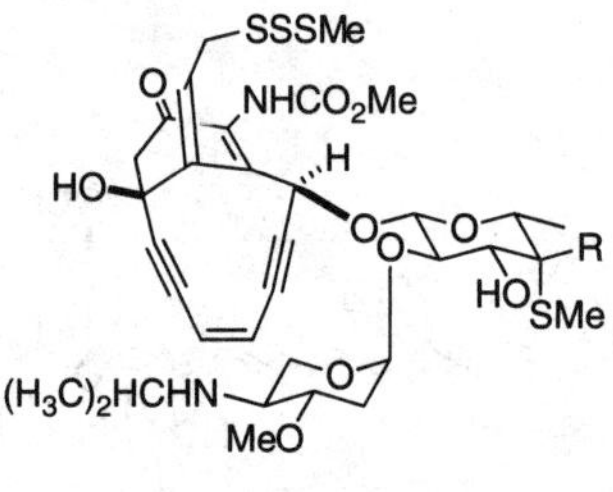

Calyculin A (**38**) is an extraordinary metabolite composed of C_{28} fatty acid and two γ-amino acids isolated from the sponge *Discodermia calyx*. It is not only highly antifungal and antitumor but also a potent cancer promoter that was found to be caused by potent inhibition of protein phosphatases 1 and 2A (1). More than 15 calyculin derivatives were isolated from several marine sponges, which indicates the involvement of symbiotic microorganisms in the production of calyculins (6).

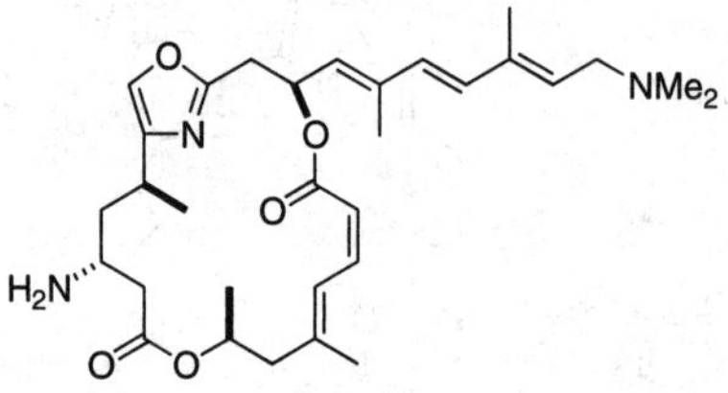

Pateamine (**39**) is a macrolide isolated from a marine sponge *Mycale* sp. Its potent cytotoxicity is attributed to inhibition of transcriptional initiation (19).

2.1.8   Aromatic Polyketides and Enediynes

Aromatic polyketides are rare metabolites in marine invertebrates. Naphtho-quinone and anthraquinone derivatives were reported as echinoderm pigments in the earlier stage of marine natural products research. The most interesting aromatic polyketides are the pentacyclic polyketides of the halenaquinone/halenaquinol class. Halenaquinone (**40**) was first isolated from the sponge *Xestospongia exigua*. Compounds of this class show a variety of biologic activities, which include inhibition of tyrosine kinase (12, 19).

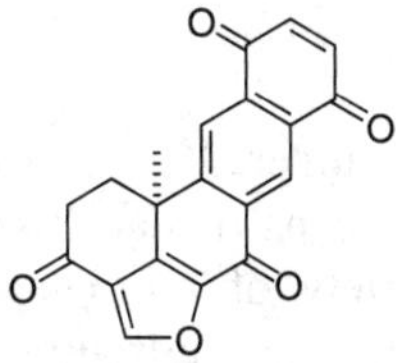

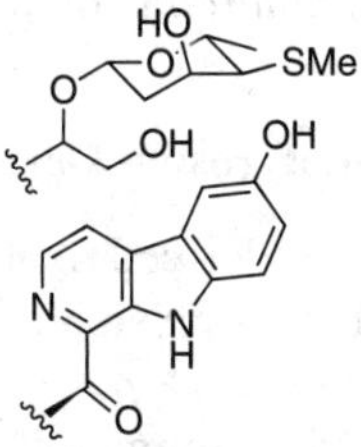

Only two groups of enediynes have been found from marine organisms; name-namicin (**41**) was first isolated the Fijian tunicate *Polysyncraton lithostrotum*, whereas shishijimicins A (**42**)-C, β-carboline-contining enediynes, were isolated from the Japanese tunicate *Didemnum proliferum*, along with **41**, which thus suggests the involvement of symbiotic microorganisms in their production (1). As expected, these compounds inhibited growth of tumor cells at pM levels.

2.2   NONRIBOSOMAL PEPTIDES AND AMINO ACID DERIVATIVES

In addition to ribosomal peptides, some of which exhibit interesting bioactivities as is the case of conotoxins, marine organisms; in particular, sponges and tuni-cates contain a wide variety of nonribosomal peptides, many of which contain unusual or unprecedented amino acids. It should be noted that these peptides show a range of biological activities.

2.2.1   Amino Acid Derivatives

Microalgae and macroalgae often contain UV-absorbing amino acids collectively dubbed "mycosporines". Mycosporine-glycine (**43**) is most widely distributed in marine organisms that use it for protection from UV irradiation. Domoic acid (**44**) is not only a helminthic agent originally isolated from a red alga, but also it is a causative agent of amnesic shellfish poisoning (14). It is a potent agonist of glutamate receptors. Dysiherbaine (**45**), which is a novel betaine isolated from the sponge *Dysidea herbacea*, is a potent non-MNDA-type agonist with very high affinity for kainate receptors (20).

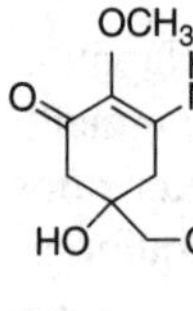

43 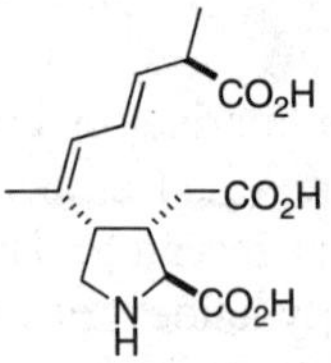   44 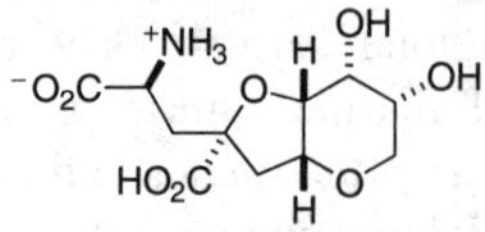   45 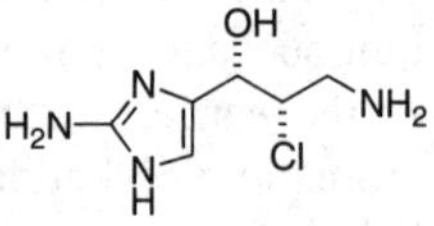   46

Girolline (**46**), which is a potent cytotoxin isolated from the sponge *Pseudaxinyssa cantharella*, inhibits protein synthesis (2).

2.2.2   Bromotyrosine Derivatives

Bromotyrosine-derived metabolites are often encountered in marine sponges of the families Aplysinidae and Pseudoceratidae, in particular *Pseudoceratina* (= *Psammaplysilla*) *purpurea*. They show a variety of biological activities, which include antimicrobial, enzyme inhibitory, and antifouling activities. Psammaplysin A (**47**) is antimicrobial, cytotoxic, and antifouling, whereas psammaplin A (**48**) is an inhibitor of histone deacetylase (2). The marine sponge *Ianthella basta* synthesizes at least 25 bastadins that are linear or cyclic peptides composed of four bromotyrosine residues [bastadin 5 (**49**)] and show antimicrobial, cytotoxic, and enzyme inhibitory activities as well as interaction with Ca^{2+} channels (21).

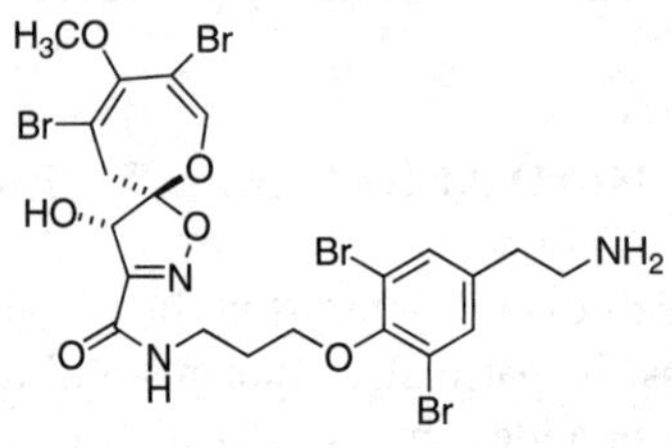

47 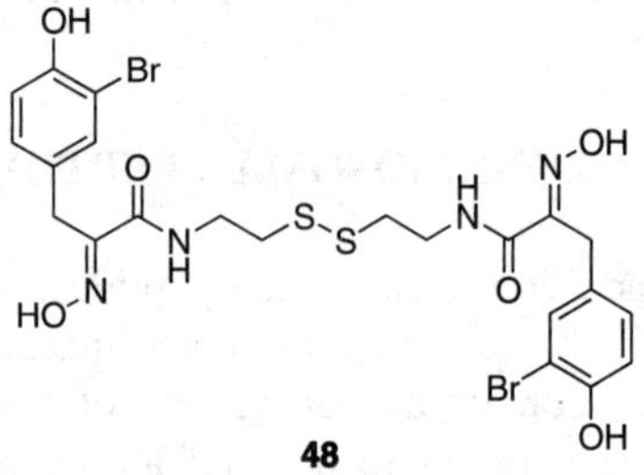   48

2.2.3   Linear Peptides

Sponges and tunicates frequently contain unusual linear and cyclic peptides; those from the former source were reviewed in 1993 and 2002 (22, 23). To avoid duplication, this review focuses on bioactive peptides isolated from other sources and new findings. Gymnangiamide (**50**) is the first described peptide from a hydroid (24). This pentapeptide from *Gymnangium regae* contains three previously unknown amino acids. Although moderately cytotoxic, its structure is reminiscent of dolastatin 10 (**51**), which is a powerful antitubulin agent of the sea hare *Dolabella auricularia* (2). Actually, dolastatins are of cyanobacterial origin. Dysinosin A (**52**) is a novel inhibitor of factor VIIa and thrombin, which was isolated from a new genus and species of Australian sponge of the family Dysideidae (12). Perhaps the most intriguing linear peptide is polytheonamide B (**53**), which is a potent cytotoxin discovered from *T. swinhoei* (25). It is composed of 48 amino acid residues, most of which are unusual amino acids. More significantly, it has the sequence of alternating D- and L-amino acids.

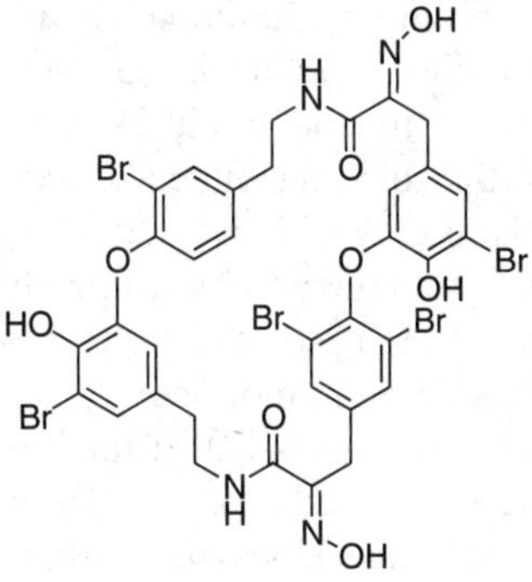

49

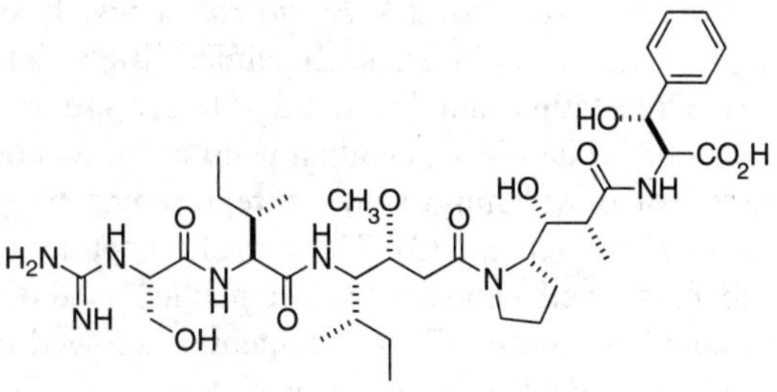

50

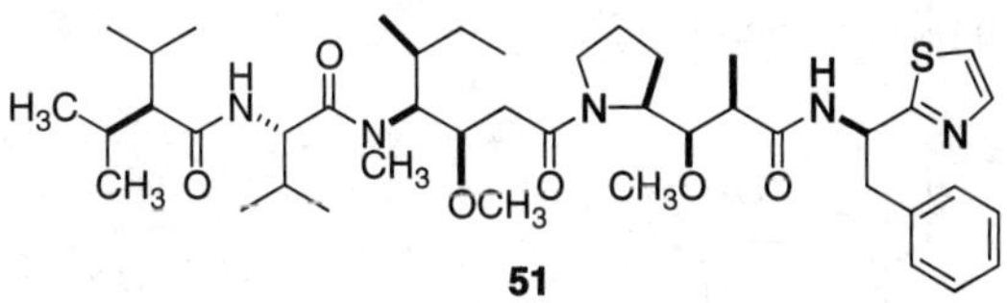

51

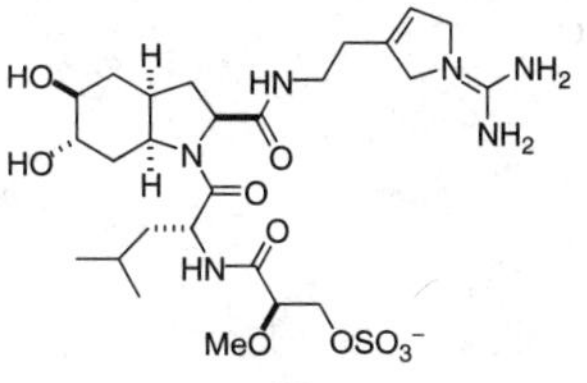

52

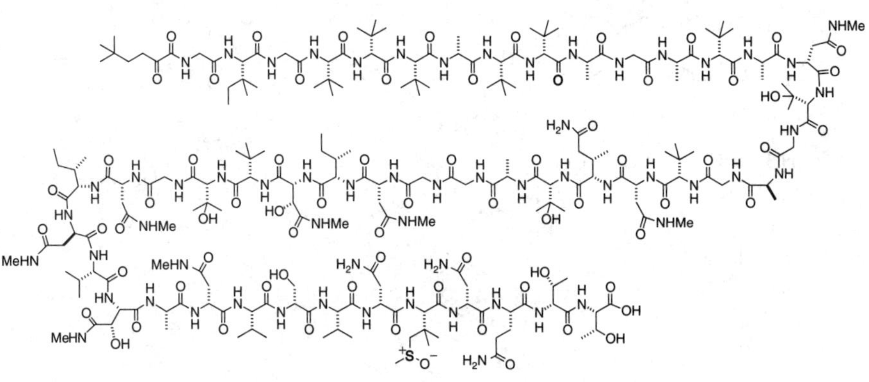

53

2.2.4   Cyclic Peptides

Most cyclic peptides have been isolated from sponges and tunicates. Azumamides A (**54**) through E, which are cyclic tetrapeptides isolated from the sponge *Mycale izuensis*, are the most recent addition to the list of cyclic peptides. They strongly inhibit histone deacetylase (12). Dolastatin 11 (**55**), which is a cyclic depsipeptide isolated from *D. auricularia*, stabilizes actin filaments as in the case of jaspakinolide/jaspamide. Kahalalide F (**56**) is a cyclic depsipeptide isolated from the sacoglossan mollusk *Elysia rufescens*. It shows promising antitumor activity and has entered Phase II clinical trials, but its mode of action is not clear (2). Dolastatins and kahalalides likely are sequestered by the mollusks from cyanobacterial diets. Ascidian peptides often contain thiazole/thiazoline and oxazole/oxazoline amino acids as represented by patellamide A (**57**) isolated from *Lissoclinum patella* (3). These cyclic peptides show cytotoxic activity. Probably the most well-known ascidian peptide is didemnin B (**58**) isolated from *Trididemnum solidum*. This depsipeptide showed remarkable antitumor activity and entered clinical trials, but it was dropped because of side effects (2). It inhibits protein synthesis. Aplidine (dehydrodidemnin B) (**59**) isolated from *Aplidium albicans* is more promising as an anticancer drug, although it shows multiple modes of action (2).

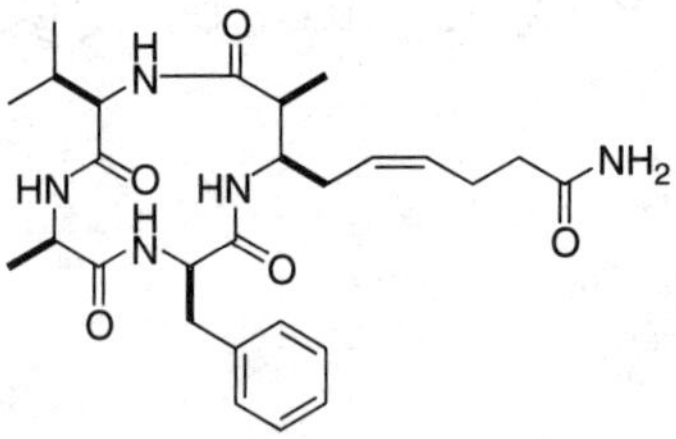

54

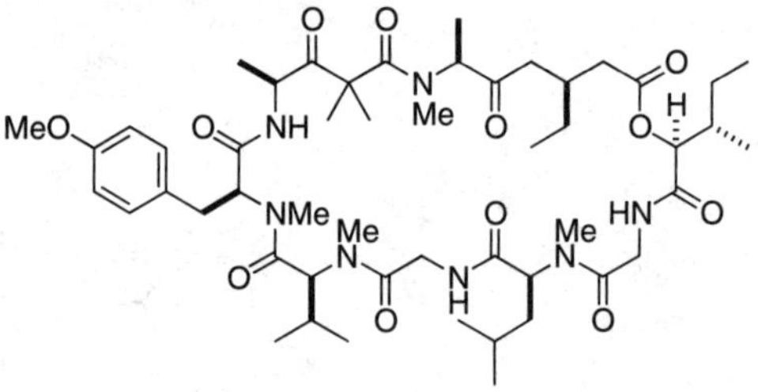

55

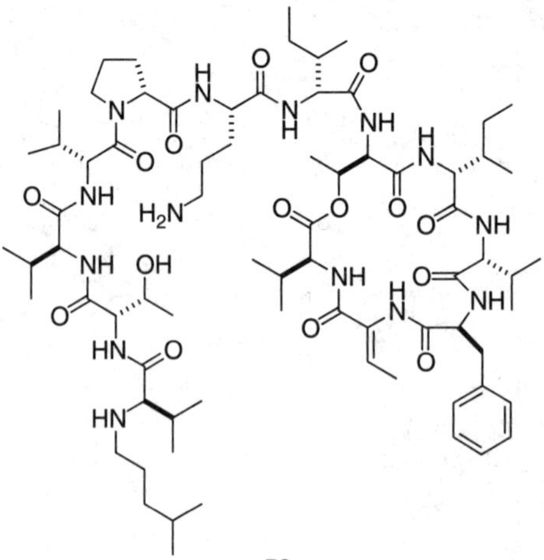

56

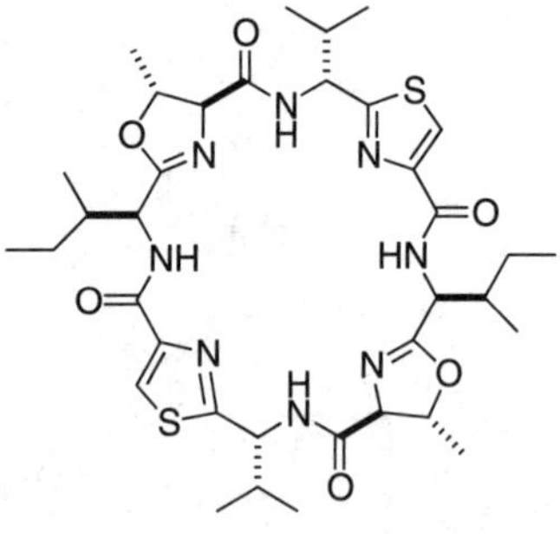

57

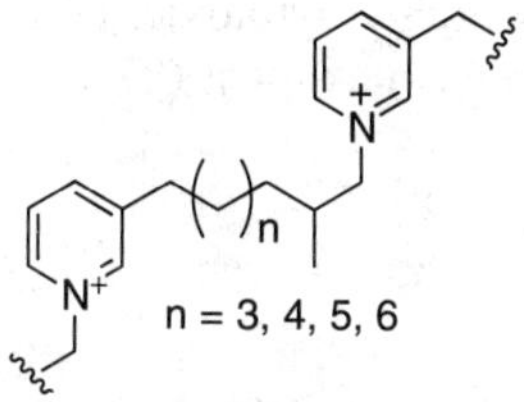

60

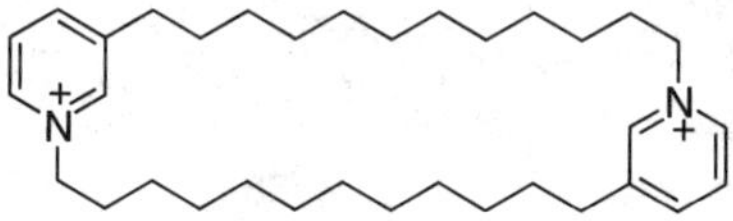

n = 3, 4, 5, 6

61

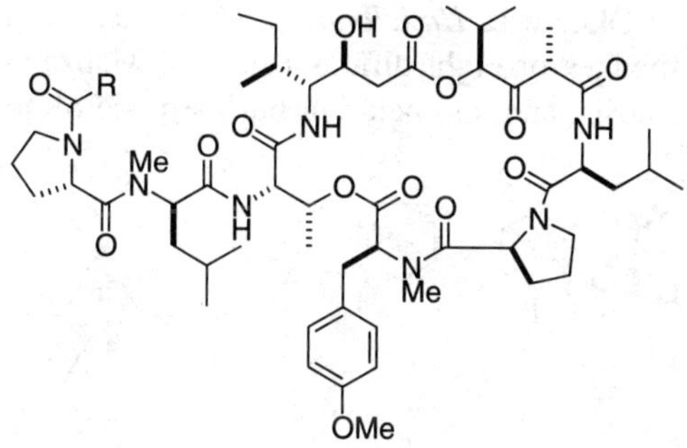

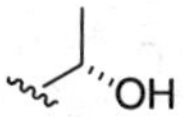

58: R = ⌇⌇⟍OH

59: R = ⌇⌇⟍O

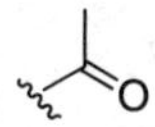

2.3 ALKALOIDS

Marine organisms produce a wide range of alkaloids with potent bioactivities, which include such specific classes as 3-alkylpiperidine, guanidine, indole, polyamine, pyridoacridine, and pyrrole-imidazole alkaloids. Their biological activities vary from antimicrobial to neurological.

2.3.1 3-Alkylpiperidines

A variety of 3-alkylpiperidine-derived compounds have been obtained from sponges belonging to five families of the order Haplosclerida (26). They show a range of bioactivities, for example, cytotoxic, antimalarial, and antifouling. It is likely that 3-alkylpiperidines are produced by sponge cells but not by symbiotic microorganisms.

The first 3-alkylpiperidine derivative reported is halitoxin (**60**), which was isolated from *Haliclona* sp. Similar polymeric alkylpyridines are also known from several sponges. In addition to the polymers, various types of metabolites of 3-alkylpyridines or 3-alkylpiperidines have been isolated, namely, macrocyclic *bis*-3-alkylpiperidine [telomerase-inhibitory cyclostellettamine A (**61**)] (12). *Bis*-quinolizadine [petrosin A (**62**)] and *bis*-1-oxaquinolizadine macrocycles [xestospongin C (**63**)] isolated from *Petrosia seriata* and *Xestospongia* spp., respectively, also belong to this group. The former is ichthyotoxic, whereas the latter is a potent vasodilator as well as an inhibitor of IP$_3$ receptor. Halicyclamine (**64**), which was isolated from *Haliclona* sp., is another group of macrocyclic *bis*-3-alkylpiperidines, whereas sarain A (**65**), which was isolated from *Reniera sarai*, has a more complex polycyclic core. These compounds are moderately cytotoxic.

The most well-known group of 3-alkylpiperidine alkaloids is the manzamines. Manzamine A (**66**), which is the first member of this group, was isolated from an Okinawan *Haliclona* sp. More than 20 manzamines have been isolated from sponges of eight different genera. Manzamine A is highly cytotoxic, antituberculosis, and antimalarial, but its mode of action remains unknown (27).

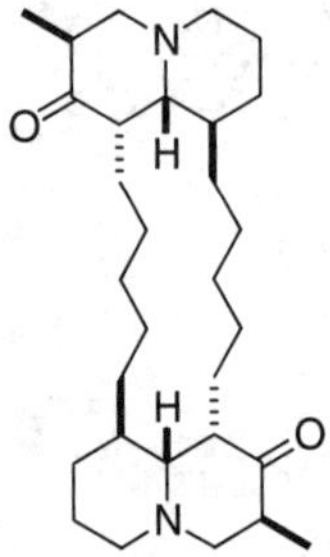

62 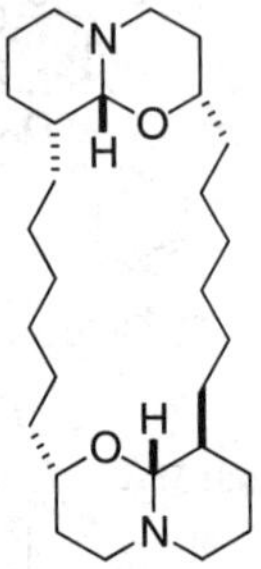   63 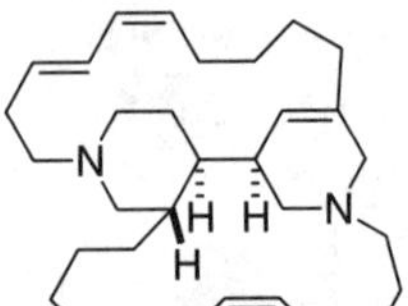   64 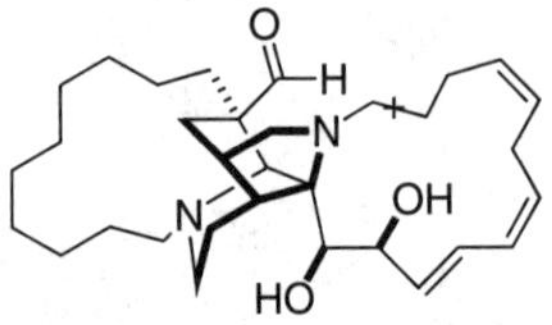   65

2.3.2 Pyrrole-Imidazoles

The pyrrole-imidazole alkaloids are found exclusively in marine sponges, in particular in the families Agelasidae, Axinellidae, and Halichondridae (28). Oroidin (67), which was discovered from *Agelas oroides*, is the building block of about 100 metabolites of this family, which can be divided into those derived from different modes of cyclization, for example, (−)-dibromophakelin (68) isolated from *Phakellia flabellate* and ageladine A (69) from *Agelas nakamurai*, and those derived from different modes of dimerization, namely sceptrin (70), which is isolated from *Agelas sceptrum*, palau'amine (71) from *Stylotella aurantium* and massadine (72) from *Stylissa massa*. Quite recently, a dimer of massadine (oridin tetramer) named stylissadine A was isolated from *Stylissa caribica*. The pyrrole-imidazole alkaloids show a range of bioactivities, which include antimicrobial, cytotoxic, antagonistic to receptors, immnosuppressive, enzyme inhibitory, and antifouling.

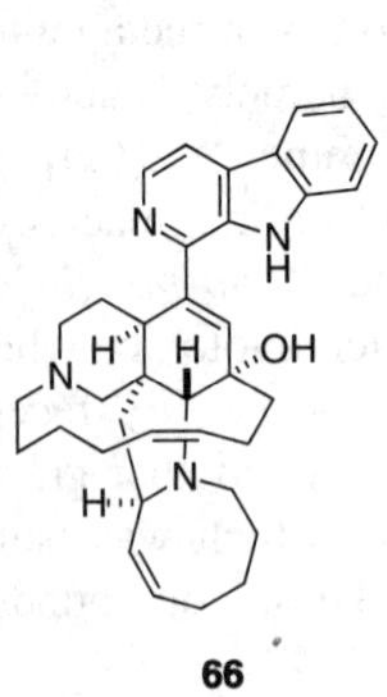

66 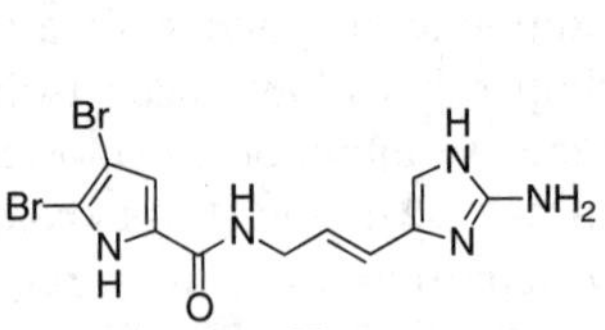   67 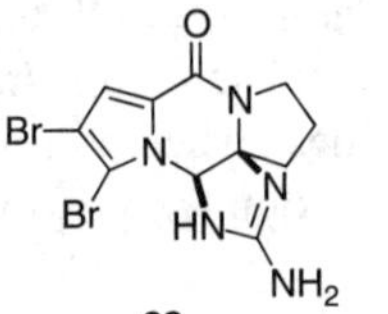   68 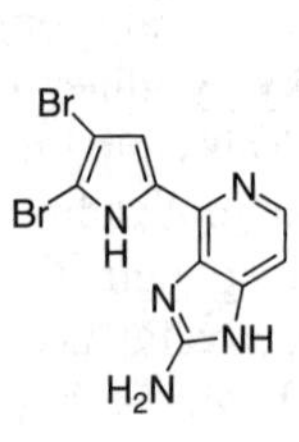   69

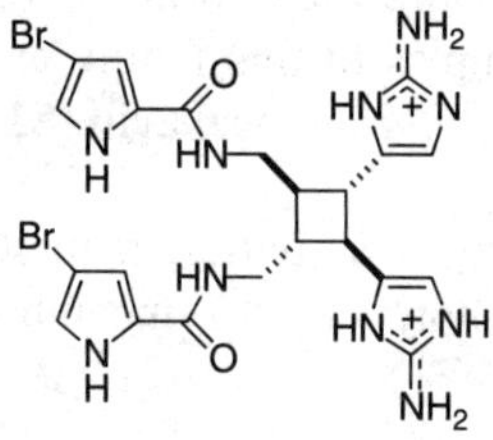

70 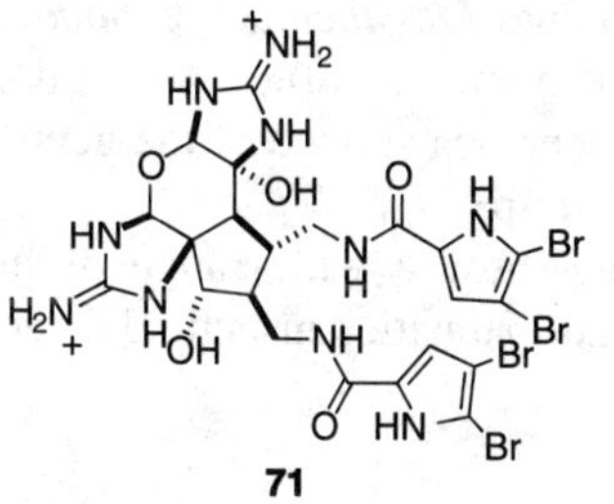   71 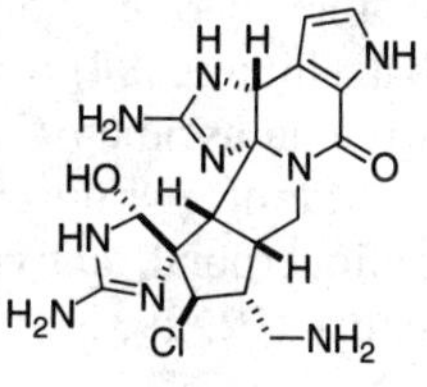   72

2.3.3 Pyridoacridines and Related Alkaloids

Pyridoacridines are highly colored polycyclic alkaloids mainly isolated from sponges and tunicates (29). They are divided into four groups, the simplest of which is styelsamine D (73), which was isolated from the ascidian *Eusynstyela latericius*. The pentacyclic pyridoacridines are also classified into two groups as represented by amphimedine (74), which was isolated from a sponge *Amphimedon* sp., and ascididemin (75), which was isolated from a tunicate *Didemnum* sp., respectively. Dercitin (76) is a thiazole-containing pentacyclic alkaloid isolated from a sponge *Dercitus* sp., whereas cyclodercitin (77), which was isolated from a sponge *Stelletta* sp., is a member of the hex acyclic group. The most complex octacyclic pyridoacridines are represented by eudistone A (78), which was isolated from *Eudistoma* sp.

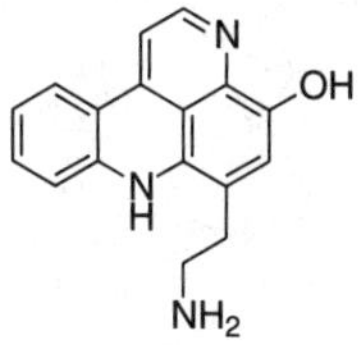

73 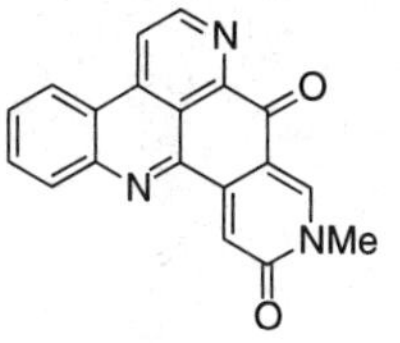   74 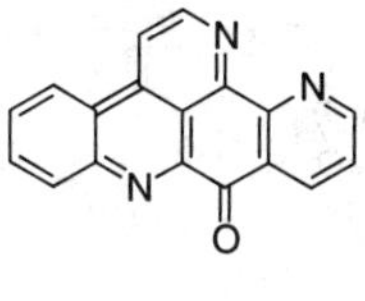   75 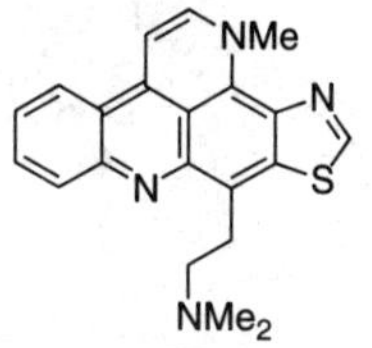   76

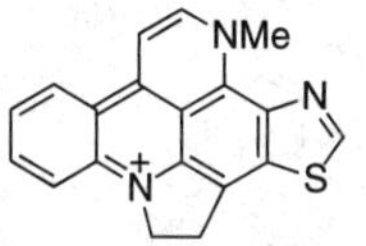

77 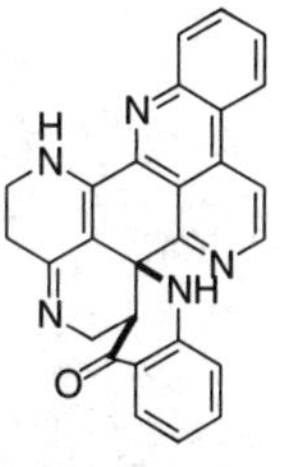   78 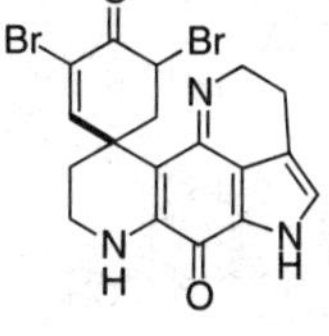   79 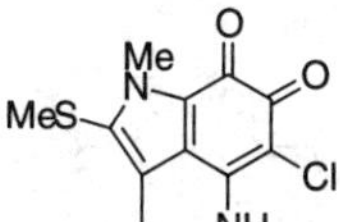   80 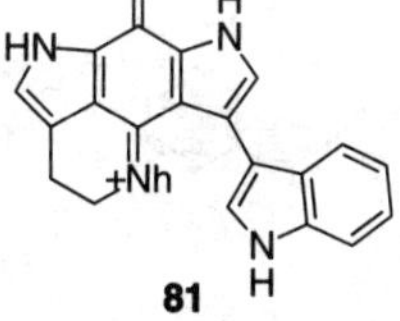   81

Alkaloids related to pyridoacridines are known also from sponges and tunicates. Discorhabdin C (79), which is the first marine pyrroloquinoline alkaloid, was isolated from the sponge *Latrunculia* cf. *bocagei*. More than 20 alkaloids of this class are known at moment (30). Additional examples of this family are batzelline A (**80**) from a deep-sea sponge of the genus *Batzella* and wakayin (**81**) from an ascidian *Clavelina* sp.

Marine pyridoacridines show a wide range of biological activities, namely antimicrobial, antiviral, antiparasitic, insecticidal, antitumor, and enzyme inhibitory (12, 29, 30).

2.3.4 Pyrrole Alkaloids

Highly substituted pyrroles are often found in sponges, ascidians, and mollusks (30). Lamellarins A (**82**)-D are the first metabolites of this group that were reported from a mollusk *Lamellaria* sp., followed by the discovery of more than 50 alkaloids of this class from sponges and ascidians. Polycitone A (**83**) was isolated from a tunicate *Polysitor* sp., whereas storniamide A (**84**) was isolated from a sponge of the genus *Cliona*. The most recent addition to this class is dictyodendrins A (**85**) through E isolated as telomerase inhibitors from the sponge *Dictyodendrilla verongiformis* (12). Various bioactivities have been reported for lamellarins and related compounds, which include antitumor, antiviral, and enzyme inhibitory.

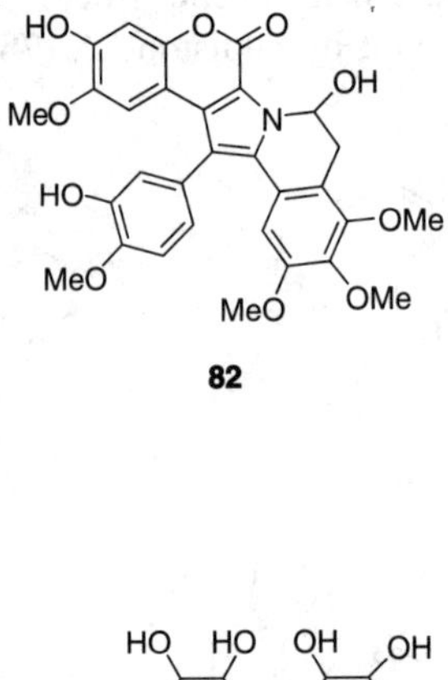

82

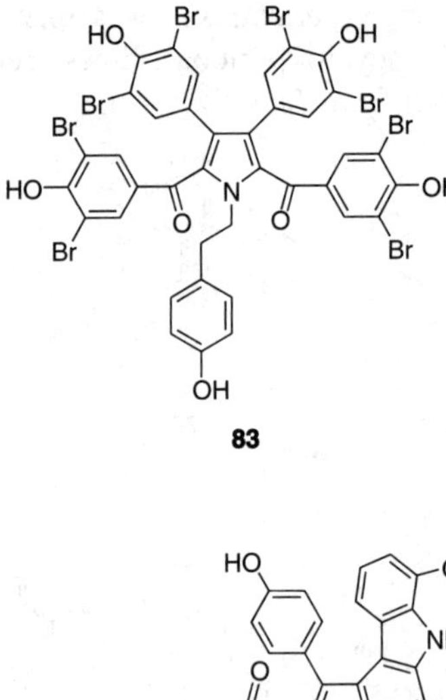

83

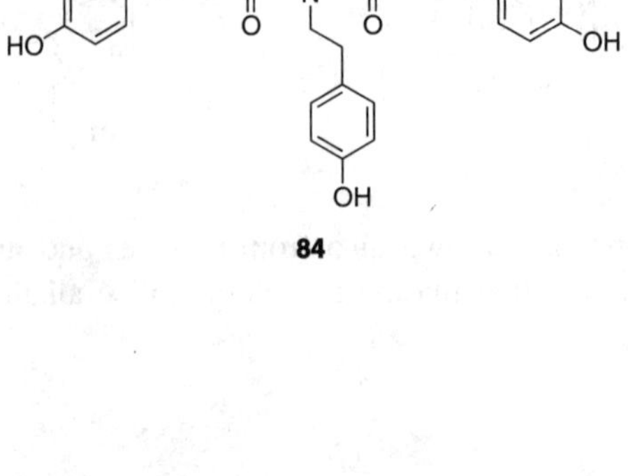

84

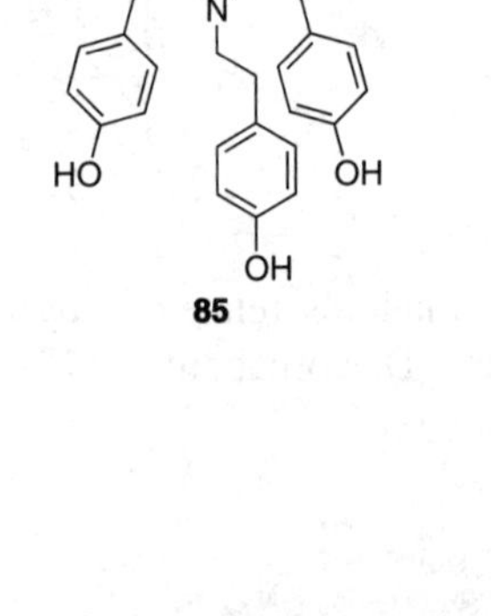

85

2.3.5 Indole Alkaloids

Many indole-containing metabolites have been reported from marine organisms, some of which were already mentioned. This section focuses on important indole-containing substances that belong to structural classes not mentioned above. These alkaloids show antimicrobial, antiparasitic, antitumor, and enzyme inhibitory activities (27). Dragmacidin (**86**) is a novel *bis*-indole isolated from a deep-sea sponge *Dragmacidon* sp., whereas another deep-sea sponge *Spongosorites ruetzleri* contains a similar *bis*-indole, topsentin (**87**).

β-Carboline-containing metabolites are known from sponges and tunicate. Eudistomins were the first β-carboline alkaloid isolated from marine organisms. Eudistomin K (**88**), which is a novel oxathiazepine ring containing β-carboline, was isolated from the tunicate *Eudistoma olivaceum*, whereas a guanidine-containing grossularine 1 (**89**) was from the tunicate *Dendrodoa grossulatia* (27).

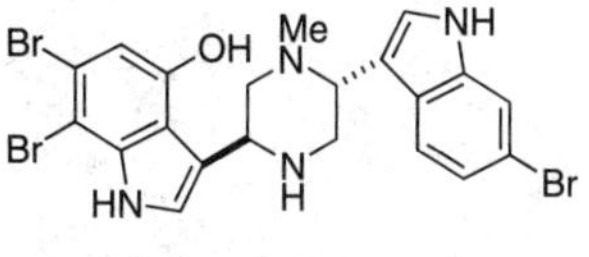

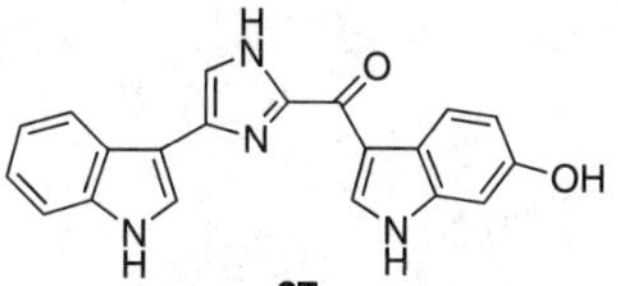

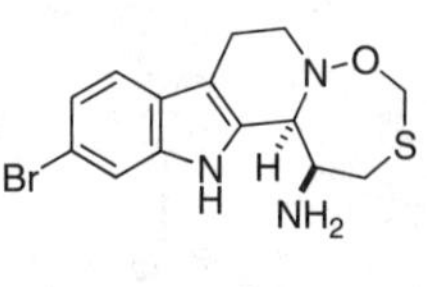

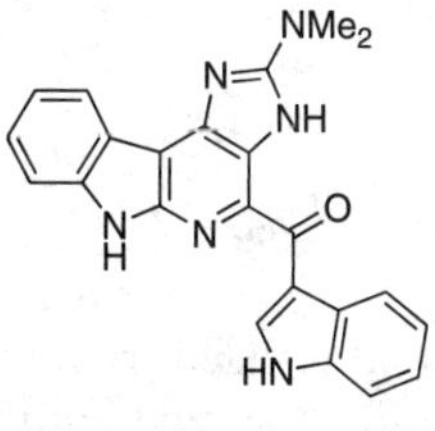

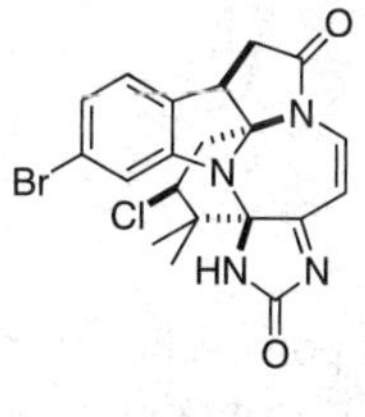

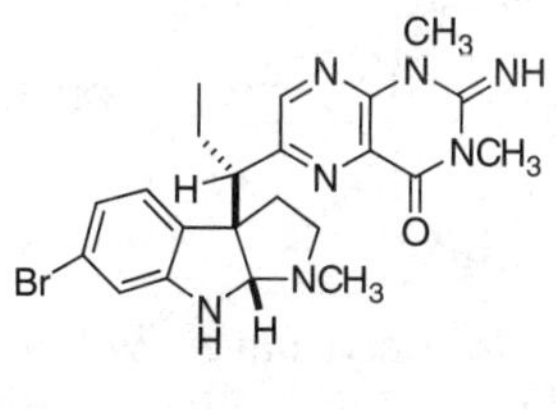

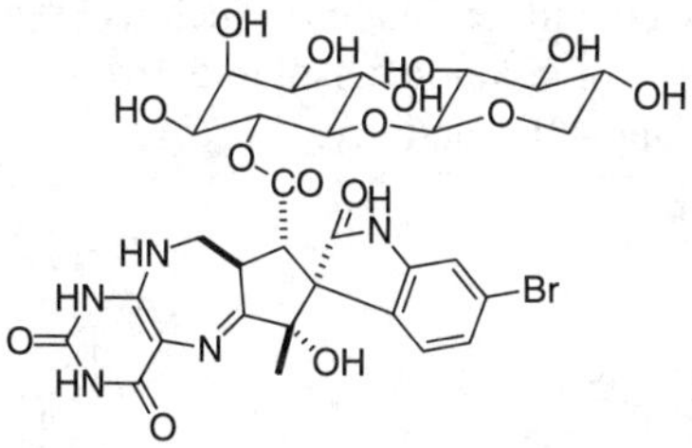

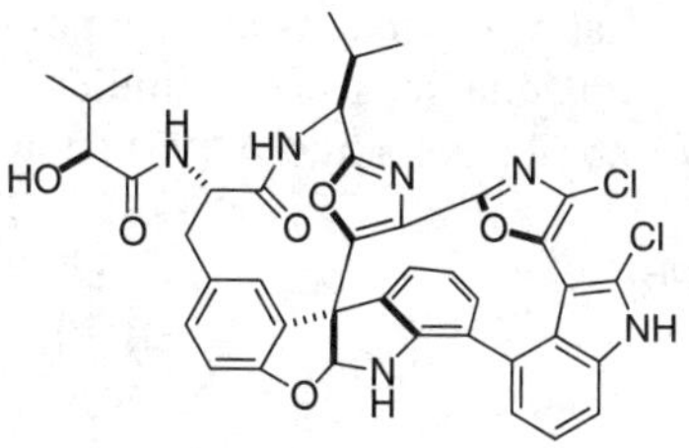

Physostigmine alkaloids are often contained in bryozoans. The highly unusual constituent is securine A (**90**) isolated from *Securiflustra securifrons* from the North Sea. Another interesting physostigmine is urochordamine A (**91**) isolated as a larval settlement inducer from the ascidian *Ciona savignyi* (30). Neosurugatoxin (**92**), which is a causative agent of "ivory shellfish poisoning" isolated from the gastropod *Babylonia japonica*, is a reversible nicotinic acetylcholine antagonist (14). Its bacterial origin has been suggested. Finally, diazonamide A (**93**) is a highly unusual cytotoxic metabolite of the tunicate *Diazona angulata*. It inhibits tubulin polymerization (2).

2.3.6 Guanidine Alkaloids

Many guanidine-containing compounds have been reported from diverse marine organisms (31). No doubt exists that the most well-known guanidine alkaloid is tetrodotoxin (**94**), which is a causative agent of puffer fish poisoning (14). It is highly toxic because of inhibition of voltage-gated Na^+ channels. Production of tetrodotoxin by bacteria of such genera as *Pseudoalteromonas* and *Vibrio* was reported. Similarly, saxitoxin (**95**) causes paralytic shellfish poisoning. Bivalves accumulate the toxin from dinoflagellates, for example, *Alexandrium catenella*, *A. tamarense*, and *Gymnodinium catenatum*. Its mode of action is similar to that of tetrodotoxin. More than 30 saxitoxin derivatives are known to date.

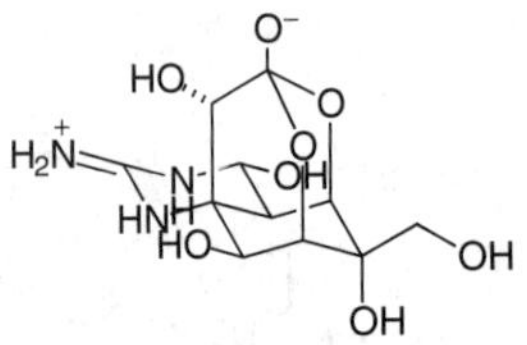

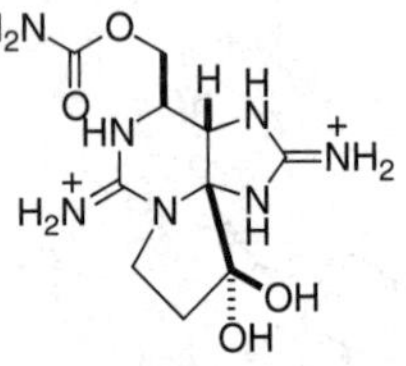

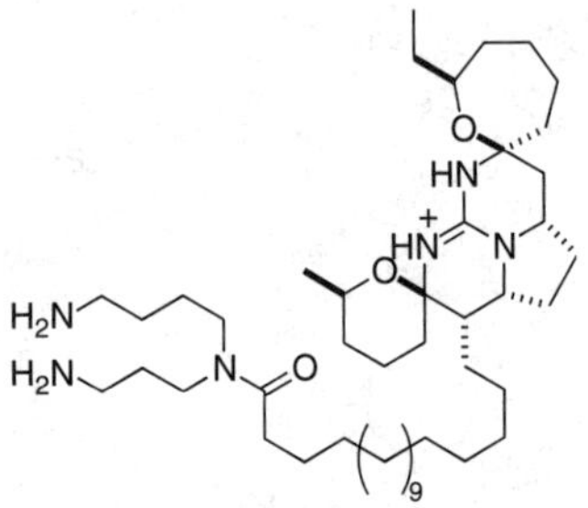

Ptilomycalin A (**96**) is a novel polycyclic guanidine alkaloid isolated from the sponge *Ptilocaulis spiculifer* (31). It is highly cytotoxic, antifungal, and antiviral. The related alkaloids were reported from the Mediterranean sponge *Crambe crambe* that also contains other types of guanidine alkaloids such as crambescin A (**97**). Batzelladines are a similar class of alkaloids isolated from a sponge *Batzella* sp.; batzelladine A (**98**) shows anti-HIV activity.

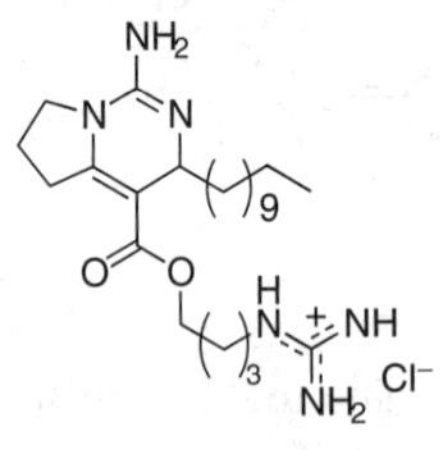

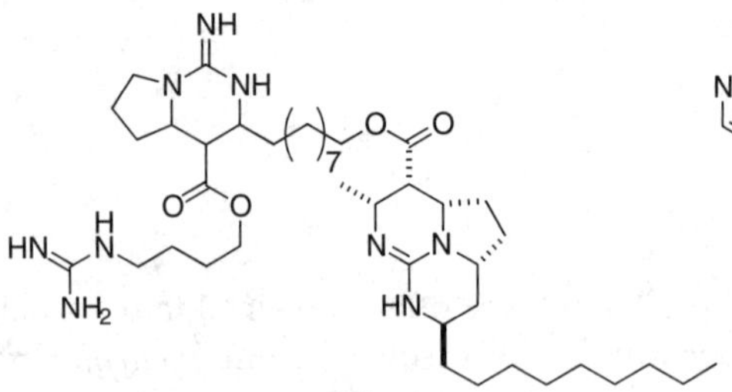

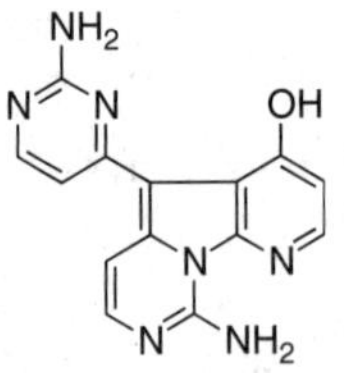

Variolin B (**99**), which is a pyridopyrropyrimidine alkaloid isolated from the sponge *Kirkpatrichia varialosa*, is strongly cytotoxic, antifungal, and antiviral (30). It inhibits cyclin-dependent kinases (12).

2.3.7 Polyamine Alkaloids

Stellettadine A (**100**), which is isolated from a sponge *Stelletta* sp., represents the first polyamine alkaloid with an arcaine backbone (32). Similar alkaloids [e.g., stellettazole A (**101**)] are also known from the same source. These compounds show larval settlement-inducing, antifungal, and enzyme inhibitory activities. Isoprenylated polyamines are encountered occasionally as metabolites of soft corals as represented by sinulamide (**102**), which inhibits H,K-ATPase (12).

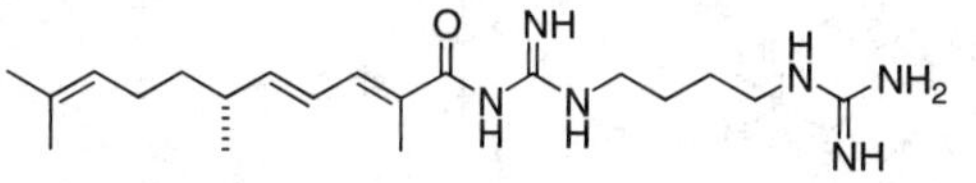

100

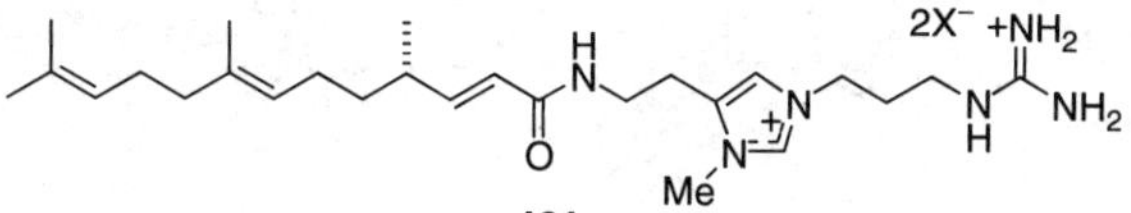

101

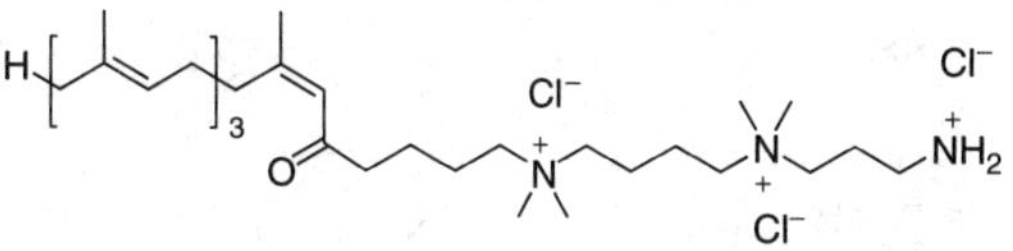

102

2.3.8 Quinolines and Isoquinolines

Aaptamine (**103**), which is a cytotoxic benzonaphthyridine alkaloid isolated from the sponge *Aaptos aaptos*, induces differentiation in chronic leukemia cells (33). Schulzeine A (**104**) is a novel dihydroquinoline alkaloids isolated from the sponge *Penares schulzei* that inhibits glycosidases (12).

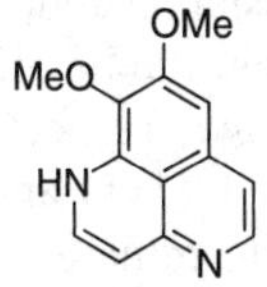

103 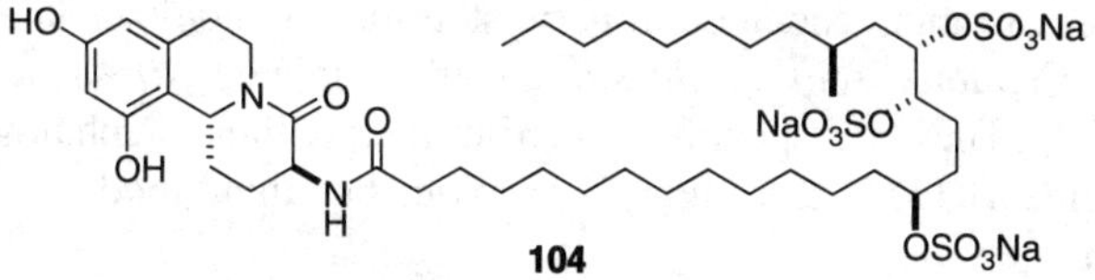   **104**

Sponges and tunicates contain tetrahydroisoquinoline alkaloids (30). Renieramycin A (**105**) from a sponge *Reniera* sp. represents the first example of this particular class of alkaloids isolated from marine organisms. Ecteinascidins, in particular ecteinascidin 743 (**106**), which was isolated from the tunicate *Ecteinascidia turbinata*, are promising as anticancer leads. Ecteinascidin 743 cleaves DNA chains and is in clinical trials (2).

2.3.9 Quinolizidines and Indolizidines

Only a few compounds of these classes of alkaloids have been reported from marine organisms. Clavepictine B (**107**), which was isolated from the ascidian

Clavelina picta, is marginally cytotoxic. Stellettamine A (**108**), which is an indolizidine derivative from a sponge *Stelleta* sp., is antifungal and cytotoxic. It also inhibits calmodulin (12). Lepadiformine (**109**), which has a similar structural feature isolated from the tunicate *Clavelina inoluccensis*, inhibits K$^+$ channels (33).

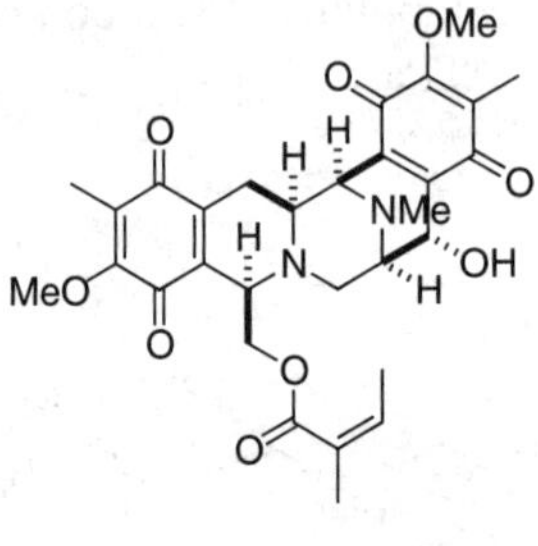

105 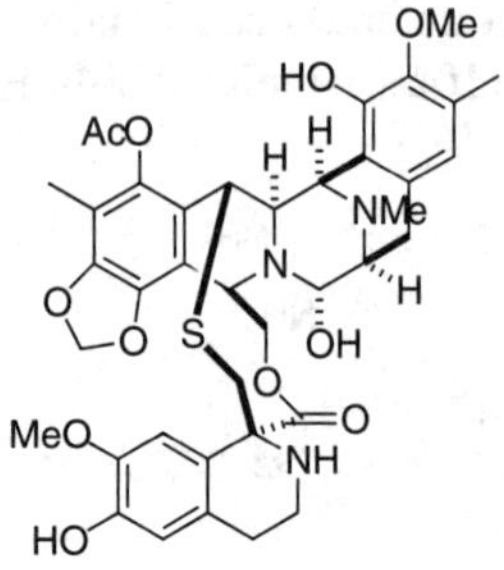   106 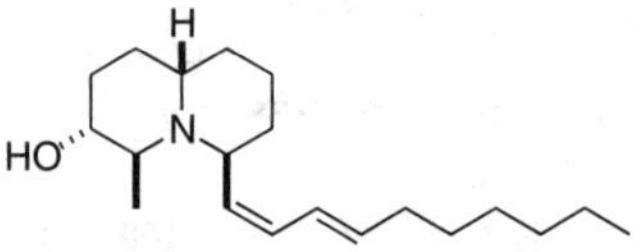   107

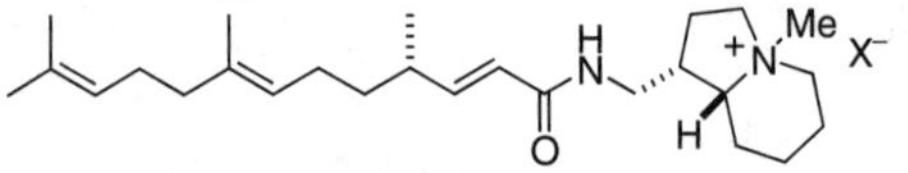

108 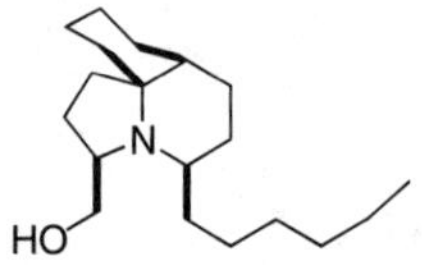   109

2.3.10   Steroidal Alkaloids

Steroidal alkaloids belong to a rare class of marine natural products. Plakinamine A (**110**), which is an antimicrobial metabolite from a sponge *Plakina* sp., is the first steroidal alkaloid isolated from marine organisms. Cephalostatins and ritterazines are unprecedented dimeric steroidal alkaloids isolated from the hemichordate *Cephalodiscus grichristii* and the tunicate *Ritterella tokioka*, respectively (1, 30). They are specific to marine metabolites. Cephalostatin 1 (**111**) and ritterazine B (**112**) are highly cytotoxic, but their mode of action remains to be elucidated.

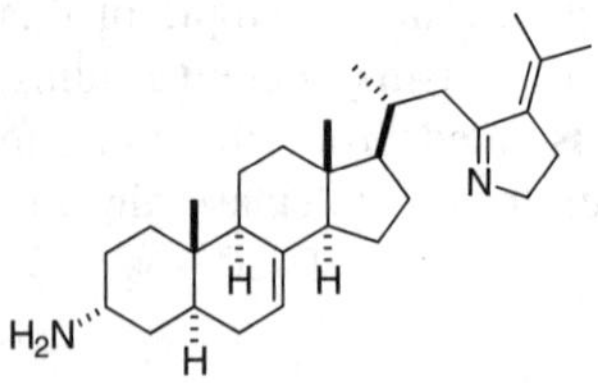

110 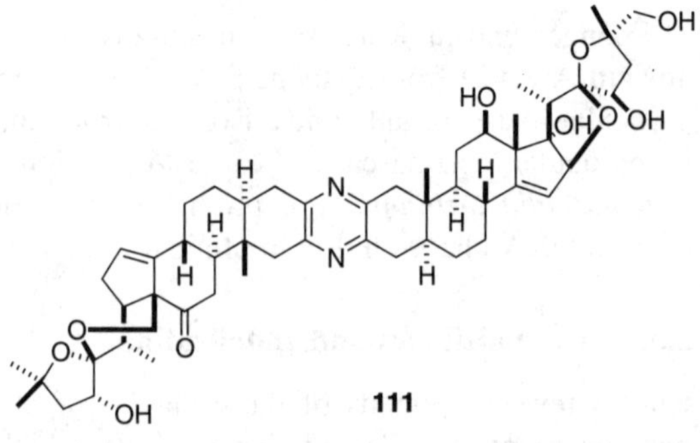   111

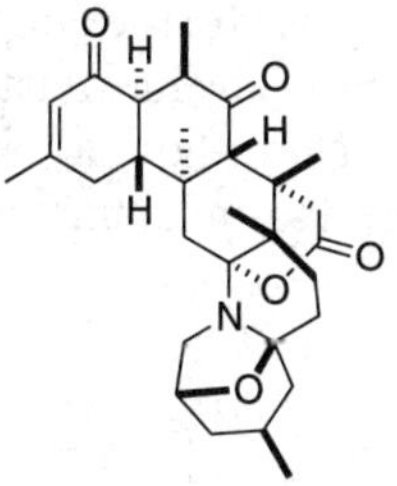

113

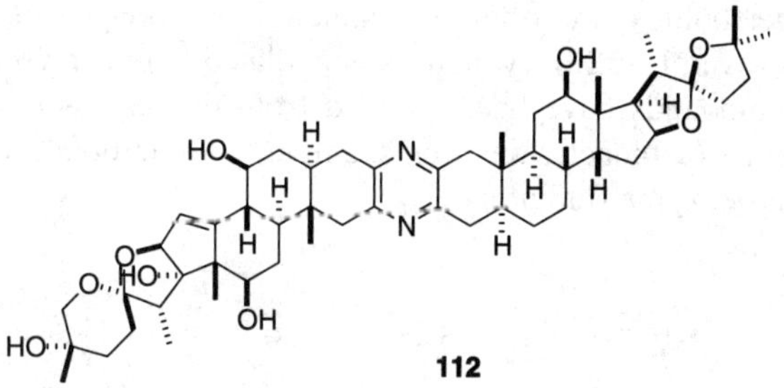

112

Although unlikely steroids or triterpenoids, zoanthamines found in zoanthids of the genus *Zoanthus* should be mentioned here. Zoanthamine (**113**) represents the first example of 10 metabolites of this class. Norzoanthamine has been best studied and found to suppress the production of IL-6 (15).

2.4 TERPENOIDS

Although those similar to terrestrial terpenoids are found commonly in marine organism, in particular in algae, several terpenoids with new or modified skeletons have been isolated frequently from algae, sponges, and cnidarians. Halogenated terpenoids are often contained in algae, whereas sulfated terpenoids and steroids are distributed widely in sponges.

2.4.1 Monoterpenoids and Sesquiterpenoids

Red algae contain highly halogenated monoterpenoids such as **114**, which is an antifeeding constituent isolated from *Plocamium cartilagineum* (34, 35).

Several halogenated sesquiterpenes with various skeletal types were reported from red algae of the genus *Laurencia*, for example, elatol (**115**), which is a chamigrane sesquiterpene that has antifeeding and antifouling activities from *L. obtusa*. A series of linear and cyclic sesquiterpenes with an 1,4-diacetoxybutadiene functionality [e.g., caulerpenyne (**116**)] have been isolated as antifeeding agents from green algae of the order Caulerpales (35). Coelenterates are a rich source of sesquiterpenes of 20 skeletal types, which include $\Delta^{9(12)}$-capnellane (**117**), an antifeeding metabolite isolated from *Capnella imbricata* (36).

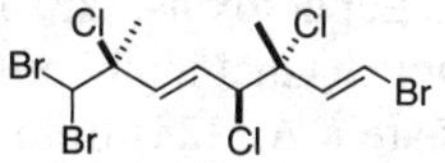

114 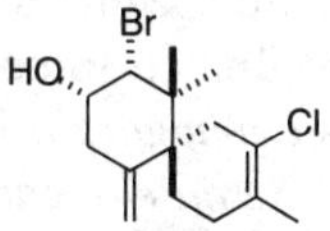   **115** 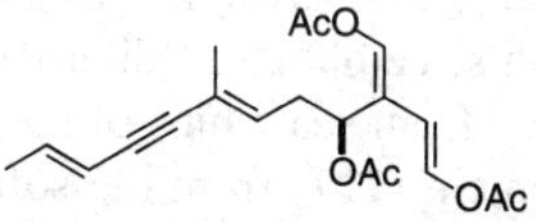   **116**

Sponges produce furanosesquiterpenes of various skeletal types, such as furodysinin (**118**) from *Dysidea herbacea* and nakafuran-8 (**119**) from *D. etheria*, the latter of which is antifeeding. Similarly, mixed shikimate-mevalonate metabolites are often encountered as sponge metabolites; the first example is avarol (**120**), which was isolated from *Dysidea avara*. Several related compounds have been isolated from dictyoceratid sponges. They show a wide range of bioactivities, for example, antimicrobial, antitumor, anti-inflammatory, and enzyme inhibitory.

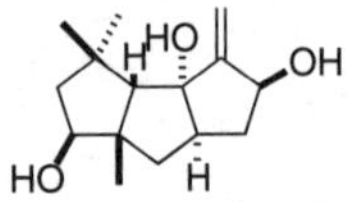

117 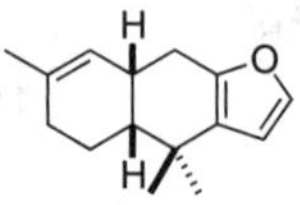   **118** 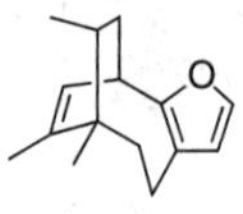   **119** 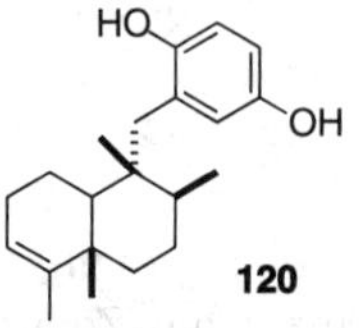   **120**

2.4.2 Diterpenoids

Brown algae are rich in diterpenoids, which can be classified into three groups, namely "xenicanes," "extended sesquiterpenes," and "dolabellanes," which are represented by dictyotadial (**121**) from *Dictyota crenulata*, spatol (**122**) from *Spatoglossum schmittii*, and amijidictyol (**123**) from *Dictyota linearis*, respectively. These terpenes are involved in chemical defense. Red algae contain brominated diterpenes of several established skeletal types.

Spongian diterpenes are a chemical marker for dictyoceratid and dendroceratid sponges (37). The first example of a spongian diterpene is isoagatholactone (**124**) isolated from *Spongia officinalis*. In addition to those possessing a basic spongian skeleton, a wide variety of rearranged spongian diterpenoids have been reported, which include gracilin B (**125**) from *Spongionella gracilis*. Many spongian diterepenes are antimicrobial and cytotoxic.

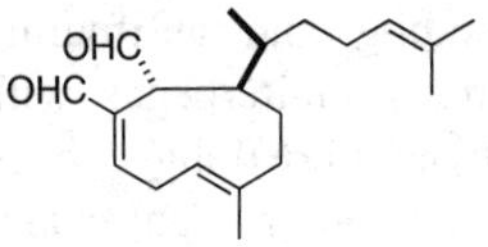

121 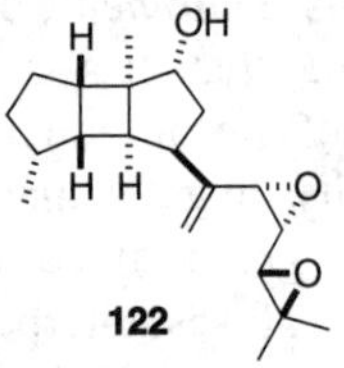   **122** 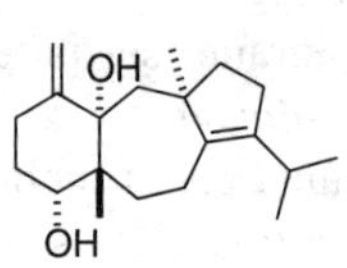   **123** 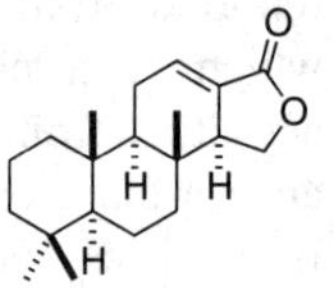   **124**

Again, soft corals and gorgonians are a rich source of diterpenoids of 19 structural classes, some of which are specific to them (35, 36). Besides tobacco plants, cembranoid diterpenes are limited to soft corals. Lophotoxin (**126**) isolated from sea whips of the genus *Lophogorgia* is a sodium channel inhibitor (3). Xenicin (**127**) from the soft coral *Xenia elongata* and briarein A (**128**) from the gorgonian *Briareum asbestinum* represent non-cembranolide diterpenes. Diterpenoids of these classes show antimicrobial, cytotoxic, and insecticidal activities. Pseudopterosin A (**129**), which is a diterpene glycoside isolated from the sea whip *Pseudopterogorgia elisabethae*, shows anti-inflammatory activity by inhibiting release of leukotriene B$_2$ from leukocytes (2, 35). A more interesting class of diterpenes includes sarcodictyn A (**130**) and eleutherobin (**131**), which were isolated from the soft corals *Sarcodictyon roseum* and *Eleutherobia* sp., respectively. These diterpenoids show potent cytotoxicity by stabilizing microtubules (2, 35).

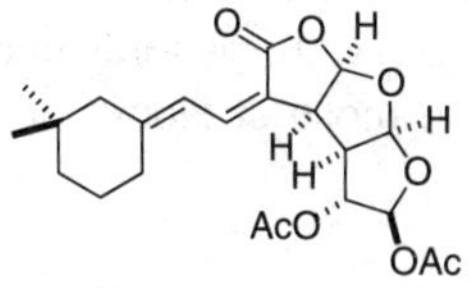

125 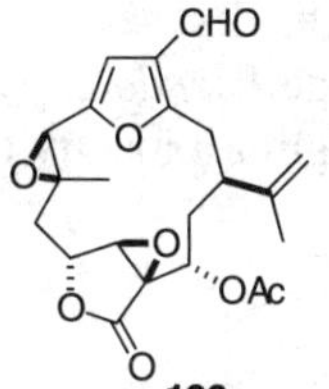 **126** 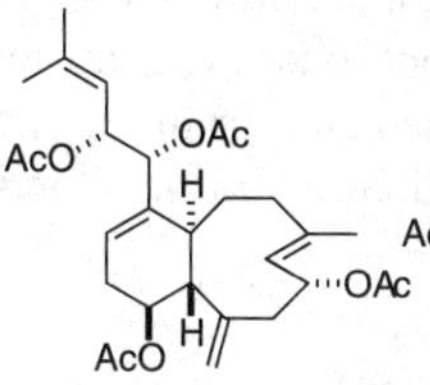 **127** 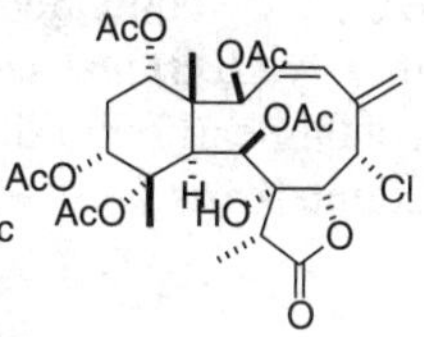 **128**

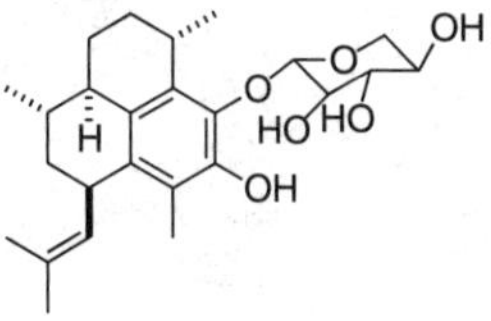

129 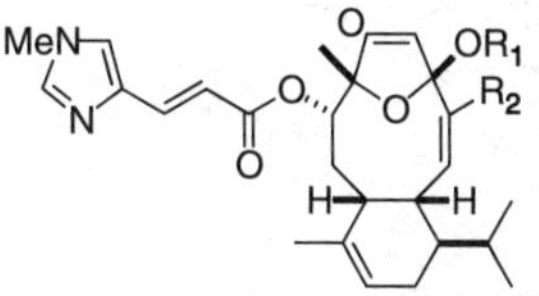 **130:** R$_1$ = H; R$_2$ = Ac **131:** R$_1$ = Me; R$_2$ = **132** 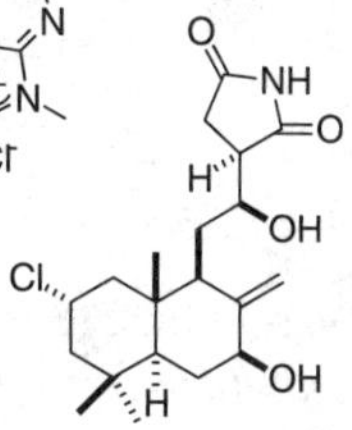 **133**

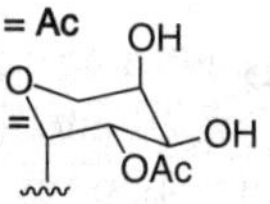

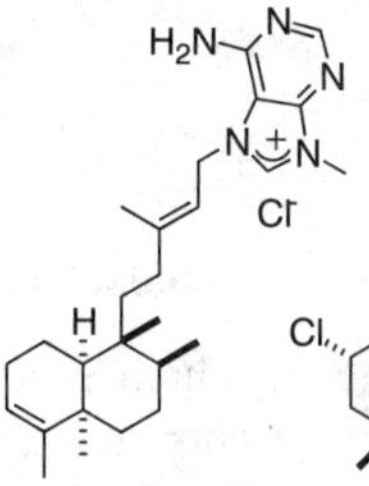

Ageladine A (**132**), which is a 9-methyladenine derivative of diterpene isolated from a sponge *Agelas* sp., is antimicrobial and inhibitory against Na, K-ATPase (12). A novel, chlorinated diterpenoid, chlorolissoclimide (**133**), which was isolated from the ascidian *Lissoclinum forskalii*, inhibits eukaryotic protein synthesis (38).

2.4.3 Isocyanoterpenes and Related Terpenoids

Isocyanide-containing natural products are rare; they have been reported only from cyanobacteria, *Penicillium* fungi, marine sponges, and nudibranchs (35, 39). Sesquiterpenoid and diterpenoid isocyanides are found in a limited species of sponges and nudibranchs that prey on these sponges. Axisonitrile-1 (**134**) isolated from the sponge *Axinella cannabina* is the first isocyanide-containing marine natural product. Isocyanopupukeanane (**135**) was isolated originally from the nudibranch *Phyllidia verrucosa* and later from a sponge *Ciocalypta* sp. Kalihinol A (**136**) and 7,20-diisocyanoadocane (**137**) were isolated from the sponges

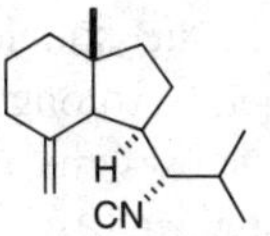

134 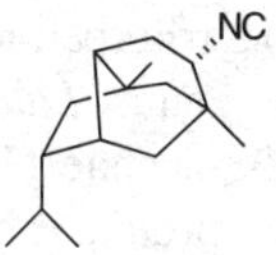 **135** 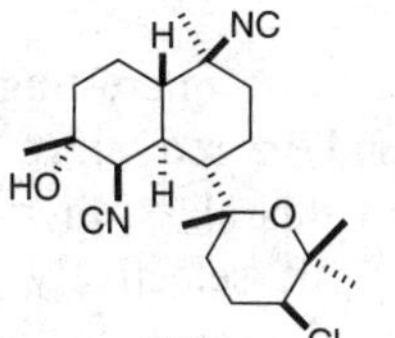 **136** 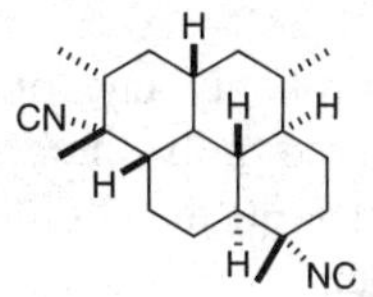 **137**

Acanthella carvenosa and *Adocia* sp., respectively. Isocyanoterpenes are often accompanied by thiocyanates, isothiocyantes, and formamides. These terpenoids show a wide range of bioactivities, which include antimicrobial, cytotoxic, ichthyotoxic, antifouling, and antimalarial. The carbonimidic dichloride group is considered to be equivalent to isocyanide, and the first sesquiterpenoid that contains this moiety (**138**) was isolated from the sponge *Pseudoaxinyssa pitys*. Terpenoids that contain carbonimidic dichloride show similar bioactivities to those of isocyanide counterparts.

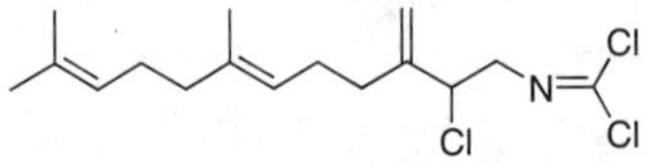

138

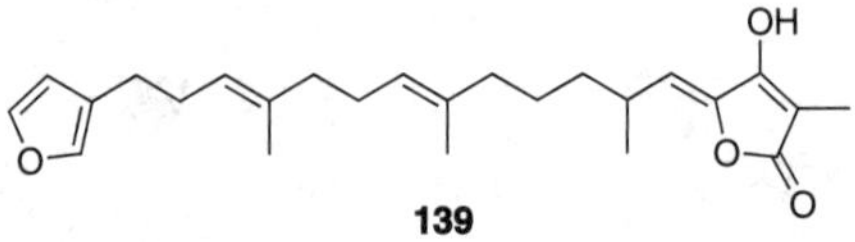

139

2.4.4   Sesterterpenoids, Triterpenoids, and Steroids

Variabilin (**139**), which is a C_{25} tetronic acid isolated from the sponge *Ircinia valiabiris*, represents the class of furano-sesterterpenes derived from sponges (34). Bioactivities include antimicrobial, cytotoxic, antifouling, and antifeedant.

Manoalide (**140**) is perhaps the most well-known marine sesterterpene isolated from the sponge *Luffariella variabilis*. It is antimicrobial, anti-inflammatory, and PLA_2 inhibitory (2, 12, 35). Several sesterterpenes of the scalarane class [e.g., scalaradial (**141**)] have been isolated from dyctioceratid sponges, many of which show anti-inflammatory activity (12). Dysidiolide (**142**), which is an unusual sesterterpene isolated from the sponge *Dysidea etheria*, is the first natural product inhibitor of cdc25A (12). Mycaperoxide A (**143**), which is a norsesterterpene peroxide isolated from *Mycale* sp., showed promising antimalarial activity (9).

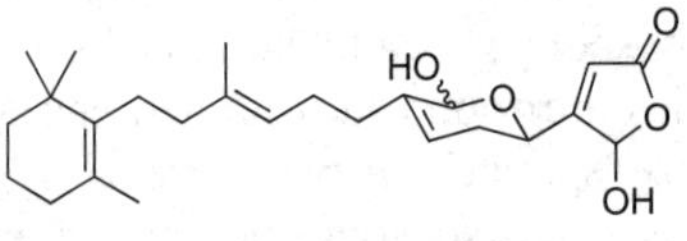

140

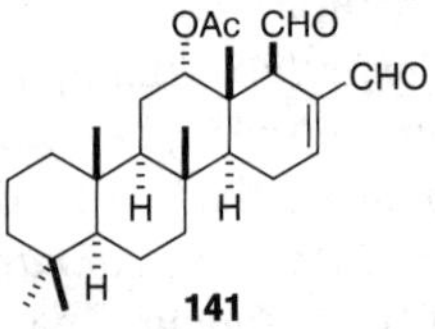

141

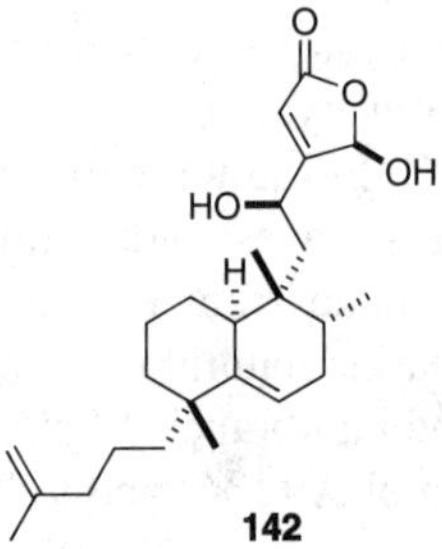

142

Probably one of the most interesting marine triterpenoids is the squalene-derived polyethers found in red algae of the genus *Laurencia*. Among the nearly 30 metabolites of this class, thyrsiferyl 23-acetate (**144**) shows the most potent cytotoxicity (40). It selectively inhibits protein phosphatase 2A. The other class of triterpenoid polyethers has been isolated from sponges of the families Halichondridae and Axinellidae (40). Sipholenone B (**145**), which was isolated from *Siphonochalina siphonella*, represents the first example of this class. Adociasulfate-2 (**146**) is a member of the triterpenoid hydroquinone sulfates isolated from a sponge *Haliclona* (aka *Adocia*) sp. that is the only known natural product inhibitor of kinesin (12, 19).

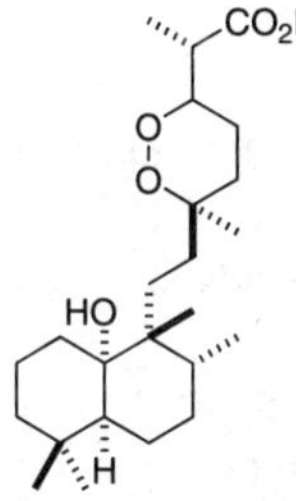

143 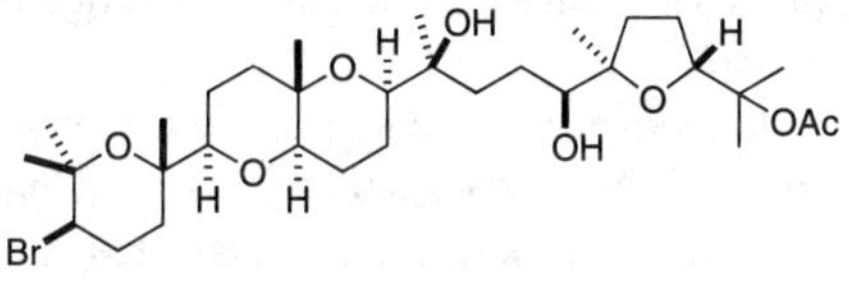   **144** 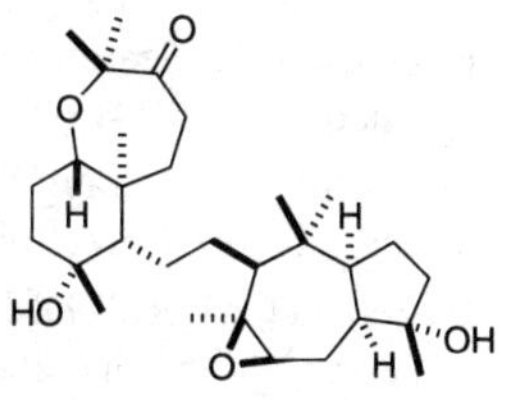   **145**

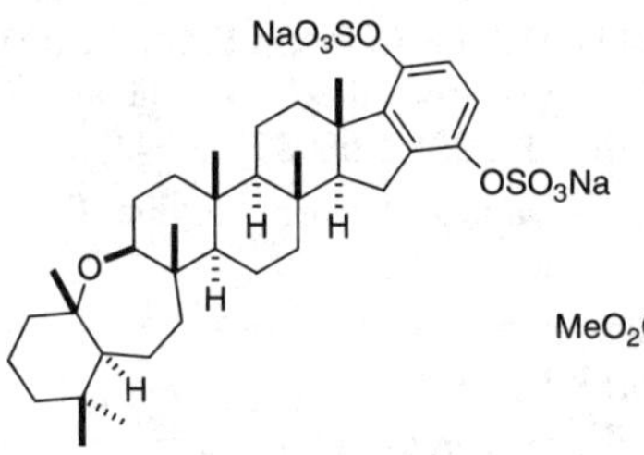

146 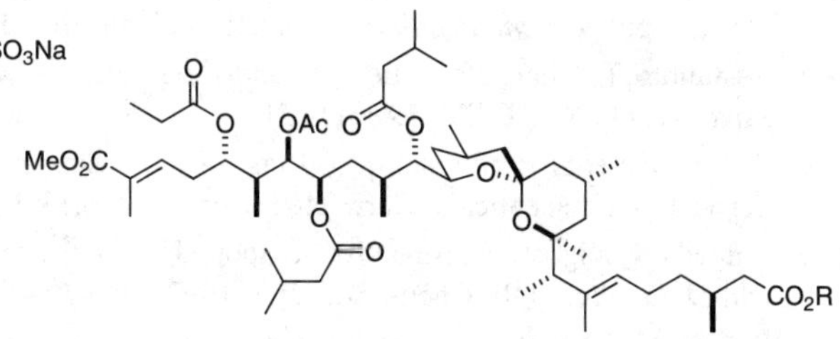   **147**: R = CH₂CH₂SO₃Na
148: R = CH₃

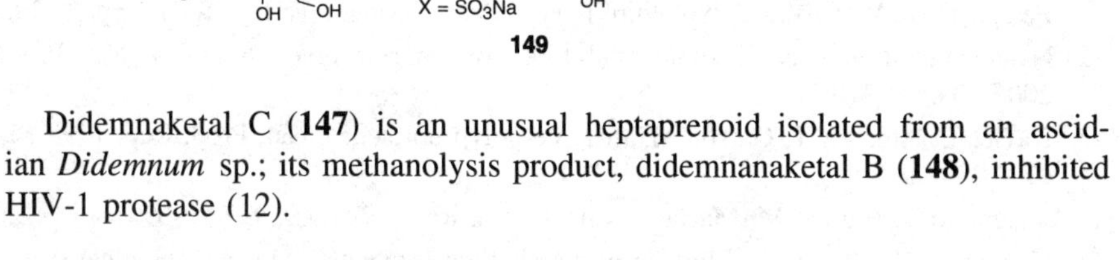

149

Didemnaketal C (**147**) is an unusual heptaprenoid isolated from an ascidian *Didemnum* sp.; its methanolysis product, didemnanaketal B (**148**), inhibited HIV-1 protease (12).

2.4.5 Sugars

Several unusual polysaccharides have been isolated from sponges (11), among which the most unusual is axinelloside A (**149**), which has been isolated recently as a potent telomerase inhibiter from *Axinella infundibula* (12). It resembles bacterial lipopolysaccharides.

2.5 ACKNOWLEDGMENTS

I would like to thank Otto Hensens for editing the manuscript.

REFERENCES

1. Fusetani N. Search for drug leads from Japanese marine invertebrates. J. Synth. Org. Chem. Jpn. 2004;62:1073–1079.
2. Newman DJ, Cragg GM. Marine natural products and related compounds in clinical and advanced preclinical trials. J. Nat. Prod. 2004;67:1216–1238.
3. Ireland CM, Roll DM, Molinski TF, McKee TC, Zabriskie TM, Swersey JC. Uniqueness of the marine chemical environment: categories of marine natural products from invertebrates. Biomedical Importance of Marine Organisms. Fautin DG, ed. 1988. California Academy of Sciences, San Francisco, CA. pp. 41–57.
4. Mancini I, Guella G, Frostin M, Hnawia E, Laurent D, Debitus C, Pietra F. On the first polyarsenic organic compound from nature: arsenicin A from the New Caledonian marine sponge *Echinochalina bargibanti*. Chem. Eur. J. 2006;12:8989–8994.
5. Yasumoto T. Chemistry, etiology, and food chain dynamics of marine toxins. Proc. Japan Acad., Ser. B. 2005;81:43–51.
6. Piel J. Bacterial symbionts: prospects for the sustainable production of invertebrate-derived pharmaceuticals. Curr. Med. Chem. 2006;13:39–50.
7. Yamada K, Kigoshi H. Bioactive compounds from the sea hares of two genera: *Aplysia* and *Dolabella*. Bull. Chem. Soc. Jpn. 1997;70:1479–1489.
8. Suzuki M, Vairappan CS. Halogenated secondary metabolites from Japanese species of the red algal genus *Laurencia* (Rhodomelaceae, Ceramiales). Curr. Topic. Phytochem. 2005;7:1–34.
9. Casteel DA. Peroxy natural products. Nat. Prod. Rep. 1999;16:55–73.
10. Costantino V, Fattorusso E, Menna M, Taglialatela-Scafati O. Chemical diversity of bioactive marine natural products: an illustrative case study. Curr. Med. Chem. 2004;11:1671–1692.
11. van Soest RWM, Fusetani N, Andersen RJ. Straight-chain acetylenes as chemotaxonomic markers of the marine Haplosclerida. In: Sponge Sciences Multidisciplinary Perspectives. Watanabe Y, Nobuhiro F, eds. 1998. Springer-Verlag, Tokyo. pp. 3–30.
12. Nakao Y, Fusetani N. Enzyme inhibitors from marine invertebrates. J. Nat. Prod. 2007;70:689–710.
13. Davies-Coleman MT, Garson MJ. Marine polypropionates. Nat. Prod. Rep. 1998;15:477–493.
14. Yasumoto T, Murata M. Marine toxins. Chem. Rev. 1993;93:1897–1909.
15. Uemura D. Bioorganic studies on marine natural products – diverse chemical structures and bioactivities. Chem. Record 2006;6:235–248.
16. Kobayashi J, Kubota T. Bioactive macrolides and polyketides from marine dinoflagellates of the genus *Amphidinium*. J. Nat. Prod. 2007;70:451–460.
17. Nishimura S, Matsunaga S, Yoshida M, Hirota H, Yokoyama S, Fusetani N. 13-Deoxytedanolide, a marine sponge-derived antitumor macrolide, binds to the 60S large ribosomal subunit. Bioorg. Med. Chem. 2005;13:449–454.

18. Molinski TF. Antifungal compounds from marine organisms. Curr. Med. Chem. Anti-Infective Agents 2004;3:197–220.

19. Nagle DG, Zhou Y-D, Mora FD, Mohammed KA, Kim Y-P, Mechanism targeted discovery of antitumor marine natural products. Curr. Med. Chem. 2004;11:1725–1756.

20. Sakai R, Kamiya H, Murata M, Shimamoto K. Dysiherbaine: a new neurotoxic amino acid from the Micronesian marine sponge *Dysidea herbacea*. J. Am. Chem. Soc. 1997;119:4112–4116.

21. Couladopuros EA, Pitsinos EN, Moutsos VI, Sarakinos G. A general method for the synthesis of bastaranes and isobastaranes: first total synthesis of bastadins 5, 10, 12, 16, 20, and 21. Chem. Eur. J. 2005;11:406–421.

22. Fusetani N, Matsunaga S. Bioactive sponge peptides. Chem. Rev. 1993;93:1793–1806.

23. Matsunaga S, Fusetani N. Nonribosomal peptides from marine sponges. Curr. Org. Chem. 2003;7:945–966.

24. Milanowski DJ, Gustafson KR, Rashid MA, Rannell LK, McMahon JB, Boyd MR. Gymnangiamide, a cytotoxic pentapeptide from the marine hydroid *Gymnangium regae*. J. Org. Chem. 2004;69:3036–3042.

25. Hamada T, Matsunaga S, Yano G, Fusetani N. Polytheonamides A and B, highly cytotoxic, linear polypeptides with unprecedented structural features, from the marine sponge, *Theonella swinhoei*. J. Am. Chem. Soc. 2005;127:110–118.

26. Andersen RJ, van Soest RWM, Kong F. 3-Alkylpiperidine alkaloids isolated from marine sponges in the order Haplosclerida. Alakloids: Chem. Biol. Perspect. 1996;10:301–355.

27. Gul W, Hamann MT. Indole alkaloid marine natural products: an established source of cancer drug leads with considerable promise for the control of parasitic, neurological and other diseases. Life Sci. 2005;78:442–453.

28. O'Malley DP, Li K, Maue M, Zografos AL, Baran PS. Total synthesis of dimeric pyrrole-imidazole alkaloids: sceptrin, ageliferin, nagelamide E, oxysceptrin, nakamuric acid, and the axinellamine carbon skeleton. J. Am. Chem. Soc. 2007;129:4762–4775.

29. Delfourne E, Bastide J. Marine pyridoacridine alkaloids and synthetic analogues as antitumor agents. Med. Res. Rev. 2003;23:234–252.

30. Urban S, Hickford SJH, Blunt JW, Munro MHG. Bioactive marine alkaloids. Curr. Org. Chem. 2000;4:765–807.

31. Berlinck RGS. Natural guanidine derivatives. Nat. Prod. Rep. 1999;16:339–365.

32. Bienz S, Bisegger P, Guggisberg A, Hesse M. Polyamine alkaloids. Nat Prod. Rep. 2005;22:647–658.

33. Folmer F, Houssen WE, Scott RH, Jaspars M. Biomedical research tools from the seabed. Curr. Opin. Drug Discov. Dev. 2007;10:145–152.

34. Faulkner DJ. Interesting aspects of marine natural products chemistry. Tetrahedron 1977;33:1421–1443.

35. Gross H, König GM. Terpenoids from marine organisms: unique structures and their pharmacological potential. Phytochem. Rev. 2006;5:115–141.

36. Coll JC. The chemistry and chemical ecology of octocorals (Coelenterata, Anthozoa, Octocorallia). Chem. Rev. 1992;92:613–631.

37. Keyzers RA, Northcote PT, Davies-Coleman MT. Spongian diterpenoids from marine sponges. Nat. Prod. Rep. 2006;23:321–334.
38. Robert F, Gao HQ, Donia M, Merrick WC, Hamann MT, Pelletier J. Chlorolissoclimides: new inhibitors of eukaryotic protein synthesis. RNA 2006;12:717–725.
39. Garson MJ, Simpson JS. Marine isocyanides and related natural products – structure, biosynthesis and ecology. Nat. Prod. Rep. 2004;21:164–179.
40. Fernández JJ, Souto ML, Norte M. Marine polyether triterpenes. Nat. Prod. Rep. 2000;17:235–246.

FURTHER READING

Allingham JS, Klenchin VA, Rayment I. Actin-targeting natural products, Properties and mechanisms of action. Cell. Mol. Life Sci. 2006;63:2119–2134.

Antunes EM, Copp BR, Davies-Coleman MT, Samaai T. Pyrroloiminoquinone and related metabolites from marine sponges. Nat. Prod. Rep. 2005;22:62–72.

Bandaranayake WM. The nature and role of pigments of marine invertebrates. Nat. Prod. Rep. 2006;23:223–255.

Blunt JW, Copp BR, Hu WP, Munro MHG, Northcote PT, Princep MR. Marine natural products. Nat. Prod. Rep. 2007;24:31–86.

Bowman EJ, Bowman BJ. V-ATPases as drug targets. J. Bioenerget. Biomembr. 2005;37: 431–435.

Bugni TS, Ireland CM. Marine-derived fungi: a chemically and biologically diverse group of microorganisms. Nat. Prod. Rep. 2004;21:143–163.

Burja AM, Banaigs B, Abou-Mansour E, Burgess JG, Wright PC. Tetrahedron 2001;57: 9347–9377.

Chang CWJ. Naturally occurring isocyano/iothiocyanato and related compounds. Fortsch. Chem. Org. Naturst. 2000;80:1–186.

Crews P, Hunter LM. The search for antiparasitic agents from marine animals. In: Marine Biotechnology, vol 1. Pharmaceutical and Bioactive Natural Products, vol. 1. Attaway DH, Zaborsky OR, eds. 1993. Plenum Press, New York. pp. 343–390.

D'Auria MV, Minale L, Riccio R. Polyoxygenated steroids of marine origin. Chem. Rev. 1993;93:1839–1895.

Daranas AH, Norte M, Fernández JJ. Toxic marine microalgae. Toxicon 2001;39: 1101–1132.

Dembitsky VM, Levitsky DO. Arsenolipids. Prog. Lipid Res. 2004;43:403–448.

Donia M, Hamann MT. Marine natural products and their potential applications as anti-infective agents. Lancet Infect. Dis. 2003;3:338–348.

Faulkner DJ. Marine natural products. Nat. Prod. Rep. 2002;19:1–48, and earlier reviews cited within.

Fenical W, Jensen PR. Developing a new resource for drug discovery: marine actinomycete bacteria. Nat. Chem. Biol. 2006;2:666–673.

Gerwick WH, Bernart MW. Eicosanoids and related compounds from marine algae. In: Marine Biotechnology, vol 1. Pharmaceutical and Bioactive Natural Products, vol 1. Attaway DH, Zaborsky OR, eds. 1993. Plenum Press, New York. pp. 101–152.

Gerwick WH, Nagle DG, Proteau PJ. Oxylipins from marine invertebrates. Topics Curr. Chem. 1993;167:117–180.

Gribble GW. The diversity of naturally occurring organobromine compounds. Chem. Soc. Rev. 1999;28:335–346.

Heys L, Moore CG, Murphy PJ. The guanidine metabolites of *Ptilocaulis spiculifer* and related compounds; isolation and synthesis. Chem. Soc. Rev. 2000;29:57–67.

Ireland CM, Copp BR, Foster MP, McDonald LA, Radisky DC, Swersey JC. Biomedical potential of marine natural products. In: Marine Biotechnology, vol 1. Pharmaceutical and Bioactive Natural Products, vol 1. Attaway DH, Zaborsky OR, eds. 1993. Plenum Press, New York. pp. 1–76.

Janin YL. Peptides with anticancer use or potential. Amino Acids 2003;25:1–40.

Jensen PR, Fenical W. Marine microorganisms and drug discovery: current status and future potential. In: Drugs from the Sea. Fusetani N, ed. 2000. Karger, Basel pp. 6–29.

Kobayashi J, Ishibshi M. Marine natural products and marine chemical ecology. Comp. Nat. Prod. Chem. 1999;8:415–649.

Kornprobst JM, Sallenave C, Barnathan G. Sulfated compounds from marine organisms. Comp. Biochem. Physiol. 1998;119B:1–51.

Lewis RJ, Garcia ML. Therapeutic potential of venom peptides. Nat. Rev. Drug Discov. 2003;2:790–802.

Llewellyn LE. Saxitoxin, a toxic marine natural product that targets a multitude of receptors. Nat. Prod. Rep. 2006;23:200–222.

Marshall KM, Barrows LR. Biological activities of pyridoacridines. Nat. Prod. Rep. 2004;21:731–751.

Minale L, Cimino G, de Stefano S, Sodano G. Natural products from Porifera. Fortsch. Chem. Org. Naturst. 1976;33:1–72.

Minale L, Riccio R, Zollo F. Steroidal oligosaccharides and polyhydorxylated steroids from echinoderms. Fortsch. Chem. Org. Naturst. 1993;62:75–308.

Moore BS. Biosynthesis of marine natural products: macroorganisms (Part B). Nat. Prod. Rep. 2006;23:615–629.

Mourabit AA, Potier P. Sponge's molecular diversity through the ambivalent reactivity of 2-aminoimidazole: a universal chemical pathway to the oroidin-based pyrrole-imidazole alkaloids and their palau'amine congeners. Eur. J. Org. Chem. 2001;2: 237–243.

Newman DJ, Cragg GM. Natural products from marine invertebrates and microbes as modulators of antitumor targets. Curr. Drug Target. 2006;7:279–304.

Northcote PT, Blunt JW, Munro MHG. Pateamine: a potent cytotoxin from the New Zealand marine sponge, *Mycale* sp. Tetrahedron Lett. 1991;32:6511–6414.

Pettit GR. The dolastatins. Forsch. Chem. Org. Naturst. 1997;70:1–79.

Piel J. Metabolites from symbiotic bacteria. Nat. Prod. Rep. 2004;21:519–538.

Pietra F. Biodiversity and Natural Product Diversity. 2002. Pergamon, Amsterdam.

Pindur U, Lemster T. Advances in marine natural products of the indole and annelated indole series: chemical and biological aspects. Curr. Med. Chem. 2001;8:1681–1698.

Rinehart KL, Shield LS, Cohen-Parsons M. Antiviral substances. In: Marine Biotechnology, vol 1. Pharmaceutical and Bioactive Natural Products, vol. 1. Attaway DH, Zaborsky OR, eds. 1993. Plenum Press, New York. pp. 309–342.

Sarma AS, Daum T, Müller WEG, eds. Secondary Metabolites from Marine Sponges. 1993. Ullstein Mosby, Berlin.

Scheuer PJ, ed. Marine Natural Products. Chemical and Biological Perspectives, vol. I–V. 1978, 1980, 1981, 1983. Academic Press, New York.

Scheuer PJ. Chemistry of Marine Natural Products. 1973. Academic Press, New York.

Schmitz FJ, Bowden BF, Toth SI. Antitumor and cytotoxic compounds from marine organisms. In: Marine Biotechnology, vol 1. Pharmaceutical and Bioactive Natural Products, vol. 1. Attaway DH, Zaborsky OR, eds. 1993. Plenum Press, New York. pp. 197–308.

Scott JD, Williams RM. Chemistry and biology of the tetrahydroisoquinoline antitumor antibiotics. Chem. Rev. 2002;102:1669–1730.

Shimuzu Y. Dinoflagellates as sources of bioactive molecules. In: Marine Biotechnology, vol 1. Pharmaceutical and Bioactive Natural Products, vol. 1 Attaway DH, Zaborsky OR, eds. 1993. pp. 391–410.

Skyler D, Heathcock CH. The pyridoacridine family tree: a useful scheme for designing synthesis and predicting undiscovered natural products. J. Nat. Prod. 2002;65: 1573–1581.

Terracciano S, Aquino M, Rodriguez M, Monti MC, Casapullo A, Riccio R, Gomez-Paloma L. Chemistry and biology of anti-inflammatory marine natural products: molecules interfering with cyclooxygenase, NF-kB and other unidentified targets. Curr. Med. Chem. 2006;13:1947–1969.

3

MICROBIAL NATURAL PRODUCTS

SERGIO SANCHEZ, SILVIA GUZMÁN-TRAMPE, MARIANA ÁVALOS,
BEATRIZ RUIZ, AND ROMINA RODRÍGUEZ-SANOJA
*Instituto de Investigaciones Biomédicas, Universidad Nacional Autónoma de México,
México*

MANUEL JIMÉNEZ-ESTRADA
Instituto de Química, Universidad Nacional Autónoma de México, México

Microorganisms produce numerous natural products that, although they may not participate directly in their growth and development, do play an important role in the organism's ecological interactions with other organisms (secondary metabolites). Many of these products have therapeutic and agricultural applications and have been crucial to the health and well-being of humans. Approximately one third of the top-selling drugs in the world that are recognized in the pharmaceutical industry as having broad structural diversity and a wide range of pharmacological activity are natural products or their derivatives. The structural diversity of these products makes them valuable as novel lead compounds against newly discovered therapeutic targets. Similar biological effects are often displayed by different chemical species. Likely, many undiscovered microbes exist that are capable of producing natural compounds exhibiting biological activities with potential commercial application. Here, we discuss several important microbial natural products, with a special emphasis on their chemical diversity.

Filamentous fungi and bacteria, particularly bacilli and actinomycetes, have been the focus of industrial and academic research for the last 80 years (1). In addition to the increase in general scientific knowledge, the study of these microorganisms has revealed that they are exceptionally rich sources of many primary and secondary metabolites. Such metabolites constitute half of the pharmaceuticals on

Natural Products in Chemical Biology, First Edition. Edited by Natanya Civjan.
© 2012 John Wiley & Sons, Inc. Published 2012 by John Wiley & Sons, Inc.

the market today and are agriculturally beneficial. Biologically active compounds have only been isolated from a very small percentage of the microbial diversity found in a given ecosystem or geographic area (2, 3). Through the use of molecular biology and other identification techniques, it is estimated that less than 1% of microbial diversity has been cultured and studied experimentally (approximately 6,000 species of prokaryotes and approximately 70,000 fungal species have been formally described) (4–7). Therefore, new species of fungi and actinomycetes, with metabolic pathways adapted for life in a wide range of terrestrial and marine environments, are being discovered every day. Microbes living in unique environments may produce a variety of unusual metabolites, allowing for the discovery and application of new chemical compounds.

The list of microbial natural products with biological activity is long and includes compounds that have had a widespread impact on agriculture, human and veterinary medicine, the food industry, and scientific research. The biologically active portion of some compounds contains unusual chemical structures that are crucial therapeutic agents with uses in oncology, immunosuppression, atherosclerosis, and the treatment of infectious diseases. Several microbial natural products also have antioxidant properties; can protect against radiation; are able to boost cellular tolerance to desiccation, salt, and heat stress; and are even used in microbial chemical intercommunication.

Approximately 22,500 metabolites reported to exhibit biological activity have been isolated from actinomycetes or fungi, with 45% of such compounds produced by actinomycetes, 38% by fungi, and 17% by unicellular bacteria (8). To date, new microbial products are in clinical development, particularly as anticancer agents and anti-infectives.

Similar biological effects are often displayed by different chemical species, giving rise to the concept of chemical diversity. This chapter focuses on the active metabolites that have been evaluated for their biological, pharmacological, and/or chemotaxonomic potential, or for the purpose of exploring chemical diversity.

3.1 ANTIBACTERIALS

Microbially produced secondary metabolites are extremely important to our health and nutrition. As a group, they have tremendous economic importance. In 2007, the market for antibiotics was US$66 billion (1) and included approximately 160 antibiotics and derivatives.

3.1.1 β-Lactams

β-lactam antibiotics act by inhibiting cell wall synthesis and are the largest and most often used group of antibiotics in the clinic because of their wide therapeutic range and low toxicity.

Penicillins (**1**) are the main exponent of the β-lactam group. This group of antibiotics was isolated from the fungus *Penicillium notatum*. Members consist of a lactam ring joined to a thiazolidine ring to form 6-aminopenicillanic acid (2). At position 6, these compounds have a lateral chain bound to an amino group that can influence their chemical and biological characteristics.

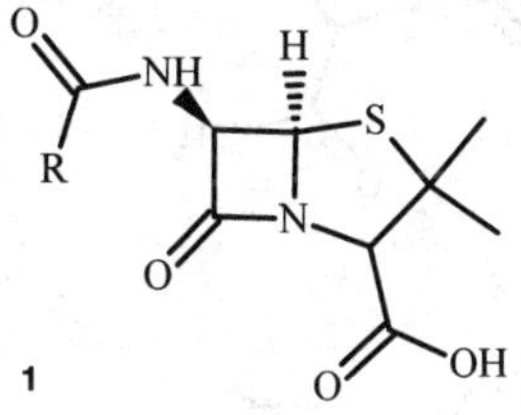

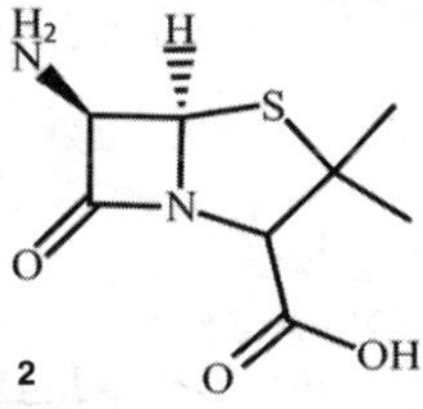

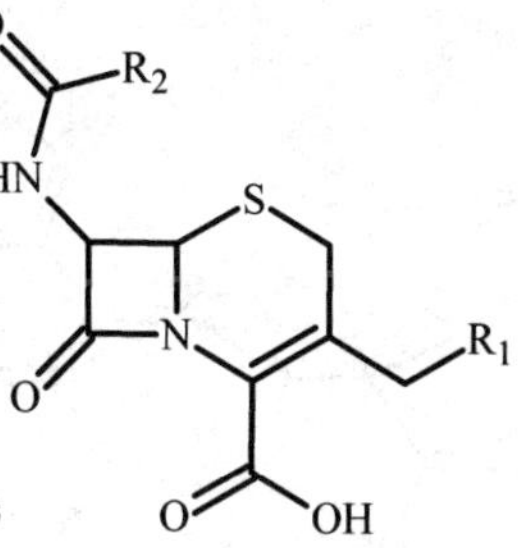

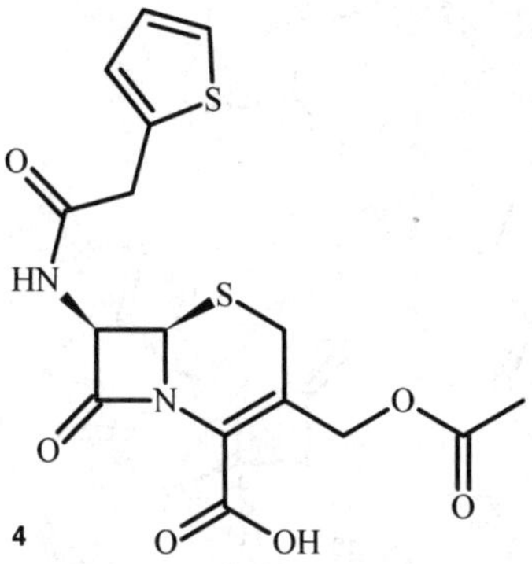

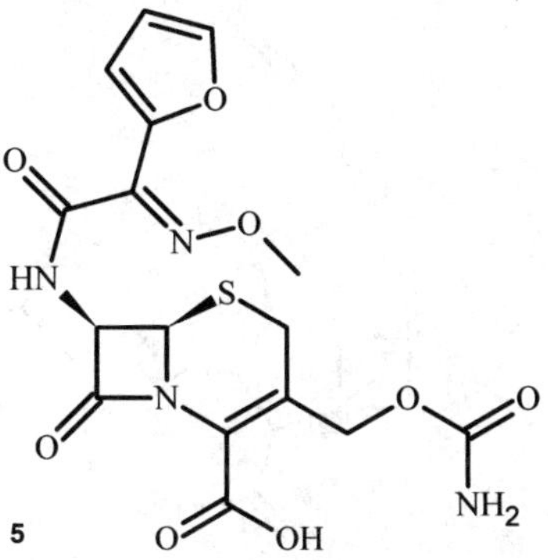

Cephalosporins have a similar structure to penicillins, but instead of the thiazolidine ring, they have a dihydrothiazine ring (3) (9). The main examples are cephalothin (4) and cefuroxime (5).

Monobactams are compounds derived from 3-aminobactamic acid. They have a monocyclic structure and have no side chain coupled to the amino group of the β-lactam ring. A classic example is aztreonam (6), which was isolated from *Chromobacterium violaceum* and is a broad-spectrum bactericidal agent that is resistant to β-lactamases.

Carbapenems were isolated from *Streptomyces cattleya*. They differ from penicillins and cephalosporins in that they have a carbon-sulfur substitution at position 1 and have an insaturation between C-2 and C-3 of the five-membered ring. The general structure can be described as a β-lactam ring fused with a pyrrolidine ring (7), with the rings sharing a nitrogen atom similar to imipenem (8) and meropenem (9) (10).

Clavulanic acid is a broad-spectrum β-lactam antibiotic with low antibacterial activity. Clavulanic acid (10) is similar to penicillanic acid (2), with a sulfur-oxygen substitution at position 3 that gives it a greater affinity for β-lactamases. Obtained from *Streptomyces clavuligerus*, clavulanic acid has been shown to be a potent inhibitor of the β-lactamases produced by staphylococci and the plasmid-mediated β-lactamases of gram-negative bacteria. Therefore, when combined with a variety of broad-spectrum semisynthetic penicillins such as amoxicillin and ticarcillin, they are effective against organisms such as *Escherichia coli, Proteus, Salmonella, Haemophilus, Pseudomonas*, and *Staphylococcus aureus*. Clavulanic acid has world sales of more than US$1 billion, and in 1995, it was the second largest selling antibacterial drug.

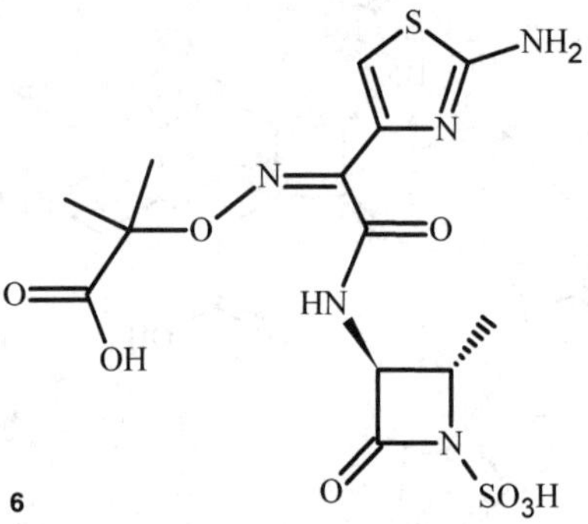

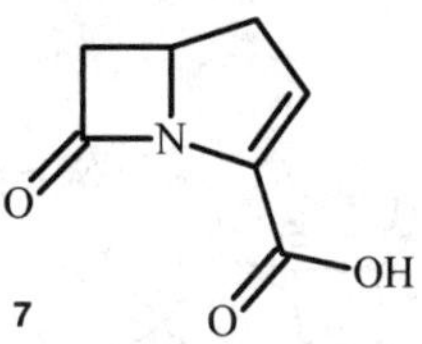

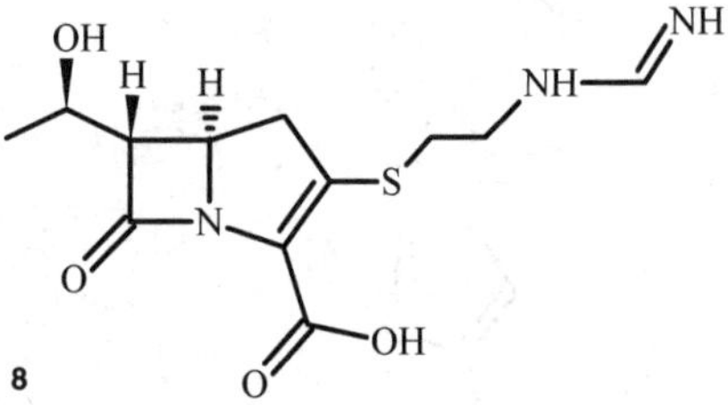

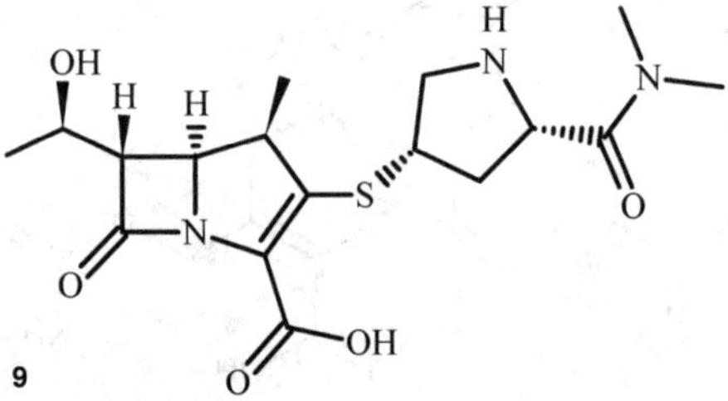

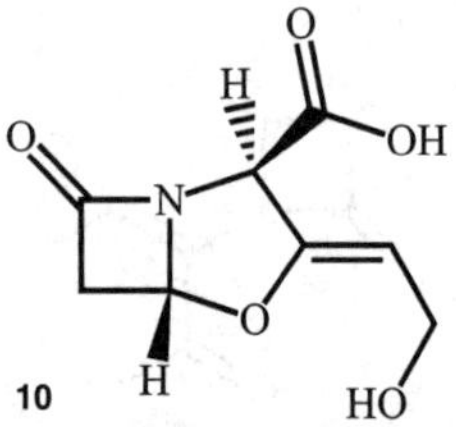

3.1.2 Antibacterial Polyketides

Macrolides are compounds characterized by a macrocyclic lactone ring of 14–16 carbon atoms bound to different sugars. Macrolides bind to the 23S subunit of the ribosome and function by blocking translation through peptidyl-tRNA release, preventing the approach of the growing peptide (11). The first macrolide antibiotic used was erythromycin (**11**), which was isolated from *Saccharopolyspora erythraea*. Additional examples are spiramycin (**12**), which is produced by *Streptomyces ambofaciens*, and kanchanamycins (**13**), which are produced by *Streptomyces olivaceus*. Kanchanamycins are classified as poliol macrolide antibiotics and have a bicyclic carbon skeleton with a terminal urea moiety as observed in kanchanamycin A (12).

Ansamycins (**14**) are a family of antibiotics with a similar structure to macrolides. These compounds contain the lactone ring, but they have an

11

12

13

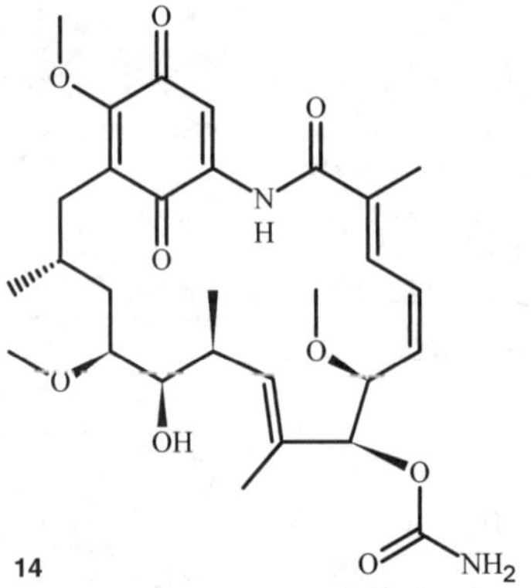

14 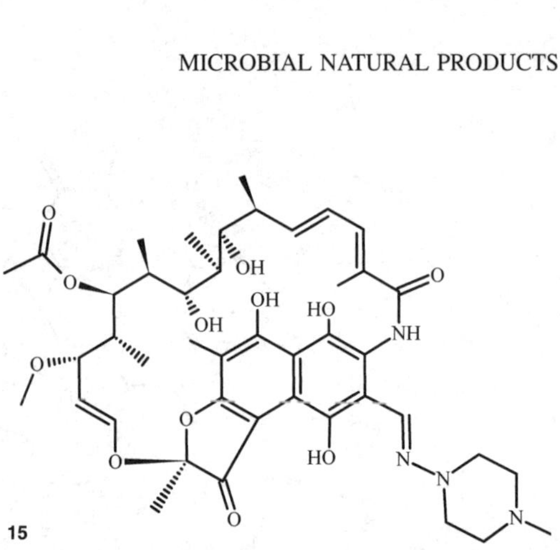   **15**

NH group within the core of this ring (13). Their antimicrobial spectrum is mainly against gram-positive bacteria. Among these compounds, the rifamycins, produced by *Amycolatopsis mediterranei*, can be cited as effective drugs in the treatment of tuberculosis and leprosy because of their antimicrobial effect against *Mycobacterium tuberculosis*. The biological activity of rifamycins is due to their ability to inhibit DNA-dependent RNA synthesis. They are also used to treat infections by *Listeria* species, *Neisseria gonorrhoeae*, *Haemophilus influenza*, and *Legionella pneumophila*. Rifamycins that are currently available include rifampicin (**15**), rifabutin (**16**), rifapentine (**17**), and rifaximin (**18**).

Streptogramins are compounds produced by some streptomycetes and can adopt one of two different chemical structures (A and B). Type A are cyclic hybrid peptide-polyketide macrolactones, whereas type B are cyclic hexa- or hepta-depsipeptides, cyclized via an ester bond between the C-terminal carboxyl group and the secondary group of a threonine residue at position 2. They have a remarkable mode of action because of the ability of the A and B types to act

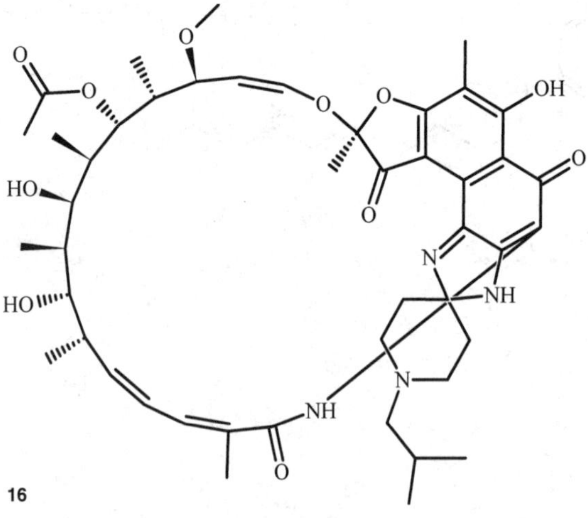

16

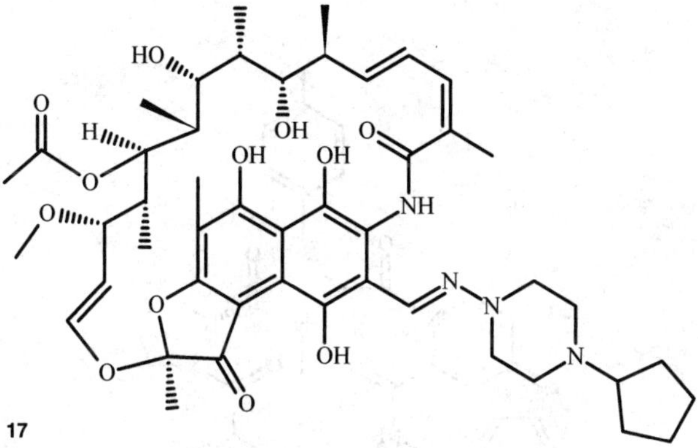

17

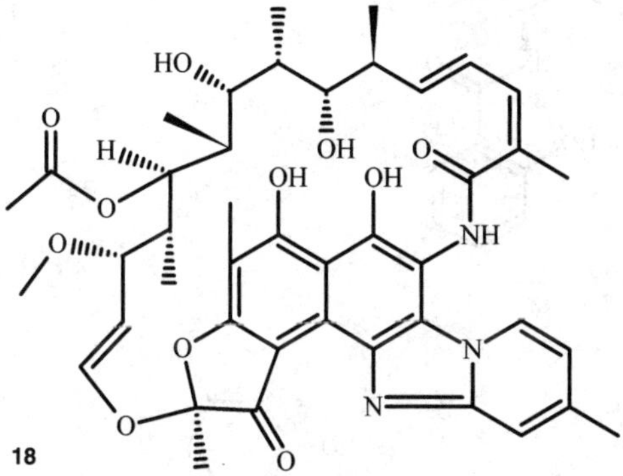

18

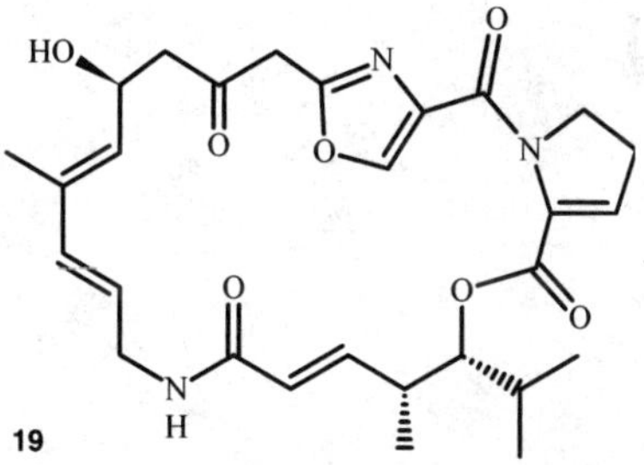

19

synergistically. The binding of type A streptogramins causes a conformational change in the 50S ribosome that increases the activity of the type B streptogramins by 100-fold, inhibiting bacterial cell growth. Each type also shows bacteriostatic effects when administered separately. Type B streptogramin prevents the extension of protein chains, causing the release of incomplete peptides, and can bind to ribosomes during any step of the protein synthesis (14). An example of a type A streptogramin is virginiamycin M_1 (19), whereas pristinamycinI$_A$ is a type B streptogramin (20).

3.1.3 Peptide Antibiotics

Peptide antibiotics are cyclic or polycyclic glycosylated structures that are active against gram-positive bacteria. They can inhibit cell wall synthesis by preventing the incorporation of N-acetylmuramic acid and N-acetylglucosamine into the peptide subunits of peptidoglycan. Common examples include vancomycin (produced by *Amycolatopsis orientalis*) (21), teicoplanin (synthesized by *Actinoplanes teichomyceticus*) (22), and ramoplanin (extracted from *Actinoplanes* sp. ATCC 33076) (23).

20

21

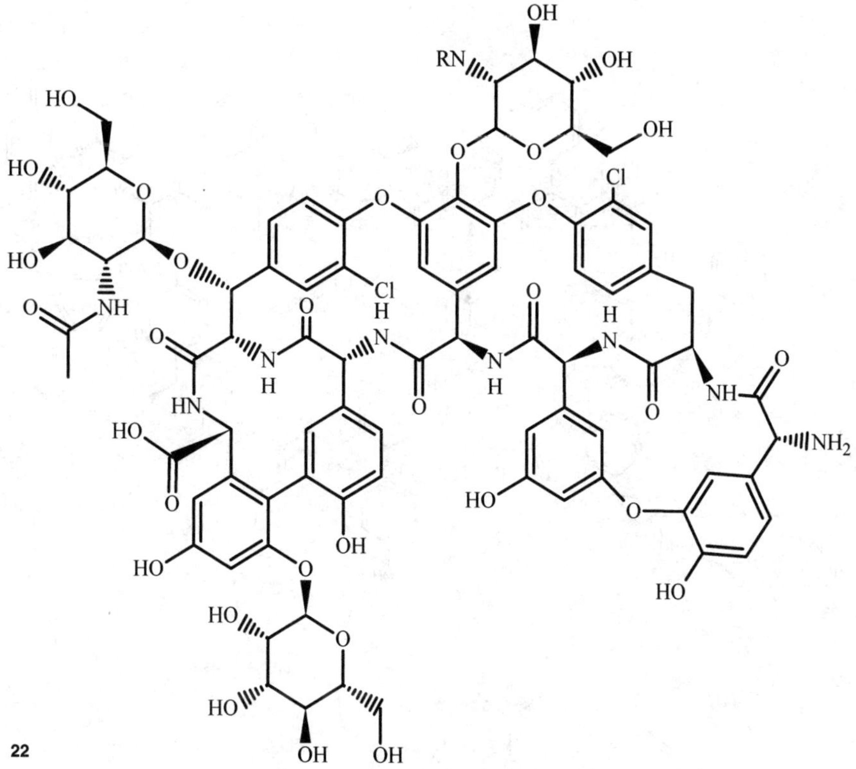

22

Gramicidins comprise several classes of peptide antibiotics, with gramicidin A, B, and C referred to collectively as gramicidin D. Gramicidin D consists of linear pentadecapeptides synthesized by *Bacillus brevis*. On the contrary, gramicidin S (**24**) is a cyclodecapeptide with the pentapeptides joined head to tail. Two uncommon amino acids form this structure: ornithine and the D-stereoisomer of phenylalanine (15).

Polymixins, produced by *Bacillus polymyxa*, are cyclic peptides with a hydrophobic chain. Polymixins are active against gram-negative bacteria because of their ability to interact with cell membrane phospholipids. The most representative antibiotic of this type is polymyxin B, which is a mixture of polymyxin B1 (**25**) and B2 (**26**).

3.2 ANTIFUNGALS

3.2.1 Polyene Antibiotics

Amphotericin B (**27**) is a macrocyclic polyene produced by *Streptomyces nodosus* with a heavily hydroxylated region on the ring opposite to the conjugated system.

23

24

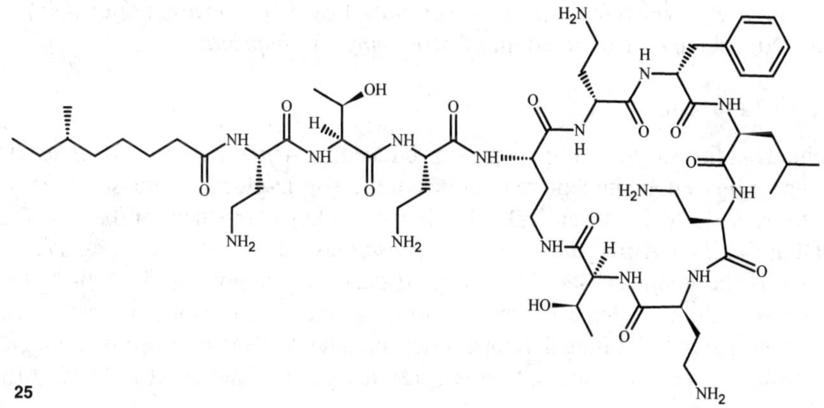

25

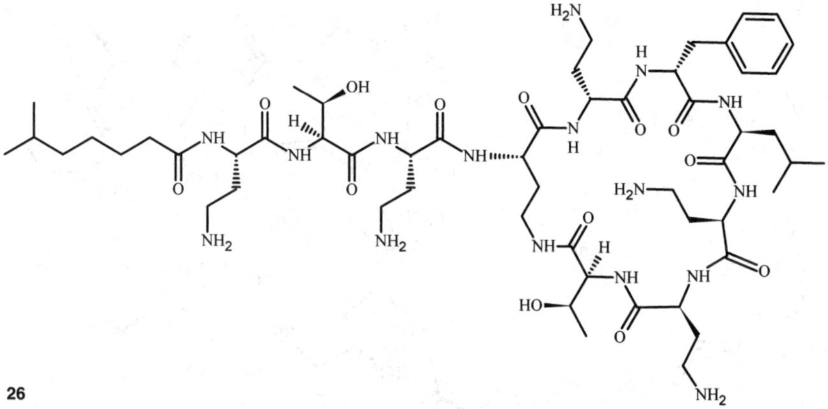

26

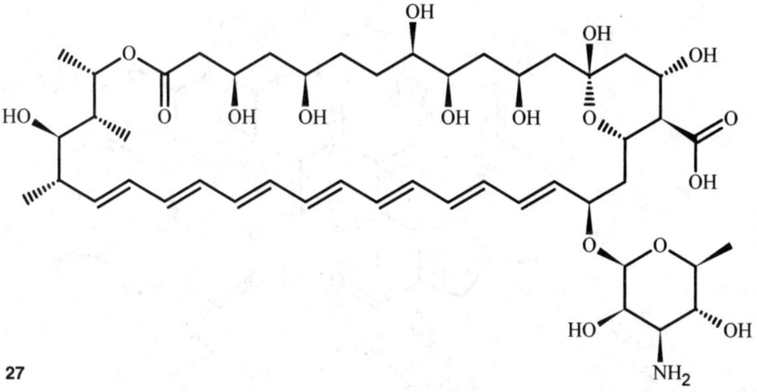

27

Amphotericin B is a fungistatic that binds to ergosterol in the fungal membrane (16), forming transmembrane channels and leading to the leakage of small organic molecules and monovalent ions (K^+, Na^+, H^+, and Cl^-). Other antifungal polyenes with a similar structure and mechanism of action include nystatin (produced by *Streptomyces noursei*), natamycin (also known as pimaricin, produced by *Streptomyces natalensis)*, filipin (produced by *Streptomyces filipinensis*), and hamycin (obtained from a strain of *Streptomyces pimprina)*.

3.2.2 Lipopeptides

Echinocandin B (**28**) is a lipopeptide produced by *Aspergillus rugulovalvus*. This compound is a noncompetitive inhibitor of (1, 3)-ß-D-glucan synthase, an enzyme that forms 1,3- β-D-glucan. 1,3-β-D-glucan is a key component of the fungal cell wall in *Candida, Aspergillus*, and *Histoplasma* as well as in *Pneumocystis carinii*, which is the major cause of HIV-related death. By inhibiting (1, 3)-ß-D-glucan synthase, echinocandin B prevents >90% of glucose incorporation into glucan.

Other natural antifungal lipopeptides include aculeacin A from *Aspergillus aculeatus*, pneumocandin B from *Glarea lozoyensis*, and FR901379 (**29**) from

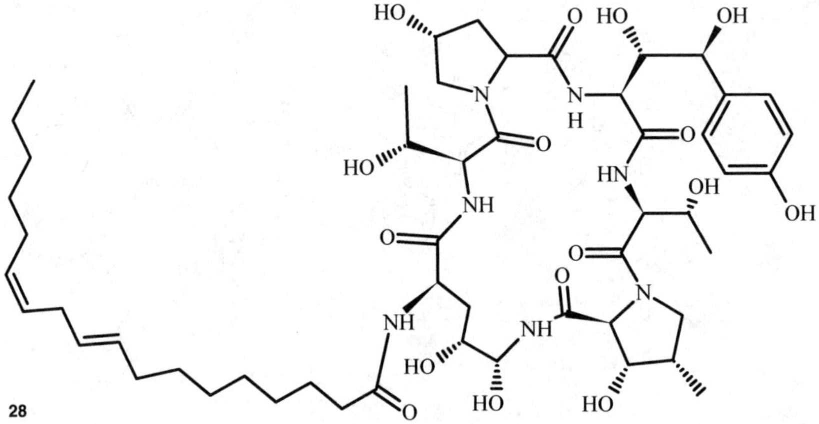

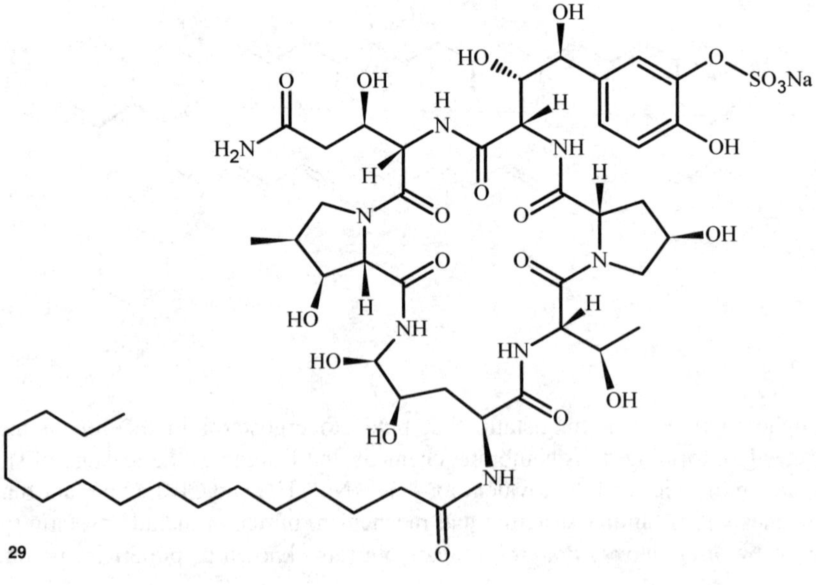

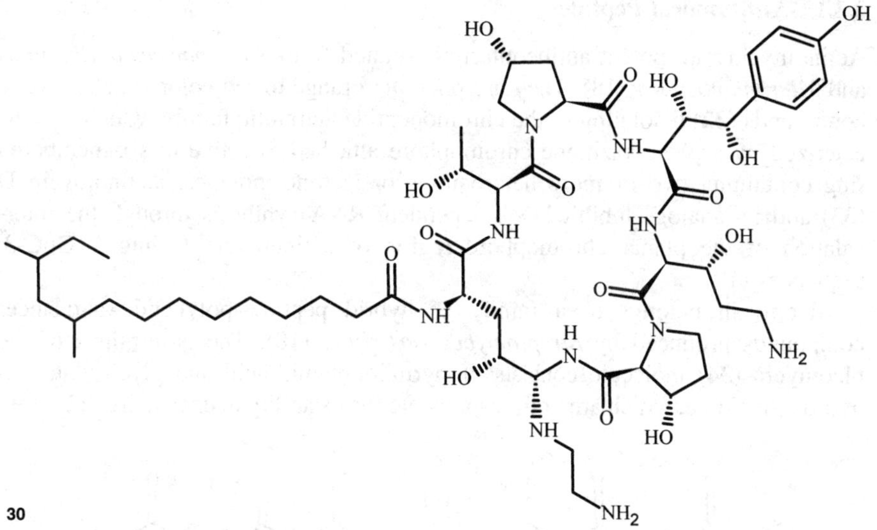

30

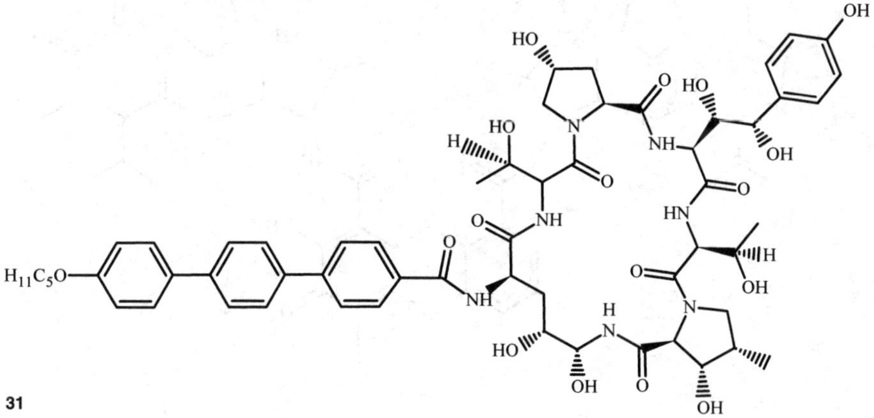

31

Coleophoma empetri. All of these antibiotics are characterized structurally by a cyclic hexapeptide that is acylated with a long side chain. The difference between FR901379 and other echinocandins is the presence of a sulfate moiety in its structure (**29**). Three echinocandins have reached the market, including caspofungin (produced by *G. lozoyensis*) (**30**), micafungin (produced from FR901379), and anidulafungin (synthesized from echinocandin B) (**31**).

3.3 ANTITUMOR COMPOUNDS

More than 60% of the anticancer drugs in clinical use are natural products derived from microorganisms (17). Antitumor metabolites exhibit broad chemical diversity and structural complexity.

3.3.1 Antitumoral Peptides

Actinomycins are potent antineoplastics isolated from *Streptomyces antibioticus* and *Micromonospora* (18). They are phenolic orange to red-colored heterocyclic compounds (**32**) belonging to the chromopeptide antibiotic family, which is characterized by a phenoxazinone chromophore attached to a five to six-membered ring containing two or more heteroatoms or lactone moieties. Actinomycin D (**33**) and its analogs inhibit DNA-dependent RNA synthesis through the intercalation of the planar chromophore region of actinomycin D into 5'-GpC-3' sequences (19).

Bleomycin belongs to a family of hybrid peptide–polyketide anticancer compounds produced by *Streptomyces verticillus* (18). The N-terminus of the bleomycin (**34**) molecule consists of pyrimidoblamic acid and glycosylated β-hydroxyhistidine, which are both responsible for its ability to chelate iron (18, 19).

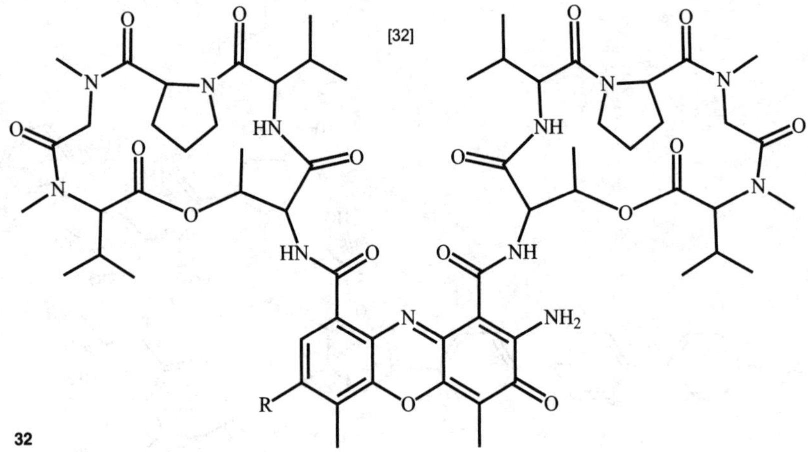

32

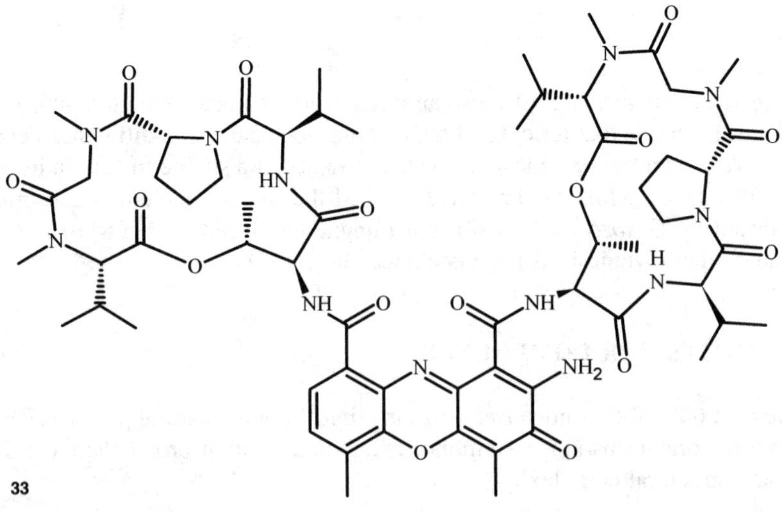

33

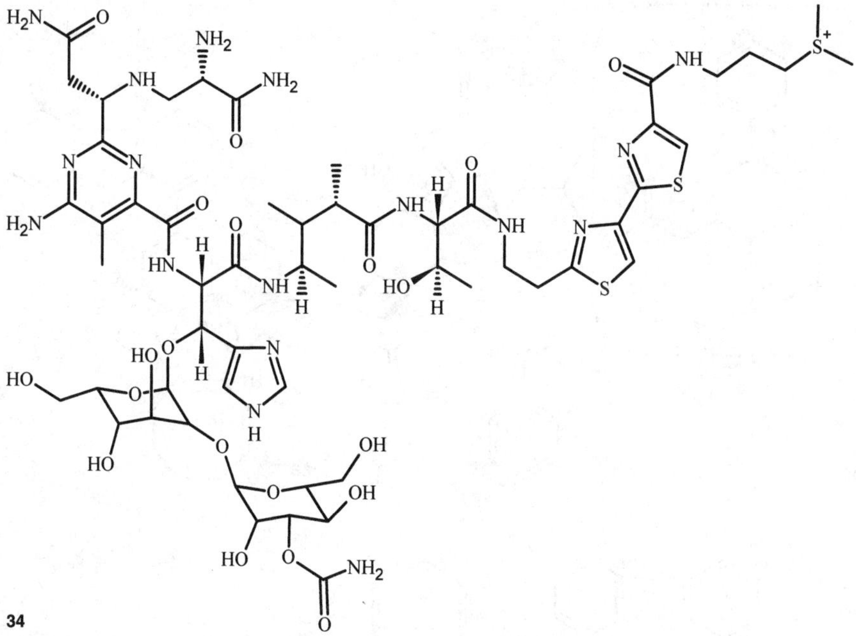

34

3.3.2 Steroid Alkaloids

Taxol is an anticancer compound produced by plants (*Taxus*) and some endophytic fungi (*Taxomyces andreanae, Taxus wallichiana*, and *Pestalotiopsis microspore*). It is a steroidal alkaloid diterpenoid that has a characteristic N-benzoylphenyl isoserine side chain (**35**), (dashed rectangle) and a tetracycline ring (18). The benzoyl group is particularly crucial for maintaining the strong bioactivity of Taxol. At elevated doses, Taxol causes mitotic arrest resulting from chromosome miss-segregation. Its unique activity has made it one of the most popular antineoplastic drugs (18, 20).

Manzamines are a family of alkaloids that exhibit a complex molecular architecture and possess antitumor, antimicrobial, antiparasitic, and insecticidal activities. *Micromonospora* sp. strain M42 produces manzamine (**36**) and 8-hydroxy-manzamine (**37**). Structurally, manzamines contain a heterocyclic unsaturated eight-membered ring. This ring forms the core in a group of opioid compounds commonly known as azocines (**38**).

3.3.3 Antitumoral Pyrrole Compounds

CC-1065, isolated from *Streptomyces zelensis*, consists of three repeating pyrroloindole subunits. CC-1065 (**39**) is a broad-spectrum antitumoral that has proven to delay death in animal cancer models.

Bizelesin is a CC-1065 analog that contains a cyclopropyl indole moiety (**40**). In addition, it contains two pyrroloindole subunits and one DNA-reactive

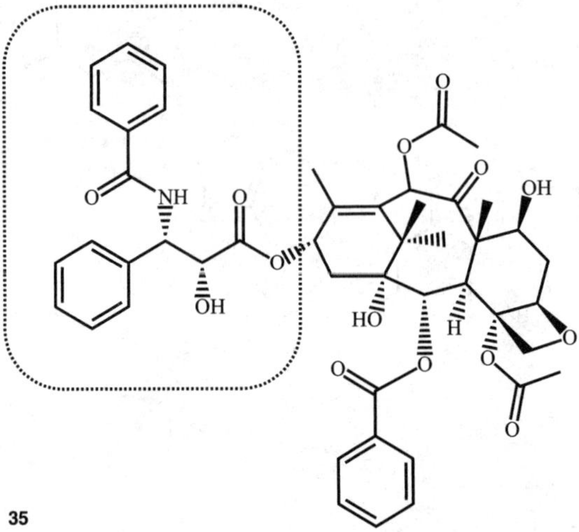

35

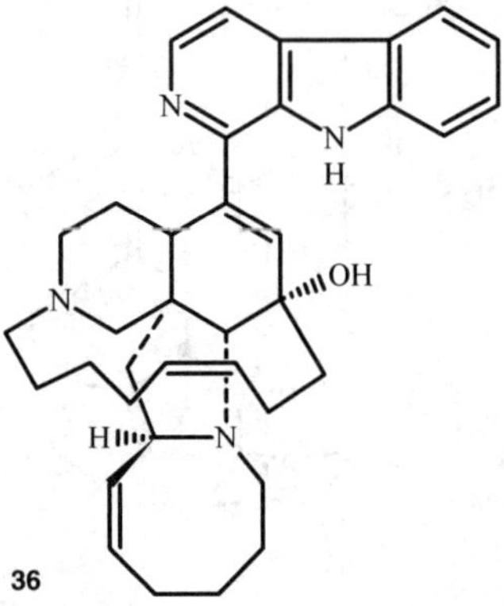

36

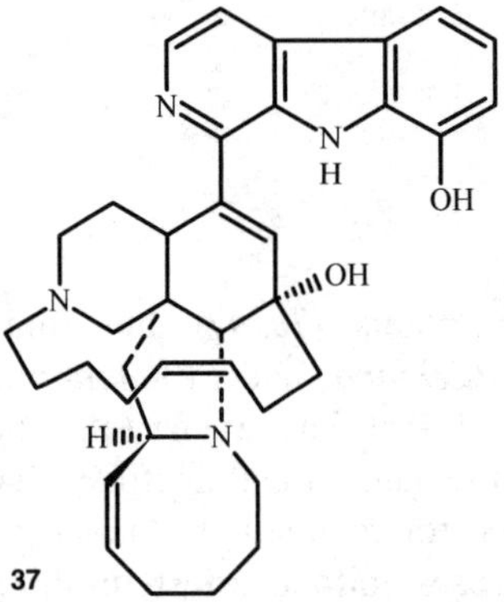

37

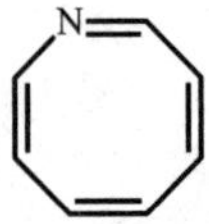

38

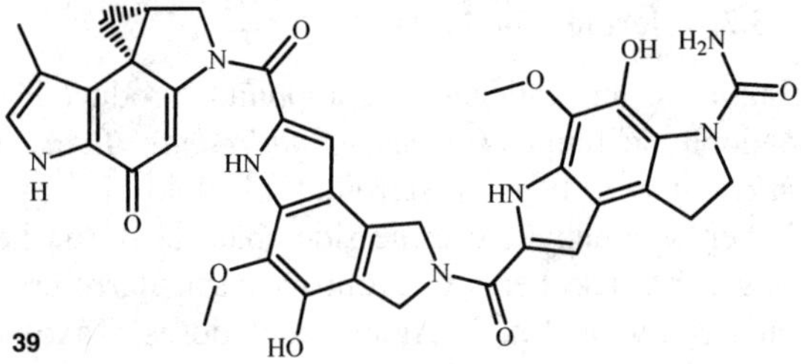

39

cyclopropylpyrroloindole moiety (18). Bizelesin cleaves doubled-stranded DNA by binding noncovalently within the minor groove with a consensus sequence of 5′-d(A/GNTTA)-3′ and 5′-d(AAAAA)-3′.

Prodigiosin is mainly isolated from *Serratia marcescens*, although it can be also found in unrelated species such as *Vibrio psychroerythrus, Streptomyces griseoviridis*, and *Hahella chejuensis*. Prodigiosin is a red pigment with a pyrrolyldipyrrolylmethene moiety and a β-ring methoxy group (**41**), (dashed rectangle) (20) that contains different alkyl substituents. In the presence of reduced metal ions such asCu^{2+}, prodigiosin is capable of cleaving single- and double-stranded DNA. Prodigiosin functions via a mechanism similar to that of CC-1065 and the bleomycins; however, it causes reductive activation of molecular oxygen, forming reactive oxygen species such as hydrogen peroxide that initiate the DNA cleavage (21).

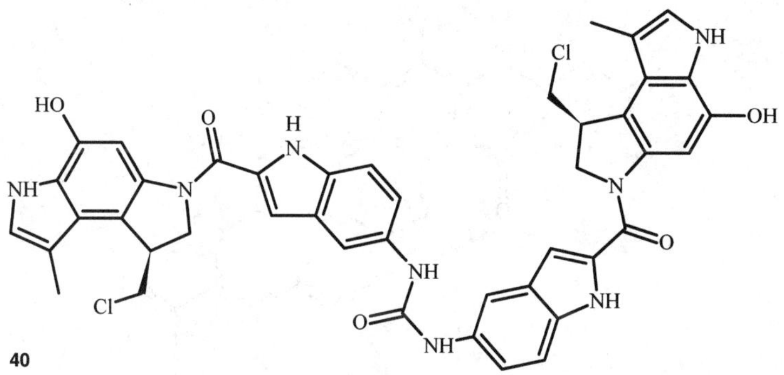

40

3.3.4 Antitumoral Polyketides

Epothilone is a colorless compound isolated from *Sorangium cellulosum* with anticancer effects. Epothilone consists of a macrolactone ring that is composed of an epoxy and a keto group in the lactone ring and a side chain exhibiting a thiazole ring (**42**), (dashed rectangles) (18). It has Taxol-like activity, but it is more effective and more water soluble. Epothilone promotes tubulin polymerization, stabilizes microtubules, and delays mitosis (18).

The kibdelones are a novel family of bioactive aromatic polyketides produced by the soil-dwelling actinomycete *Kibdelosporangium* sp. They comprise isoquinolinone and tetrahydroxanthone ring systems and display potent cytotoxicity toward a range of human cancer cell lines. Complete relative stereostructures have been assigned to kibdelones A–C (**43**) (shows type C).

3.3.5 Anthracyclines

These *Streptomyces*-derived compounds are some of the most effective anticancer agents ever developed and are effective against a wide range of cancers. Their main adverse effect is cardiotoxicity, which considerably limits their usefulness. Daunorubicin, produced by *Streptomyces peucetius*, was the first anthracycline discovered (**44**). Doxorubicin (**45**) (Adriamycin) was isolated shortly after from a *S. peucetius*-derived strain. Doxorubicin is a 14-hydroxylated version of daunorubicin.

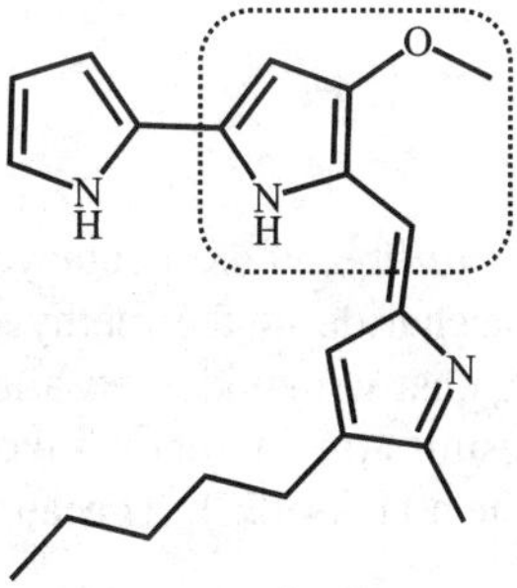

41 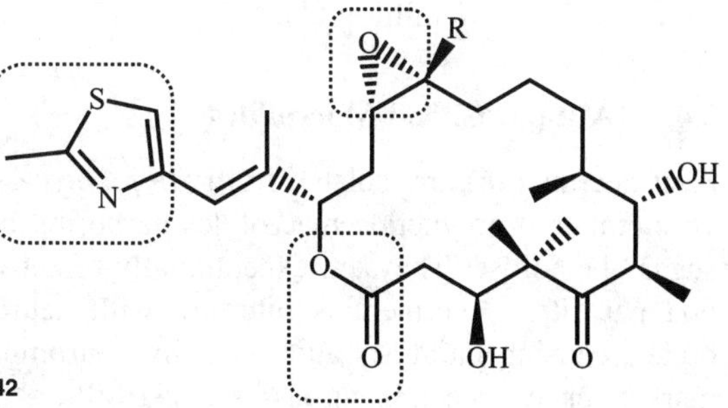   42

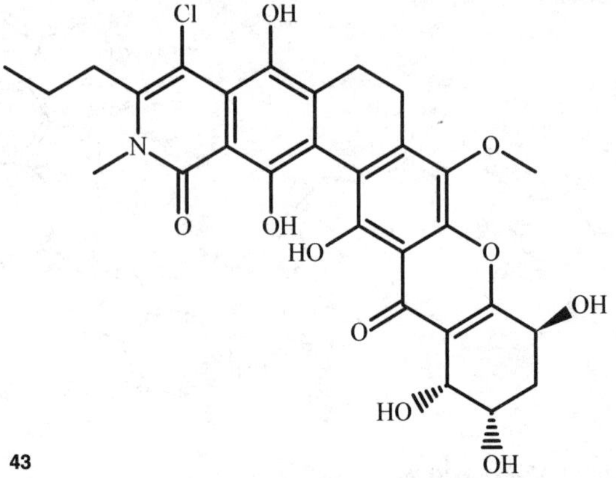

43

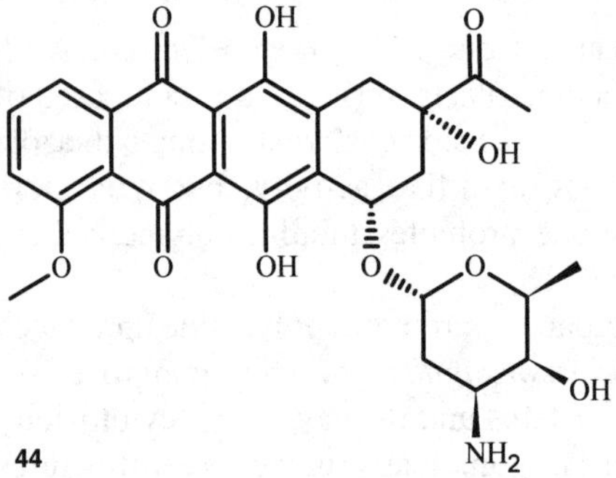

44

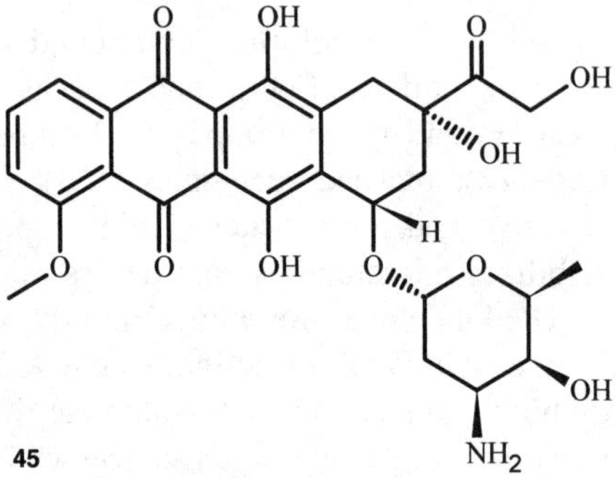

45

The antineoplastic activity of these drugs has been mainly attributed to their strong interactions with DNA in the target cells.

3.4 ANTIPARASITICS

Parasitic diseases caused by protozoa such as *Leishmania, Trypanosome*, or *Plasmodium* result in more than three million human deaths annually in developing countries. There are 3200 varieties of parasites, and occasionally, 2 or more can cause infection at the same time. In 2006, the parasiticide animal market was valued at US$464 million (1).

3.4.1 Antiparasitical Macrolides

Avermectins (**46**) are isolated from *Streptomyces avermitilis* and are pentacyclic, 16-membered polyketide macrolides harboring a disaccharide of the methylated sugar oleandrose. They are exceptionally effective against nematode and arthropod parasites. Avermectins interfere with neurotransmission in many invertebrates, causing paralysis and death by neuromuscular attacks (22). The annual market for avermectins surpasses US$1 billion.

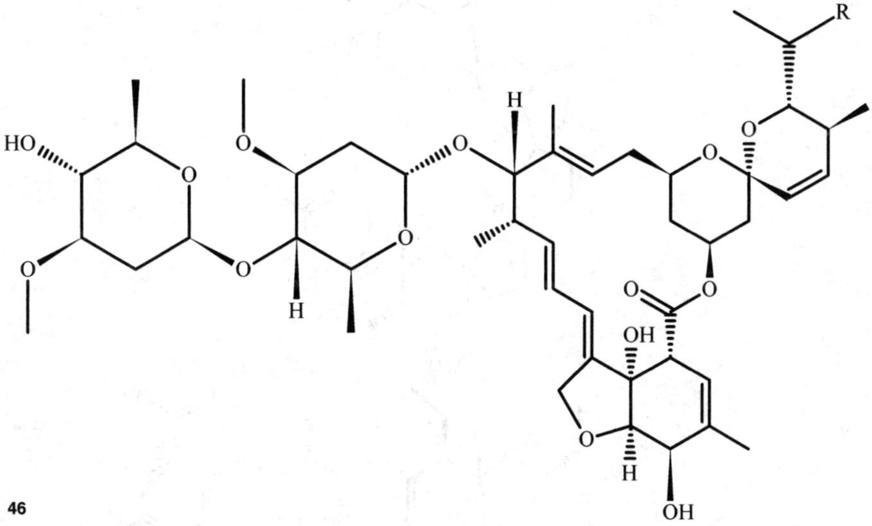

46

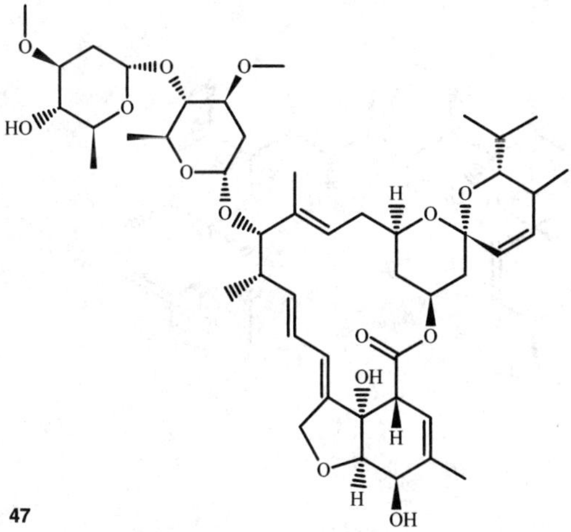

47 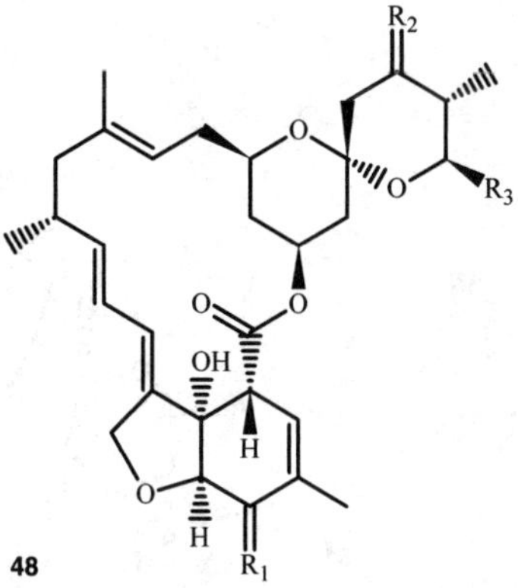 48

A semisynthetic derivative of avermectine, 22,23-dihydroavermectin B1, called ivermectin (47), is 1000 times more active than thiabendazole and is a commercial veterinary product.

The avermectins are closely related to the milbemycins (48), a group of nonglycosylated macrolides produced by *Streptomyces hygroscopicus* subsp. *aureolacrimosus* (23). These compounds possess activity against worms and insects.

3.4.2 Polyethers

Monensin is a polyether ionophore antibiotic (49) produced by *Streptomyces cinnamonensis*. It is a narrow-spectrum antibiotic with extreme potency against the

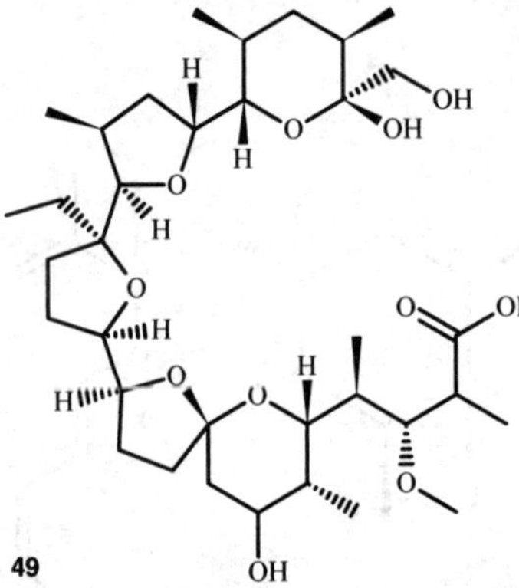

49

50

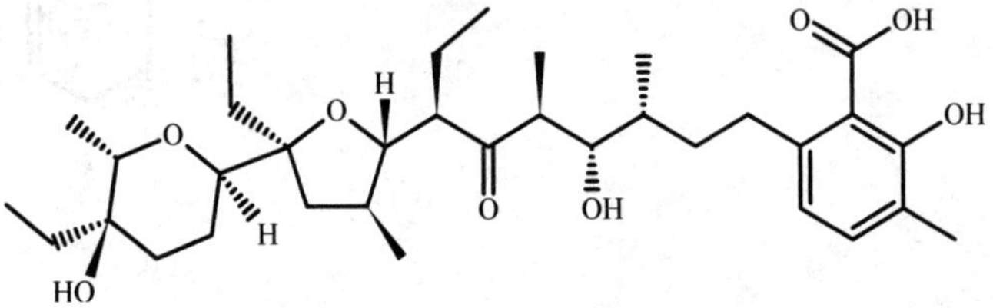

51

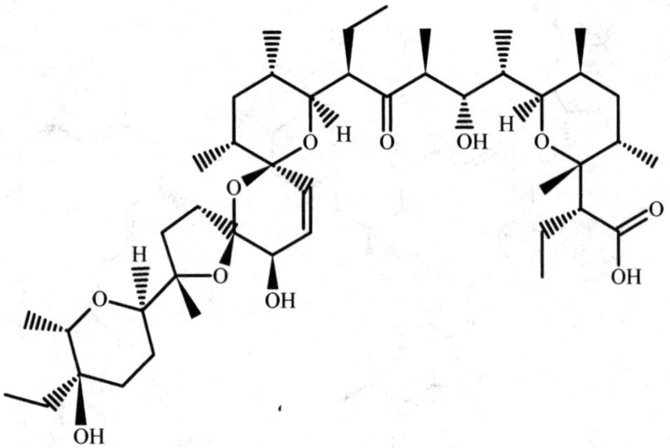

coccidia (24). The discovery of monensin led to the development of additional microbial ionophoric antibiotics such as lasalocid (**50**), narasin (**51**), and salinomycin (**52**). All are produced by various *Streptomyces* species and have similar structures to monensin. This group of compounds forms complexes with the polar cations K⁺, Na⁺, Ca²⁺, and Mg²⁺, severely affecting the osmotic balance in the parasitic cells and thus causing their death (25).

3.5 ENZYME INHIBITORS

Enzyme inhibitors are compounds that bind to enzymes and decrease their activity. They have applications in medicine as well as in agriculture as herbicides

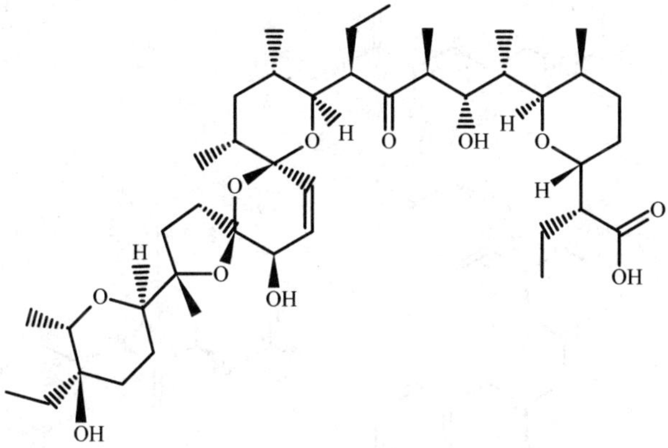

52

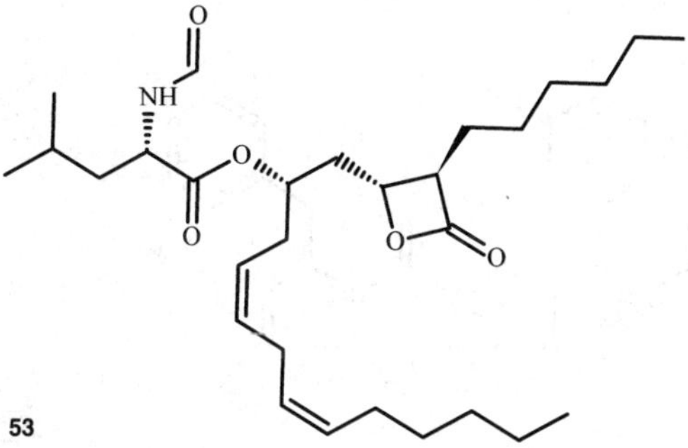

53

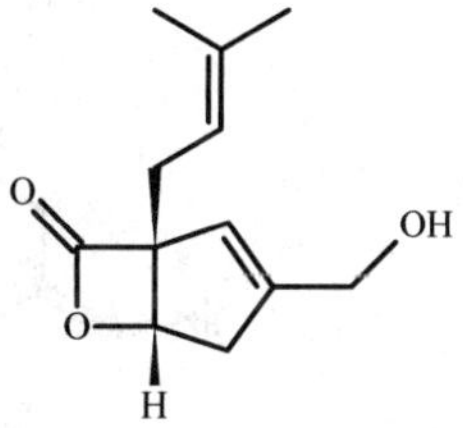

54

and pesticides. Several enzyme inhibitors with various industrial applications have been isolated from microorganisms (26).

3.5.1 Enzyme Inhibitory Lactones

Lipstatin (**53**) is a pancreatic lipase inhibitor produced by *Streptomyces toxytricini* that is used to combat obesity and diabetes. It interferes with the gastrointestinal absorption of fat (27). Lipstatin contains a β-lactone structure that is likely responsible for its irreversible binding to the active site on lipase. It is an irreversible inhibitor of several lipases, including pancreatic, gastric, carboxyl ester, and bile-stimulated lipases present in milk.

Vibralactone is a lipase inhibitor obtained from cultures of *Boreostereum vibrans* (28). Vibralactone (**54**) is an unusual fused β-lactone-type metabolite that can inhibit pancreatic lipase.

3.5.2 Enzyme Inhibitory Peptides

Antipain (**55**) is a reversible inhibitor of serine/cysteine proteases, such as plasmin and thrombin, as well as some trypsin-like serine proteases. It is a bioactive peptide produced by *Streptomyces yokosukaensis* and acts by forming a hemicetal adduct between its aldehyde group and the active serine of the protease.

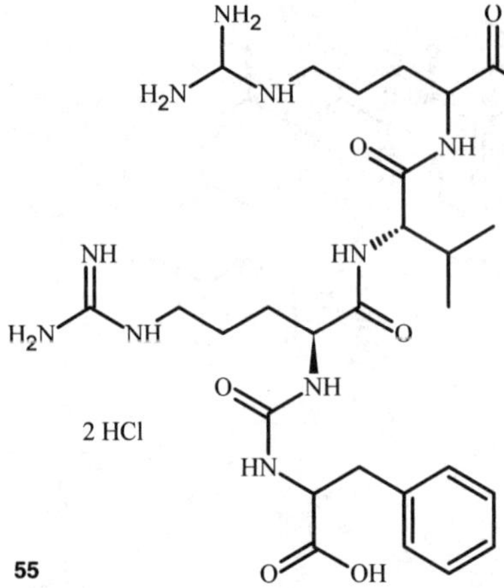

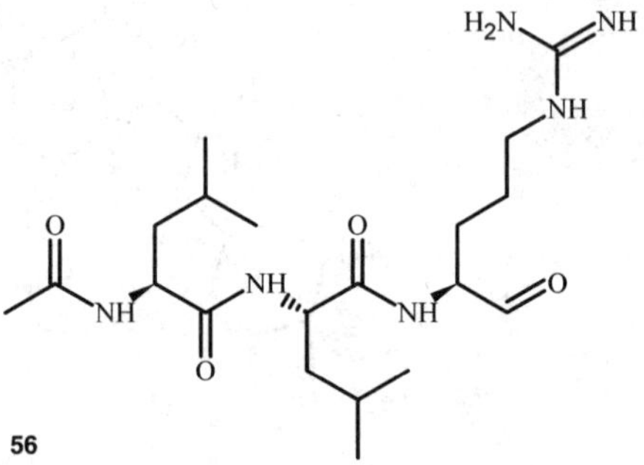

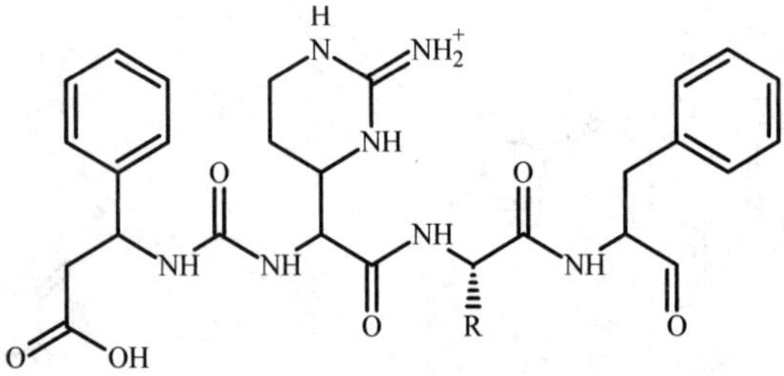

Leupeptins (**56**) are a group of naturally occurring protease inhibitors produced by more than 17 different species of actinomycetes (26). The main compounds in this group are acetyl-L-leucyl-L-leucyl-argininal (AC-LL) and propionyl-L-leucyl-L-leucyl-argininyl (Pr-LL). The aldehyde of the L-argininal moiety is thought to be essential for the antiprotease activity. Leupeptin inhibits plasmin and bradykinin formation.

Chymostatin (**57**) is another protease inhibitor that is isolated from different actynomycetes. It can inhibit serine proteases that exhibit a chymotrypsin-like specificity, including α, β, γ, and δ chymotrypsin and most cysteine proteases. Mechanistically, the tripeptide sequence Gly-Leu-Phe is necessary to position the inhibitor at the active site of the protease.

Pepstatin (**58**) shows antipepsin activity and specifically inhibits proteases and cathepsin D. More than five pepstatins have been isolated, differing from each other in the fatty acid moiety. At least one of two hydroxyl groups of pepstatin seems to be important for pepstatin's activity.

Other enzyme inhibitory compounds are the angiotensin-I-converting enzyme (ACE) inhibitors commonly used for the treatment of hypertension. Many research groups have been screening for novel ACE inhibitors of microbial origin, focusing on fungi, actinomycetes, and baker's yeast. WF-10129 (**59**), obtained from *Doratomyces putredinis*, is one of the most potent ACE inhibitors. The ACE inhibitor from *Saccharomyces cerevisiae* is considered to be a good candidate for antihypertensive drugs and functional foods (29). The amino acid sequence of the *S. cerevisiae* ACE inhibitor was determined to be Tyr-Asp-Gly-Gly-Val-Phe-Arg-Val-Tyr-Thr (30).

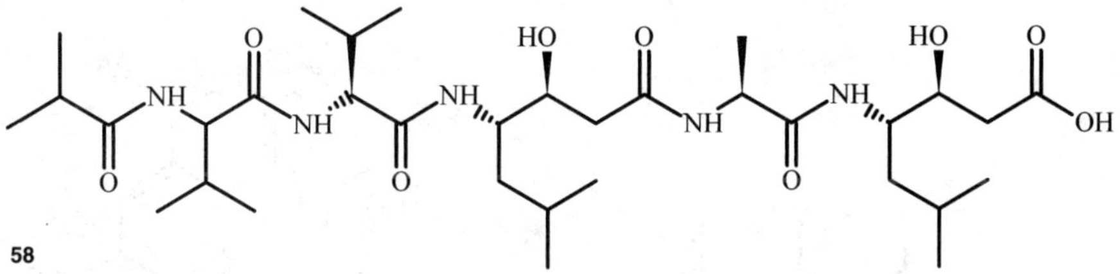

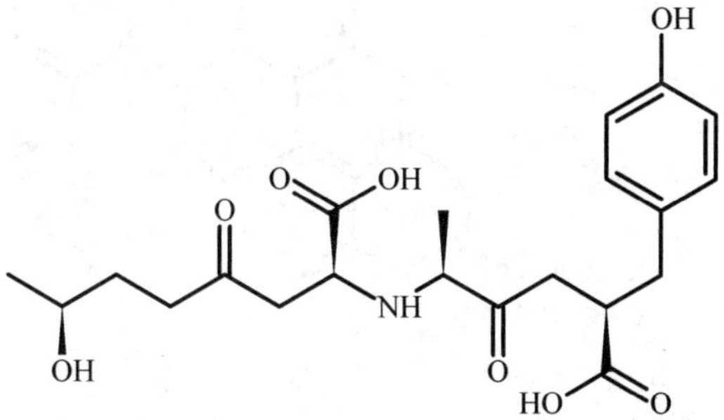

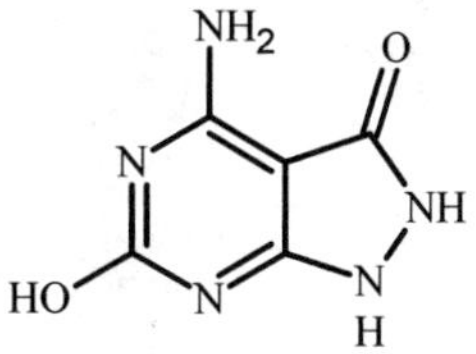

3.5.3 Purine Analogs

One additional enzyme inhibitor target is xanthine oxidase (XO). XO catalyzes the oxidation of hypoxanthine to uric acid through xanthine. An excessive accumulation of uric acid in the blood, called hyperuricemia, causes gout (31). Inhibitors of XO have been shown to decrease uric acid levels and thus have an anti-hyperuricemic effect.

A potent inhibitor of XO, hydroxyakalone (**60**), was purified from the fermentation broth of *Agrobacterium aurantiacum* sp. (32). The structure of hydroxyakalone was determined to be 4-amino−1*H*-pyrazolo[3, 4-d]pyrimidine-3-one-6-ol.

3.5.4 Naphtho γ-Pyrones

Fractionation of an endophyte (*Aspergillus niger* IFB-E003) extract from *Cynodon dactylon* resulted in four known compounds. Among them, rubrofusarin B (**61**) and aurasperone A (**62**) are strong co-inhibitors of XO (33).

Fungal products can be also used as enzyme inhibitors in the treatment of cancer, diabetes, Alzheimer's, and other diseases. Enzyme inhibitors used for

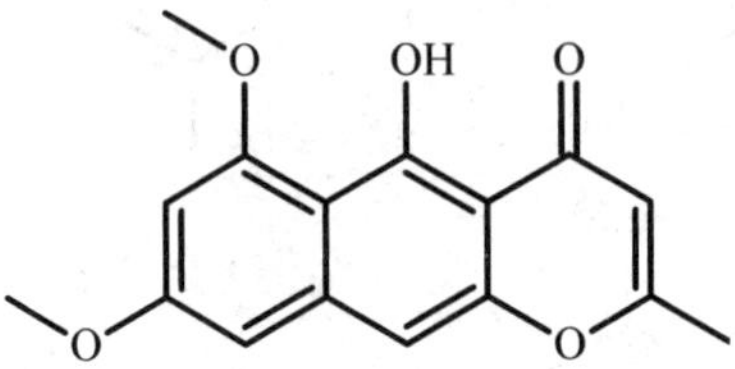

61

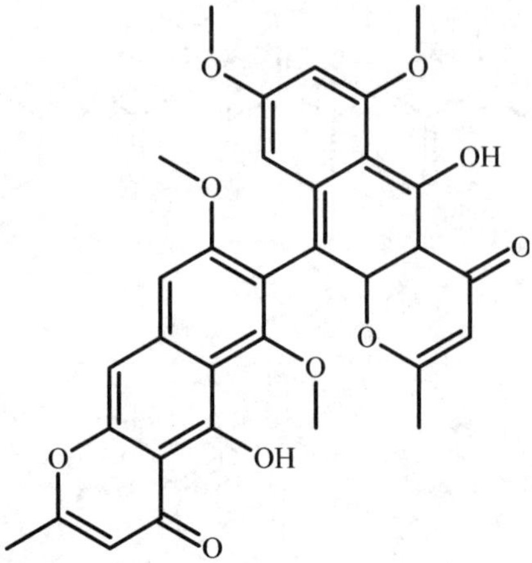

62

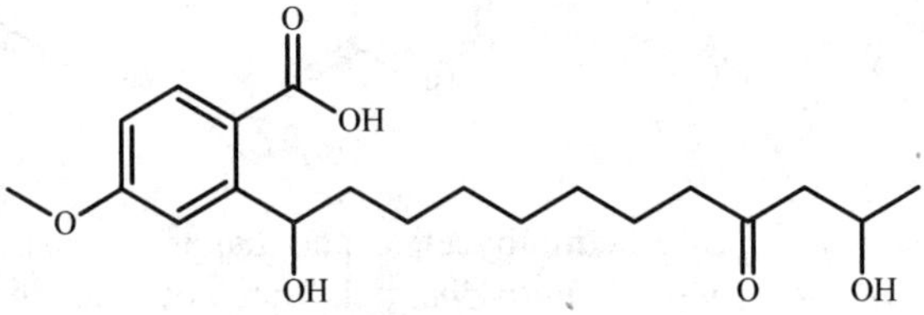

63

treatment include the acetylcholinesterase inhibitor sporotricholone (**63**) (isolated from *Sporotrichum* sp.); balanol (**64**) (from *Verticillium balanoides*), which inhibits protein kinases; and genistein (**65**) (from *Kitasatospora kifunensis*), which inhibits tyrosine kinases (34).

In animals, norepinephrine is a sympathetic system neurotransmitter synthesized from tyrosine. Tyrosine hydroxylase is involved in the rate-limiting step of norepinephrine biosynthesis. Deoxy-frenolicin (**66**) is an antibiotic produced by actinomycetes and is reported to inhibit tyrosine hydroxylase. A portion of this structure (surrounded by the dotted line) is intriguing, as these types of compounds are known to inhibit tyrosine hydroxylase or dopamine hydroxylase.

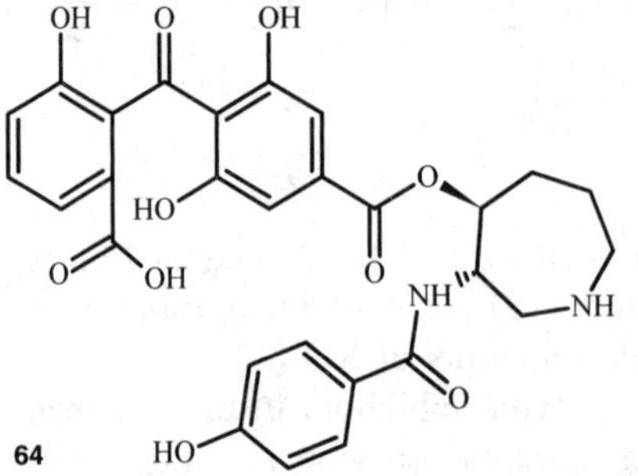

64

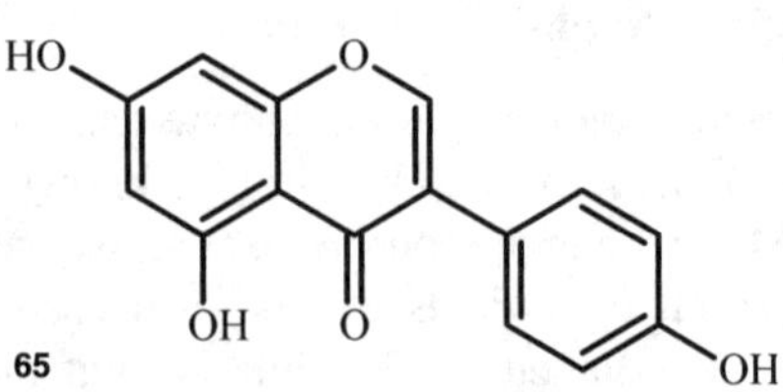

65

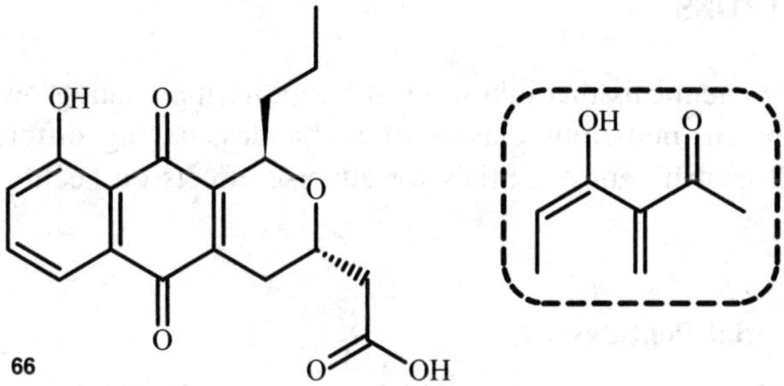

66

3.6 GASTROINTESTINAL MOTOR STIMULATORS

Drugs affecting gastrointestinal motility have become valuable in the management of several diseases and are useful for some postoperative patients. Some macrolides that can enhance the transit of material through the gastrointestinal tract are called prokinetics.

3.6.1 Gastrointestinal-Active Macrolides

Erythromycin (11) is a commonly used macrolide antibiotic produced by *Sacch. erythraea*. Erythromycin (11) causes diarrhea, nausea, and vomiting in part as a result of the action of erythromycin on motilin receptors in the gut. Erythromycin is thus an attractive option for use in patients with gastrointestinal motility problems. Stimulation of motilin receptors results in contraction and improved emptying of the stomach. Some erythromycin analogs have been developed that, although lacking an antibiotic effect, retain their ability to act on motilin receptors; these are referred to as "motilides."

Similar effects are observed with the use of oleandomycin, which is produced by *S. antibioticus*. Both erythromycin (11) and oleandomycin (67) have a 14-membered lactone ring to which a dimethylamino sugar and a neutral sugar are attached at positions C3 and C5 in a parallel glycosidic linkage.

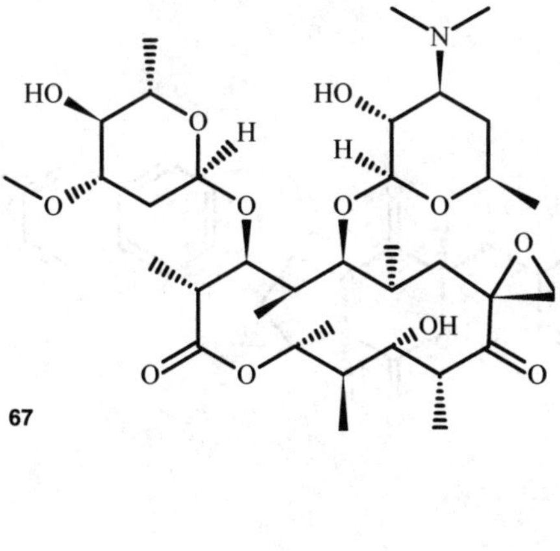

67

3.7 HERBICIDES

Herbicides are chemicals that inhibit or interrupt normal plant growth and development. There are numerous classes of herbicides, having different modes of action as well as different potentials for adverse effects on health and the environment.

3.7.1 Herbicidal Peptides

Bialaphos (**68**) is a tripeptidic herbicide named N(4-[hydroxy(methyl)phosphinoyl]-L-homoalanyl-L-alanyl-L-alanine) that is produced by *Streptomyces hygroscopicus*. Bialaphos is a protoxin which, upon metabolism in plants, is converted to the active form called L-phosphinothricin (**69**). Phosphinothricin is a structural analog of glutamic acid and acts a potent competitive and irreversible inhibitor of the enzyme glutamine synthetase. Phosphinothricin has bactericidal (against gram-positive and gram-negative bacteria), fungicidal (against *B. cinerea*), and herbicidal properties (35). The inhibition of glutamine synthetase causes the accumulation of ammonia and the inhibition of photorespiration, resulting in glyoxylate accumulation in chloroplasts. In turn, this inhibition causes the inhibition of ribulose bisphosphate carboxylase. Phosphinothricin is also produced by the bacteria *S. hygroscopicus* and *S. viridochromogenes*.

Maculosin (**70**) is a cyclic dipeptide produced by *Alternaria alternata* and is host-specific to spotted knapweed (*Cantaurea maculosa*). Maculosin-binding activity has been detected toward receptors from spotted knapweed leaves.

Tentoxin (**71**) is a cyclic tetrapeptide by-product of *Alternaria alternate*. It causes phytotoxic damage to both monocotyledonous and dicotyledonous weed species. This secondary metabolite functions by inhibiting CF1 ATPase activity.

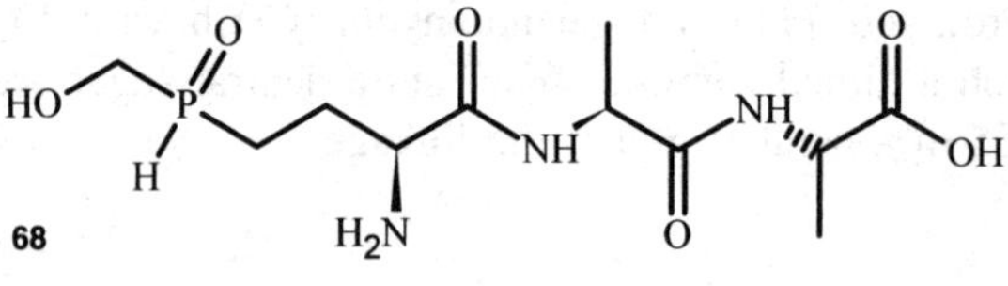

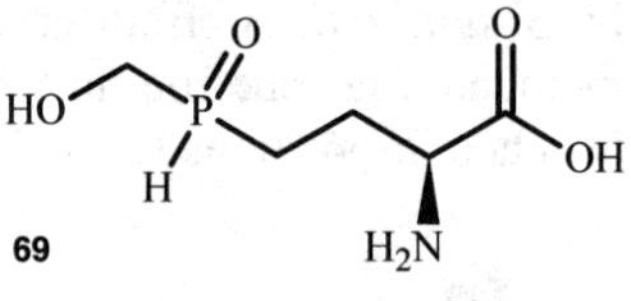

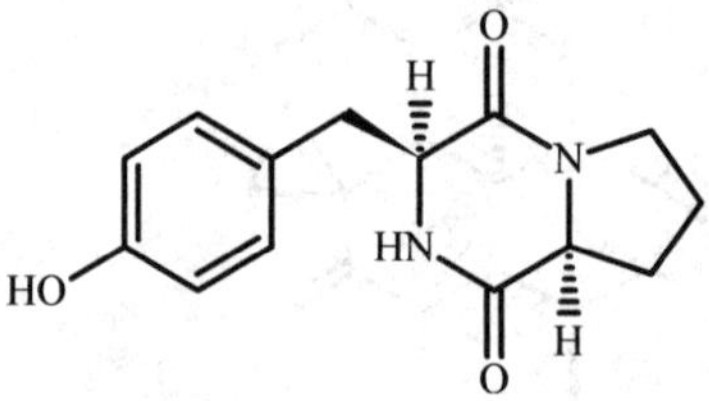

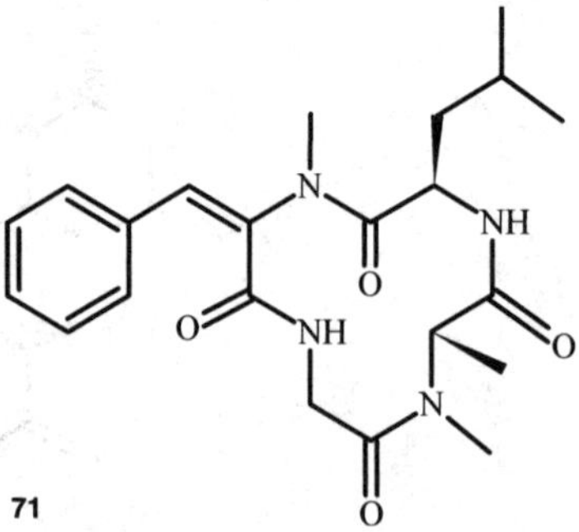

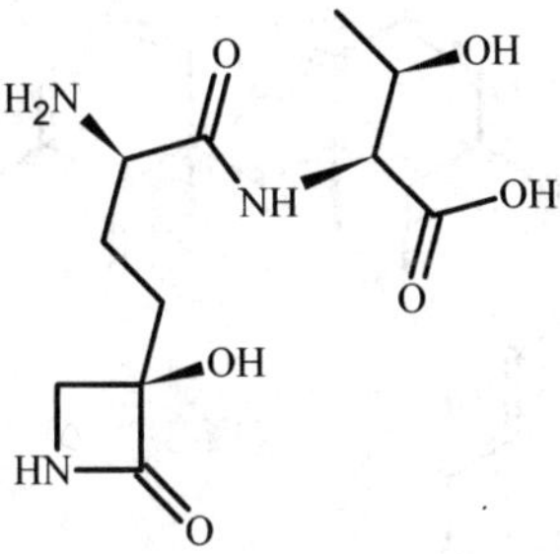

72

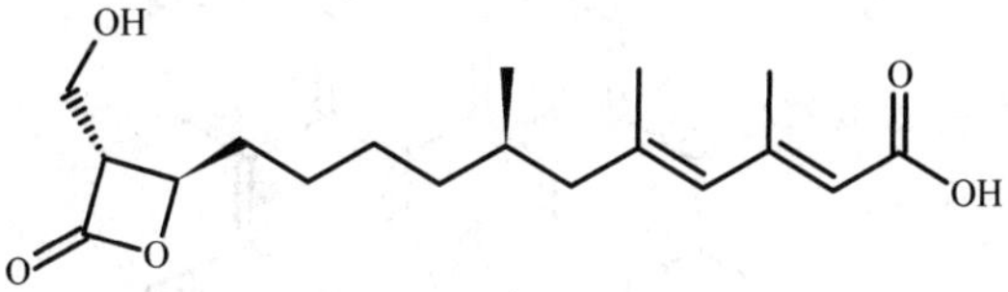

73

3.7.2 Monobactams

Tabtoxin (**72**) is a simple monobactam biotoxin produced by *Pseudomonas syringae* var. *tabaci*. This compound acts by inhibiting glutamine synthetase activity; however, the toxin is a less potent herbicide than is bialaphos.

Antibiotic 1233A (**73**) is a monobactam produced by *Scopulariopsis candida, Cephalosporium* sp., and *Fusarium* sp. It is a potent phytotoxin that inhibits the 3-hydroxy-3-methylglutaryl coenzyme A synthetase.

3.7.3 Herbicidal Lactones

Cyanobacterin (**74**) (diaryl-substituted γ-lactone) has a chlorine in one of its aromatic rings. It is produced by the cyanobacterium *Scytonema hofmanni* and is a potent inhibitor of cyanobacteria, green algae, and higher plants. The herbicide functions via an inhibition of the Hill reaction when p-benzoquinone, K3[Fe(CN)6], dichlorophenolindophenol, or silicomolybdate are used as electron acceptors. Cyanobacterin inhibits O_2-evolving photosynthetic electron transport in all plants. The most probable site of action is photosystem II.

3.7.4 Herbicidal Macrolides

Herbimycin (**75**) is a new benzaquinoid ansamycin herbicidal agent produced by *S. hygroscopicus*. It is a broad-spectrum herbicide against both monocotyledonous and dicotyledonous weeds and is particularly effective against *Cyperus microiria* STEUD. Herbimycin irreversibly and selectively inhibits tyrosine kinases by

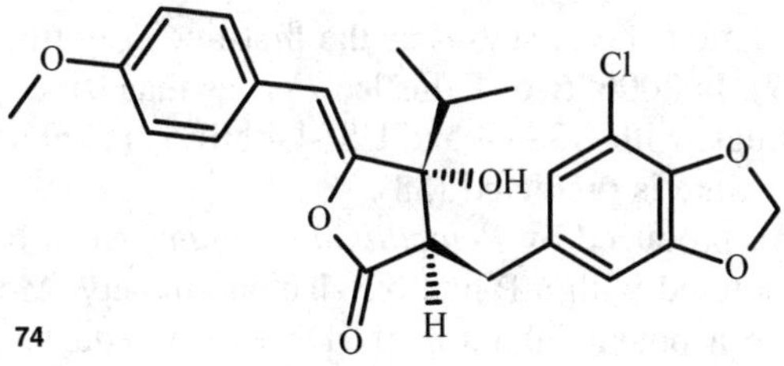

74

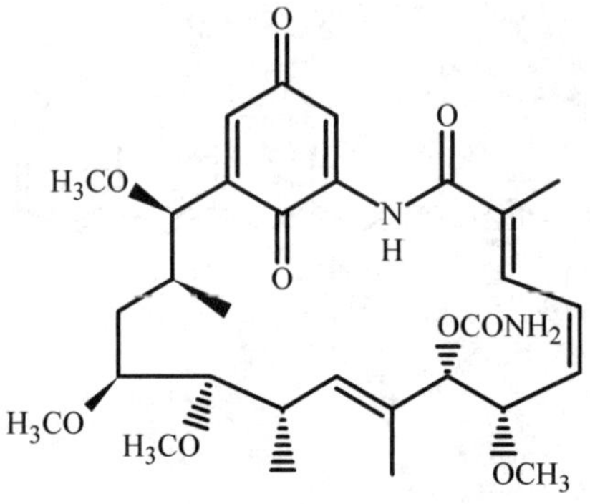

reacting with thiol groups. It also binds to Hsp90 (Heat Shock Protein 90) and alters its function.

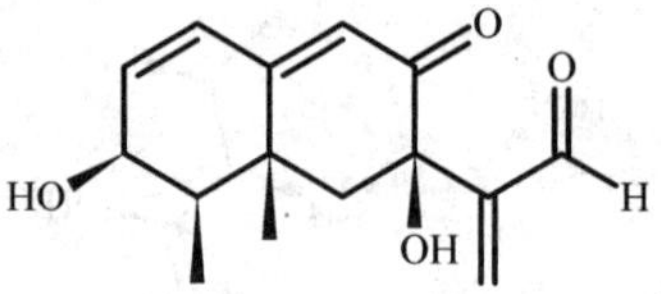

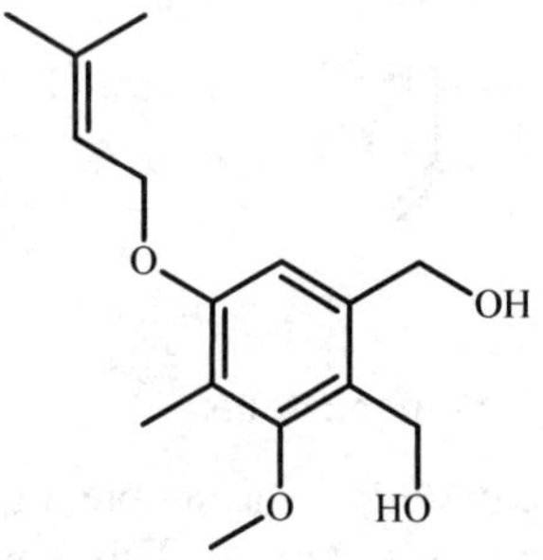

3.7.5 Terpenes

Bipolaroxin (**76**) is a sesquiterpene obtained from *Bipolaris cynodontis*, a fungal pathogen of Bermuda grass (*C. dactylon*). It has been found to be host-selective in low concentrations.

Zinniol (**77**) is produced by several *Alternaria* spp. and *Phoma macdonaldii* Boerma. It causes necrosis in tissues, likely by disrupting calcium regulation (36).

3.8 HYPOCHOLESTEROLEMICS

Hypercholesterolemia is a causative factor of coronary artery disease, myocardial infarction, and ischemic brain infarction. Current guidelines for cholesterol treatment target low-density cholesterol (LDL-C). To reduce plasma cholesterol levels, one option is to inhibit HMG-CoA reductase, the rate-limiting enzyme in the synthesis of cholesterol.

3.8.1 Naphtalenes

Recent recommendations favor statins as the first-line drug treatment for hypercholesterolemia (37). In 2006, two statins led *Forbes* magazine's list of America's 20 Best Selling Drugs, with US$8.4 and US$4.4 billion in annual sales. Increased use of statins over time is predicted (38).

Mevastatin (**78**), produced by *Penicillium citrinum*, has a hexahydronaphthalene skeleton substituted with a P-hydroxy-lactone moiety. Mevastatin has been demonstrated to be a potent inhibitor of HMG-CoA reductase. The search for

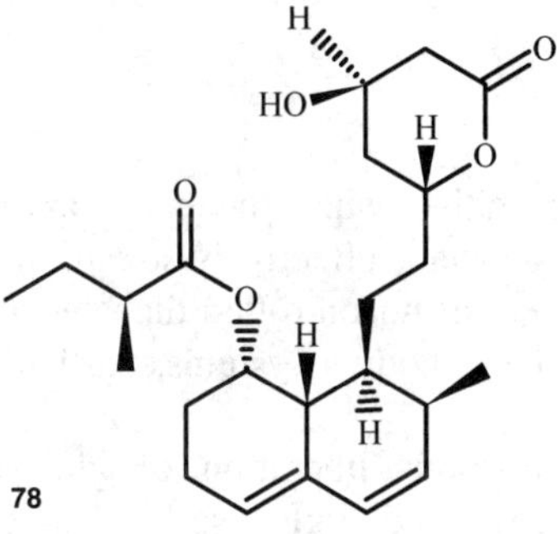

78 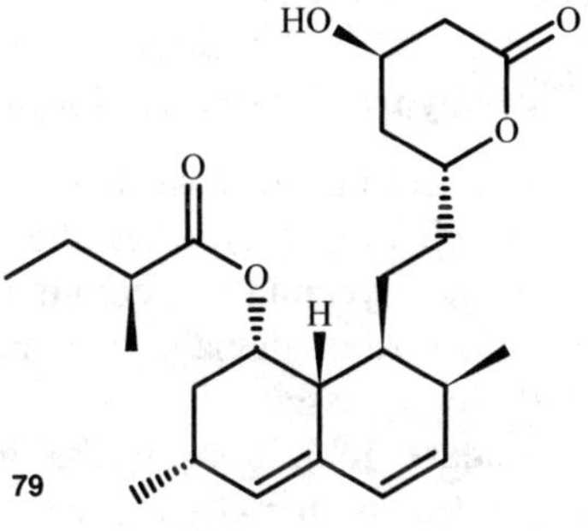   79

additional HMG-CoA reductase inhibitors was continued for another 10 years after mevastatin's discovery, leading to the isolation of several additional compounds of the mevastatin family (39).

Alberts et al. (40) isolated a strain of *Aspergillus terreus* that produced a more efficient statin: mevinolin or lovastatin (**79**). The chemical structure of lovastatin is identical to mevastatin, except for an additional methyl group. Merck later developed a second-generation semisynthetic derivative of lovastatin, which is still now the second best marketed statin. This derivative was named simvastatin (**80**). Although it can be produced synthetically from lovastatin, its synthesis can be also accomplished with an engineered *A. terreus* strain.

In addition to these more widely known compounds, several lovastatin- or mevastatin-related metabolites with a different composition of the C8 side chain have been isolated and characterized. An *in silico* analysis of the lovastatin biosynthetic cluster protein was recently performed. The study hypothesized that two previously undescribed proteins in the cluster may function as tailoring enzymes for the glycosylation machinery, and they may be useful for the preparation of new lovastatin analogs. It is predicted that incorporation of these enzymes with cytochrome P450s into heterologous hosts may help in the large-scale production of lovastatin in microbial fermentations (41).

Interestingly, statins are also emerging as drugs for the treatment of cancer (42), osteoporosis (43). Alzheimer's disease (44), Parkinson's disease (45), and rheumatoid arthritis (46).

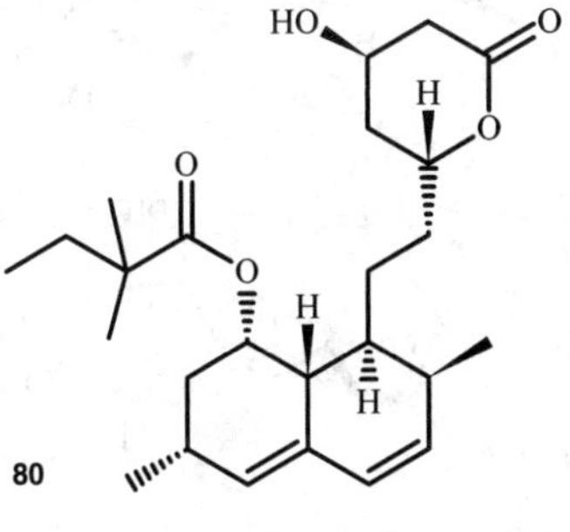

80

3.9 HYPOGLYCEMICS

3.9.1 Carbohydrate-Based Inhibitors

Diabetes is a chronic disease that occurs either when the pancreas does not produce enough insulin or when the body cannot effectively use the insulin it produces. Hyperglycemia is a common effect of uncontrolled diabetes and, over time, leads to serious damage to many of the body's systems, particularly the nerves and blood vessels.

α-Glucosidase inhibition in the gut restrains liberation of glucose from oligosaccharides and thereby reduces the postprandial glucose levels and insulin responses (47).

Acarbose (**81**) is a pseudotetrasaccharide made by *Actinoplanes* sp. SE50. It contains an aminocyclitol moiety, valienamine, which inhibits intestinal α-glucosidase and sucrose absorption, resulting in a decrease in starch breakdown in the intestine that is useful in combating diabetes in humans (48).

Recently, several acarviosin-containing α-amylase inhibitors have been reported. One of them, acarviostatin III03 (**82**), isolated and purified from *Streptomyces coelicoflavus*, is the most effective α-amylase inhibitor known. It is 260 times more potent than acarbose (49).

Other naturally occurring carbasugars such as validamine, valienamine, and hydroxyvalidamine were isolated from the fermentation broth of the antibiotic validamycin A fermentation in *S. hygroscopicus* culture. Their activity is attributable to their ability to mimic the ground- and/or transition-state structures of glucopyanosyl cations, which have been postulated to be formed during the hydrolysis of α-glucopyranosides.

Miglitol, produced by *Gluconobacter oxydans* through biotransformation of D-sorbitol, was recently selected as the most favorable inhibitor out of a large number of agents active *in vitro*. For the synthesis of miglitol, an N-hydroxyethyl derivative of 1-amino-1-deoxy-d-sorbitol is used (50).

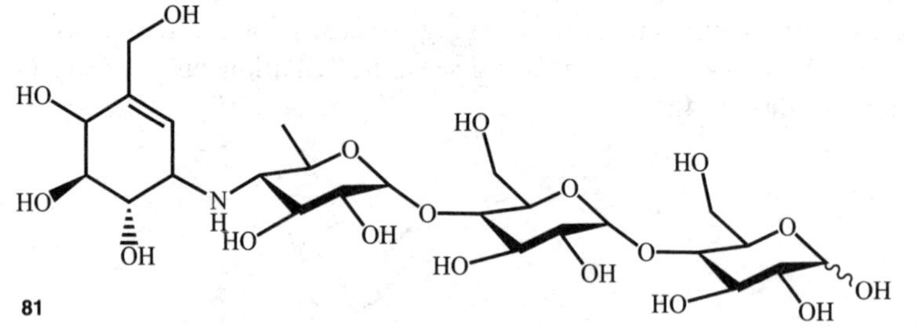

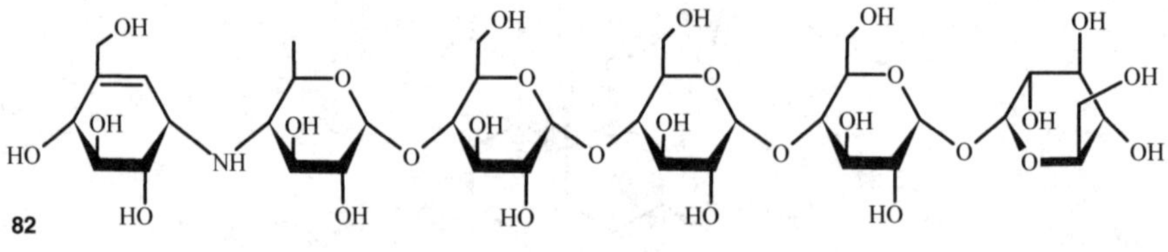

3.9.2 Hypoglycemic Peptides

Tendamistat, a 74-residue-long protein produced by *Streptomyces tendae* 4158 and *Streptomyces lividans*, is a potent α-amylase inhibitor (Ki = 0.2 nM). Its ability to bind α-amylase is localized to three residues: Trp18, Arg19, and Tyr20. These residues are held within a reverse-turn that blocks the catalytic site (51). Because of its resistance against most hydrolytic enzymes, Tendamistat may be orally available for diabetes mellitus treatment. However, its immunogenicity could prevent its further development. Therefore, developmental research is ongoing, with a focus on small peptides that are based on the essential binding elements of the natural protein inhibito (52–54).

3.9.3 Fatty Acids

HOD (10-hydroxy-8(*E*)-octadecenoic acid) exhibits strong anti-α-glucosidase (EC 3.2.1.20) activity. HOD (**83**) is an intermediate in the bioconversion of oleic acid to 7,10-dihydroxy-8(*E*)-octadecenoic acid (DOD) by *Pseudomonas aeruginosa* (PR3). Compared with acarbose, HOD is a stronger inhibitor of α-glucosidase and shows competitive inhibition against yeast α-glucosidase (55).

3.9.4 Diterpenoids

Platensimycin (**84**) (PTM) is a novel tetracyclic enone produced by *Streptomyces platensis* and is a broad-spectrum antibiotic against gram-positive bacteria. PTM is a selective inhibitor of mammalian fatty acid synthase (FAS), the discovery of which has led to improved liver steatosis and the amelioration of diabetes in db/db mice (56).

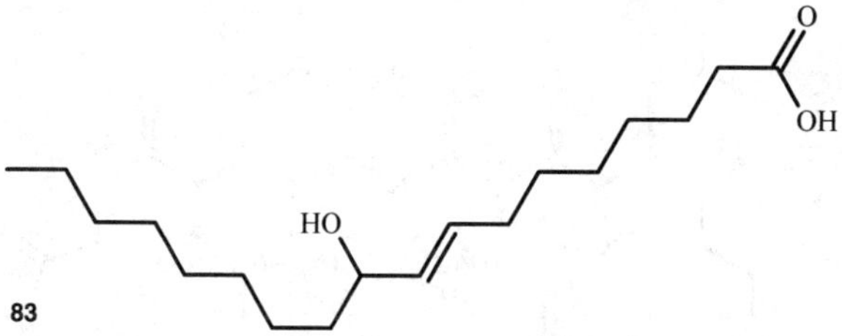

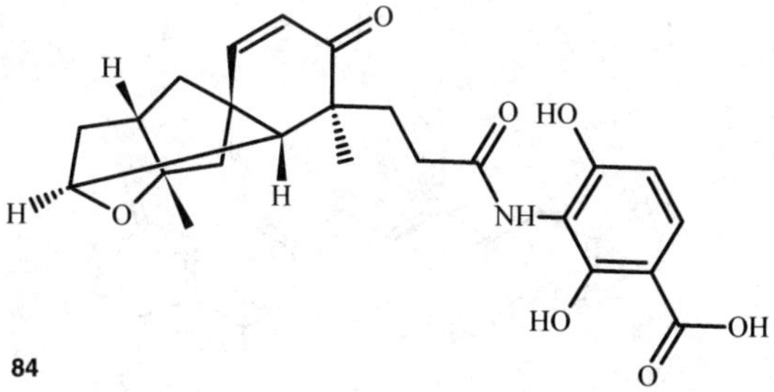

3.10 IMMUNOSUPPRESSANTS

The induction of immunosuppression is generally used to prevent the body from rejecting an organ transplant or for the treatment of autoimmune diseases such as rheumatoid arthritis or Crohn's disease. Several microbial compounds have been discovered that can suppress the immune response.

3.10.1 Immunosuppressive Peptides

Cyclosporine A (**85**) is a cyclic undecapeptide produced by the mold *Tolypocladium nivenum*. Mechanistically, it suppresses the immune system by interrupting calcium-mediated activating events (calcineurin) through complex formation with cyclophilins and by impairing cytoplasmic and nuclear responses of lymphocytes to alloantigens (57). With regard to its structure, the amino acid (4R)-4-[(E)-2-butenyl]-4-methyl-L-threonine (Bmt or C-9 amino acid) seems to be essential for immunosuppression.

3.10.2 Terpenoids with Immunosuppressive Activity

Mycophenolic acid can be produced by *Penicillium brevicompactum* (58) and *Penicillium echinulatum*. Its chemical structure (**86**) consists of a single phenolic hydroxyl, one CH_3O group, a lactone group, and a double bond (58, 59). Mycophenolic acid has been used in the treatment of psoriasis and other inflammatory dermatoses. It was never commercialized as an antibiotic because of its

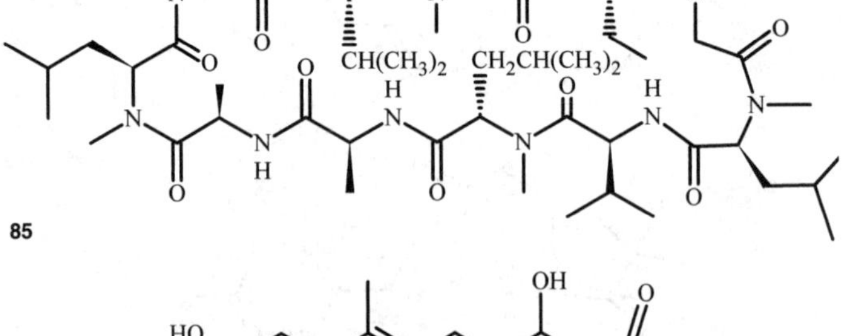

85

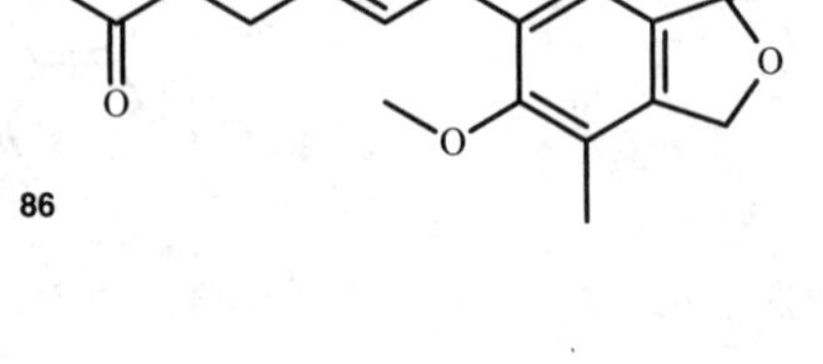

86

toxicity, but its 2-morpholinoethylester was approved as a new immunosuppressant for kidney and heart transplantation (1). This compound acts as an inhibitor of inosine monophosphate dehydrogenase, which is rate limiting in the synthesis of guanosine nucleotides and therefore affects T- and B-lymphocytes that are dependent on this pathway.

Myriocin (**87**) (ISP-1, thermozymocidine) is an acyclic diterpene that was shown to be 10- to 100-fold more potent than cyclosporine A. This compound is produced by *Isaria sinclairii, Myriococcum albomyces* and *Mycelia sterilia* (60). Myriocin potently inhibits the enzyme serine palmitoyltransferase, thereby decreasing sphingolipid content. Myriocin shows high toxicity and low solubility compared with cyclosporine A. Thus, this compound has been chemically modified to create a less toxic drug named fingolimod (**88**). In addition, fingolimod has been approved recently for the treatment of multiple sclerosis (60). It is the first orally administered disease-modifying treatment for multiple sclerosis to be approved in the United States, and it disrupts the disease's attack on the central nervous system (CNS) through a unique mechanism.

3.10.3 Immunosuppressive Macrolides

Tacrolimus (**89**) and rapamycin (**90**) are macrolides linked to amino acids that were isolated from *Streptomyces tsukubaensis* and *S. hygroscopicus*, respectively.

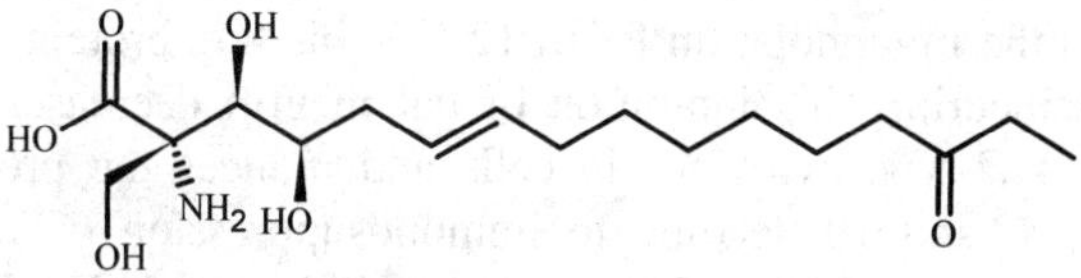

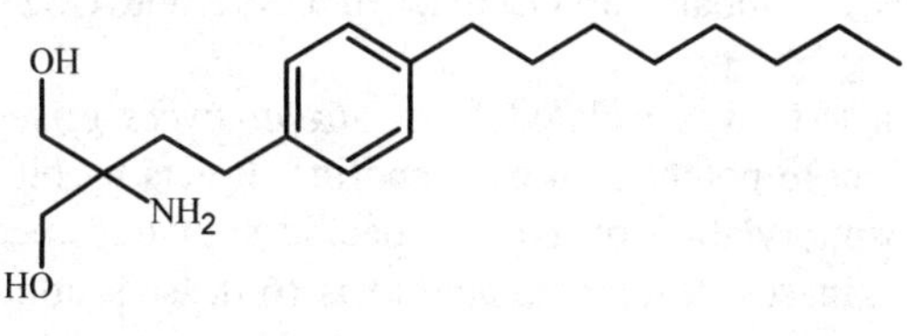

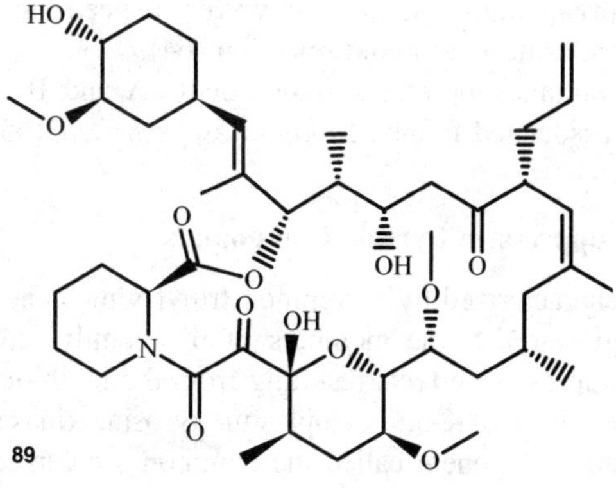

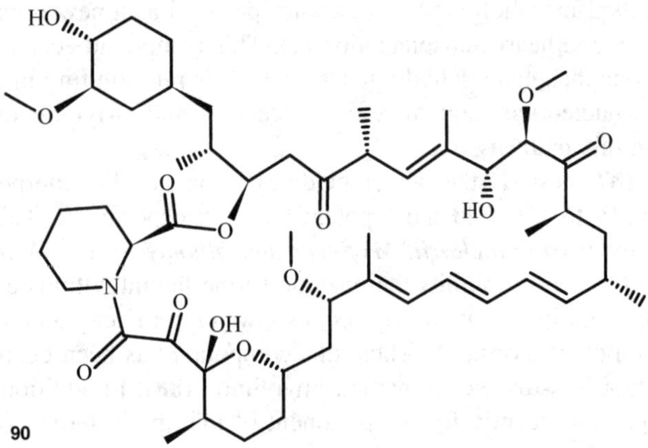

Both compounds are 100-fold more potent than cyclosporin as immunosuppressants and are less toxic. Their mechanism of action centers on its formation of a complex with the immunophilin FKBP12 (FK-binding protein), causing an inhibition of calcineurin. This inhibition of calcineurin decreases the presentation of Th1- and Th2-type cytokines in cells and reduces the production of interleukins IL-2, 4, and 10, leading to immunosuppression (61). Tacrolimus and rapamycin sales in global markets reached US$2 and US$1.5 billion in 2007, respectively (1).

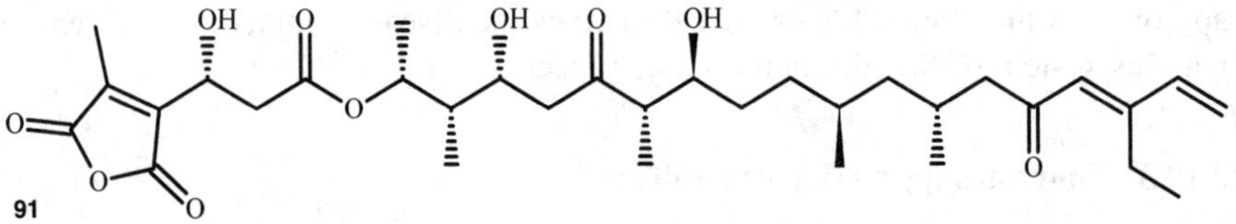

Tautomycetin (91) was isolated from *Streptomyces griseochromogenes* (62) and is 100-fold more potent than cyclosporine. It acts by blocking the induction of tyrosine phosphorylation of T-cell–specific signaling mediators downstream of Src tyrosine kinases, leading to apoptosis (63). This immunosuppressant has been assayed in organ transplantation; however, it needs to be synergistically combined to cyclosporine A to avoid rejection (64).

Recently, the immunosuppressants dalesconols A and B (92) were isolated from the amantis-associated fungus *Daldinia eschscholtzii* (65).

3.10.4   Immunosuppressive Pyrrole Compounds

Prodiginines are characterized by a common tripyrrylmethene skeleton (a structure called prodigiosene). Some members of this family have been shown to possess immunosuppressive effects resulting from the inhibition of phosphorylation and the activation of JAK-3, a cytoplasmic tyrosine kinase associated with a cell surface receptor component called the common γ-chain. This component is

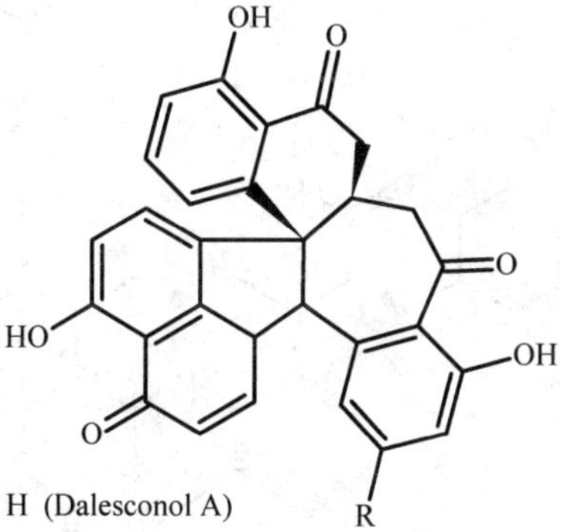

92 R= H (Dalesconol A)
    R= OH (Dalesconol B)

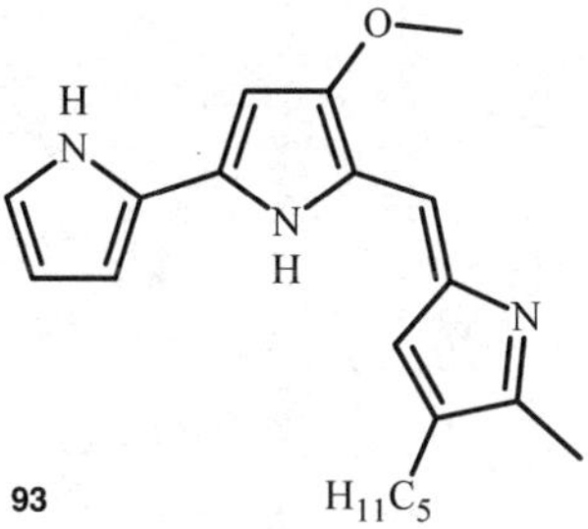

93

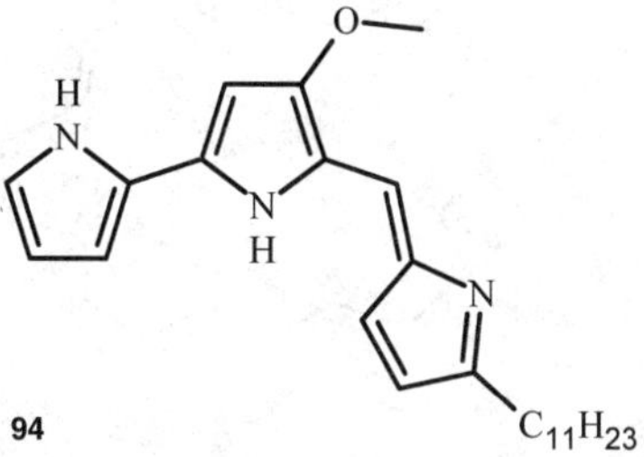

94

exclusive to IL-2 cytokine family receptors. Among prodiginines, prodigiosin (**93**) is synthesized by *S. marcescens, Serratia plymuthica, H. chejuensis, Zooshikella rubidus, Pseudomonas magnesiorubra*, and *V. psychroerythreus* (66–68).

Undecylprodigiosin (**94**) is produced by several actinomycetes such as *Streptomyces coelicolor, Streptomyces longisporus rubber, Actinomadura madurae, Streptoverticillium rubrireticuli*, and *Saccharopolyspora* sp. (67, 69, 70). Butyl-meta-cycloheptylprodiginine (**95**), produced by *Saccharopolyspora* sp. nov. and *S. coelicolor* (67, 69), has a ring that is formed between positions 2 and 4 of pyrrole ring C in prodigiosene (67). Cyclo-prodigiosin (**96**), produced by *Vibrio gazogenes, Alteromonas rubra*, and *Pseudoalteromonas denitrificans*, has a ring between positions 3 and 4 of pyrrole ring C and a methyl group at position C-2. Cyclononylprodigiosin (**97**), produced by *Actinomadura pelletieri* and *Actinomadura madurae* (67), is a macrocyclic prodiginine with a ring formation between position 2 of pyrrole C and position 10 of pyrrole A.

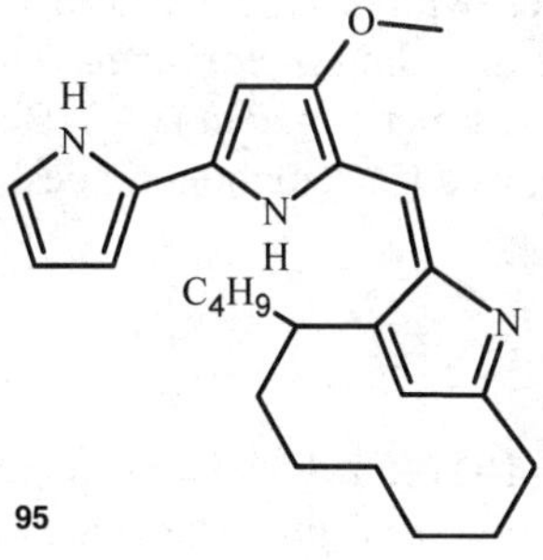

95

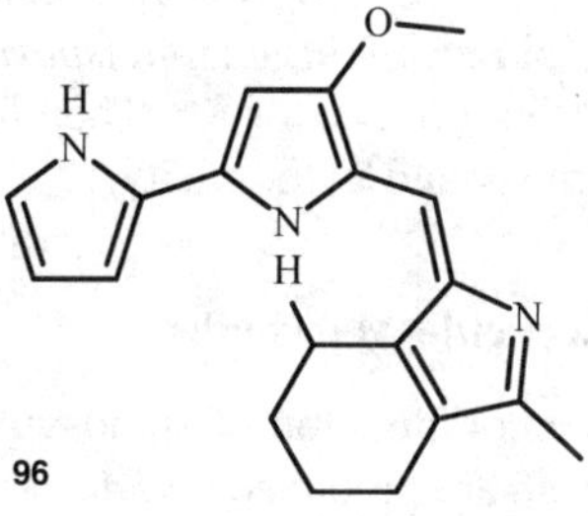

96

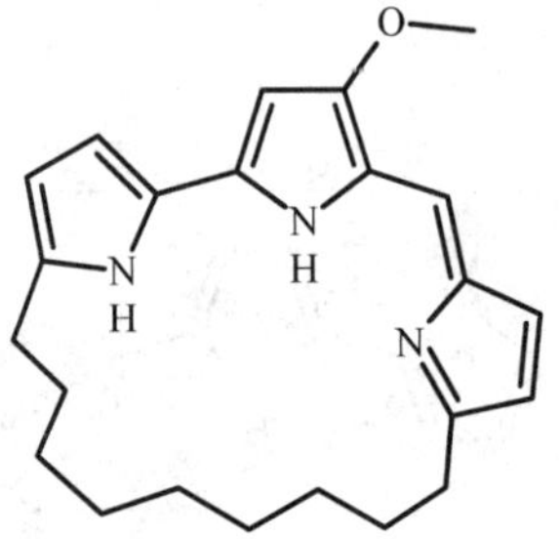

97 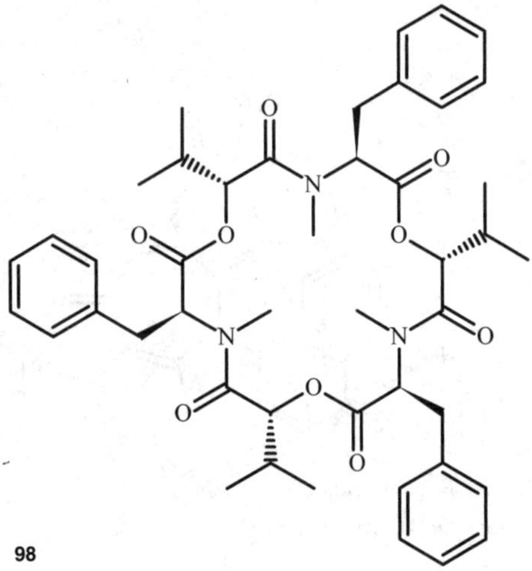   98

3.11 INSECTICIDES

Microbial insecticides are valuable because they are essentially nontoxic and nonpathogenic to wildlife, humans, and potentially beneficial insects apart from the target pest.

3.11.1 Insecticide Peptides

Beauvericin (**98**) is a cyclic hexadepsipeptide with alternating methyl-phenylalanyl and hydroxyisovaleryl residues. It is produced by *Beauveria bassiana* (72), a soil fungus that infects the larvae and adults of many species such as beetles, fire ants, and termites. Beauvericin forms a complex with ions, thereby permitting its transport across cell membranes.

Bassianolide (**99**) is a nonribosomal cyclodepsipeptide isolated from the cultured mycelia of *B. bassiana*. Bassianolide causes paralysis in a broad range of host insects that are of agricultural, veterinary, and medical significance. It selectively inhibits the tonic component of the contraction that is induced by acetylcholine (73).

Destruxins are cyclic hexadepsipeptides with a type I β-turn conformation (**100**). They are produced by several fungi, including *Metarhizium anisopliae* var. *anisopliae* and *Lecanicillium longisporum* (formerly *lecanii*) (74). Destruxins can chelate Ca^{2+} but not Na^+ or K^+. Therefore, as a Ca^{2+} ionophore, destruxins cause muscle paralysis and cytotoxicity.

3.11.2 Insecticide Macrolides

Spinosad is a combination of spinosyn A (**101**) (65–95%) and spinosyn D (**102**) (5–35%). Both are produced by *Saccharopolyspora spinosa* and are active against

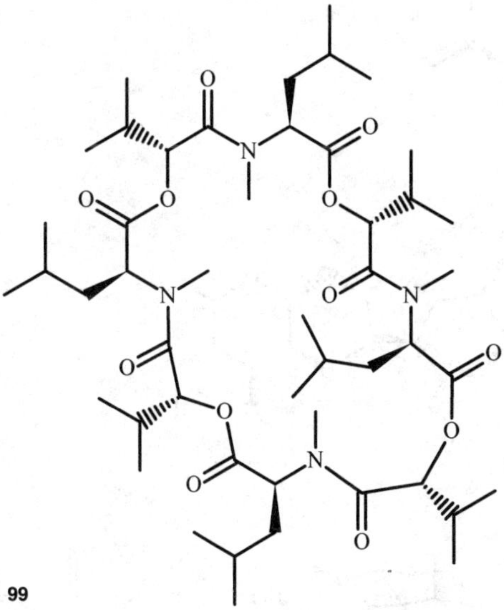

99

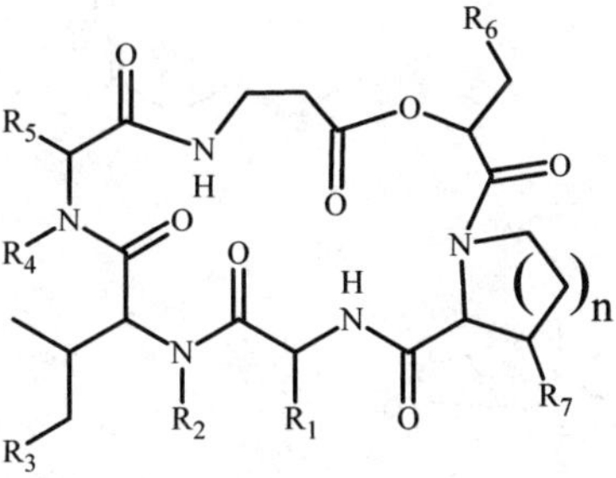

100

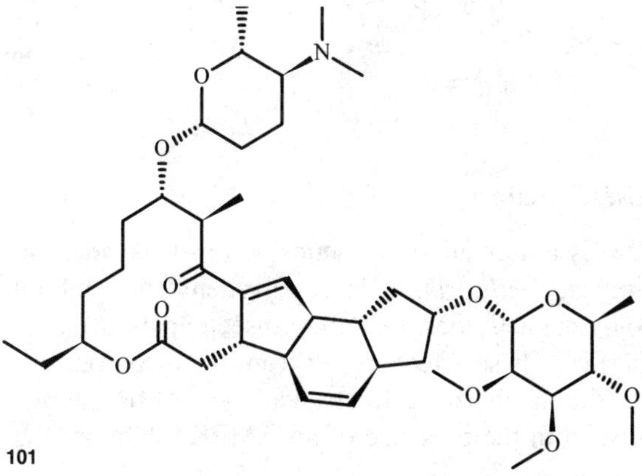

101

a wide variety of insects from *Lepidoptera* and *Diptera*. Spinosyn A comprises a tetracyclic polyketide aglycone (21 carbon atoms) to which is attached a neutral saccharide substituent (2,3,4-tri-O-methyl-α-L-rhamnosyl) on the C-9 hydroxyl group and an amino sugar moiety (β-D-forosaminyl) on the C-17 hydroxyl group (75). Spinosyn D is 6-methyl-spinosyn A. The spinosyns are highly efficient killers when they are ingested or contacted, attacking the insect's nervous system by a unique mechanism of action involving the disruption of nicotinic acetyl-choline receptors (76).

Recently, derivatives of spinosyns have been isolated from *Saccharopolyspora pogona* NRRL 30141. They differ in side chain length at C-21, having a butenyl substitution (butenyl-spinosyns or pogonins) (**103**) (77).

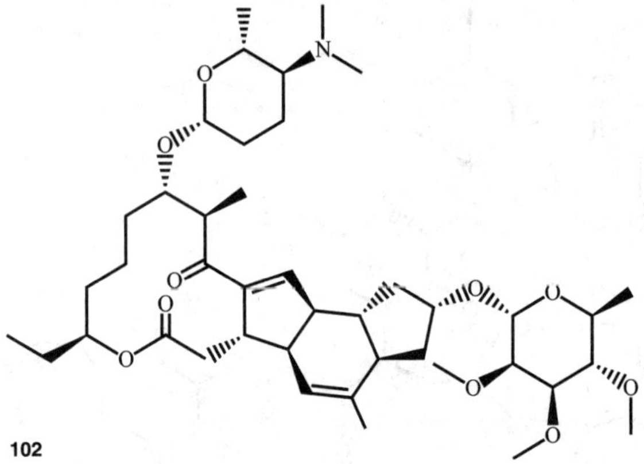

102

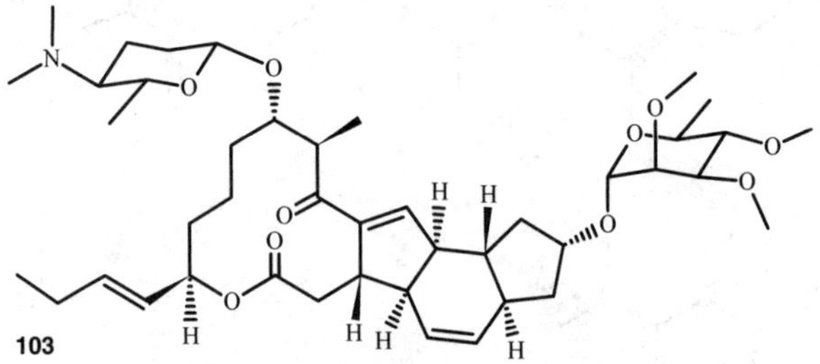

103

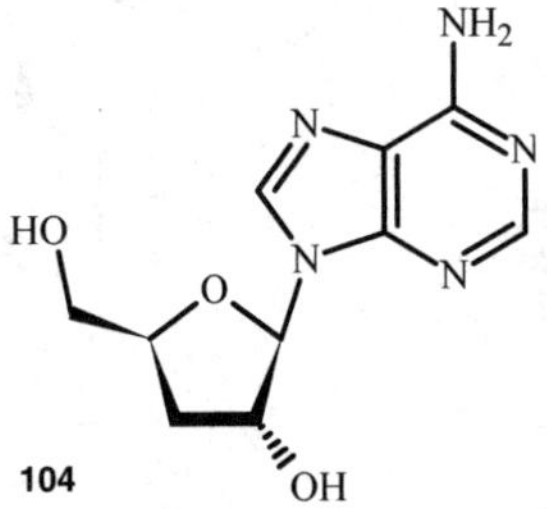

104

3.11.3 Nucleoside Analogs

Cordycepin (**104**) is a derivative of adenosine (3′-deoxyadenosine) produced by the fungus *Cordyceps militaris*. It is a polyadenylation inhibitor with a large spectrum of biological activities, including insecticidal (78), antifungal, and anti-neoplastic properties. These effects result from the incorporation of the molecule into RNA. In addition, the drug likely activates AMP kinase (AMPK), as its effects are abolished in the presence of an AMPK inhibitor.

3.12 CONCLUDING REMARKS

Throughout history, microbial natural products have been used as the active agents in many traditional medical preparations and they continue to play an important role as pharmaceuticals. With the continuing need for novel drug-like lead compounds with antibiotic, anticancer, and immunosuppressant effects (among other pharmacological activities), the chemical diversity derived from microbial products will be increasingly relevant to the future of drug discovery.

As only a small fraction of the world's microbial biodiversity has been tested for biological activity, it can be assumed that microbial products will continue to offer novel leads for therapeutic agents, provided that the microorganisms are available for screening.

Microbial bioactive secondary metabolites with pharmacological applications have inspired many developments in organic chemistry, leading to advances in synthetic methodologies and to the possibility of making analogs of the original lead compound with improved pharmacological or pharmaceutical properties. With the application of various techniques to create analogs and derivatives of these microbial products, it becomes possible to derive novel compounds that can be patented, even when the original structure has been previously disclosed.

The use of terrestrial bacteria has been limited by difficulties in culturing the vast majority of species. This has led to an interest in various genetic manipulation techniques, such as combinatorial genetics. The application of molecular biological techniques is increasing the availability of novel compounds that can be conveniently produced in bacteria or yeasts, and combinatorial chemistry approaches are being based on natural product scaffolds to create screening libraries that closely resemble drug-like compounds. More recently, a metagenomics approach has been used to access a wider range of synthetic capabilities from bacteria, leading to the discovery of novel compounds with antibiotic activity, such as the turbomycins. Therefore, the chemical diversity derived from microbial secondary metabolites will be increasingly relevant to the future of drug discovery.

3.13 ACKNOWLEDGMENTS

This work was supported by grants CB2008-100564-IIBO from CONACYT, Mexico, and PAPIIT, IN209210 from Dirección General de Asuntos del Personal Académico, UNAM, México.

REFERENCES

1. Demain AL, Sánchez S. Microbial drug discovery. J. Antibiot. 2009;62:5–16.

2. Handelsman J, Rondon MR, Brady S, Clardy J, Goodman RM. Molecular biological provides access to the chemistry of unknown soil microbes: a new frontier for natural products. Chem. Biol. 1998;5:R245–R249.

3. Pearce C. Biologically active fungal metabolites. Adv. Appl. Microbiol. 1997;44: 1–80.

4. Pedros-Alio C. Marine microbial diversity: can it be determined? Trends Microbiol. 2006;14:257–263.

5. Hawksworth DL. The fungal dimension of biodiversity, magnitude, significance and conservation. Mycol. Res. 1991;95:641–655.

6. Davies J. Millennium bugs. Trends Biochem. Sci. 1999;24:M2–M5.

7. Priscu JC, Fritsen CH, Adams EE, Giovannoni SJ, Paerl HW, McKay CP, Doran PT, Gordon DA, Lanoil BD, Pinckney JL. Perennial Antarctic lake ice: an oasis for life in a polar desert. Science 1998;280:2095–2098.

8. Berdy J. Bioactive microbial metabolites a personal review. J. Antibiot. (Tokyo) 2005; 58:1–26.

9. Stork CM. Antibiotics, antifungals, and antivirals. In: Goldfrank's Toxicologic Emergencies. Nelson LH, Flomenbaum N, Goldfrank LR, Hoffman RL, Howland MD, Lewin NA, eds. 2006. McGraw-Hill, New York.

10. Craig WA. The pharmacology of meropenem, a new carbapenem antibiotic. Clin. Infec. Dis. 1997;24:S266–S275.

11. Alvarez-Elcoro S, Enzler MJ. The macrolides: erythromycin, clarithromycin, and azithromycin. 1999;74: 613–634.

12. Stephan H, Kempter C, Metzger JW, Jung G, Potterat O, Pfefferle C, Fiedler HP. Kanchanamycins, new polyol macrolide antibiotics produced by *Streptomyces olivaceus* Tü 4018 II. Structure elucidation. J. Antibiot. 1996;49:765–769.

13. Wehrli W. Ansamycins. chemistry, biosynthesis and biological activity. Top. Curr. Chem. 1977;72:21–49.

14. Mukhtar TA, Wright GD. Streptogramins, oxazolidinones, and other inhibitors of bacterial protein synthesis. Chem. Rev. 2005;105:529–542.

15. Dutton JC, Haxwell M, McArthur H, Wax RG. Peptide Antibiotics: Discovery, Modes of Action, and Applications. 2002. Marcel Dekker, New York.

16. Georgopapadakou NH. Antifungals targeted to cell wall: focus on β-1,3-glucan synthase. Expert Opin. Investig. Drugs 2001;10:269–280.

17. Van Lanen S, Ben S. Microbial genomics for the improvement of natural product discovery. Curr. Opin. Microbiol. 2006;9:252–260.

18. Cragg GM, Kingston DGI, Newman DJ. Anticancer Agents from Natural Products. 2005. CRC Press. Boca Raton, FL.

19. Qu X, Ren J, Riccelli P, Benight A, Chaires J. Enthalpy/Entropy compensation: influence of DNA flanking sequence on the binding of 7-amino actinomycin D to its primary binding site in short DNA duplexes. Biochemistry 2003;42:11960–11967.

20. Strobel G, Bryn D. Bioprospecting for microbial endophytes and their natural products. Microbiol. Mol. Biol. Rev. 2003;67:491–502.

21. Park G, Tomlinson J, Misenheimer J, Kucera G, Manderville R. Photo-induced cytotoxicity of prodigiosin analogues. Bull. Korean Chem. Soc. 2007;28:49–52.

22. Tooth JA, Davis MW, Avery LA. *avr*-15 encodes a chloride channel subunit that mediates inhibitory glutamatergic neurotransmission and ivermectin sensitivity in *Caenorhabditis elegans*. EMBO J. 1997;16:5867–5879.

23. Mishima H, Ide J, Muramatsu S, Ono M. Milbemycins, a new family of macrolide antibiotics. Structure determination of milbemycins D, E, F, G, H, J and K. J. Antibiot. 1983;36:980–990.

24. Westley JW. Polyether antibiotics: versatile carboxylic acid ionophores produced by *Streptomyces*. Adv. Appl. Microbiol. 1977;22:177–223.

25. Matabudul DK, Lumley ID, Points JS. The determination of 5 anticoccidial drugs (nicarbazin, lasalocid, monensin, salinomycin and narasin) in animal livers and eggs by liquid chromatography linked with tandem mass spectrometry (LC-MS-MS). Analyst 2002;127:760–768.

26. Umezawa H. Low-molecular-weight enzyme inhibitors of microbial origin. Ann. Rev. Microbiol. 1982;36:75–99.

27. Weibel EK, Hadvary P, Hochuli E, Kupfer E, Lengsfeld H. Lipstatin, an inhibitor of pancreatic lipase, produced by *Streptomyces toxytricini*. J. Antibiot. 1987;40: 1081–1085.

28. Liu DZ, Wang F, Liao TG, Tang JG, Steglich W, Zhu HJ, Liu JK. Vibralactone: a lipase inhibitor with an unusual fused beta-lactone of the basidiomycete *Boreostereum vibrans*. Org. Lett. 2006;8:5749–5752.

29. Ondetti MA, Rubin B, Cushman DW. Enzyme of the rennin-angiotensin system and their inhibitors. Annu. Rev. Biochem. 1982;51:283–308.

30. Rhee KH. Purification and identification of an antifungal agent from *Streptomyces* sp. KH-614 antagonistic to rice blast fungus, *Pyricularia oryzae*. J. Microbiol. Biotechnol. 2003;13:984–988.

31. Borges F, Fernandes E, Roleira F. Progress towards the discovery of xanthine oxidase inhibitors. Curr. Med. Chem. 2002;9:195–217.

32. Izumida H, Miki W, Sano H, Endo M. Agar plate method, a new assay for chitinase inhibitors using a chitin-degrading bacterium. J. Marine Biotechnol. 1995;2:163–166.

33. Song YC, Li H, Ye YH, Shan CY, Yang YM, Tan RX. Endophytic naphthopyrone metabolites are co-inhibitors of xanthine oxidase, SW1116 cell and some microbial growths. FEMS Microbiol. Lett. 2004;241:67–72.

34. Paterson RRM. Fungal enzyme inhibitors as pharmaceuticals, toxins and scourge of PCR. Curr. Enzyme Inhib. 2008;4:46–59.

35. Duke SO, Cantrell CL, Meepagal KM, Wedge DE, Tabanca N, Schrader KK. Natural toxins for use in pest management. Toxins 2010;2:1943–1962.

36. Strobel G, Sugawara F, Hershenhorn J. Pathogens and their products affecting weedy plants. Phytoparasitica 1992;20:307–323.

37. Lebenthal Y, Horvath A, Dziechciarz P, Szajewska H, Shamir R. Are treatment targets for hypercholesterolemia evidence based? Systematic review and meta-analysis of randomised controlled trials. Arch. Dis. Child. 2010;95:673–680.

38. Barrios-González J, Miranda RU. Biotechnological production and applications of statins. Appl. Microbiol. Biotechnol. 2010;85:869–883.

39. Endo A. The discovery and development of HMG-CoA reductase inhibitors. J. Lipid Res. 1992;33:1569–1582.

40. Alberts AW, Chen J, Kuron G, Hunt V, Huff J, Hoffman C, Rothrock J, Lopez M, Joshua H, Harris E, Patchett A, Monaghan R, Currie S, Stapley E, Albers-Schonberg G, Hensens O, Hirshfield J, Hoogsteent K, Liesch J, Springer J. Mevinolin: a highly potent competitive inhibitor of hydroxymethylglutaryl-coenzyme A reductase and a cholesterol-lowering agent. Proc. Natl. Acad. Sci. U.S.A. 1980;77:3957–3961.

41. Subazini TK, Kumar GR. Characterization of lovastatin biosynthetic cluster proteins in *Aspergillus terreus* strain ATCC 20542. Bioinformation 2011;6:250–254.

42. Vaklavas C, Chatzizisis YS, Tsimberidou AM. Common cardiovascular medications in cancer therapeutics. Pharmacol. Ther. 2011;130:177–190.

43. Bakhireva LN, Shainline MR, Carter S, Robinson S, Beaton SJ, Nawarskas JJ, Gunter MJ. Synergistic effect of statins and postmenopausal hormone therapy in the prevention of skeletal fractures in elderly women. Pharmacotherapy 2010;30:879–887.

44. Barone E, Cenini G, Di Domenico F, Martin S, Sultana R, Mancuso C, Murphy MP, Head E, Butterfield DA. Long-term high-dose atorvastatin decreases brain oxidative and nitrosative stress in a preclinical model of Alzheimer disease: a novel mechanism of action. Pharmacol. Res. 2011;63:172–180.

45. Roy A, Pahan K. Prospects of statins in Parkinson disease. Neuroscientist 2011;17: 244–255.

46. Tang TT, Song Y, Ding YJ, Liao YH, Yu X, Du R, Xiao H, Yuan J, Zhou ZH, Liao MY, Yao R, Jevallee H, Shi GP, Cheng X. Atorvastatin upregulates regulatory T cells and reduces clinical disease activity in patients with rheumatoid arthritis. J. Lipid Res. 2011;52:1023–1032.

47. Casirola DM, Ferraris RP. Alpha-glucosidase inhibitors prevent diet induced increases in intestinal sugar transport in diabetic mice. Metabolism 2006;55:832–841.

48. Van de Laar F. Alpha-glucosidase inhibitors in the early treatment of type 2 diabetes. Vasc. Health Risk Manag. 2008;4:1189–1195.

49. Geng P, Qiu F, Zhu Y, Bai G. Four acarviosin-containing oligosaccharides identified from *Streptomyces coelicoflavus* ZG0656 are potent inhibitors of α-amylase. Carbohydr. Res. 2008;343:882–892.

50. Deppenmeier U, Hoffmeister M, Prust C. Biochemistry and biotechnological applications of *Gluconobacter* strains. Appl. Microbiol. Biotechnol. 2002;60:233–242.

51. Camerino M, Thompson P, Chalmer D. New drugs for diabetes: the design and molecular modelling of α-amylase peptidomimetics. Proc. 4th Int. Peptide Symp. Australian Peptide Association, Australia, 2007. pp. 1–2.

52. Heyl DL, Fernandes S, Khullar L, Stephens J, Blaney E, Opang-Owusu H, Stahelin B, Pasko T, Jacobs J, Bailey D, Brown D, Milletti MC. Correlation of LUMO localization with the alpha-amylase inhibition constant in a tendamistat-based series of linear and cyclic peptides. Bioorg. Med. Chem. 2005;13:4262–4268.

53. Heyl DL, Tobwala S, Lucas LS, Nandanie AD, Himm RW, Kappler J, Blaney EJ, Groom J, Asbill J, Nzoma JK, Jarosz C, Palamma H, Schullery SE. Peptide inhibitors of α-amylase based on tendamistat: development of analogues with φ-amino acids linking critical binding segments. Prot. Pept. Lett. 2005;12:275–280.

54. Rehm S, Han S, Hassani I, Sokocevic A, Jonker HRA, Engels JW, Schwalbe H. The high resolution NMR structure of parvulustat (Z-2685) from *Streptomyces parvulus* FH-1641: comparison with tendamistat from *Streptomyces tendae* 4158. Chem. Biol. Chem. 2009;10:119–127.

55. Souren P, Hou CT, Kang SC. α-Glucosidase inhibitory activities of 10-hydroxy-8(*E*)-octadecenoic acid: an intermediate of bioconversion of oleic acid to 7,10-dihydroxy-8(*E*)-octadecenoic acid. New Biotechnol. 2010;27:419–423.

56. Wu M, Singh SB, Wang J, Chung CC, Salituro S, Karanam VB, Lee SH, Powles M, Ellsworth KP, Lassman ME, Miller C, Myers RW, Tota MR, Zhang BB, Li C. Antidiabetic and antisteatotic effects of the selective fatty acid synthase (FAS) inhibitor platensimycin in mouse models of diabetes. Proc. Natl. Acad. Sci. U.S.A. 2011;108:5378–5383.

57. Kahan BD. Forty years of publication of transplantation proceedings—the second decade: the cyclosporine revolution. Transplant Proc. 2009;41:1423–1437.

58. Bentley R. Mycophenolic acid: a one hundred year odyssey from antibiotic to immunosuppressant. Chem. Rev. 2000;100:3801–3825.

59. Sneader W. Drug Discovery: A History. 2005. John Wiley and Sons, Ltd., England.

60. Strader CR, Pearce CJ, Oberlies NH. Fingolimod (FTY720): a recently approved multiple sclerosis drug based on a fungal secondary metabolite. J. Nat. Prod. 2011;74: 900–907.

61. Kino T, Hatanaka H, Miyata S, Inamura N, Nishiyama M, Yajima T, Goto T, Okuhara M, Kohsaka M, Aoki H, Ochia T. FK-506, a novel immunosuppressant isolated from a *Streptomyces*. J. Antibiot. 1987;60:1256–1265.

62. Cheng X-C, Kihara T, Ying X, Uramoto M, Osada H, Kusakabe H, Wang B-N, Kobayashi Y, Ko K, Yamaguchi I, Shen Y-C, Isono K. A new antibiotic, tautomycetin. J. Antibiot, 1989;42:141–144.

63. Shim J-H, Lee H-K, Chang E-J, Chae W-J, Han J-H, Han D-J, Morio T, Yang J-J, Bothwell A, Lee S-K. Immunosuppressive effects of tautomycetin *in vivo* and *in vitro* via T cell-specific apoptosis induction. Proc. Natl. Acad. Sci. U.S.A. 2002;99: 10617–10622.

64. Han DJ, Jeong YL, Wee YM, Lee AY, Lee HK, Ha JC, Lee SK, Kim SC. Tautomycetin as a novel immunosuppressant in transplantation. Transplant Proc. 2003; 35:547.

65. Snyder S, Sherwood T, Ross A. Total syntheses of *Dalesconol* A and B. Angew. Chem. Int. Ed. 2010;49:5146–5150.

66. Kim D, Lee JS, Park YK, Kim JF, Jeong H, Oh TK, Kim BS, Lee CH. Biosynthesis of antibiotic prodiginines in the marine bacterium *Hahella chejuensis* KCTC 2396. J. Appl. Microbiol. 2007;102:937–944.

67. Williamson NR, Fineran PC, Leeper FJ, Salmond GPC. The biosynthesis and regulation of bacterial prodiginines. Nat. Rev. Microbiol. 2006;4:887–899.

68. Lee JS, Kim YS, Park S, Kim J, Kang SJ, Lee MH, Ryu S, Choi JM, Oh TK, Yoon JH. Exceptional production of both prodigiosin and cycloprodigiosin as major metabolic constituents by a novel marine bacterium, *Zooshikella rubidus* S1-1. Appl. Environ. Microbiol. 2001;77:4967–4973.

69. Liu R, Cui CB, Duan L, Gu QQ, Zhu VW. Potent *in vitro* anticancer activity of metacycloprodigiosin and undecylprodigiosin from a spong-derived actinomycete *Saccharopolyspora* sp. nov. Arch. Pharm. Res. 2005;28:1341–1344.

70. Luti KJ, Matuvina F. *Streptomyces coelicolor* increases the production of undecylprodigiosin when interacted with *Bacillus subtilis*. Biotechnol. Lett. 2011;33: 113–118.

71. Kawauchi K, Shibutani K, Yagisawa H, Kamata H, Nakatsuji S, Anzai H, Yokoyama Y, Ikegami Y, Moriyama Y, Hirata H. A possible immunosuppressant, cycloprodigiosin hydrochloride, obtained from Pseudoalteromonas denitrificans. Biochem. Biophys. Res. Commun. 1997;237:543–547.

72. Nakajyo S, Shimizu K, Kametani A. On the inhibitory mechanism of bassianolide a cyclodepsipeptide, in acetylcholine-induced contraction in guinea-pig taenia coli. Jpn. J. Pharmacol. 1983;33:573–582.

73. Xu Y, Orozco R, Wijeratne EMK, Espinosa-Artiles P, Gunatilaka AAL, Stock P, Molnár l. Biosynthesis of the cyclooligomer depsipeptide beauvericin, a virulence factor of the entomopathogenic fungus *Beauveria bassiana*. Chem. Biol. 2008;15:898–907.

74. Butt TM, Ben El Hadj N, Skrobek A, Ravensberg WJ, Wang C, Lange CM, Vey A, Shah UK, Dudley E. Mass spectrometry as a tool for the selective profiling of destruxins; their first identification in *Lecanicillium longisporum*. Rapid Commun. Mass Spectrom. 2009;23:1426–1434.

75. Waldron C, Matsushima P, Rosteck PR, Broughton MC, Turner J, Madduri K. Cloning and analysis of the spinosad biosynthetic gene cluster of *Saccharopolyspora spinosa*. Chem. Biol. 2001;8:487–499.

76. Kirst HA. The spinosyn family of insecticides: realizing the potential of natural products research. J. Antibiot. 2010;63:101–111.

77. Hahn DR, Gustafson G, Waldron C, Bullard B, Jackson JD, Mitchell J. Butenyl spinosyns, a natural example of genetic engineering of antibiotic biosynthetic genes. J. Ind. Microbiol. Biotechnol. 2006;33:94–104.

78. Kim J-R, Yeon S-H, Kim H-S, Ahn Y-J. Larvicidal activity against *Plutella xylostella* of cordycepin from the fruiting body of *Cordyceps militaris*. Pest Manag. Sci. 2002;58:713–717.

FURTHER READING

Enoch DA, Ludlam HA, Brown NM. Invasive fungal infections: a review of epidemiology and management options. J. Med. Microbiol. 2006;55:809–818.

Eschenauer G, DePestel DD, Carver PL. Comparison of echinocandin antifungals. Ther. Clin. Risk Manag. 2007;3:71–97.

Firn RD, Jones CG. Natural products-a simple model to explain chemical diversity. Nat. Prod. Rep. 2003;20:382–391.

Harvey AL. Natural products in drug discovery. Drug Discov. Today 2008;13:894–901.

Hübel K, Leßmann T, Waldmann H. Chemical biology—identification of small molecule modulators of cellular activity by natural product inspired synthesis. Chem. Soc. Rev. 2008;37:1361–1374.

Klekota J, Roth FP. Chemical substructures that enrich for biological activity. Bioinformation 2008;24:2518–2525.

Osada H, Hertweck C. Exploring the chemical space of microbial natural products. Curr. Opin. Chem. Biol. 2009;13:133–134.

Schmidt EW. The hidden diversity of ribosomal peptide natural products. BMC Biol. 2010;8:83.

Schuffenhauer A, Brown N. Chemical diversity and biological activity. Drug Discov. Today Technol. 2006;3:387–395.

Sing SB, Pelaez F. Biodiversity, chemical diversity and drug discovery. Prog. Drug Res. 2008;65:142–174.

PART II

BIOSYNTHESIS OF NATURAL PRODUCTS

4

NONRIBOSOMAL PEPTIDES

GEORG SCHOENAFINGER AND MOHAMED A. MARAHIEL

Philipps-University of Marburg, FB Chemie/Biochemie, Marburg, Germany

Many microorganisms have evolved an unusual way of producing secondary peptide metabolites. Large multidomain enzymatic machineries, the so-called nonribosomal peptide synthetases (NRPSs), are responsible for the production of this structurally diverse class of peptides with various functions, such as cytostatic, immunosuppressive, antibacterial, or antitumor properties. These secondary metabolites differ from peptides of ribosomal origin in several ways. Their length is limited to a mere 20 building blocks, roughly, and mostly a circular or branched cyclic connectivity is found. Furthermore, aside from the proteinogenic amino acids, a larger variety of chemical groups is found in these bioactive compounds: D-configured amino acids, fatty acids, methylated, oxidized, halogenated, and glycosylated building blocks. These functional and structural features are known to be important for bioactivity, and often natural defense mechanisms are thus evaded. In this chapter, we describe the enzymatic machineries of NRPSs, the chemical reactions catalyzed by their subunits, and the potential of redesigning or using these machineries to give rise to new nonribosomal peptide antibiotics.

Nonribosomal peptide synthetases (NRPSs) compose a unique class of multidomain enzymes capable of producing peptides (1–4). In contrast to the ribosomal machinery where the mRNA template is translated, the order of catalytically active entities within these synthetases intrinsically determines the sequence of building blocks in the peptide product (Fig. 4.1). As a consequence, generally speaking, each NRPS can only produce one defined peptide product. This is chemically implemented by the fact that all substrates and reaction intermediates are spatially fixed to the synthetase by covalent linkage—thereby eliminating

Natural Products in Chemical Biology, First Edition. Edited by Natanya Civjan.
© 2012 John Wiley & Sons, Inc. Published 2012 by John Wiley & Sons, Inc.

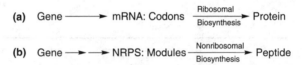

Figure 4.1 Comparison of ribosomal and nonribosomal peptide synthesis. (a) In the ribosomal information pathway, the sequence of codons in the mRNA determines the sequence of amino acids in the protein, whereas (b) the sequence of modules in the nonribosomal peptide synthetases intrinsically determines the primary sequence of the peptide product.

side product formation caused by diffusion. The catalytic entities responsible for the incorporation of a distinct building block into the product are called modules. Each module carries out several chemical steps required for the synthesis of nonribosomal peptides: Recognition of the building block, activation, covalent attachment, translocation, and condensation. In several cases, additional modifications are found, such as epimerization [Tyrocidine, (5)], cyclization [Gramicidin S, (6)], oxidation [Epothilone, (7)], reduction [linear gramicidin (8)], methylation [Cyclosporin, (9)], and formylation [linear gramicidin, (10)]. *In vitro* studies have shown that each module can be subdivided into catalytically active domains to which the different reactions mentioned above can be assigned (1–4). Thus, the so-called adenylation (A), peptidyl carrier protein (PCP), and the condensation (C) domains were identified as being essential to all NRPSs. In addition, a second group of so-called optional domains exists: the epimerization (E), cyclization (TE or Cy), oxidation (Ox), reduction (R), N-methylation (Mt), and formylation (F) domains. Aside from NRPSs themselves, several enzymes are known to act on some peptides while they are still bound to the synthetase or even after their release. These modifying enzymes can glycosylate [Vancomycin, (11)], halogenate [Vancomycin, (11)], or reduce [linear gramicidin, (8)] the peptides *in trans*. With several hundred different building blocks found in nonribosomal peptide products, it becomes evident that their diversity is vast (Fig. 4.2). This chapter addresses the biologic background of these secondary metabolites, the enzymatic machineries of NRPSs, and the chemical reactions catalyzed by their domains. Furthermore, the possibility of manipulating NRPSs and using certain domains to produce novel compounds is discussed.

4.1 BIOLOGIC BACKGROUND

Nonribosomal peptides are produced by a large number of bacteria, fungi, and lower eucaryotes. For most of these compounds, their biologic role is unknown. One might suspect that these secreted molecules are used for unknown forms of communication or simply to critically increase the chance of survival for the producing cell in its habitat, because the metabolic cost of their production is enormous. However, the function of some nonribosomal compounds has

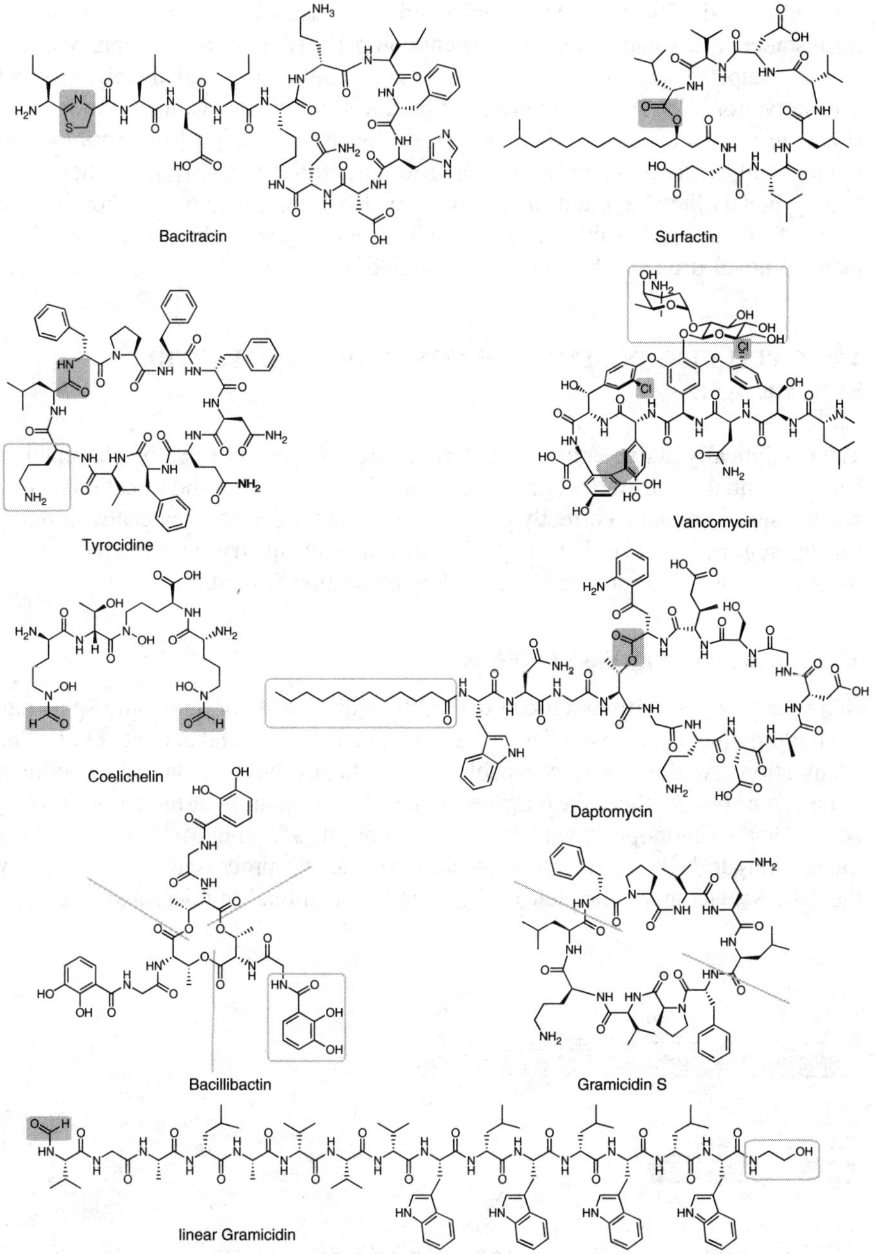

Figure 4.2 A selection of nonribosomal peptides. Chemical and structural features that contribute to the vast diversity of this class of metabolites are highlighted: Heterocycle (bacitracin), lactone (surfactin, daptomycin), ornithine and lactam (Tyrocidine), sugar, chlorinated aromats, C–C crosslink (Vancomycin), N-formyl groups (Coelichelin and linear gramicidin), fatty acid (daptomycin), dihydroxybenzoate and trimeric organization (bacillibactin), dimeric organization (gramicidin S), and ethanolamine (linear gramicidin).

been identified: The well-studied penicillin produced by *Penicillium notatum*, for instance, is a weapon against nutrient competitors, and the siderophore bacillibactin helps its producer, *Bacillus subtilis*, acquire iron and thereby prevent iron starvation. For us, natural products produced by microorganisms attract considerable attention because their observed bioactivities range from antibiotic to immunosuppressive and from cytostatic to antitumor. Not only have these secondary metabolites been optimized for their dedicated function over millions of years of evolution, but they also represent promising scaffolds for the development of novel drugs with improved or altered activities.

4.2 CATALYTIC DOMAINS OF NONRIBOSOMAL PEPTIDE SYNTHETASES

The catalytically active entities that NRPSs are composed of can be classified as being essential to all NRPSs or being responsible for special modifications. Only when a set of domains correctly acts in appropriate order, the designated product can be synthesized (3) (Fig. 4.3). The function, chemistry, and interactions of these domains are discussed in the following section (Fig. 4.4).

4.2.1 Essential Domains of NRPSs

Before any peptide formation can occur, the amino acids or, generally speaking, the building blocks to be condensed need to be recognized and activated (12). The adenylation (A) domains are capable of specifically binding one such building block. Once bound, the same enzyme catalyses the formation of the corresponding acyl-adenylate-monophosphate by consumption of ATP (Fig. 4.4). The resulting mixed anhydride is the reactive species that can be processed additionally by the NRPS machinery. Sequence alignments, mutational studies, and structural

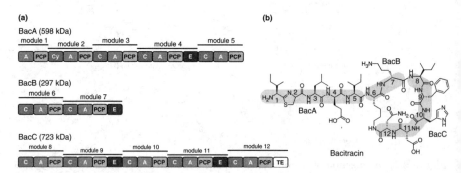

Figure 4.3 The nonribosomal machinery (a) needed to produce bacitracin (b) consists of three separate synthetases: BacA, BacB, and BacC, composing a total of 12 modules and 40 catalytic domains. These synthetases specifically interact with each other to produce the nonribosomal peptide bacitracin.

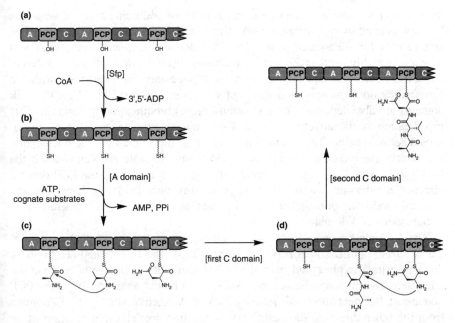

Figure 4.4 Single reactions in NRPSs and their timing. (a) After ribosomal synthesis of the *apo*-enzymes, the PCP domains are postsynthetically modified with 4′-Phosphopantetheine cofactors by a 4′Ppan transferase, e.g., Sfp. (b) In a second step, the A domains bind their cognate substrates as well as ATP and form the corresponding acyl-adenylate intermediates. These are transferred onto the cofactor of the neighboring PCP domains. (c) The C domains catalyze the condensation of two building blocks. The specificities of C domains and the affinities of aminoacyl-/peptidyl-*holo*-PCP domains ensure that no internal start reactions occur: (d) Only after the first condensation domain has acted does the second C domain seem to process the intermediate. During synthesis, the growing product chain is continuously translocated toward the C-terminal end of the enzyme.

data have revealed that amino acid residues at certain positions in the enzyme determine the specificity of an A domain (13). This result can be explained by the thereby generated chemical and physical environment of the substrate binding site. Some A domains, however, are known to have a relaxed substrate specificity. In these cases, chemically or sterically similar amino acids also are recognized, processed analogously, and thus found at that very position in the product. For example, the A domain of the first module of the gramicidin synthetase LgrA (10) not only activates valine but also activates isoleucine, which is found in 5% of linear gramicidins extracted from producing strains.

When the first building block has been recognized and activated by the A domain, the next essential domain comes into play: The peptidyl carrier protein (PCP) domain. Like the acyl carrier protein (ACP) in fatty acid biosynthesis, this PCP domain is responsible for keeping the reaction intermediates bound to the enzymatic machinery. Thus, a directed order of additional reaction steps can be implemented by controlled translocation, and NRPSs are thus often described as assembly line-like machineries. The PCP domain consists of 90 amino acid residues, roughly, and is known to rearrange itself to at least three different tertiary structures in aqueous solution, as is necessary for interaction with the surrounding domains at certain stages of synthesis (14). Just like ACPs, the PCP domains are also dependent on a posttranslational modification to function. This modification is the attachment of a 4′-phosphopantetheine cofactor to a conserved serine residue. The terminal thiol group of this cofactor is the nucleophile that attacks the mixed anhydride (acyl-AMP) and therefore covalently binds the NRPS substrates *via* a thioester bond. After such an acylation, the PCP domain directs the substrate toward the next processing domain. If we leave out any optional modifying domains at this point, this next domain would generally be a condensation (C) domain.

The C domain is needed for the condensation of two biosynthetic intermediates during nonribosomal peptide assembly (15). The PCP-bound electrophilic donor substrate is presented from the N-terminal side of the synthetase. On the other side, the nucleophilic acceptor substrate—bound analogously to the PCP domain of the next module—reaches back to the active site of the C domain from the other direction (downstream). In the first condensation reaction of an NRPS, both of these substrates would typically be aminoacyl groups connected to their PCP domains. Condensation is initiated by the nucleophilic attack of the α-amino group of the acceptor substrate onto the thioester group of the donor substrate. The cofactor of the upstream PCP domain is released, and the resulting amide bond now belongs to the dipeptide, which remains bound to the downstream PCP domain. Thus, a translocation of the condensation product toward the next module has occurred. All condensation reactions are strictly unidirectional—always transporting the growing product chain toward the module closer to the C-terminus of the machinery. The elongated peptide then serves as the donor in a subsequent condensation step on the next module. Usually, there are as many condensation domains in an NRPS as there are peptide bonds in the linear peptide product. This general translocation model implies that the biosynthesis is linear—altogether dependent on delicate, situationally changing affinities that guarantee correct timing for each reaction and that prevents side reactions (Fig. 4.4). Even though this model successfully puts the biosynthetic enzymes in relation with their products for most known NRPS systems, some exceptions are known: The structures of syringomycin (16) or coelichelin (17) cannot be sufficiently explained by merely deciphering the buildup of their NRPSs when using this model. Obviously, other regulatory mechanisms and forms of inter-domain communication are not yet fully understood.

When the last condensation reaction has occurred, the linear precursor needs to be released from the enzyme. For this important last step, several mechanisms are known: simple hydrolysis of the thioester (balhimycin, vancomycin), intramolecular cyclization leading to a lactam (tyrocidine, bacitracin) or a lactone (surfactin), or even reductive thioester cleavage (linear gramicidin). In some cases, the linear precursor is dimerized (gramicidin S) or even trimerized (bacillibactin, enterobactin) before cyclization (Fig. 4.2). Even though these reactions are critical for the compound's bioactivity, the catalytic domains responsible for the release are not found in all NRPS systems and will therefore be called "modifying" domains.

4.2.2 Modifying Domains of NRPSs

Apart from the essential domains in NRPSs, several so-called modifying domains are not found in every NRPS system. Nevertheless, they are required for proper processing of their designated substrate within their synthetase. Deletion or inactivation of these modifying domains usually results in the production of compounds with bioactivities severely reduced or altogether abolished.

Most nonribosomal peptides have a cyclic connectivity. In these cases, a C-terminal, so-called thioesterase (TE) domain, is often found in the synthetase. These TE domains all share an invariant serine residue belonging to a catalytic triad (Asp-His-Ser), which is known to be acylated with the linear peptide before cyclization (18). Once the substrate is translocated from the PCP domain onto the TE domain, the regiospecific and stereospecific intramolecular attack of a nucleophile onto the C-terminal carbonyl group of the substrate is directed by the enzyme. This nucleophile can be the N-terminal α-amino group of the linear peptide (tyrocidine, gramicidin S), a side-chain amino (bacitracin) or hydroxyl group (surfactin). Since the ester bond between the substrate and the TE domain is cleaved by these cyclization reactions, the resulting lactams or lactones are released from the synthetic machinery by this step. In a few cases, the modular arrangement of NRPSs suggests that only one half (gramicidin S) or one third (bacillibactin, enterobactin) of the extracted peptide product can be produced by one assembly line-like synthesis (Fig. 4.2). These synthetases are considered iterative (19) because they have to complete more than one linear peptide synthesis before one molecule of the secondary metabolite can be released. According to a proposed model, the first precursor is translocated onto the TE domain, the second monomer is then produced and transferred to the TE domain-bound first monomer leading to a dimer. An analogous trimerization occurs—if applicable—and finally the product is released by cyclization.

Another modifying reaction that is commonly found in NRPSs is the epimerization (E) of an amino acid (5). E domains that are always situated directly downstream of a PCP domain catalyze these reactions. The most C-terminal amino acid of the reaction intermediate is racemized by an E domain, no matter whether the substrate is an aminoacyl group alone or a peptidyl group. The mechanism of these E domains is so far unclear, even though a catalysis that involves one or more catalytic bases to deprotonate the α-carbon atom as a first step seems likely. The resulting planar double-bond species then needs to be reprotonated from the other side to invert the absolute configuration of the building block. This result can be accomplished by a nearby protonated catalytic base in the enzyme or water, which is positioned opposite of the first catalytic base. Nevertheless, a mixture of both stereoisomers always can be detected when the substrate bound to the enzyme is analyzed, which is indicative for either a nonstereoselective or a reversible reaction. Once the epimerized substrate undergoes the subsequent condensation reaction, only the species with an inverted stereocenter is found in the elongated product. Thus, the C domain only processes the inverted species. In some rare cases, the C domain also exhibits epimerization activity besides its normal function, and it is then called the "dual C/E" domain [Arthrofactin, (20)].

Another structural feature often found in NRPS products is N-methylated amide bonds. The domain that introduces this C_1 unit, the so-called methyltransferase (Mt) domain, is situated between the A and the PCP domain (21). By consumption of S-Adenosyl-methionine, the α-amino group of the acceptor substrate is methylated before condensation with the donor.

In the case of linear gramicidin, the N-terminus of the nonribosomal peptide carries a formyl group (10). Just like in the bacterial ribosomal synthesis, only a formylated first building block is processed additionally by the corresponding enzymatic machinery. Thus, one can find a distinct formylation (F) domain at the very N-terminus of the synthetase. Another formylated NRPS product is coelichelin whose N-terminal ornithine residue is believed to be N_δ-formylated *in trans* by a formyltransferase genetically associated with the NRPS (17). Formyltetrahydrofolate is used as source of the formyl group by these enzymes.

The essential condensation domain mentioned above can, in some cases, not only condensate but also catalyze a side-chain cyclization. It is then called cyclization (Cy) domain. The cyclization is initiated by a nucleophilic attack of the side-chain heteroatom on the carbonyl group of the amide bond formed by the same domain. When water is eliminated, a stable pentacycle is integrated into the peptide chain without altering the rest of the backbone. The nucleophiles known to be reactants in these Cy domain reactions are either threonine/serine [mycobactin A, (22)] or cysteine [bacitracin, (23)]. The former leads to (methyl-)oxazoline heterocycles, whereas the latter gives rise to thiazoline-like units. Another domain sometimes associated with this heterocyclization is the oxidation (Ox) domain [Epothilone, (7)]. It is located between A and PCP domains, and it catalyzes the oxidation of oxa/thiazoline intermediates, which leads to oxazoles or thiazoles, respectively.

4.3 TECHNIQUES FOR THE PRODUCTION OF NOVEL NONRIBOSOMAL PEPTIDES

With a constantly growing number of pathogenic bacterial strains resistant to the known antibiotics, the demand for novel antibiotics or, more generally speaking, therapeutic agents is evident. Because many NRPS products already have such activities and their chemical and structural diversity is so huge, efforts have been made to use NRPSs to broaden the known spectrum of therapeutics. In this section, the possibilities of using NRPS machineries or parts of them to produce new bioactive compounds are addressed.

4.3.1 Module or Domain Exchange

When considering the modular buildup of NRPSs, the possibility of altering the peptide product by insertion, deletion, or exchange of modules seems to be an obvious approach for the production of new compounds. Because the A domains determine the specificity of each module, even an exchange of fractions smaller than whole modules in a synthetase could lead to an altered product. In the past, various attempts have succeeded using these strategies (24, 25). For instance, the exchange of an A domain in the surfactin NRPS with other A domains of both bacterial and fungal origin lead to the formation of the expected variants of surfactin (26). However, in all of these early studies, the apparent turnover rates were significantly lower than in the wild-type systems. According to common understanding of NRPSs, two explanations for the drastically slowed down synthetic process can be given. First, the borders chosen to dissect and to fuse the catalytic domains might have been unsuitable. Even though the reoccurring, highly variable so-called linker regions between each pair of catalytic domains seem not to exhibit secondary structures, their sequence and length might be critical for proper inter-domain communication. So far, no structure of any enzyme consisting of two or more NRPS domains has been published, which makes it difficult to define the right domain border when preparing a cloning strategy for fusion or for dissection. Second, the specificity of the C domains might result in a reduced product turnover. Even though a relaxed specificity for the donor substrate has been reported, the acceptor site seems to be highly specific, which discriminates against artificial substrates (27). Both the mode of catalytic action and the molecular and structural basis for the selectivity are not fully understood for C domains so that a straightforward approach for overcoming these low turnover rates currently cannot be given.

4.3.2 Changing the Specificity Code for A Domains

Sequence alignments of A domains have revealed that domains activating the same type of building block share a set of conserved residues in the primary protein sequence (13). With the A domain's crystal structure, one can find that these residues form the substrate binding pocket (28). These residues are therefore referred to as the "selectivity-conferring code" of NRPSs (13). One can now rationally exchange these sets of residues and can obtain fully functional A domains with altered substrate recognition. For example, this process has been done for the first module of the surfactin synthetase srfAA, which activated glutamate (29). In this case, the only difference in the selectivity-conferring code compared with a glutamine activating A domain lies in one residue. Thus, the single mutation of Lys239 to Gln239 in the enzyme leads to the desired and predicted shift of specificity. In another experiment, three residues were altered to change the substrate recognition of the aspartate activating A domain in srfB2 to asparagine. The corresponding bacterial strain was shown to produce the expected variant of surfactin containing asparagine at position 5. Even though this elegant way of manipulating NRPSs works, a few drawbacks are worth mentioning. On the one hand, product turnover rates are predicted to be very low. As discussed, the C domains that have to process the artificial substrates are predicted to discriminate against nonnatural substrates, which kinetically impede product formation drastically. On the other hand, this method is limited to the building blocks that other known A domains activate. Yet, the vast diversity and bioactivity of nonribosomal peptides mainly arises from their unusual connectivities and a large number of postsynthetic modifications, which one cannot address when merely changing the A domains' specificities.

4.3.3 Chemoenzymatic Approaches

A very powerful method for producing novel antibiotics is the chemoenzymatic approach (30). The idea behind this strategy is to leave out the enzymatic buildup of the linear peptide scaffolds and replace it by solid-phase peptide synthesis (Fig. 4.5). Once the desired peptide is produced, its C terminus needs to be synthetically activated (usually as a thioester) before the substrate can be subjected to enzymatic cyclization using a TE domain. The advantages of solid-phase synthesis (SPPS) are obvious: Virtually any oligo-peptide can be made in a short time and in large quantities. Even though this is true for most oligo-peptides, some amino acid sequences seem very difficult to synthesize, and the popular Fmoc protective group strategy always imposes the risk of racemization. Many different building blocks (already modified with protective groups necessary for SPPS) can be purchased, and by automated parallel peptide synthesis whole libraries can be produced very quickly. The reason why one can use such peptides as substrates lies in the relaxed substrate specificity of many TE domains. The TE domain of the Tyrocidine synthetase, which carries out a head-to-tail cyclization of the decapeptide DPhe-Pro-Phe-DPhe-Asn-Gln-Tyr-Val-Orn-Leu, for instance, only recognizes the two N-terminal and C-terminal amino acids of the natural substrate. The side chains of the amino acids at other positions are not recognized by the enzyme, and experiments with substrates that carry substitutions to alanine in these positions still lead to analogous cyclodecamers (31). The major advantage of using TE domains for cyclization reactions is their regiospecificity, stereospecificity, and chemospecificity. Thus, no protective groups are needed during these enzymatic cyclization reactions, and undesired side product formation is minimized. Additionally, these reactions are carried out under mild aqueous conditions, usually pH 7–8.

To follow this chemoenzymatic approach, the synthetic substrates must be transferred onto the catalytically active serine residue of the TE domain. This transfer can either be done directly or with the help of a PCP domain. In the natural system, translocation is realized by the interaction between the PCP and the TE domain. The substrate, which is bound to the 4′Ppan cofactor of the PCP domain as a thioester, acylates the hydroxyl group of the serine. Chemically speaking, the acylation of the TE is a result of a *trans*-esterification. When using

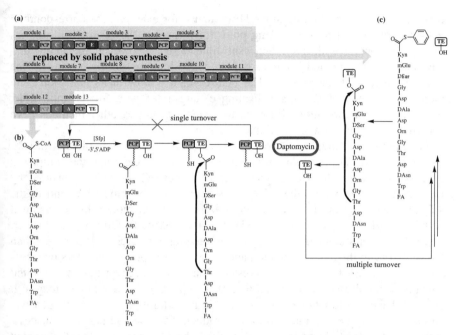

Figure 4.5 Chemoenzymatic approaches for the production of novel bioactive compounds. In this example, the enzymatic buildup of the linear precursor of daptomycin by its NRPSs (DptA, DptBC, and DptD) is substituted by solid-phase synthesis (a). By using the 4′Ppan transferase Sfp and the CoA-thioester of the linear peptide, the *apo*-enzyme PCP-TE and be modified, and after *trans*-esterification cyclized by the TE domain (b). Because the resulting *holo*-enzyme cannot be modified again, this is a single turnover reaction. Another strategy uses thiophenole-esters of the linear peptides to be cyclized (c). When these compounds are used, no PCP domain is necessary. The TE domain is readily acylated, and regiospecific and stereospecific cyclization toward daptomycin or, depending on the linear peptide provided, toward variants thereof occurs. Because the enzyme is not altered in any way after product release, this setup results in a multiple turnover.

TE domains, the substrate provided *in trans* must also have an appropriate acylation potential. Several key techniques have been developed to covalently attach synthetic substrates to PCP and TE domains. In the first method, the relaxed substrate specificity of the 4′Ppan transferase Sfp is used to load acyl moieties onto PCP domains enzymatically. Just like in the natural priming reaction in which the 4′Ppan part of CoA is transferred onto the conserved serine residue of the *apo*-PCP domain, Sfp does analogously attach S-acylated 4′Ppans, which originates from S-acylated CoA substrates (32). These CoA substrates can readily be obtained by one coupling reaction directly after solid phase synthesis. With this technique, virtually any substrate can be brought to a desired position in recombinant NRPS enzymes containing an *apo*-PCP domain. This result is of great value when elucidating the catalytic properties and substrate specificities of other domains. When investigating TE domains, for example, the corresponding *apo*-PCP-TE would be the starting point for screening the cyclization abilities of the TE domain with a synthetic substrate library. However, the major disadvantage of this method becomes evident when looking closely at the enzyme after the release of the product: The PCP domain is now in its *holo*-state, and therefore, subsequent enzymatic loading of substrates with Sfp is impossible. Because product formation is limited to a single turnover, other methods have been developed to allow for multiple turnover.

For multiple turnover reactions, the TE domains must be supplied with substrates *in trans*; yet the acylation potential must be sufficient and the compound must be recognized by the enzyme. The first approach made was inspired by the natural system where the substrate is activated as a thioester bound to the 4′Ppan cofactor. The idea was to minimize the 4′Ppan moiety by replacing it with N-acetyl-cysteamine (SNAc) (33). This method works fairly well; however, it seems in later studies that the acylation potential is of greater importance than the similarity to the natural situation and a variety of peptidyl-thioesters with better leaving groups than SNAc was tested. The fastest turnover rates were found when thiophenol-esters were used (30, 34). Thiophenol has several advantages: It does not have any functional groups other than the thiol group, it is inexpensive, and it can easily be separated from the product. This method has been successfully used to shed light on the promiscuity of the TE domain of the daptomycin NRPS (35).

Even though these techniques allow for the production of new potentially bioactive compounds, they are usually closely related to known substances, which basically implies an analogous mode of action, and the deviations normally alter quantitative parameters such as solubility and affinity. Nevertheless, hundreds of nonribosomal systems still need to be explored, and the discoveries of new ones are frequently reported. The basic understanding of the catalytic functions of nonribosomal domains and modules that we have today is a good starting point for additional exploration and use of systems that are not yet understood fully.

REFERENCES

1. Walsh CT. Polyketide and nonribosomal peptide antibiotics: modularity and versatility. Science 2004;303:1805–1810.
2. Finking R, Marahiel MA. Biosynthesis of nonribosomal peptides1. Annu. Rev. Microbiol. 2004;58:453–488.
3. Fischbach MA, Walsh CT. Assembly-line enzymology for polyketide and nonribosomal Peptide antibiotics: logic, machinery, and mechanisms. Chem. Rev. 2006; 106:3468–3496.
4. Keller U, Schauwecker F. Combinatorial biosynthesis of non-ribosomal peptides. Comb. Chem. High Throughput Screen. 2003;6:527–540.
5. Stein DB, Linne U, Hahn M, Marahiel MA. Impact of epimerization domains on the intermodular transfer of enzyme-bound intermediates in nonribosomal peptide synthesis. ChemBioChem. 2006;7:1807–1814.

6. Kohli RM, Trauger JW, Schwarzer D, Marahiel MA, Walsh CT. Generality of peptide cyclization catalyzed by isolated thioesterase domains of nonribosomal peptide synthetases. Biochemistry 2001;40:7099–7108.

7. Chen H, O'Connor S, Cane DE, Walsh CT. Epothilone biosynthesis: assembly of the methylthiazolylcarboxy starter unit on the EpoB subunit. Chem. Biol. 2001;8: 899–912.

8. Schracke N, Linne U, Mahlert C, Marahiel MA. Synthesis of linear gramicidin requires the cooperation of two independent reductases. Biochemistry. 2005;41(44): 8507–8513.

9. Velkov T, Lawen A. Mapping and molecular modeling of S-adenosyl-L-methionine binding sites in N-methyltransferase domains of the multifunctional polypeptide cyclosporin synthetase. J. Biol. Chem. 2003;278:1137–1148.

10. Schoenafinger G, Schracke N, Linne U, Marahiel MA. Formylation domain: an essential modifying enzyme for the nonribosomal biosynthesis of linear gramicidin. J. Am. Chem. Soc. 2006;128:7406–7407.

11. Hubbard BK, Walsh CT. Vancomycin assembly: nature's way. Angew. Chem. Int. Ed. Engl. 2003;42:730–765.

12. Luo L, Burkart MD, Stachelhaus T, Walsh CT. Substrate recognition and selection by the initiation module PheATE of gramicidin S synthetase. J. Am. Chem. Soc. 2001;123:11208–11218.

13. Stachelhaus T, Mootz HD, Marahiel MA. The specificity-conferring code of adenylation domains in nonribosomal peptide synthetases. Chem. Biol. 1999;6:493–505.

14. Koglin A, Mofid MR, Lohr F, Schafer B, Rogov VV, Blum MM, et al. Conformational switches modulate protein interactions in peptide antibiotic synthetases. Science. 2006;14(312):273–276.

15. Keating TA, Marshall CG, Walsh CT, Keating AE. The structure of VibH represents nonribosomal peptide synthetase condensation, cyclization and epimerization domains. Nat. Struct. Biol. 2002;9:522–526.

16. Bender CL, Alarcon-Chaidez F, Gross DC. Pseudomonas syringae phytotoxins: mode of action, regulation, and biosynthesis by peptide and polyketide synthetases. Microbiol. Mol. Biol. Rev. 1999;63:266–292.

17. Lautru S, Deeth RJ, Bailey LM, Challis GL. Discovery of a new peptide natural product by Streptomyces coelicolor genome mining. Nat. Chem. Biol. 2005;1:265–269.

18. Kohli RM, Trauger JW, Schwarzer D, Marahiel MA, Walsh CT. Generality of peptide cyclization catalyzed by isolated thioesterase domains of nonribosomal peptide synthetases. Biochemistry 2001;40:7099–7108.

19. Gehring AM, Mori I, Walsh CT. Reconstitution and characterization of the Escherichia coli enterobactin synthetase from EntB, EntE, and EntF. Biochemistry 1998;37:2648–2659.

20. Balibar CJ, Vaillancourt FH, Walsh CT. Generation of D amino acid residues in assembly of arthrofactin by dual condensation/epimerization domains. Chem. Biol. 2005; 12:1189–1200.

21. Billich A, Zocher R. N-Methyltransferase function of the multifunctional enzyme enniatin synthetases. Biochemistry 1987;26:8417–8423.

22. Quadri LE, Sello J, Keating TA, Weinreb PH, Walsh CT. Identification of a Mycobacterium tuberculosis gene cluster encoding the biosynthetic enzymes for assembly of the virulence-conferring siderophore mycobactin. Chem. Biol. 1998;5:631–645.

23. Eppelmann K, Doekel S, Marahiel MA. Engineered biosynthesis of the peptide antibiotic bacitracin in the surrogate host Bacillus subtilis. J. Biol. Chem. 2001;276: 34824–34831.

24. Mootz HD, Kessler N, Linne U, Eppelmann K, Schwarzer D, Marahiel MA. Decreasing the ring size of a cyclic nonribosomal peptide antibiotic by in-frame module deletion in the biosynthetic genes. J. Am. Chem. Soc. 2002;124:10980–10981.

25. Schauwecker F, Pfennig F, Grammel N, Keller U. Construction and in vitro analysis of a new bi-modular polypeptide synthetase for synthesis of N-methylated acyl peptides. Chem. Biol. 2000;7:287–297.

26. Stachelhaus T, Schneider A, Marahiel MA. Rational design of peptide antibiotics by targeted replacement of bacterial and fungal domains. Science 1995;269:69–72.

27. Belshaw PJ, Walsh CT, Stachelhaus T. Aminoacyl-CoAs as probes of condensation domain selectivity in nonribosomal peptide synthesis. Science 1999;284:486–489.

28. May JJ, Kessler N, Marahiel MA, Stubbs MT. Crystal structure of DhbE, an archetype for aryl acid activating domains of modular nonribosomal peptide synthetases. Proc. Natl. Acad. Sci. U.S.A. 2002;99:12120–12125.

29. Eppelmann K, Stachelhaus T, Marahiel MA. Exploitation of the selectivity-conferring code of nonribosomal peptide synthetases for the rational design of novel peptide antibiotics. Biochemistry. 2002;41:9718–9726.

30. Grunewald J, Marahiel MA. Chemoenzymatic and template-directed synthesis of bioactive macrocyclic peptides. Microbiol. Mol. Biol. Rev. 2006;70:121–146.

31. Trauger JW, Kohli RM, Mootz HD, Marahiel MA, Walsh CT. Peptide cyclization catalysed by the thioesterase domain of tyrocidine synthetase. Nature 2000;407: 215–218.

32. Sieber SA, Walsh CT, Marahiel MA. Loading peptidyl-coenzyme A onto peptidyl carrier proteins: a novel approach in characterizing macrocyclization by thioesterase domains. J. Am. Chem. Soc. 2003;125:10862–10866.

33. Ehmann DE, Trauger JW, Stachelhaus T, Walsh CT. Aminoacyl-SNACs as small-molecule substrates for the condensation domains of nonribosomal peptide synthetases. Chem. Biol. 2000;7:765–772.

34. Sieber SA, Tao J, Walsh CT, Marahiel MA. Peptidyl thiophenols as substrates for nonribosomal peptide cyclases. Angew. Chem. Int. Ed. Engl. 2004;43:493–498.

35. Grunewald J, Sieber SA, Mahlert C, Linne U, Marahiel MA. Synthesis and derivatization of daptomycin: a chemoenzymatic route to acidic lipopeptide antibiotics. J. Am. Chem. Soc. 2004;126:17025–17031.

FURTHER READING

Baltz RH. Molecular engineering approaches to peptide, polyketide and other antibiotics. Nat. Biotechnol. 2006;24:1533–1540.

Cane DE. Introduction: Polyketide and Nonribosomal Polypeptide Biosynthesis. From Collie to Coli. Chem. Rev. 1997;97:2463–2464.

Du L, Shen B. Biosynthesis of hybrid peptide-polyketide natural products. Curr. Opin. Drug Discov. Devel. 2001;4:215–228.

Grunewald J, Kopp F, Mahlert C, Linne U, Sieber SA, Marahiel MA. Fluorescence resonance energy transfer as a probe of peptide cyclization catalyzed by nonribosomal thioesterase domains. Chem. Biol. 2005;12:873–881.

Hahn M, Stachelhaus T. Selective interaction between nonribosomal peptide synthetases is facilitated by short communication-mediating domains. Proc. Natl. Acad. Sci. U.S.A. 2004;101:15585–15590.

Salomon CE, Magarvey NA, Sherman DH. Merging the potential of microbial genetics with biological and chemical diversity: an even brighter future for marine natural product drug discovery. Nat. Prod. Rep. 2004;21:105–121.

Sieber SA, Linne U, Hillson NJ, Roche E, Walsh CT, Marahiel MA. Evidence for a monomeric structure of nonribosomal Peptide synthetases. Chem. Biol. 2002;9: 955–956.

5

PLANT TERPENOIDS

CHRISTOPHER I. KEELING

Michael Smith Laboratories, University of British Columbia, British Columbia, Canada

JÖRG BOHLMANN

Michael Smith Laboratories and Departments of Botany and Forest Sciences,
University of British Columbia, British Columbia, Canada

Terpenoids are the largest class of all known natural products. Plants produce a variety of terpenoid compounds that number in the thousands. Some terpenoids are involved in plant growth and development directly (i.e., in primary metabolism), but most plant terpenoids are thought to function in interactions of plants with their biotic and abiotic environment and have traditionally been referred to as secondary metabolites. In addition to the isolation and identification of plant terpenoids, research has concentrated on the biosynthesis, the biological function, and the exploitation of plant terpenoids for human use as biomaterials and pharmaceuticals. Plant terpenoids are biosynthesized from C_5 precursors by the action of prenyl transferases and terpenoid synthases. Often, terpenes are acted on by cytochromes P450 and other enzymes to increase their functionalization. Terpenoid biosynthesis in plants involves several subcellular compartments. The accumulation of terpenoids requires efficient transport systems and specialized anatomical structures. Using isoprene (a hemiterpene), menthol (a monoterpene), artemisinin (a sesquiterpene), and paclitaxel [better known under the registered trademark Taxol (Bristol Myers Squibb, New York)] and diterpene resin acids (diterpenes) as examples, we highlight some strategies, techniques, and results of plant terpenoid research with a strict focus on the low-molecular-weight (C_5–C_{20}) terpenoids of specialized plant metabolism.

Natural Products in Chemical Biology, First Edition. Edited by Natanya Civjan.
© 2012 John Wiley & Sons, Inc. Published 2012 by John Wiley & Sons, Inc.

All plant terpenoids are derived from C_5 precursors, and most plant terpenoids can be grouped according to their number of C_5 building blocks as hemiterpenoids (C_5), monoterpenoids (C_{10}), sesquiterpenoids (C_{15}), diterpenoids (C_{20}), or polyterpenoids (C_{5xn}). In addition, condensation of C_{15}- and C_{20}-intermediates give rise to triterpenoids (C_{30}) and tetraterpenoids (C_{40}), respectively. Many irregular terpenoids or terpenoid derivates are also found in plants. Terpenoids are involved in all forms of plant interactions with other organisms including plant reproduction, defense, and signaling. They are the most diverse group of plant chemicals and have been used by humans for centuries in both traditional and modern industrial applications. As such, plant terpenoids are used widely as pharmaceuticals, flavor and aroma chemicals, vitamins, pigments, and large-volume biological feedstock for the production of a suite of industrial materials such as, for example, industrial resins and print inks. Plant terpenoids are an important group of natural product chemicals that are actively being explored as alternatives for petroleum-based materials.

5.1 BIOLOGICAL BACKGROUND

Over 20,000 terpenoids have been identified (1), and more are being discovered continuously. Plant terpenoids are important in both primary and secondary (specialized) metabolism. Their importance in primary metabolism includes physiological, metabolic, and structural roles such as plant hormones, chloroplast pigments, roles in electron transport systems, and roles in the posttranslational modification of proteins. In secondary metabolism, the roles of plant terpenoids are incredibly diverse but are associated most often with defense and communication of sessile plants interacting with other organisms. Examples include terpenoid chemicals that form physical and chemical barriers, antibiotics, phytoalexins, repellents and antifeedants against insects and other herbivores, toxins, attractants for pollinators or fruit-dispersing animals, host/nonhost selection cues for herbivores, and mediators of plant–plant and mycorrhiza interactions (2, 3).

Some plants produce terpenoids in specialized cells or tissues such as the glandular trichomes on the surface of peppermint leaves, scent-releasing epidermal cells of certain flowers, or the epithelial cells that surround the resin ducts of conifers. These structures place high concentrations of terpenoids in areas most likely to be encountered by the interacting organism.

The biosynthetic pathways of plant terpenoids are highly regulated and highly spatially organized in subcellular compartments and sometimes in specialized cells. Terpenoid biosynthesis can be regulated by plant hormones, developmental programs, diurnal cycles, herbivory, or pathogen infection. Identifying what role specific terpenoids play in plants, how and where they are biosynthesized, and how their biosyntheses are regulated allows us to better understand their importance to the survival of the plant and thus make use of this knowledge in crop improvement or in the production of terpenoids for medicinal or industrial uses.

5.2 CHEMISTRY

Much chemistry research in plant terpenoids has been to elucidate the structure, define the biosynthetic pathways, characterize the enzymes involved, and develop systems for the large-scale production of medicinally or industrially important terpenoids (4). Progress on the identification and the study of plant terpenoids is reviewed regularly in the journal *Natural Product Reports*, and the biosynthesis of terpenoids, including plant terpenoids, has been reviewed comprehensively (5). Plant terpenoids can be volatile or nonvolatile, lipophilic or hydrophilic, cyclic or acyclic, chiral or achiral, and they often have double-bond stereochemistry. The chemical diversity of terpenoid structures originates largely from the terpenoid synthase enzymes that stabilize different carbocation intermediates, allow rearrangements or water termination, and direct stereochemistry; the diversity also originates from the many different terpenoid-modifying enzymes.

Two major complementary approaches to studying plant terpenoids have been established. One approach involves the isolation and the structural identification of terpenoid chemicals of interest from plant tissues based on traditional natural products research followed by targeted search for the relevant enzymes and genes that control biosynthesis. The second approach explores the emerging plant genome sequences to discover complete sets of genes that encode terpenoid biosynthetic enzymes. The combination of these two approaches is the most powerful approach to a comprehensive understanding of plant terpenoid chemistry and its biosynthetic origins.

The diversity of plant terpenoids reflects the complexity and the diversity of the pathways that biosynthesize them. The recent sequencing of the genomes of four different plant species and large collections of expressed sequence tags (ESTs) from many other plants may indicate the diversity of pathways and chemicals we might expect in any one species. For example, the genes that encode putatively active terpenoid synthases (TPS) comprise at least 32 in the Arabidopsis (*Arabidopsis thaliana*) genome (6), at least 15 in the rice (*Oryza sativa*) genome (7), at least 47 in the poplar (*Populus trichocarpa*) genome (8), and at least 89 in the genome of a highly inbred grapevine (*Vitis vinifera*) Pinot Noir variety (9). The large gene family of TPS, which is important to generating structural diversity of terpenoid chemicals in plants, apparently results from repeated gene duplication and subsequent neofunctionalization or subfunctionalization (10, 11). Most TPS produce more than one product from a single substrate, and these products are often modified by the action of additional enzymes such as cytochromes P450 and reductases. Thus, the number of distinct terpenoids found in any one plant species is predicted to be manifold higher than the number of TPS genes present in that species. Genomics approaches, which can identify the candidate genes for terpenoid production, together with functional characterization of heterologously expressed enzymes and the identification of the resulting plant terpenoids, can enhance the discovery of the biochemical pathways substantially *in planta* as has been demonstrated in recent years with research in Arabidopsis (12), rice (13), and grapevine (14). Ideally, the functional genomics approach is combined with classical and modern approaches of isolation, identification, and metabolite profiling of terpenoids from plant tissue.

The combined genomics and chemical approaches to plant terpenoid research are not restricted to the few plant species for which more or less complete genome sequences are now available. The discovery of many of the genes and enzymes for the formation of terpenoids such as menthol and related monoterpenes in peppermint (*Mentha x piperita*) (15), artemisinin in *Artemisia annua* (16), Taxol in the yew tree (*Taxus*) (17), or conifer diterpene resin acids in species of spruce (*Picea*) and pine (*Pinus*) (18) have been possible on the foundation of highly specialized efforts of EST and full-length cDNA sequencing combined with characterization of recombinant enzymes and analysis of the terpenoid metabolome of the target plant species.

5.2.1 Common Steps in Plant Terpenoid Biosynthesis

The universal precursors to terpenoids, the C_5-compounds dimethylallyl pyrophosphate (DMAPP) and isopentenyl pyrophosphate (IPP), originate from two pathways in plants (Fig. 5.1). The mevalonate (MEV) pathway is well described in many eukaryotic organisms. This pathway is present in the cytosol/endoplasmic reticulum of plants. More recently, another pathway has been described, the $2C$-methyl-D-erythritol-4-phosphate (MEP) pathway, which is found in the plastids of plants (19). The localization of the different pathways and the plastid-directing transit peptides found in hemi-TPS, mono-TPS, and di-TPS, but not in sesqui-TPS, result in the production of terpenoids from at least two different precursors pools.

Hemiterpenoids are produced from the isoprenyl diphosphate DMAPP. All other terpenoids are produced from DMAPP and IPP via longer-chain prenyl diphosphate intermediates formed by prenyl transferases. Prenyl transferases (20) catalyze the formation of geranyl diphosphate (GPP), farnesyl diphosphate (FPP), and geranylgeranyl diphosphate (GGPP) from one molecule of DMAPP and one, two, or three molecules of IPP, respectively (Fig. 5.1). Isoprenyl diphosphates are the substrates for all TPS, which lead to the hemiterpenoids, monoterpenoids, sesquiterpenoids, and diterpenoids, which will be highlighted with selected examples in the following sections.

5.2.2 Isoprene (C_5, Hemiterpene)

Isoprene (2-methyl 1,3-butadiene, Fig. 5.2) is the simplest terpenoid produced by many plants, and is produced abundantly by some tree species such as poplars. The mechanism of isoprene biosynthesis is a straightforward diphosphate ionization of the DMAPP precursor followed by deprotonation of the allylic cation (Fig. 5.2). Only a few isoprene synthases have been identified and characterized in plants, from *Populus* spp. (poplar) (21, 22) and *Pueraria montana* (kudzu vine) (23). These TPS contain a transit peptide that targets the plastids, and thus isoprene biosynthesis is derived from products of the MEP pathway. Isoprene is

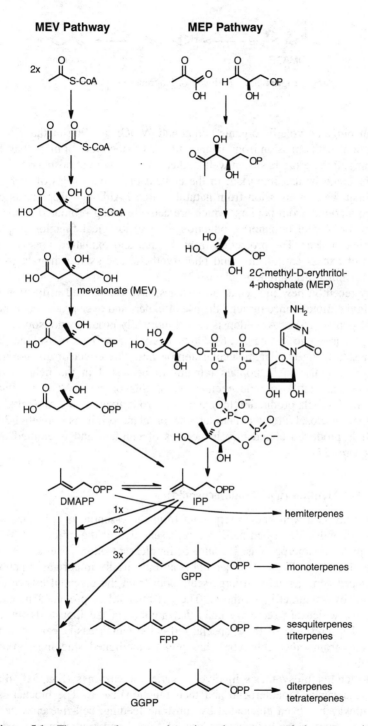

Figure 5.1 The two pathways to the universal precursors of plant terpenoids.

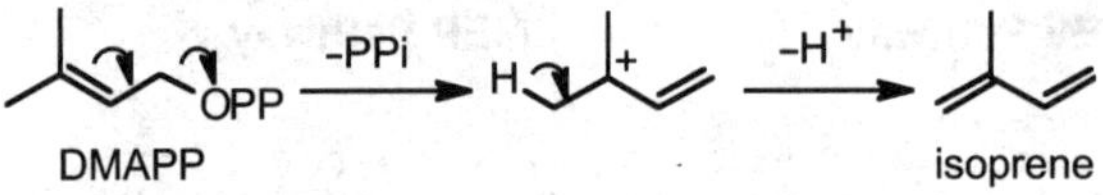

Figure 5.2 Pathway of isoprene biosynthesis.

a major biogenic volatile organic compound (VOC; rivalling methane in global production) with emission from plants estimated in the order of more than 10^{12} kg per year and therefore has been well studied for its role in atmospheric chemistry (24). Isoprene is also important in the context of global cycles of carbon fixation versus carbon emission from natural sources. Although isoprene is emitted in large amounts from poplars, which are actively being promoted as plantation species for biofuel (ethanol) production, its physiological function in plants is somewhat unclear. The protection from thermal and oxidative stress as well as release of excess carbon flux and photosynthetic energy are thought to be the main functions of isoprene in plants (21, 22).

Very recently, new molecular approaches have established a function of isoprene in thermotolerance through the use of under- and over-producing transgenic lines of poplar (25) or Arabidopsis (which normally produces no isoprene) using the poplar isoprene synthase (26). Other possible functions of isoprene remain to be tested using similar molecular approaches. The effect of down-regulation of isoprene emission in poplars remains to be tested in the field to explore whether it is viable to reduce emission of this biogenic VOC in plantation forests for biofuels production to maximize carbon fixation and minimize carbon emission. A closely related hemiterpene of plant origin is 2-methyl-3-buten-2-ol. It is produced abundantly in needles of conifers and is emitted into the atmosphere (27).

5.2.3 (–)-Menthol (C_{10}, Monoterpenoid)

(–)-Menthol is a well-known terpenoid from the essential oil of mint (*Mentha* spp.) (15), and is described here as a representative of the different acyclic and cyclic plant monoterpenoids. Because of its pleasant odor, taste, and anesthetic and antimicrobial effects, (–)-menthol is an industrially important terpenoid and is produced commercially in large scale both from the essential oils of *Mentha* spp. and by asymmetric synthesis. The essential oil is produced in glandular trichomes, which are secretory cells that number in the thousands on *Mentha* leaves. The presence of these specialized cells, which easily can be separated physically from other cell types, has greatly facilitated studying (–)-menthol biosynthesis.

(–)-Menthol biosynthesis involves a series of enzymes (Fig. 5.3) that first generate a cyclic monoterpene and then functionalize it. The biochemistry of this pathway has been elucidated by substrate feeding, cell-free enzyme assays with plant extracts, and characterization of cloned and recombinantly expressed

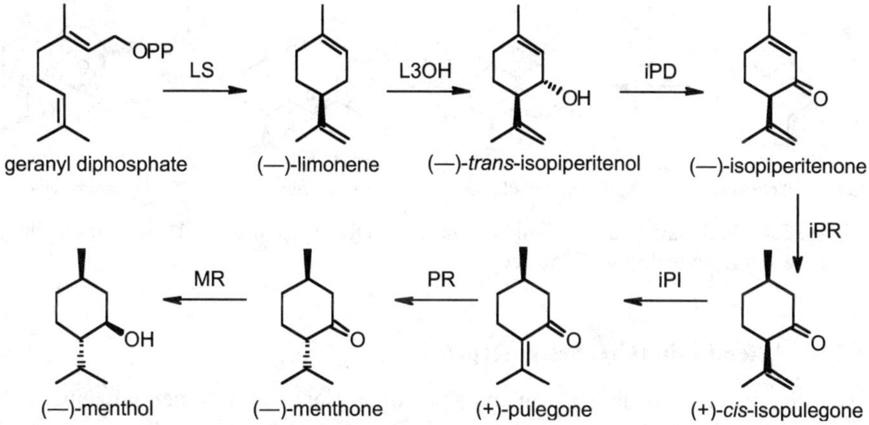

Figure 5.3 Pathway of (−)-menthol biosynthesis in *Mentha*. *LS*, (−)-limonene synthase; *L3OH*, (−)-limonene-3-hydroxylase; *iPD*, (−)-*trans*-isopiperitenol dehydrogenase; *iPR*, (−)-isopiperitenone reductase; *iPI*, (+)-*cis*-isopulegone isomerase; *PR*, (+)-pulegone reductase; *MR*, (−)-menthone reductase.

enzymes (28). The biosynthesis of menthol, which has been studied for more than two decades by Croteau et al. (reviewed in Reference 15), is arguably the best-characterized pathway of a functionalized monoterpenoid and can serve as a paradigm for many monoterpenoids in plants in general. First, using geranyl diphosphate as substrate, (−)-limonene synthase generates (−)-limonene and minor amounts of myrcene, (−)-alpha-pinene, and (−)-beta-pinene. (−)-Limonene then undergoes a series of transformations ultimately to yield (−)-menthol. These modifications involve first the allylic hydroxylation to (−)-*trans*-isopiperitenol by the cytochrome P450 limonene-3-hydroxylase (CYP71D13). Allylic oxidation of this alcohol to (−)-isopiperitenone is then catalyzed by the NAD-dependent isopiperitenol dehydrogenase. Subsequently, NADPH-dependent (−)-isopiperitenone reductase catalyzes the formation of (+)-*cis*-isopulegone. (+)-*cis*-Isopulegone is enzymatically isomerized to the more stable alpha,beta-unsaturated ketone—(+)-pulegone—by (+)-*cis*-isopulegone isomerase. (−)-Menthone and (+)-isomenthone (in a 2:1 to 10:1 ratio) are then formed by the action of the NADPH-dependent (+)-pulegone reductase. Finally, (−)-menthone reductase reduces (−)-menthone to (−)-menthol. All of these enzymes and their corresponding genes in *Mentha* have been isolated, functionally characterized, and their enzymology studied (15). For some enzymes, substrate specificities or product outcomes after directed mutations as well as enzyme localization have been investigated. The structure of (−)-limonene synthase has been determined recently (29). Based on the detailed knowledge of (−)-menthol and its biosynthetic pathway, it has become possible to improve the composition of the monoterpene-rich essential oil of *Mentha* through metabolic engineering (30, 31).

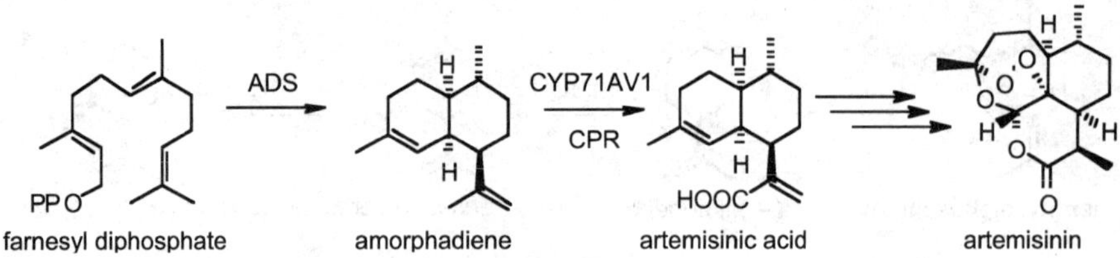

Figure 5.4 Pathway of artemisinin biosynthesis in *Artemisia annua*. *ADS*, amorphadiene synthase; *CPR*, cytochrome P450 reductase.

5.2.4 Artemisinin (C_{15}, Sesquiterpenoid)

Artemisinin is used here as an example of a plant sesquiterpenoid with both traditional value as well as with medicinal and social value in the twenty-first century. Research on artemisinin has also established new benchmarks for bio-chemical engineering and functional genomics of plant terpenoids. Artemisinin is a functionalized sesquiterpene with a unique peroxide linkage from the sweet wormwood (*Artemisia annua*). Chinese herbalists have used it since ancient times, and it is now used for its unique efficacy to treat multidrug-resistant strains of the malaria parasite *Plasmodium falciparum*. Its medicinal importance has prompted studies into its biosynthesis and its biochemical engineering so that cost-effective methods for producing it in large scale and in consistent quality may be realized.

Biosynthesis of artemisinin *in planta* begins with the formation of the sesquiterpene amorpha-4,11-diene in glandular tricomes of *A. annua* leaves (Fig. 5.4) by amorpha-4,11-diene synthase (32, 33). Amorphadiene is oxidized to artemisinic acid in three steps by a multifunctional cytochrome P450 (CYP71AV1) (16). The remaining steps have not yet been established but are predicted to include nonenzymatically catalyzed photooxidation reactions (34). The application of a semi-synthetic route from artemisinic acid to artemisinin, along with the availability of the characterized plant enzymes to produce artemisinic acid described above from *A. annua*, have permitted the complete synthesis of artemisinin via microbial host cells (16). Introducing these enzymes into *Escherichia coli* or *Saccharomyces cerevisiae* and engineering an unnatural, or fine-tuning the natural, mevalonate pathway in these microorganisms have resulted in significant production of artemisinic acid in fermentations (16, 35, 36).

5.2.5 Taxol (C_{20}, Diterpenoid)

Taxol is another example of a medicinally important functionalized plant ter-pene, in this case a diterpenoid, with large pharmaceutical and economic value (Fig. 5.5). Taxol is a potent anticancer drug that was isolated and identified from the bark of the Pacific yew (*Taxus brevifolia*) more than 35 years ago (37). The name Taxol is now a registered trademark, but the literature commonly uses this

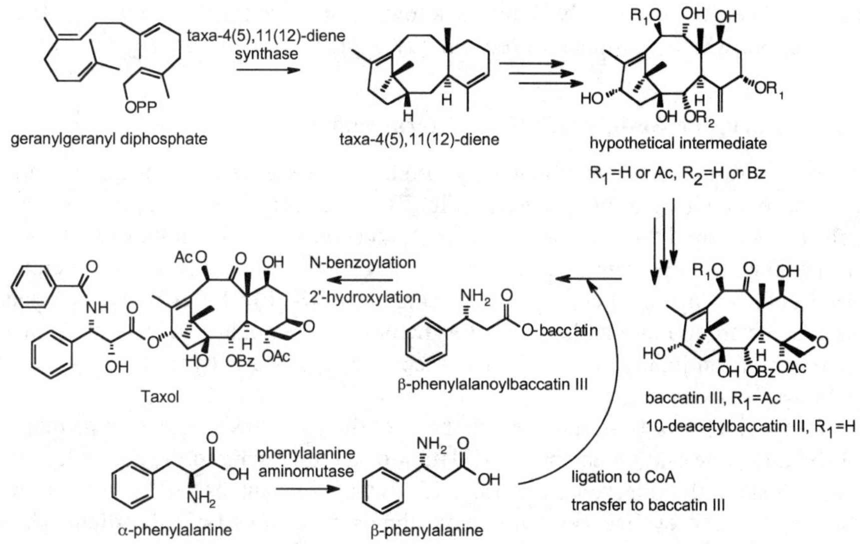

Figure 5.5 Pathway of Taxol biosynthesis in *Taxus* spp.

name rather than the generic name paclitaxel. The total synthesis of Taxol is possible (38), but it is not economically feasible currently because of the challenges of stereochemistry, low yield, and high cost. The study of Taxol by Croteau et al. (17) is an exceptional example of how a terpenoid biosynthetic pathway was rationalized, and the synthesis and testing of various hypothetical precursors with cell-free extracts have yielded the discovery of many enzymes and genes in this complex pathway. Because Taxol is only found at very low levels in slow growing trees, and is one of hundreds of Taxol-like compounds produced in a metabolic grid, the use of an inducible *Taxus* cell culture system has accelerated this research.

Taxol is biosynthesized in 19 steps from GGPP that originates from precursors of the MEP pathway (Fig. 5.5). The biosynthesis of Taxol begins with the formation of the tricyclic diterpene skeleton of taxa-4(5),11(12)-diene (17, 39). All genes for the enzymes in this early pathway have been identified in *Taxus cuspidata* and a taxa-4(5),11(12)-diene synthase has been identified in several *Taxus* species. The mechanism of this di-TPS has been explored in detail (39). Taxa-4(5),11(12)-diene is then hydroxylated by several cytochrome P450 taxoid oxygenases to yield a putative intermediate decorated with seven alcohol or ester groups. Many cytochrome P450 enzymes that catalyze these transformations have been identified and characterized, but two enzymes remain uncharacterized (17). The biosyntheses of the ester functionalities have been studied, and several acyl and aroyl transferases have been identified and characterized. Finally, studies on the steps in the aromatic side chain assembly and attachment has yielded several enzymes that include a phenylalanine aminomutase, a C13-phenylpropanoyl-CoA transferase, and an *N*-benzoyl transferase. Although the pathway of biosynthesis has not been resolved fully, what is known can be applied to improve Taxol production, which continues to rely on *Taxus* plants or cell cultures.

5.2.6 Diterpene Resin Acids (C_{20}, Diterpenoids)

Diterpene resin acids are abundantly produced in conifers of the pine family (Pinaceae) and in other plant species (Fig. 5.6). They are produced in the epithelial cells that surround the resin ducts that are found constitutively, or they are induced in the xylem upon wounding and are important for the physical and chemical plant defenses against herbivores and pathogens (18, 40). Industrially, diterpene resin acids are important chemicals for the naval stores industry, in printing inks, as potential antimicrobials and pharmaceuticals, and are byproducts of wood pulping processes.

Two major steps exist in the biosynthesis of diterpene resin acids: the formation of the diterpene and the stepwise oxidation of the diterpene to the corresponding acid. Most conifer diterpenes are tricyclic; they are biosynthesized by bifunctional di-TPS that first cyclize geranylgeranyl diphosphate to (+)-copalyl diphosphate and then cyclize this intermediate even more to form the various diterpenes (11, 18) (Fig. 5.6). Some di-TPS produce multiple products, whereas others produce only single products. These diterpenes are then oxidized stepwise to the resin acids by multifunctional, multisubstrate cytochromes P450 (41).

5.3 CHEMICAL TOOLS AND TECHNIQUES

The study of plant terpenoids shares many of the same tools for their isolation, identification, characterization, and synthesis that are required in other natural product research. Advances in separation science, analytical chemistry, spectroscopic tools, and synthetic organic chemistry all affect the study of terpenoids in plants.

5.3.1 Extraction and Separation

Isolation of the plant terpenoids usually begins with some form of extraction from the plant source. Certain tissue types such as roots, leaves, or flowers are often extracted. Often, the terpenoids of interest are produced in very specialized tissues on plant surfaces or within the plant such as the glandular trichomes of *Mentha* spp. or the resin ducts of conifer species. If known, the increased concentration and reduced complexity gained by selectively extracting these specific cell types may outweigh the increased difficulty of isolating them. Homogenization of the tissue in an appropriate solvent usually is sufficient to extract the terpenoids. If the terpenoids of interest are relatively volatile, then vapor-phase collection can reduce sample complexity; this can be done without the use of solvent in some instances. Recently, Tholl et al. (42) have reviewed the analytical methods of headspace sampling, solid phase micro-extraction (SPME), and

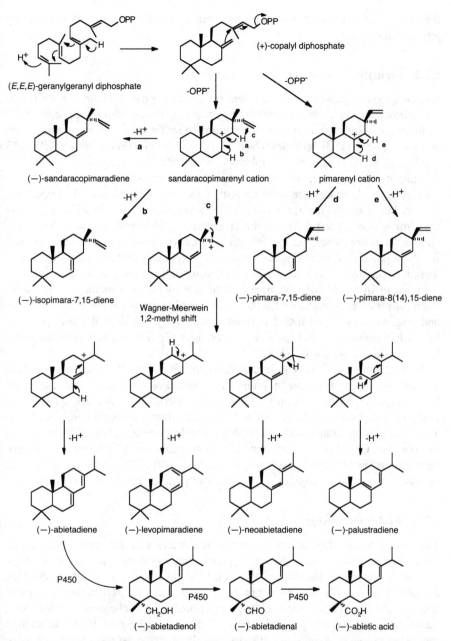

Figure 5.6 Pathway of conifer diterpene resin acid biosynthesis. Bifunctional di-TPS convert geranylgeranyl diphosphate to various diterpenes, which are oxidized stepwise by multisubstrate and multifunctional cytochromes P450 to the corresponding diterpene acid. The oxidation of (−)-abietadiene to (−)-abietic acid is shown as an example.

the capture of plant volatiles onto absorbent media for subsequent analysis by gas chromatography.

5.3.2 Structural Determination

Hanson (43) has published an excellent review of the methods and the strategies for rigorous structural determination of terpenoids. Routine survey-style analyses of plant terpenoids often are not so rigorous in structural assignment. Databases of mass spectra [such as Wiley (New York) and NIST (Gaithersburg, MD) MS databases] or databases that combine mass spectra and retention indices (44, 45) facilitate assignment of some commonly encountered plant terpenoids. Of course, these databases are only helpful if they include the specific terpenoids that are being analyzed and the likely alternatives. Sometimes, in addition to having similar mass-spectral fragmentation patterns, two terpenoids may share very similar retention indices on the 5% phenyl methylpolysiloxane GC column used for these databases, which makes structural assignment without additional information impossible. The availability of retention index information on at least two columns of different polarity increases the confidence in structural assignment. Such information is available for the more volatile terpenoids involved in flavors and fragrances (46), and insect semiochemicals (47). For nonvolatile terpenoids, LC-MS is often used, but there are no comparable databases of terpenoids for LC-MS as there are for GC-MS.

Stereochemistry often is an integral component to both the chemical structure and the biological function of plant terpenoids. For volatile terpenoids, chiral GC stationary phases (48) provide the enantiomeric separation for quantitative analysis, and, provided an authentic standard of known absolute configuration is available, elution comparison can establish absolute configuration of the unknown sample (49). Chiral phases for liquid chromatography can also be used to resolve enantiomers, both analytically and preparatively (50). Often, chiral synthesis is necessary to clearly establish the absolute configuration of an unknown terpenoid.

5.3.3 Molecular Biology

The recent availability of plant genome sequences and the methods to clone homologous genes from different plant species using molecular biology approaches has provided the ability to identify the capacity of plants to produce additional terpenoids that may not have been detected in that specific plant species before. Rather than isolate, fractionate, and chemically identify a particular plant terpenoid profile, plant TPS or cytochromes P450 are cloned into a heterologous expression system, and the enzyme assay products of these recombinant enzymes with isoprenoid substrates are analyzed directly. This approach has proven useful in situations in which the terpenoids themselves may not be detectable in the plant under normal growing conditions, and the inducer or environmental conditions required for their production are not yet known (12).

5.3.4 Synthesis

Independent synthesis of the identified terpenoid is often required to confirm structural assignment (51) and to test the biological and/or pharmacological functions. In addition, biosynthetic studies often require the synthesis of putative precursors for functional characterization of enzymes. Terpenoid structures challenge chemists in many of the same ways that other natural products do. Their structural diversity (complicated by stereochemistry, carbocyclic skeletons, and often multiple functionalization) provides opportunities for synthetic organic chemists to develop new methodologies for synthesis. Recently, Maimone and Baran (52) have reviewed some synthetic challenges terpenoids present and the solutions employed.

5.3.5 Biochemical and Metabolic Engineering

Often, the original natural plant sources of medicinally or industrially important terpenoids cannot supply sufficient material for their demand. In some instances, such as *Taxus* for the production of Taxol (53), the original plant species is amenable to growth in plant cell culture for commercial production. For several plant terpenoids, an understanding of their biosynthetic pathway has advanced such that many, if not all, enzymes and genes involved have been identified and characterized. These developments provide the opportunity to engineer bacteria or yeast to produce these terpenoids *de novo* or from more readily available precursors in large-scale fermentations (54). In addition, the increased, reduced, or altered biosynthesis of a terpenoid in a plant can be engineered, which results in plant products with greater benefit for human use (55). The enhanced production of terpenoids both *in planta* and in microbial fermentations may provide a renewable alternative feedstock for petroleum-based industrial materials and fuels.

An active area of research is the directed evolution of TPS by site-directed mutagenesis of amino acids that are important in influencing product outcomes. These studies not only allow us to understand how the diversity of terpenoids evolved from gene duplication and neofunctionalization, but also allow us to modify multiproduct TPS to favor one of the wild type products or to force the enzyme into producing unnatural products. Both techniques have use in the large-scale production of medicinally or industrially important terpenoids.

REFERENCES

1. Conolly JD, Hill RA. Dictionary of Terpenoids. 1991. Chapman & Hall, London.
2. Gershenzon J, Dudareva N. The function of terpene natural products in the natural world. Nat. Chem. Biol. 2007;3:408–414.
3. Keeling CI, Bohlmann J. Genes, enzymes and chemicals of terpenoid diversity in the constitutive and induced defence of conifers against insects and pathogens. New Phytol. 2006;170:657–675.

4. Dudareva N, Pichersky E, Gershenzon J. Biochemistry of plant volatiles. Plant Physiol. 2004;135:1893–1902.

5. Cane DE, ed. Isoprenoids, Including Carotenoids and Steroids. 1999. Elsevier, London.

6. Aubourg S, Lecharny A, Bohlmann J. Genomic analysis of the terpenoid synthase (AtTPS) gene family of *Arabidopsis thaliana*. Mol. Genet. Genomics 2002;267: 730–745.

7. Goff SA, Ricke D, Lan T-H, Presting G, Wang R, Dunn M, Glazebrook J, Sessions A, Oeller P, Varma H. et al. A draft sequence of the rice genome (*Oryza sativa* L.ssp. *japonica*). Science 2002;296:92–100.

8. Tuskan GA, Difazio S, Jansson S, Bohlmann J, Grigoriev I, Hellsten U, Putnam N, Ralph S, Rombauts S, Salamov A. et al. The genome of black cottonwood, *Populus trichocarpa* (Torr. & Gray).Science 2006;313:1596–1604.

9. Jaillon O, Aury J-M, Noel B, Policriti A, Clepet C, Casagrande A, Choisne N, Aubourg S, Vitulo N, Jubin C. et al. The grapevine genome sequence suggests ancestral hexaploidization ini major angiosperm phyla. Nature 2007;449:463–467.

10. Bohlmann J, Meyer-Gauen G, Croteau R. Plant terpenoid synthases: Molecular biology and phylogenetic analysis. Proc. Natl. Acad. Sci. U.S.A. 1998;95:4126–4133.

11. Martin DM, Fäldt J, Bohlmann J. Functional characterization of nine Norway spruce TPS genes and evolution of gymnosperm terpene synthases of the TPS-d subfamily. Plant Physiol. 2004;135:1908–1927.

12. Ro D-K, Ehlting J, Keeling CI, Lin R, Mattheus N, Bohlmann J. Microarray expression profiling and functional characterization of AtTPS genes: Duplicated *Arabidopsis thaliana* sesquiterpene synthase genes At4g13280 and At4g13300 encode root-specific and wound-inducible (Z)- γ-bisabolene synthases. Arch. Biochem. Biophys. 2006;448:104–116.

13. Peters RJ. Uncovering the complex metabolic network underlying diterpenoid phytoalexin biosynthesis in rice and other cereal crop plants. Phytochemistry 2006;67: 2307–2317.

14. Martin DM, Bohlmann J. Identification of *Vitis vinifera* (-)-alpha-terpineol synthase by *in silico* screening of full-length cDNA ESTs and functional characterization of recombinant terpene synthase. Phytochemistry 2004;65:1223–1229.

15. Croteau RB, Davis EM, Ringer KL, Wildung MR. (−)-Menthol biosynthesis and molecular genetics. Naturwissenschaften 2005;92:562–577.

16. Ro D-K, Paradise EM, Ouellet M, Fisher KJ, Newman KL, Ndungu JM, Ho KA, Eachus RA, Ham TS, Kirby J. et al. Production of the antimalarial drug precursor artemisinic acid in engineered yeast. Nature 2006;440:940–943.

17. Croteau R, Ketchum REB, Long RM, Kaspera R, Wildung MR. Taxol biosynthesis and molecular genetics. Phytochem. Rev. 2006;5:75–97.

18. Keeling CI, Bohlmann J. Diterpene resin acids in conifers. Phytochemistry 2006;67: 2415–2423.

19. Lichtenthaler HK. The 1-deoxy-D-xylulose-5-phosphate pathway of isoprenoid biosynthesis in plants. Ann. Rev. Plant Physiol. Plant Mol. Biol. 1999;50:47–65.

20. Takahashi S, Koyama T. Structure and function of *cis*-prenyl chain elongating enzymes. Chem. Rec. 2006;6:194–205.

21. Miller B, Oschinski C, Zimmer W. First isolation of an isoprene synthase gene from poplar and successful expression of the gene in *Escherichia coli*. Planta. 2001;213: 483–487.

22. Sasaki K, Ohara K, Yazaki K. Gene expression and characterization of isoprene synthase from *Populus alba*. FEBS Letters 2005;579:2514–2518.

23. Sharkey TD, Yeh S, Wiberley AE, Falbel TG, Gong D, Fernandez DE. Evolution of the isoprene biosynthetic pathway in kudzu. Plant Physiol. 2005;137:700–712.

24. Guenther A, Karl T, Harley P, Wiedinmyer C, Palmer PI, Geron C. Estimates of global terrestrial isoprene emissions using MEGAN (Model of Emissions of Gases and Aerosols from Nature). Atmos. Chem. Phys. Dis. 2006;6:3181–3210.

25. Behnke K, Ehlting B, Teuber M, Bauerfeind M, Louis S, Hänsch R, Polle A, Bohlmann J, Schnitzler JP, Transgenic, non-isoprene emitting poplars don't like it hot. Plant J. 2007;51:485–499.

26. Loivamäki M, Gilmer F, Fischbach RJ, Sörgel C, Bachl A, Walter A, Schnitzler JP. Arabidopsis, a model to study biological functions of isoprene emission? Plant Physiol. 2007;144:1066–1078.

27. Fisher AJ, Baker BM, Greenberg JP, Fall R. Enzymatic synthesis of methylbutenol from dimethylallyl diphosphate in needles of Pinus sabiniana. Arch. Biochem. Biophys. 2000;383:128–134.

28. Wise ML, Croteau R. Monoterpene biosynthesis. In: Isoprenoids, Including Carotenoids and Steroids. Cane DE, ed. 1999. Elsevier, London. pp. 97–153.

29. Hyatt DC, Youn B, Zhao Y, Santhamma B, Coates RM, Croteau RB, Kang C. Structure of limonene synthase, a simple model for terpenoid cyclase catalysis. Proc. Natl. Acad. Sci. U.S.A. 2007;104:5360–5365.

30. Mahmoud SS, Croteau RB. Metabolic engineering of essential oil yield and composition in mint by altering expression of deoxyxylulose phosphate reductoisomerase and menthofuran synthase. Proc. Natl. Acad. Sci. U.S.A. 2001;98:8915–8920.

31. Mahmoud SS, Croteau RB. Menthofuran regulates essential oil biosynthesis in peppermint by controlling a downstream monoterpene reductase. Proc. Natl. Acad. Sci. U.S.A. 2003;100:14481–14486.

32. Bertea CM, Freije JR, van der Woude H, Verstappen FW, Perk L, Marquez V, De Kraker JW, Posthumus MA, Jansen BJ, de Groot A. et al. Identification of intermediates and enzymes involved in the early steps of artemisinin biosynthesis in *Artemisia annua*. Planta Med. 2005;71:40–47.

33. Mercke P, Bengtsson M, Bouwmeester HJ, Posthumus MA, Brodelius PE. Molecular cloning, expression, and characterization of amorpha-4,11-diene synthase, a key enzyme of artemisinin biosynthesis in *Artemisia annua* L. Arch. Biochem. Biophys. 2000;381:173–180.

34. Wallaart TE, Bouwmeester HJ, Hille J, Poppinga L, Maijers NC. Amorpha-4,11-diene synthase: cloning and functional expression of a key enzyme in the biosynthetic pathway of the novel antimalarial drug artemisinin. Planta 2001;212:460–465.

35. Chang MCY, Eachus RA, Trieu W, Ro D-K, Keasling JD. Engineering *Escherichia coli* for production of functionalized terpenoids using plant P450s Nat. Chem. Biol. 2007;3:274–277.

36. Shiba Y, Paradisea EM, Kirbya J, Ro D-K Keasling JD. Engineering of the pyruvate dehydrogenase bypass in *Saccharomyces cerevisiae* for high-level production of isoprenoids. Metabol. Eng. 2007;9:160–168.

37. Wani MC, Taylor HC, Wall ME, Coggan P, McPhail AT. The isolation and structure of taxol, a novel antileukemic and antitumor agent from *Taxus brevifolia*. J.Am. Chem. Soc. 1971;93:2325–2327.

38. Xiao Z, Itokawa H, Lee K-H. Total synthesis of taxoids. In: *Taxus –The genus Taxus*. Itokawa H, Lee K-H, eds. 2003. Taylor & Francis, London. pp. 245–297.

39. Walker K, Croteau R. Taxol biosynthetic genes. Phytochemistry 2001;58:1–7.

40. Langenheim JH. Plant Resins: Chemistry, Evolution, Ecology and Ethnobotany. 2003. Timber Press, Portland, OR.

41. Ro D-K, Arimura G, Lau SY, Piers E, Bohlmann J. Loblolly pine abietadienol/ abietadienal oxidase PtAO (CYP720B1) is a multifunctional, multisubstrate cytochrome P450 monooxygenase. Proc. Natl. Acad. Sci. U.S.A. 2005;102:8060–8065.

42. Tholl D, Boland W, Hansel A, Loreto F, Röse US, Schnitzler JP. Practical approaches to plant volatile analysis. Plant J. 2006;45:540–560.

43. Hanson Jr. The development of strategies for terpenoid structure determination. Nat. Prod. Rep. 2001;18:607–617.

44. Adams RP. Identification of Essential Oil Components by Gas Chromatography/Mass Spectrometry. 2007. Allured Publishing Co., Carol Stream, IL.

45. Hochmuth DH. MassFinder 3. 2007. Hamburg, Germany.

46. Acree T, Arn H. Flavornet and Human Odor Space. *http://www.flavornet.org*.

47. El-Sayed AM. The Pherobase: Database of Insect Pheromones and Semiochemicals. *http://www.pherobase.com*.

48. Shurig V. Separation of enantiomers by gas chromatography. J. Chrom. A 2001; 906:275–299.

49. König WA, Hochmuth DH. Enantioselective gas chromatography in flavor and fragrance analysis: strategies for the identification of known and unknown plant volatiles. J. Chrom. Sci. 2004;42:423–439.

50. Keeling CI, Ngo HT, Benusic KD, Slessor KN. Preparative chiral liquid chromatography for enantiomeric separation of pheromones. J. Chem. Ecol. 2001;27:487–497.

51. Nicolaou KC, Snyder SA. Chasing molecules that were never there: misassigned natural products and the role of chemical synthesis in modern structure elucidation. Angew. Chem. Int. Ed. Engl. 2005;44:1012–1044.

52. Maimone TJ, Baran PS. Modern synthetic efforts toward biologically active terpenes. Nat. Chem. Biol. 2007:3(7):396–407.

53. Tabata H. Paclitaxel production by plant-cell-culture technology. Adv. Biochem. Engin. Biotechnol. 2004;7:1–23.

54. Withers ST, Keasling JD. Biosynthesis and engineering of isoprenoid small molecules. Appl. Microbiol. Biotechnol. 2007;73:980–990.

55. Townsend BJ, Llewellyn DJ. Reduced terpene levels in cottonseed add food to fiber. TRENDS in Biotech. 2007;25:239–241.

6

POLYKETIDES IN FUNGI

THOMAS J. SIMPSON AND RUSSELL J. COX

School of Chemistry, University of Bristol, Bristol, United Kingdom

Fungi produce a wide variety of biologically active compounds. Among these compounds, the polyketides form a large and structurally diverse group. These compounds are synthesized by highly programmed, large iterative multifunctional proteins, which are called the polyketide synthases. This review describes the structure and biosynthesis of polyketide fungal metabolites and highlights recent work on the links between gene sequence, protein architecture, and biosynthetic programming for fungal polyketide synthases.

Polyketides have long been recognized as one of the most important classes of secondary metabolites (1). They occur in plants, bacteria, and marine organisms as well as in fungi. Fungal polyketides vary from the simplest monocyclic aromatic compounds, for example, orsellinic (**1**) and 6-methylsalicylic (6-MSA) (**2**) acids to polycyclic aromatics such as citrinin (**3**), alternariol (**4**), islandicin (**5**), deoxyherqueinone (**6**), and norsolorinic acid (**7**). Although initially associated with the formation of aromatic compounds, many polyketides are nonaromatic (e.g., the macrolide decarestrictine D (**8**), long-chain polyfunctional molecules exemplified by T-toxin (**9**) and the decalins, lovastatin (**10**), and compactin (**11**)). Many other metabolites consist of an aromatic ring attached to a more highly reduced moiety (e.g., zeralanenone (**12**), dehydrocurvularin (**13**), and monocerin (**14**)). Additional diversity results from extensive oxidative metabolism of preformed polyketide structures (e.g., penicillic acid (**15**) and patulin (**16**)), which are formed from cleavage and rearrangement of 6-MSA and orsellinic acid, respectively, and indeed ring cleavage is a very common feature with

---

Natural Products in Chemical Biology, First Edition. Edited by Natanya Civjan.
© 2012 John Wiley & Sons, Inc. Published 2012 by John Wiley & Sons, Inc.

the potent hepatotoxin aflatoxin B_1 (17) being derived by extensive ring cleavages and rearrangements of norsolorinic acid (7). Other metabolites contain a polyketide-derived moiety as part of a larger molecule whose biosynthesis is other than polyketide. A classic example is mycophenolic acid (18), in which the branched carboxylic side chain is derived via a cleaved farnesyl moiety. Other compounds of mixed terpenoid-polyketide origin include the mycotoxin viridicatumtoxin (19). Again the origins of the polyketide-derived moiety may be disguised as a result of extensive metabolism as observed in the meroterpenoid metabolites austin (20) and paraherquonin (21), which are all derived via 3,5-dimethylorsellinic acid (2). The xenovulenes (22)–(24) are an interesting group where it has been shown that the cyclopentananone, benzenoid, and tropolone moieties all have a common biosynthetic origin via ring expansion and ring contaction of 3-methylorcinaldehyde. Other groups contain Krebs' cycle intermediates, for example, the tetronic acid, carlosic acid (25), and the squalestatins (29) or amino-acid derived moieties (e.g., fusarin C (26)).

6.1 BIOLOGICAL PROPERTIES

Another important feature of fungal polyketides is their vast range of biological activities both beneficial and harmful. Thus, griseofulvin (27) was one of the first effective antifungal agents. Penicillic acid (15), which was discovered shortly after the penicillins, is also a powerful antibiotic, but unfortunately it proved too toxic for clinical use. Mycophenolic acid (18) has been "rediscovered" as an immunosuppressive agent. The strobilurins (e.g., (28)), although not themselves used in the field formed the basis for the development of the widely used methoxyacrylate group of antifungal agents and, of course the statins, as represented by lovastatin (10), are among the most widely prescribed drugs for control of cholesterol levels and associated heart disease. As a cursory inspection of their structures would suggest, the squalestatins (e.g., (29)), are effective inhibitors of squalene synthase though their early promise as cholesterol-lowering clinical candidates declined because of inherent toxicity. They contain two separate polyketide chains linked to oxaloacetate. Overall, fungal metabolites and their pharmaceutical and agrochemical derivatives have total sales of many tens of billions of pounds annually. In addition to these beneficial effects, the large group of mycotoxins represented first and foremost by aflatoxin B_1 (17) and others such as the fuminosins (30), zearaleneone (12), citrinin (3), ochratoxins (e.g., (31)), are the cause of many problems in both animal and human health, and spoilage of both growing crops and stored foodstuffs through contamination by mycotoxin producing fungi is a cause of major economic losses worldwide.

6.2 BIOSYNTHESIS

Although diverse in structure, the class is defined by the common biosynthetic origin of the carbon atoms: These atoms are derived from the CoA thiolesters of small carboxylic acids, such as acetate and malonate. As long ago as 1953, Birch realized that polyketide biosynthesis is related to fatty acid biosynthesis and some of the earliest applications of radioisotopes to natural product biosynthesis were to fungal polyketide metabolites, where the ease of fermentation and isolation of metabolites in pure form, and relatively efficient uptake of simple labeled precursors facilitated the work. In more recent years, fungal metabolites in general, and polyketides in particular, were the focus of the rapidly expanding applications of stable isotope labeling in the 1970s and 1980s beginning with incorporations of singly ^{13}C-labeled precursors with analysis of regiospecifity of labeling being greatly facilitated by the contemporaneous development of Fourier Transform methods and their application to ^{13}C NMR. The application of doubly ^{13}C-labeled precursors led to the concept of bond labeling, which allowed *inter alia* the mode of cyclization of linear polyketide precursors into polycyclic molecules, bond fragmentation, and rearrangements processes to be detected for the first time through analysis of the resulting ^{13}C-^{13}C coupling patterns. This method was then rapidly followed by applications of isotope induced shifts in ^{13}C NMR which allowed indirect detection of ^{2}H and ^{18}O labels and the use of direct ^{2}H NMR. These new methods (2) allowed stereochemistry and regiochemistry of labeling to be detected and in particular permissible levels of oxidation and reduction in otherwise undetectable biosynthetic intermediates to be determined. Along with similar work with bacterial polyketides, this method laid the basis for the ideas of the processive mode of polyketide biosynthesis to be established. It was a major step because it changed fundamentally the idea that polyketide chains were assembled in their entirety and then subjected to necessary reductive modifications to the simple, in retrospect, idea that these changes occur concomitant with chain elongation rather than post-elongation. Although it is still not uncommon for these classic "Birch" fully oxygenated polyketide intermediates to be invoked it is now evident that in most polyketide metabolites, these have no reality. The concept of processive polyketide assembly and modification brought polyketide biosynthesis even closer to the process of fatty acid biosynthesis in which full reductive processing in each cycle of chain condensation and elongation is the norm. The rapid developments in understanding of the molecular genetics of polyketide biosynthesis particularly in bacteria in the 1990s were fully consistent with the processive mode. Understanding of the genetics of fungal polyketide biosynthesis still lags behind that of bacterial polyketides and the remainder of this chapter will provide a brief overview of current understanding.

6.2.1 Polyketide Assembly

The basic assembly cycle for both polyketide and fatty acid biosynthesis is shown in Fig. 6.1 in which a starter unit, normally acetate is transferred to the ketosynthase (KS) or condensing enzyme which catalyzes a decarboxylative condensation with malonate, bound after after malonyl tranferase (MT) catalyzed malonylation to the acyl carrier protein (ACP). During fatty acid biosynthesis, the resulting β-ketothiolester (**32**) is subjected to additional chemical processing while attached

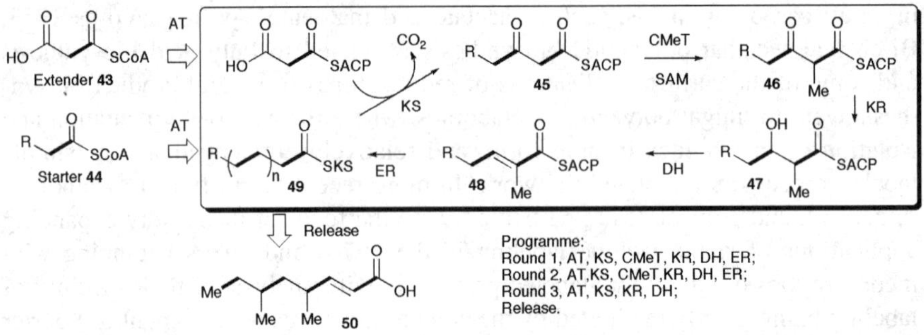

Figure 6.1 Generic polyketide assembly pathway reactions catalyzed by iterative fungal polyketide synthases. The assembly sequence for the squalesatin tetraketide intermediate (37) is shown for illustration.

to the terminal thiol of the ACP: first: it is reduced by a β-ketoacyl reductase (KR) to a secondary alcohol (34), which then undergoes a dehydratase (DH)-catalyzed dehydration to form an αβ-unsaturated thiolester (35), and final enoyl reduction (ER) yields a fully saturated thiolester (36). Fungal PKS deploy all these reactions, but additionally the chain can be methylated, using a methyl group from S-adenosylmethionine (SAM). This reaction probably occurs after KS, which gives an α-methyl-β-ketothiolester (33). During the biosynthesis of palmitic acid (C_{16}), there are seven cycles of these reactions. The final reaction of FAS is hydrolysis of the thiolester by a dedicated thiolesterase (TE). Apart from the capacity for C-methylation, fungal PKSs in common with other PKSs have the ability to use the condensation and reductive cycle in a highly controlled manner to produce polyketide intermediates in which no reductive modification has occurred to give a a classic poly-β-ketothiolester, or more generally "reduced" or "processive" polyketide intermediates in which a complete spectrum of reduction and/or C-methylation has occurred in each condensation cycle as indicated in Fig. 6.1 for the squalesatin tetraketide (37) synthase. Another variation is that a range of alternate starter units can be used [e.g., hexanoate in the case of nor-solorinic acid (7) or benzoate in the case of the strobilurins or squalestatins (28) and (29)].

6.2.2 Enzymology

The understanding of the relationship between FAS and PKS proteins, the application of molecular genetics, and more latterly genomics, has greatly facilitated the discovery and understanding of polyketide synthases from diverse sources. The homology in catalytic function between FAS and PKS enzymes is preserved in their respective gene sequences. It is now clear that fungal PKSs belong to the class of Type 1 iterative syntheses represented by mammalian FASs. Type 1 FAS proteins are large multifunctional proteins in which single (or occasionally two) peptides contain the sequences for KS, ACP, AT, KR, DH, ER, and TE activities—these catalytic functions are carried out by particular functional

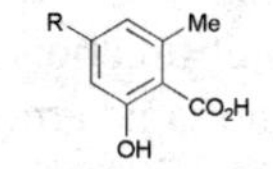

orsellinic acid (R = OH) 1
6-methylsalicylic acid (R = H) 2

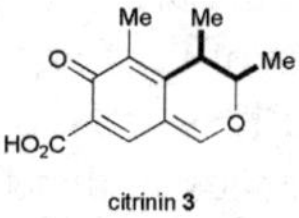

citrinin 3

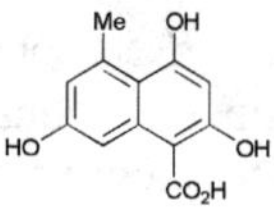

alternariol 4

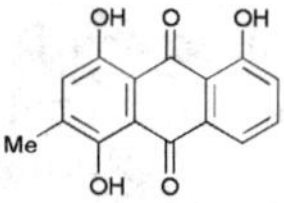

islandicin 5

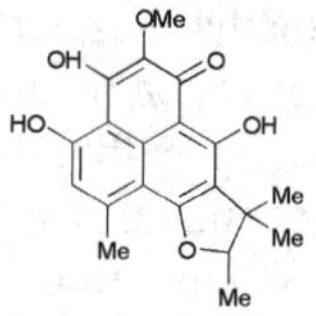

deoxyherqueinone 6

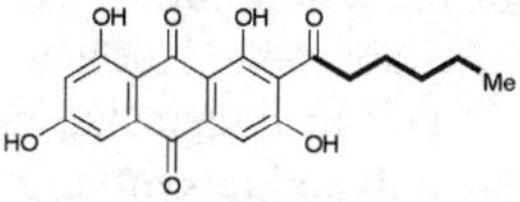

norsolorinic acid 7

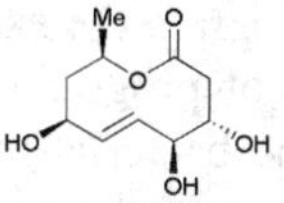

decarestrictine D 8

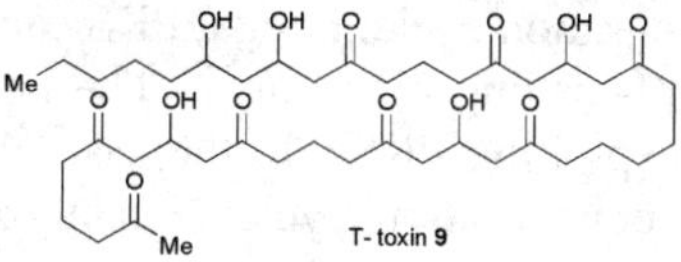

T- toxin 9

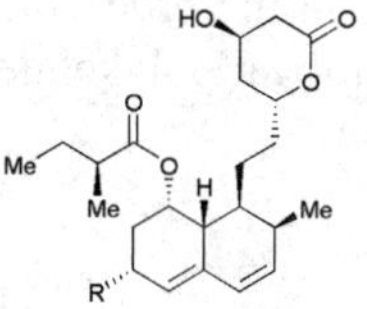

lovastatin 10 (R = Me)
compactin 11 (R = H)

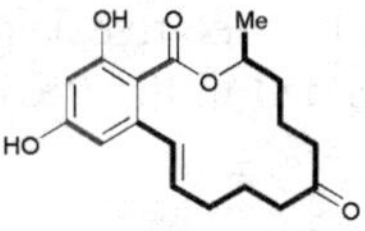

zearalenone 12

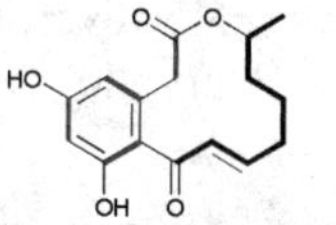

dehydrocurvularin 13

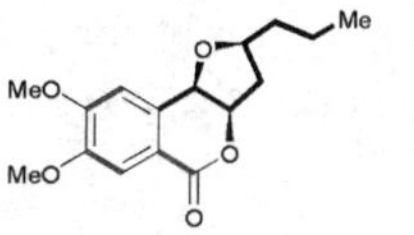

monocerin 14

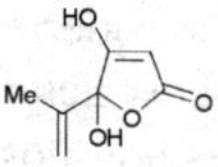

penicillic acid 15

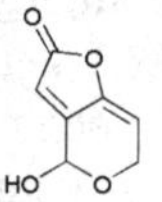

patulin 16

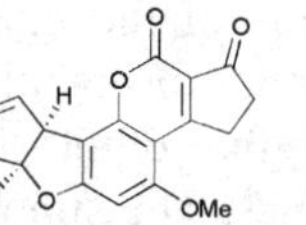

aflatoxin B₁ 17

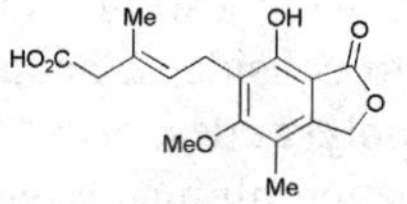

mycophenolic acid 18

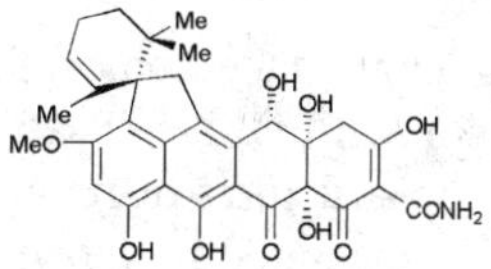

viridicatumtoxin 19

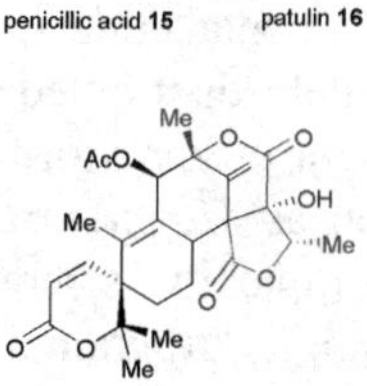

austin 20

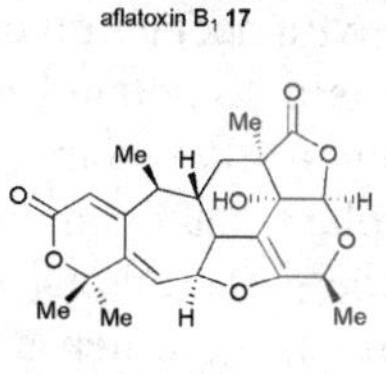

paraherquonin 21

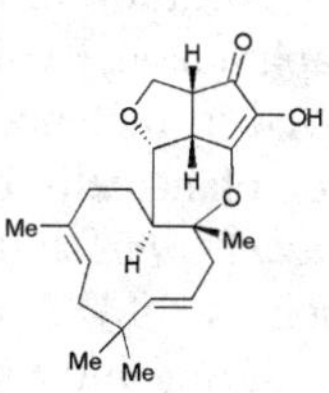

xenovulene A 22

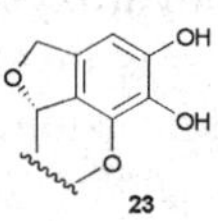

23

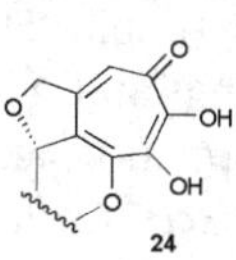

24

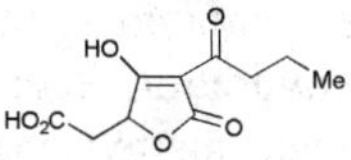

carlosic acid 25

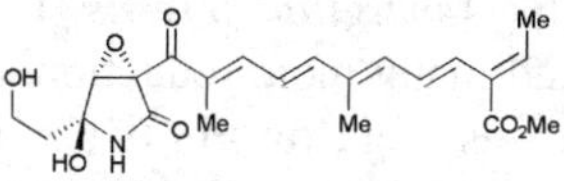

fusarin C 26

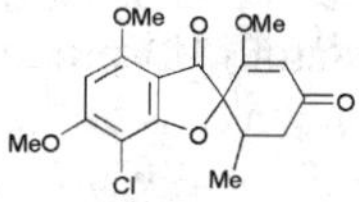

griseofulvin 27

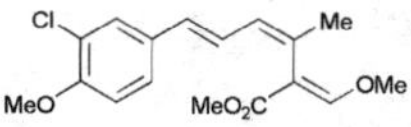

strobilurin B 28

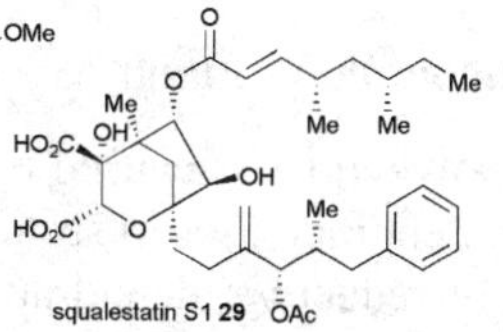

squalestatin S1 29

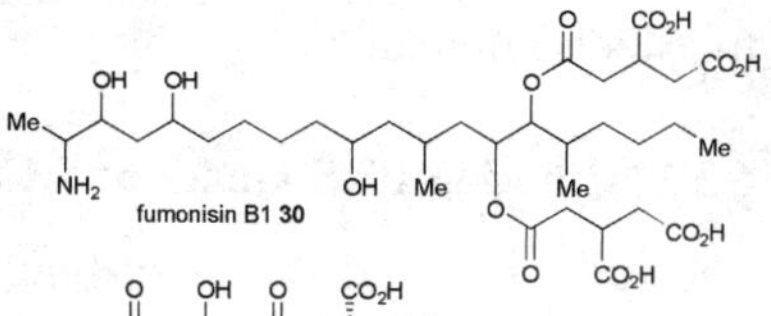

fumonisin B1 30

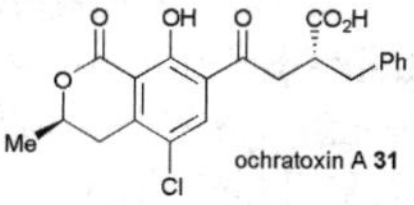

ochratoxin A 31

domains. Similarly, the genes for Type 1 FAS proteins are correspondingly large single open reading frames, and Type 1 PKSs consist of very large multifunctional proteins with individual functional domains. Thus, PKSs use much the same array of chemical reactions as FAS—but the key differences is that of *programming*: FASs have to control chain length (i.e., the number of extensions), but PKS can additionally control starter unit selection and the extent of reduction during each condensation cycle. Fungal PKS can program the extent of chain methylation and the off-loading mechanism. The issue of programming is key to understanding and exploiting PKSs. In the case of the bacterial modular polyketide synthases, each condensation cycle is catalyzed by a discreet module containing all the catalytic domains required. In this case, the program is explicit in the order and composition of the modules. However, for the iterative Type 1 fungal polyketide synthases, the programme is cryptic—encoded in the PKS itself.

6.3 FUNGAL POLYKETIDE SYNTHASES

Fungi make some of the simplest and some of the most complex polyketides known (3). It is useful to consider a heirarchy of complexity when considering the structures of fungal polyketides, because the complexity in chemical structure is generated by enzymes and ultimately genes. The simplest structures are those such as orsellinic acid (**1**). Addition of an extra acetate gives a pentaketide such as 1,3,6,8-tetrahydroxynaphthalene, which is a compound widely distributed in fungi and which is involved in melanization—a key component of apressorium formation and invasion of plant cells by plant pathogen such as *Magnaporthe grisea, Colletotrichum lagenarium*, and others. Additional complexity is represented by compounds such as 6-MSA acid (**2**), in which a single, programmed reduction reaction occurs during biosynthesis. More reduction is then observed in compounds such as T-toxin (**19**) produced by *Cochliobolus heterostrophus* and lovastatin (**10**) produced by *Aspergillus terreus*. In these compounds, many more carbon atoms are used and many more reduction and dehydration reactions occur. Additionally in many cases (e.g., fusarin C (**26**)), pendant methyl groups have been added from the *S*-methyl of methionine, catalyzed by a *C*-methyl transferase domain.

6.3.1 Linking PKS Genes and Compounds in Fungi

The huge structural variety of fungal polyketides is caused by differences in programming of their PKS proteins—apparent increases in structural complexity are caused by increasing use and control of reductive, dehydrative, and methylating steps by the PKS. This inrease must be beacuse of differences in PKS protein sequence and structure. This fact has been exploited in the development of rapid methods for the cloning of fungal PKS genes associated with the biosynthesis of particular fungal polyketide types.

Bingle et al. (3) realized that these subtle protein sequence differences should be reflected in DNA sequence, and polymerase chain reaction (PCR) primers could be designed to amplify fragments of fungal PKS genes selectively from fungal genomic DNA (or cDNA). In early work in this area, they hypothesized that fungal polyketides could be grouped into two classes: nonreduced (NR) compounds such as orsellinic acid (**1**), norsolorinic acid (**7**), and 1,3,6,8-tetrahydroxynaphthalene, and partially reduced (PR) compounds, such as 6-MSA (**2**). At the time, very few fungal PKS genes were known, and based on very limited sets of sequences, they designed degenerate PCR primers that were complimentary to conserved DNA sequences in the KS domains in fungal PKS responsible for the biosynthesis of NR and PR compounds. Later, the same analysis was extended to the KS domains of highly reduced (HR) compounds, such as lovastatin (**10**), when DNA sequence data became available for the lovastatin nonaketide and diketide synthases (LNKS and LDKS, respectively) (4). The availability of these sequences also allowed the development of selective PCR primers for CMeT domains.

This sequence analysis has been significantly extended as genomic approaches have been applied to fungi recently (5). Full genome sequences have now been obtained for more than a dozen fungi. In each organism, many PKS genes have been discovered. For example, *Aspergillus niger* contains 34 PKS genes, so several hundred fungal PKS genes are known. Sequence comparison of all these new PKS genes, however, shows that the three classes of fungal PKS genes predicted by Bingle et al. (3) are the same three classes observed in the most recent sequence comparisons (5). Despite the fact that so many fungal PKS genes have been discovered, however, relatively few genes have been definitively linked to the biosynthesis of specific metabolites. Because the NR, PR, and HR nomenclature is useful for describing both the chemical products and their cognate genes, the state of knowledge of fungal PKS is reviewed in this way.

6.4 FUNGAL NR-PKS

The tetraketide orsellinic acid (**1**) is the simplest tetraketide, which requires no reductions during its biosynthesis. One of the first discovered fungal PKS, orsellinic acid synthase (OSAS) was isolated from *Penicillium madriti* in 1968. Despite the early work with the protein, however, the OSAS-encoding gene has not yet been discovered, and nothing is known of the catalytic domains or their organization. However, genes involved in the biosynthesis of several other nonreduced polyketides are now known, and a general pattern of domain organization has emerged. In all known cases, these genes encode Type 1 iterative PKS proteins. At the N-terminus, a domain is present that seems to mediate the loading of a *starter unit* (Fig. 6.2a). It seems that the starter unit can derive from either a dedicated FAS, another PKS, or an acyl CoA. The starter unit loading domain (SAT) is followed by typical KS and AT domains responsible for chain extension and malonate loading. Beyond the AT is a conserved domain known as the product template (PT) domain. This domain is followed by an ACP. Some NR-PKS seem to terminate after the ACP, but many feature a diverse range of different

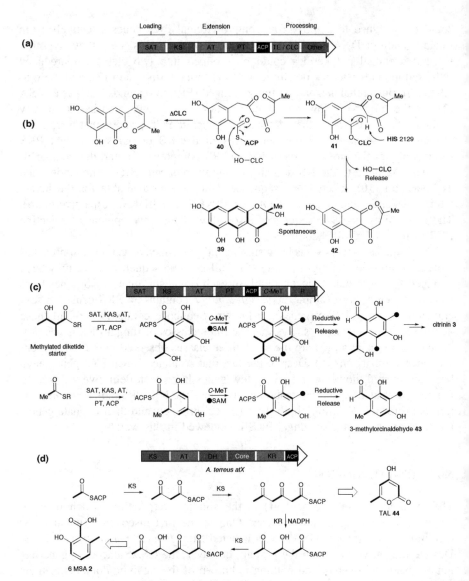

Figure 6.2 (a) General architecture of NR PKS genes; proposed mechanism of, (b) the WAS CLC domain, (c) the *Monascus purpureus* citrinin PksCT and *A. strictum* MOS synthases; and (d) domain architecture of *A. terreus* and *P. patulum* MSAS, and proposed mechanism of 6-MSA (**2**) and TAL (**44**).

domains that include cyclases, methyl transferases, and reductases. Thus, it seems that these synthases are arranged with an N-terminal *loading component*; a central *chain extension component* that consists of KS, AT, PT and ACP domains; and a *C*-terminal *processing component*.

6.4.1 NR-PKS Loading Component

Feeding experiments with isotopically labeled precursors have shown that many NR fungal polyketides are formed by the use of "advanced" starter units. In the classic case of norsolorinic acid (**7**) biosynthesis, it has long been known that hexanoate forms the starter unit. Differential specific incorporation of acetate into the early and late positions in compounds such as citrinin (**3**) have been used to argue that these compounds may have been formed by more than one PKS so that one PKS makes an advanced starter unit, which is passed to a second PKS for additional extension.

Evidence for this suggestion is growing. Townsend has defined the molecular basis for the ability of NR-PKS to use starter units derived from other FAS or PKS systems. Two genes in the aflatoxin biosynthetic gene cluster of *Aspergillus parasiticus* (*stcJ* and *stcK*) encode the α and β components of a typical fungal FAS (HexA and HexB). Clustering of these FAS genes with the NSAS PKS suggested that HexA and HexB probably produced hexanoate for use as the norsolorinic acid starter unit. The protein complex formed between NSAS, HexA, and HexB, which is known as NorS, was isolated and characterized (6). This 1.4-MDa protein complex synthesizes norsolorinic acid (**7**) from malonyl CoA, acetyl CoA, and NADPH. Townsend showed that hexanoyl CoA is not a free intermediate produced by NorS, which suggests that hexanoate produced as an ACP derivative by the HcxA/HcxB FAS must be passed directly to NSAS. In the absence of NADPH, hexanoate cannot be formed by the FAS components; thus, no norsolorinic acid is formed.

The N-terminal domain of NSAS posesses canonical acyl transferase sequence motifs so that this domain could be a candidate for the required starter unit transferase. This domain was cloned and expressed along with the ACP and was shown to catalyze the transfer of hexanoate from CoA onto the the ACP. Site-directed mutagenesis experiments to remove the proposed catalytic cysteine of the transferase resulted in loss of catalytic activity. The N-terminal transferase showed significant selectivity for the transfer of hexanoate over longer or shorter acyl chains (7). Thus, the N-terminal domain of NSAS acts as a starter unit:ACP transacylase (SAT) component.

Sequence comparison with other known NR PKS suggests that such SAT domains are common. In the few cases where the PKS sequence has been cor-related with product structure the presence of SAT domains now explains prior results from feeding experiments, which suggested the use of advanced starter units. For example it is now known (8) that two PKS genes are involved in the biosynthesis of zearalenone (**12**)—one of these is a HR-PKS and probably provides a highly reduced hexaketide as a starter unit. The second zearalenone PKS is a NR-PKS possessing an N-terminal SAT domain, which likely loads the hexaketide ready for three further extensions. Most NR PKS seem to possess potential SAT domains whether they require an acetate starter unit or not. For example, polyketide synthases involved in the biosynthesis of YWA1 (**39**) (WAS) and tetrahydroxynaphthalene (THNS). In the case of THNS from *C. lagenarium*, *in vitro* experiments have implied that the purified protein uses malonyl CoA as the starter unit. The THNS SAT domain may therefore be involved with loading and decarboxylation of malonate to use as a starter unit, much as the bacterial Type 2 KS_β component does.

6.4.2 NR PKS Chain Extension Component

The extension components of NR polyketide synthases consist of KS, AT, PT, and ACP domains. Sequence analysis of PT domains (450–550 residues) from a range of NR PKS genes in which the chemical products are known suggests that it is conceiveable that the PT domain is involved in chain-length determination. Comparison of the PT domains from the *Acremonium strictum* PKS1, citrinin PKS, zearalenone PKS-B, NSAS, sterigmatocystin PKS, dothistromin PKS, THNS from *C. lagenarium* and *Wangiella dermatitidis*, and WAS suggests that these domains group into clades that correspond with chain length (9). Thus the citrinin, ASPKS1, and zearalenone-B groups correspond to tetraketide synthases; the NSAS, sterigmatocystin, and dothistromin PKS are all octaketide synthases; the THNS are hexaketide synthases; and the wA synthase (WAS) forms an outgroup, which is a heptaketide.

The domains found after the chain extension components at the C-termini of fungal NR-PKS are highly varied. These components include putative Claisencyclase/thiolesterases (CLC/TE), *C*-methyl transferases (*C*-MeT), reductases (R), and additional ACP domains. Recent work indicates that these processing components act *after* chain assembly to modify either a poly-keto or a cyclized intermediate.

6.4.3 CLC/TE Domains

The first fungal NR PKS gene to be cloned was *Aspergillus nidulans* wA (encoding WAS). Limited domain analysis was carried out to determine the presence of KAS, AT, and ACP domains. Later, it was shown that *wA* also possess a TE domain, which form one of the most common processing components of NR PKS. The TE domain can either operate as a standard thiolesterase or be involved in a cyclization-release mechanism. Fujii et al. (10) expressed *A. nidulans wA* (encoding WAS) in *Aspergillus oryzae*. Initially, the expression strain produced the isocoumarin (**38**), which indicated that WAS is a heptaketide synthase (Fig. 6.2b). It was then realized that the expression construct used in this experiment had a deletion that resulted in the expressed WAS missing the final 67 amino acids of the C-terminal TE domain. When the complete *wA* gene was expressed, however, the heptaketide naphthopyrone YWA1 (**39**) was produced (10). A series of experiments that involve step-wise shortening of the C-terminus of WAS showed that deletion of as few as 32 amino acids resulted in production of the isocoumarin (**38**). Site-directed mutagenesis of a conserved serine and histidine in the C-terminal domain also resulted in a switch from naphthopyrone production to citreoisocoumarin production.

Isotopic feeding experiments using ^{13}C-labeled acetate indicated the folding pattern shown in Fig. 6.2b for the naphthopyrone (**39**). Thus, both the isocoumarin (**38**) and the naphthopyrone (**39**) must result from the cyclization of the common intermediate (**40**). This finding suggests that the WAS chain extension component produces a heptaketide and catalyzes the cyclization and aromatization of the first ring. The C-terminal domain must therefore catalyze a second (Claisen) cyclization reaction to form (**42**). The involvement of conserved serine and histidine residues suggests involvement of the CLC-bound intermediate (**41**) shown in Fig. 6.2b. Thus, the TE domain has been renamed as Claisen Cyclase (CLC). These domains also occur in the known NSAS and THNS proteins in which the same chemistry must occur to provide the observed products.

6.4.4 C-MeT Domains

Few NR PKS are known to possess C-methylation domains, although many known fungal nonreduced polyketides are C-methylated, such as 3,5-dimethylorsellinic acid. A small group of NR PKS have been identified in genome sequences (5), which feature a C-MeT domain located after the ACP (Fig. 6.2c). The first correlation between a gene sequence and a compound came from citrinin (**3**), in which the PKS involved in citrinin biosynthesis in *Monascus ruber* was reported (11). Here, the C-MeT domain must be programmed because it acts twice during polyketide biosynthesis when a probable methylated diketide starter unit is extended. It is not yet clear whether the C-MeT domain acts during extension, after chain extension but before aromatization, or after aromatization. 1,3-Dihydroxyaromatics are known to tautomerize easily to keto forms, and it is conceivable that it could act as the nucleophile for the reaction with SAM.

6.4.5 R Domains

Reductases are currently rare as part of the processing component of NR PKS. Evidence for the role of these reductase domains has been obtained by the isolation of the PKS responsible for the formation of the tetraketide component found in xenovulene A (**22**). The PKS gene (MOS) was found to have SAT, KS, AT, C-MeT, and R domains (Fig. 6.2c), and heterologous expression of the gene in *A. oryzae* resulted (9) in a high yield of 3-methylorcinaldehyde (**43**). Although not described in the literature, sequence analysis of the citrinin PKS sequence discussed above shows that it also possesses a C-terminal thiolester reductase domain.

Similar domains are known from NRPS systems in which reductase domains are sometimes used as chain release mechanisms, which release an aldehyde or primary alcohol. In the case of MOS and citrinin biosynthesis the reductive release mechanism makes good sense as this provides the products with C-1 at the correct oxidation state (Fig. 6.2c).

6.5   FUNGAL PR-PKS

The domain structure of PR PKSs is much closer to mammalian FAS, with an N-terminal KS followed by AT, and DH domains (Fig. 6.2d). A so-called "core" domain follows the DH, which is followed by a KR and the PKS terminates with an ACP domain. The domain structure differs considerably from the NR PKS—no SAT domain or PT domain exists, and the PKS terminates after the ACP and seems not to require a TE/CLC domain. No obvious catalytic machinery exists for off-loading the product. The DNA sequence of the KS domain is distinguishable from NR-PKS and FAS KAS domains using selective PCR primers and DNA probes (3). Overall, however, catalytic domains closely match those of the mamalian FAS, although the "core" domain is different. Remarkably, the fungal MSAS genes are closely related to recently discovered *bacterial* genes for the synthesis of the nonreduced tetraketide orsellinic acid (1) (12) apart from lacking a ~450-amino acid region that encompasses the KR domain.

Although many PR PKS genes are known from genome sequencing projects, only three genes have been matched to chemical products—in all cases, the tetraketide 6-MSA (2) MSAS from *Penicellium patulum* is one of the smallest Type 1 PKS at 188 KDa. It proved relatively easy to isolate, and many of the earliest *in vitro* studies of the enzymology of any PKS were carried out with this enzyme (13, 14). The PKS is evidently programmed—acetate must be extended three times, and the KR must only act once, after the second extension. In the absence of NADPH, the KR reaction cannot occur, the tetraketide is not produced, and triketide lactone (TAL) (44) is produced instead (Fig. 6.2d). This reaction reveals that chain extension and reduction are linked, and it indicates that the KR must act during chain extension. Mammalian FAS also produces TAL (44) under the same circumstances.

Chain length determination seems to use a "counting mechanism" as in the case of the Type 2 actinorhodin PKS. Incubation of various acyl CoA starter units with malonyl CoA with MSAS in the absence of NADPH and acetyl CoA resulted in two chain extensions to produce the corresponding substituted TALs.

Moriguchi et al. (15) have carried out an interesting series of expression experiments in the yeast *Saccharoveyces cerivisiae* in which two copies of the MSAS gene can be expressed simultaneously. This inspection has allowed complementation experiments in which they show that up to 44 amino acids from the N-terminus can be removed and activity retained. However, at the *C*-terminus, deletion of as few as nine amino acids caused loss of activity because of removal of key ACP residues. Use of combinations of deletion mutants, with the knowledge that MSAS forms homotetramers, provided evidence that the ACP of one peptide chain must interact with the KS of another chain. In addition, a short core domain region was identified (15). This finding was essential for successful complementation and suggested that it acts as a motif required for subunit-subunit recognition similarly to the core region of mammalian FAS, which has been shown to mediate assembly of the synthase.

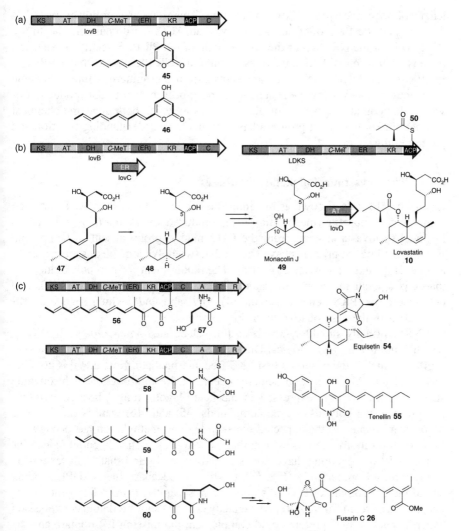

Figure 6.3 Expression of *lovB* in *A. nidulans*, (a) in the absence and, (b) the presence of lovC; (c) *lovD* catalyzed acyl transfer of *lovB* diketide (**50**) to monacolin J (**49**); (d) proposed roles of FUSS domains in the biosynthesis of "pre-fusarin" (**60**).

6.6   FUNGAL HR-PKS

The final class of fungal PKS produces complex, highly reduced compounds such as lovastatin (**10**), T-toxin (**19**), fumonisin B1 (**30**), and squalestatin (**29**). In all cases, these fungal compounds are produced by iterative Type 1 PKS. These PKSs have an N-terminal KS domain, followed by AT and DH domains (Fig. 6.3). In many cases, the DH is followed by a *C*-MeT domain. Some HR PKSs possess an ER domain, but those that lack it have a roughly equivalent length of sequence with no known function. This domain is followed by a KR domain, and the PKS often terminates with an ACP. No domain similar to the PT domain of the NR PKS or the core domain of the PR PKS seems to exist, and there is no N-terminal SAT domain as found in the NR PKS. As with the NRs and PR PKSs, many HR PKS genes are known from the numerous fungal genome sequences, but as yet few gene sequences have been linked to the production of known compounds. However, of the few cases in which both gene and chemical product are known, some progress has been made in understanding function and programming.

6.6.1 The Lovastatin Polyketide Synthases

Lovastatin (**10**) (also known as mevinolin) is produced by *A. terreus*. The related compound compactin (mevastatin) (**11**), produced by *Penicillium citrinum* is identical to lovastatin apart from the C-12 methyl group absent in compactin. Isotopic feeding experiments have shown that two polyketide chains are required: a nonaketide and a methylated diketide. The requirement for two polyketide syn-thases is evident in the gene clusters associated with biosynthesis of (**10**) and (**11**) where two PKS genes are found (4), *lovB* and *lovF*, which encode LNKS and LDKS in the case of lovastatin (Fig. 6.3b).

LNKS formally should synthesize a fully elaborated nonaketide such as (**47**), which could undergo a biological Diels-Alder reaction to form dihydromonacolin L (**48**)—which is the observed first PKS-free intermediate. Note that it is possible that the Diels Alder reaction actually occurs at the more activated hexaketide stage. When *lovB* was expressed in the heterologous fungal host *A. nidulans* (4) however, the polyunsaturated compounds (**45**) and (**46**) were isolated. These pyrones are related to the expected nonaketide (e.g., methylation has occurred at the correct position), but it is evident that reductions, specifically enoyl reductions and later keto reductions, have not occurred correctly and that chain extension has terminated prematurely (Fig. 6.3a), which is becauase the NADPH binding site of the ER domain of LNKS is impared. A separate ER encoding gene, *lovC*, occurs downstream from *lovB*, and Kennedy et al. (4) showed that in coexpression experiments the lovC protein could complement the missing ER domain and the expected dihydromonacolin L (**48**) was produced (Fig. 6.3b). It is evident that the lovC protein must interact with LNKS and control programming by ensuring enoyl reduction at the correct positions and allowing complete chain extension. It now seems that lovC type proteins are a common feature of HR PKS systems in fungi.

In the bacterial modular PKS systems, usually a C-terminal TE domain is involved in off-loading the product—either to another PKS or to solution (see Chapter 7). LNKS seems to possess part of an NRPS condensation (C) domain immediately downstream of the ACP, and this domain has been proposed to be involved in product release, presumably by either activating water as a nucle-ophile or the C-5 hydoxyl rather than the nitrogen of an aminothiolester as activated by most NRPS C domains. LDKS is closely related to LNKS, but its ER domain seems to be intact. LDKS is unusual among fungal PKS because it is not iterative—a single round of extension and processing affords the diketide (**50**) (Fig. 6.3b). In this respect, LDKS closely resembles a single module of a bacterial modular PKS (1). LDKS again lacks an obvious product release domain, which ends immediately after the ACP. A specialized acyl transferase, encoded by *lovD*, transfers (**50**) from LDKS onto the C-10 hydroxyl of monacolin J (**49**) (16).

6.6.2 The Squalestatin S1 Polyketide Synthases

Squalestatin S1 (**29**) is a potent inhibitor of mammalian squalene synthase. It is produced by *Phoma* species, and like lovastatin, consists of two polyketide chains: a main chain hexaketide and a sidechain tetraketide. Like lovastatin, both chains are methylated, but unusually for a fungal HR polyketide, the main chain is formed from a non-acetate starter unit—benzoate is incorporated at this position.

Cox et al. (17) cloned a HR PKS gene from *Phoma sp*, *PhPKS1*. This gene was expressed in the heterologous fungal host *A. oryzae*, which produced the tetraketide (**37**). *PhPKS1* thus encodes the squalestatin tetraketide synthase (SQTKS). SQTKS is highly homologous to LDKS, which lacks the C-terminal NRPS condensation domain of LNKS. Like LDKS, SQTKS posseses a functional ER domain, but SQTKS carries out three extensions. Like LDKS, all modification reactions occur after the first extension. The stereochemistry of the branching methyl group is the same in each case. Two more extensions occur—all modifying reactions occur again after the first of these, but neither ER or C-MeT are used after the final extension (Fig. 6.1).

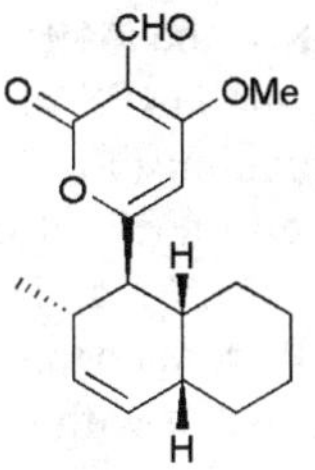

Solanapyrone A **51**

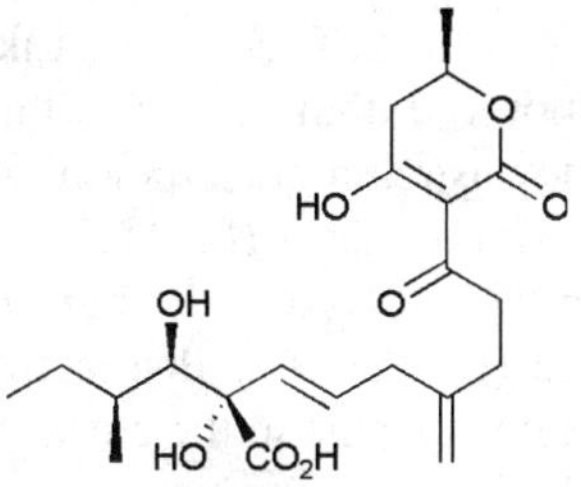

Alternaric acid **52**

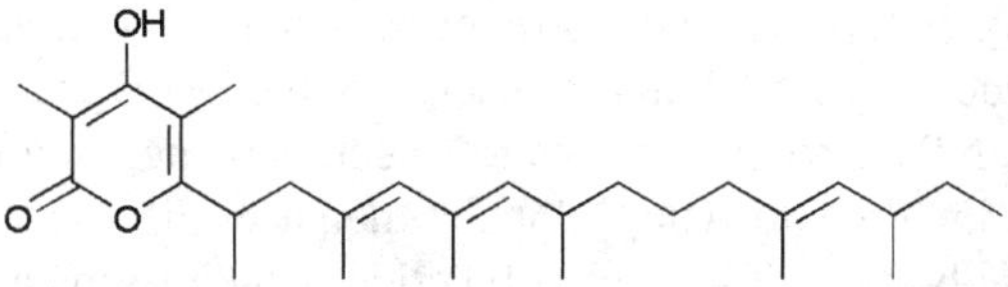

Alternapyrone **52**

6.6.3 HR PKS from *Alternaria Solani*

Alternaria solani is a plant pathogen and the causitive agent of early blight in solanum species. It produces numerous polyketides, such as solanopyrone A (**51**) and alternaric acid (**52**), and it is thus an ideal target species for speculative PKS gene-fishing expeditions. Fujii et al. (18) have conducted just such investigations, using PCR primers based on conserved PKS sequences as probes with genomic DNA libraries (18). An early investigation yielded two hits—one a HR PKS gene named *alt5*, and another, a NR PKS named *pksA*. The *alt5* gene encodes a typical HR PKS, which is known as PKSN, with the usual array of catalytic domains. Inspection of the ER sequence suggested that it should be functional like those from LDKS and SQTKS. Expression of *alt5* showed this to be correct—a single compound was synthesized in good yield ($\sim$15mgl^{-1}), which proved to be the octamethylated decaketide pyrone (**53**), named alternapyrone. This compound is the most complex polyketide yet reported to be produced by an iterative PKS, which shows programmed chain length control, keto-reduction, methylation, and enoyl reduction.

6.6.4   HR PKS-NRPS

Fungi produce a wide range of bioactive compounds derived from polyketides fused to amino acids. Examples include fusarin C (**26**), equisetin (**54**), and tenellin (**55**). Fusarin C (**26**) consists of a tetramethylated heptaketide fused to homoserine and is produced by strains of the plant pathogens *Fusarium moniliforme* and *Fusarium venenatum*. Genomic DNA libraries from these organisms were used to isolate a gene cluster centered around a 12-Kb ORF encoding a HR PKS fused to a nonribosomal peptide synthetase (NRPS) module (19). The PKS region is homologous to LNKS: KAS, AT, and DH domains are followed by CMeT, a defective ER, KR, and ACP domains. Like LNKS, the ACP is upstream of an NRPS condensation (C) domain, but in this case the NRPS module is complete, which features downstream adenylation (A), thiolation (T), and C-terminal thiolester reductase (R) domains (Fig. 6.3c). Directed knockout of the PKS-NRPS gene proved it to be involved in the biosynthesis of (**26**), and it was thus named fusarin synthetase (FUSS). The disfunctional ER domain and the fact that no *lovC* homolog seems to exist in the cluster, is consistent with the polyunsaturated nature of the polyketide moiety.

It is probable that FUSS assembles a tetramethylated heptaketide (**56**) attached to the ACP (Fig. 6.3c)—the structure of which is similar to the heptaketide pyrone (**46**) produced by LNKS in the absence of the LovC protein. In parallel, the A domain of the NRPS module seems to select, activate, and attach homoserine (**57**) to the thiolation domain. The C domain then uses the amide of homoserine to form an amide with the ACP-bound polyketide, which forms a covalently bound intermediate peptide (**58**). The final reaction catalyzed by FUSS may be reductive release of the thiolester, which forms peptide aldehyde (**59**). Finally, Knoevenagel cyclization would give the putative prefusarin (**60**). Other genes in the FUSS cluster are presumably responsible for the required additional transformation of (**60**) to fusarin C (**26**) itself: epoxidation; oxidation of a pendant methyl to a carboxylate and esterification; and hydroxylation a to nitrogen.

A highly homologous PKS-NRPS gene has been shown to be involved in the biosynthesis of equisetin (**54**) in *Fusarium heterosporum* (20). EQS posseses the same catalytic domains as FUSS, but examination of the structure of (**54**) indicates that the pyrollidinone carbon derived from the carboxylate of the amino acid (serine in this case) is not reduced, which indicates either a reoxidation mechanism, or the fact that the R domain does not produce an aldehyde intermediate in this case.

6.7 CONCLUSION

Recognition that the fungal PKS can be categorized into three subsets has allowed more detailed consideration of the programming elements. The NR PKSs are arranged into loading, extension, and processing components, and to some extent this hypothesis has been verified by expression and study of individual catalytic domains and by the construction of hybrid NR PKS genes. However, the programming elements of the PR and HR PKS remain obscure. The first experiments to probe programming in HR PKS have involved domain swaps, but few conclusions have yet been drawn. Little information is known about the three-dimensional structure of fungal PKS—whereas sequence and domain organization similarity with mammalian FAS mean that broad descriptions of the architecture of the catalytic domains can be modeled, detailed hypotheses that involve individual domains or peptide motifs cannot yet be linked with programming.

REFERENCES

1. Staunton J, Weissman KJ. Polyketide biosynthesis: a millennium review. Nat. Prod. Rep. 2001;18:380–416.

2. Simpson TJ. Applications of multinuclear NMR to structural and biosynthetic studies of polyketide microbial metabolites. Chem. Soc. Rev. 1987;16:123–160.

3. Bingle LEH, Simpson TJ, Lazarus CM. Ketosynthase domain probes identify two subclasses of fungal polyketide synthase genes. Fung. Genet. Biol. 1999;26:209–223.

4. Kennedy J, Auclair K, Kendrew SG, Park C, Vederas JC, Hutchinson CR. Modulation of polyketide synthase activity by accessory proteins during lovastatin biosynthesis. Science 1999;284:1368–1372.

5. Kroken S, Glass SNL, Taylor JW, Yoder OC, Turgeon BG. Phylogenomic analysis of type 1 polyketide synthase genes in pathogenic and saprobic ascomycetes. Proc. Natl. Acad. Sci. U.S.A. 2003;100:15670–15675.

6. Watanabe CMH, Townsend CA. Initial characterisation of a type 1 fatty acid synthase and polyketide synthase multienzyme complex NorS in the biosynthesis of aflatoxin B_1. Chem. Biol. 2002;9:981–988.

7. Crawford JM, Dancy BCR, Hill EA, Udwary D, Townsend CA. Identification of a starer unit–acyl carrier protein transacylase domain in an iterative type 1 polyketide synthase. Proc. Natl. Acad. Sci. U.S.A. 2006;103:16728–16733.

8. Gaffoor I, Trail F. Characterisation of two polyketide synthase genes involved in zearalenone biosynthesis in *Gibberella zeae*. App. Env. Microbiol. 2006;72:1793–1799.

9. Bailey AM, Cox RJ, Harley K, Lazarus CM, Simpson TJ, Skellam E. Characterisation of 3-methylorcinaldehyde synthase (MOS) in *Acremonium strictum*: first observation of a reductive release mechanism during polyketide biosynthesis. Chem. Commun. 2007;4053–4055.

10. Fujii I, Watanabe A, Sankawa U, Ebizuka Y. Identification of a claisen cyclase domain in fungal polyketide synthase WA, a naphthpyrone synthase of *Aspergillus nidulans*. Chem. Biol. 2001;8:189–197.

11. Shimizu T, Kinoshita H, Ishihara S, Sakai K, Nagai S, Nihira T. App. Env. Microbiol. 2005;71:3453–3457.

12. Ahlert J, Shepard E, Lomovskaya N, Zazopoulos E, Staffa Bachmann BO, Huang K, Fonstein L, Czisny A, Whitwam RE, Farnet CM, Thorson JS. The calicheamicin gene cluster and its iterative type 1 enediyne PKS. Science 2002;297:1173–1176.

13. Beck J, Ripka S, Siegner A, Schiltz E, Schweizer. The multifunctional 6-methylsalicylic acid synthase gene of *Penicillium patulum*. Its gene structure relative to that of other polyketide synthases. Eur. J. Biochem. 1990;192:487–498.

14. Child CJ, Spencer JB, Bhogal P, Shoolingin-Jordan PM. Structural similarities between 6-methylsalicylic acid synthase from *Penicillium patulum* and vertebrate type 1 fatty acid synthase: evidence from thiol modification studies. Biochemistry 1996;35:12267–12274.

15. Moriguchi T, Ebizuka Y, Fujii I. Analysis of subunit interactions in the iterative type 1 polyketide synthase ATX from *Aspergillus terreus*. ChemBioChem. 2006;7:1869–1874.

16. Xie X, Watanabe K, Wojcicki WA, Wang CCC, Tang Y. Biosynthesis of lovastatin analogues with a broadly specific acyl transferase. Chem. Biol. 2006;13:1161–1169.

17. Cox RJ, Glod F, Hurley D, Lazarus CM, Nicholson TP, Rudd BAM, Simpson TJ, Wilkinson B, Zhang Y. Rapid cloning and expression of a fungal polyketide synthase gene involved in squalestatin biosynthesis. Chem. Commun. 2004;2260–2261.

18. Fujii N, Shimomaki S, Oikawa H, Ebizuka Y. An iterative type 1 polyketide synthase PKSN catalyses synthesis of the decaketide alternapyrone with regiospecific octamethylation. Chem. Biol. 2005;12:1301–1309.

19. Song ZS, Cox RJ, Lazarus CM, Simpson TJ. Fusarin C biosynthesis in *Fusarium moniliforme* and *Fusarium venenatum*. ChemBioChem. 2004;5:1196–1203.

20. Fillmore JP, Warner DD, Schmidt EW, Sims JW. Equisetin Biosynthesis in *Fusarium heterosporum*. Chem. Commun. 2005;186–188.

FURTHER READING

For general review of polyketides:
O'Hagan D. The Polyketide Metabolites. Ellis Horwood Ltd. Chichester, UK, 1991.

For comprehensive listings of fungal polyketide molecules:
Turner WB. Fungal Metabolites. Academic Press, London, 1971.
Turner WB, Aldridge DC. Fungal Metabolites Part II. Academic Press, London, 1983.

For a review of mycotoxins: Steyn PS, Vleggar R. Mycotoxins and Phycotoxins, Elsevier, Amsterdam, 1986.

For further discussion of fungal PKS genes: Cox RJ. Polyketides, proteins and genes in Fungi: programmed nano-machines begin to reveal their sectors. Org. Biomol. Chem. 2007;5:2010–2026.

7

MODULAR POLYKETIDE SYNTHASES

TONIA J. BUCHHOLZ, JEFFREY D. KITTENDORF, AND DAVID
H. SHERMAN

University of Michigan, Ann Arbor, Michigan

Polyketides constitute a large class of microbial and plant-derived secondary metabolites that displays a vast array of structural diversity. These organic molecules vary in molecular weight, functional group modification, and include linear, polycyclic, and macrocyclic structural forms. Currently, polyketide natural products find clinical use as antibiotics, antiparasitic agents, antifungals, anticancer drugs, and immunosuppressants. Given these impressive and wide-ranging pharmacologic activities, an ever-increasing demand is placed on natural products research to uncover novel polyketide metabolites for the benefit of human and animal health. Modular polyketide synthases are nature's platform for the expansion of chemical diversity. This review provides new perspectives on important biosynthetic mechanisms that contribute to this variety. This includes control of double-bond configuration and regiochemistry, introduction of β-branching during polyketide chain assembly, and other processes that contribute to introduction of unique chemical functionality into these fascinating systems.

Despite the promise of modern synthetic technologies to enhance pharmaceutical discovery significantly, natural products continue to be the greatest source of all new drug leads (1). Currently, many examples of natural product-derived pharmaceuticals are employed to benefit human health (2, 3); polyketides constitute a large class of microbial and plant-derived secondary metabolites that displays a vast array of structural diversity. These organic molecules vary in molecular weight and functional group modification; they include linear, polycyclic, and macrocyclic structural forms. Currently, polyketide natural products

Natural Products in Chemical Biology, First Edition. Edited by Natanya Civjan.
© 2012 John Wiley & Sons, Inc. Published 2012 by John Wiley & Sons, Inc.

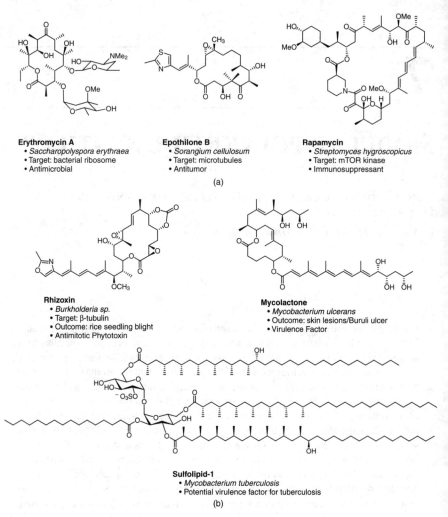

Erythromycin A
- *Saccharopolyspora erythraea*
- Target: bacterial ribosome
- Antimicrobial

Epothilone B
- *Sorangium cellulosum*
- Target: microtubules
- Antitumor

Rapamycin
- *Streptomyces hygroscopicus*
- Target: mTOR kinase
- Immunosuppressant

(a)

Rhizoxin
- *Burkholderia sp.*
- Target: β-tubulin
- Outcome: rice seedling blight
- Antimitotic Phytotoxin

Mycolactone
- *Mycobacterium ulcerans*
- Outcome: skin lesions/Buruli ulcer
- Virulence Factor

Sulfolipid-1
- *Mycobacterium tuberculosis*
- Potential virulence factor for tuberculosis

(b)

Figure 7.1 Examples of clinically-relevant polyketide natural product drugs (a) and virulence factors (b).

find clinical use as antibiotics, antiparasitic agents, antifungals, anticancer drugs, and immunosuppressants (Fig. 7.1a). Given these impressive and wide-ranging pharmacological activities, an ever-increasing demand is placed on natural prod-ucts research to uncover novel polyketide metabolites for the benefit of human health.

Although the clinical use of polyketide-inspired pharmaceuticals has been appreciated for decades, polyketide-derived metabolites have been recognized recently for their role in bacterial virulence. For example, the pathogenesis of *Mycobacterium ulcerans*, the causative agent of the devastating skin disease known as Buruli ulcer, is the result of the secretion of polyketide-derived toxins known as the mycolactones (Fig. 7.1b) (4). These polyketide toxins are responsible largely for the necrotic lesions that are characteristic of this debilitating condition. As such, the disruption of mycolactone biosynthesis may lead to an effective chemotherapy for Buruli ulcer. Recent findings also suggest that the virulence of another mycobacterial species, *Mycobacterium tuberculosis*, could be partially dependent on polyketide biosynthesis. The cell surface sulfolipid-1 (SL-1) is among several virulence-associated molecules produced by *M. tuberculosis*. SL-1 consists of a sulfated disaccharide core (trehalose-2-sulfate) that displays four lipidic substituents; all but one substituent seems to be polyketide-derived (Fig. 7.1b) (5). Finally, the toxic agent in rice seedling blight, which is a highly destructive fungal disease that inflicts severe agricultural losses worldwide, has been identified recently as the polyketide metabolite, rhizoxin (Fig. 7.1b) (6). Interestingly, rhizoxin is not produced directly by the fungus (*Rhizopus*), but rather by the endosymbiotic bacteria *Burkholderia* that thrives within the fungus. Together, these three examples suggest that inhibition of polyketide biosynthesis may lead to effective chemotherapy for controlling certain human and plant bacterial diseases.

7.1   PROTOTYPICAL POLYKETIDE BIOSYNTHESIS

The biosynthesis of many important polyketide compounds occurs via a stepwise, assembly-line type mechanism that is catalyzed by type I modular polyketide synthases (PKSs). These modular PKSs are composed of several large, multifunctional enzymes that are responsible for catalyzing the initiation, elongation, and processing steps that ultimately give rise to the characteristic macrolactone scaffold (Fig. 7.2) (7–11). Structural studies have been critical in developing a sophisticated understanding of the overall architecture and mechanism of type I PKSs and their homologs in recent years (12–18). A review from the perspective of the 6-deoxyerythronolide B synthase (a well-studied type I PKS) was published recently by Khosla et al. (19).

It is well established that the sequential arrangement of modules within a PKS system serves effectively as a biosynthetic program, which is responsible for dictating the final size and structure of the polyketide core. Typically, initiation of polyketide biosynthesis begins by the acyltransferase (AT) catalyzed linkage of a coenzyme A (CoA) priming unit (e.g., methylmalonyl-CoA, malonyl-CoA, propionyl-CoA) to the acyl carrier protein (ACP) of the loading module. Once initiated, downstream elongation modules carry out repetitive extensions of the starter unit. In most PKS systems, each elongation module contains at minimum an AT domain, an ACP domain, and a ketosynthase (KS) domain (Fig. 7.2a). The AT domain is responsible for loading the appropriate CoA extender unit onto the ACP domain (i.e., malonyl-CoA, methylmalonyl-CoA, etc.). The KS domain then, catalyzes a decarboxylative condensation of the extender unit with the growing polyketide chain obtained from the preceding module to generate an ACP-bound β-ketoacyl product. In addition to the three core domains,

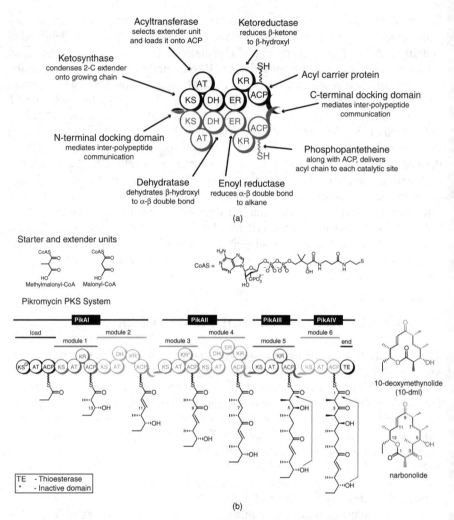

Figure 7.2 Conventional modular type I PKS paradigm. (a) Individual domains in a full type I polyketide synthase extension module. Homodimeric contacts are made in the N-terminal docking, ketosynthase, dehydratase, enoyl reductase, and C-terminal docking domains. (b) PKS system for 10-deoxymethynolide and narbonolide generation.

each elongation module may contain up to three additional domains [ketoreductase (KR), dehydratase (DH), enoyl reductase (ER)] that are responsible for the reductive processing of the β-keto functionality prior to the next extension step (Fig. 7.2a). These reductive steps contribute greatly to the overall structural diversity that is observed among polyketide natural products. The presence of a KR domain alone generates a β-hydroxyl functionality, the presence of both a KR and a DH domain generates an alkene, whereas the combination of KR, DH, and ER results in complete reduction to the alkane. Finally, termination of polyketide biosynthesis is catalyzed by a thioesterase (TE) domain located at the carboxy terminus of the final elongation module. The activity of this domain results in the cleavage of the acyl chain from the adjacent ACP; typically, intramolecular cyclization results in the formation and release of a macrolactone ring. Tailoring enzymes, such as hydroxylases and glycosyl transferases, often serve to further modify the polyketide to yield the final bioactive compound.

The modular organization of type I PKSs has made them particularly attractive targets for rational bioengineering. Combinatorial biosynthetic efforts centered on prototypical modular PKSs have been the topic of many recent outstanding review articles (20–23). Currently, several strategies are being pursued that attempt to leverage PKS systems for the generation of structurally diverse polyketides. For example, it has been demonstrated that alterations of individual catalytic domains (i.e., inactivation, substitution, addition, deletion) within a PKS module can result in predicted structural alterations of the final PKS product. Likewise, the addition, deletion, or exchange of intact modules can also impart structural variety into polyketide metabolites. Using these and other approaches, hundreds of novel polyketide structures have been generated, which established the tremendous potential of these applications. However, these successes seem to be more the exception rather than the rule, as many efforts result in trace levels, or they fail to provide the desired metabolite. This finding suggests that much remains to be learned regarding the molecular intricacies of these complex biosynthetic machines. This review provides new perspectives on important mechanisms that contribute to structural diversity in modular PKSs. These mechanisms include control of double-bond configuration and regiochemistry, introduction of β-branching during polyketide chain assembly, and other processes that contribute to introduction of unique chemical functionality into these fascinating systems.

7.2 POLYKETIDE DOUBLE BONDS

7.2.1 Trans Double Bonds

The presence of unsaturated carbon–carbon bonds within most polyketide compounds exemplifies the overall structural diversity that is a hallmark of this class of important natural products. Typically, the installation of double bonds into nascent polyketide chains relies on the two-step processing at the β-keto group by the successive activity of KR and DH domains that are embedded within a given PKS elongation module. After KS catalyzed chain elongation that extends the growing chain by two carbon atoms, the KR domain, when present, directs the NADPH-dependent reduction of the β-ketone to yield a 3-hydroxyacyl intermediate. Subsequently, an embedded DH domain within the elongation module catalyzes dehydration of the 3-hydroxyacyl intermediate, normally which results in the incorporation of an *(E)*-trans α,β unsaturated bond into the growing polyketide chain (Fig. 7.3a).

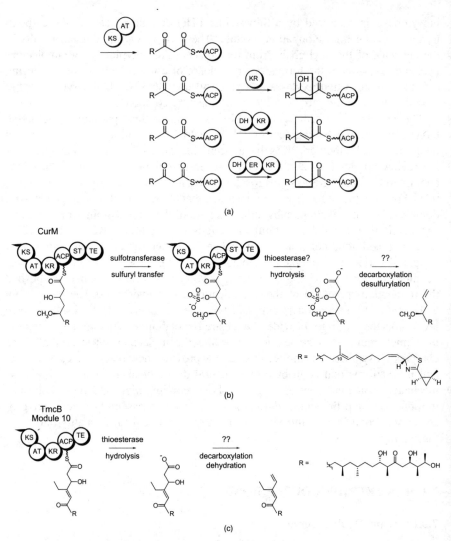

Figure 7.3 (a) Traditional view of reductive processing at the β-ketone position in the growing polyketide chain. Presence of a ketoreductase domain leads to formation of an alcohol. An active dehydratase domain can further process the alcohol moiety to an alkene. Complete saturation to the alkane is accomplished by an enoyl reductase domain. (b) Proposed terminal double bond formation for curacin biosynthesis. (c) Proposed terminal double bond formation in tautomycetin biosynthesis.

It should be noted that the KR catalyzed reduction of a β-ketoacyl intermediate has stereochemical consequences because a new chiral center is introduced into the growing oligoketide. Aside from serving to enhance the structural diversity of the final polyketide product even more, the stereochemical outcome of this reaction can have profound effects on any subsequent processing or elongation reactions. As such, an understanding of how KR domains exert stereochemical control of their hydroxylated product is a critical aspect to deciphering the mechanism of DH-mediated double bond formation. Through bioinformatic and biochemical analyses, an appreciation of ketoreductase-influenced stereochemistry has emerged (24, 25). Thus, Caffrey (25) has proposed that KR domains can be divided into two classes, depending on the final configuration of the β-hydroxyl moiety. The so-called "A" class generates an L-3-hydroxy product, whereas the "B" class produces the D-3-hydroxy polyketide intermediate. Although little difference exists between these two putative classes at the amino acid sequence level, the presence of a conserved aspartate residue within an LDD motif correlates well with "B" class. This motif is absent in the defined "A" class of KR domains. An additional diagnostic feature of the "A" class of KR domains is the presence of a conserved tryptophan residue. Recently, Keatinge-Clay (18) has proposed a refinement of the KR class descriptions as originally suggested by Caffrey (25), effectively increasing the number of possible KR types from two to six (18). This new classification takes into consideration whether a given KR domain (either reductively competent or incompetent) is located in an epimerization-competent module. Although this new classification offers a more complete description of PKS KR domains, for simplicity we will continue to use Caffrey's KR nomenclature throughout our discussion of double-bond formation.

While examples of both D- and L-hydroxyl group configurations can be found within polyketide natural products, recent evidence suggests that DH domains require a stereospecific 3-hydroxyacyl intermediate. Bioinformatic analyses performed on 71 KR domains for which the stereochemical outcome of the reduction is cryptic because of subsequent dehydration revealed that all belong to the "B" class of KR domains (25). As such, it appears that the generally preferred substrate for DH domains is a D-3-hydroxyacyl chain. However, direct experimental evidence has been difficult to obtain because the 3-hydroxyacyl intermediate is transient in modules that contain a DH domain. Recently, biochemical studies of the DH domain found in module 2 of the pikromycin PKS system (Fig. 7.2b) (26) have supported this hypothesis; inactivation of the DH domain resulted in the exclusive generation of the D-3-triketide acylthioester intermediate from a diketide substrate (27). Aside from this study, no other reports probe the substrate preference or catalytic mechanism of DH domains within PKS systems; therefore, much of what is known has been elucidated from studies of fatty acid biosynthesis (28). Previous studies on the dehydration step that is catalyzed by the yeast fatty acid synthase confirmed the *syn* elimination of water from a D-(3 R)-hydroxyacylthioester substrate (29). This result is consistent with the stereospecificity of the PKS DH domain and may suggest that *trans* unsaturated bonds, typically found in polyketides, are likewise formed via *syn* water elimination.

7.2.2 Cis Double Bonds

Although rare, several PKS biosynthetic systems can install *cis* double bonds into the final polyketide product (Fig. 7.4). Several possible mechanisms could

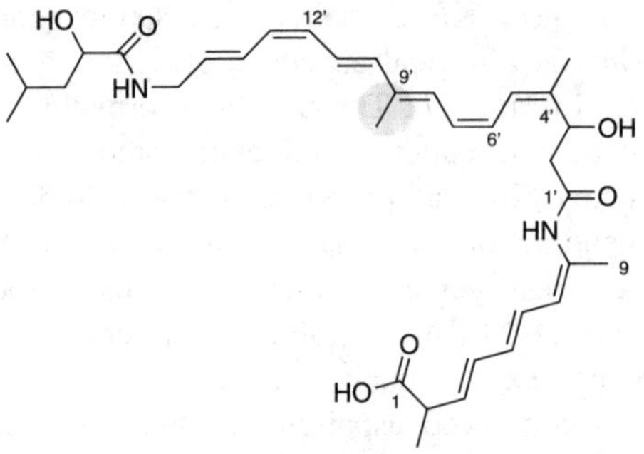

1 - Bacillaene

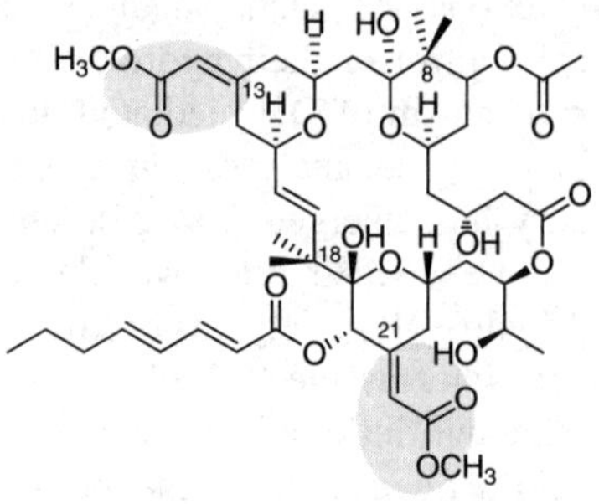

2 - Bryostatin 1

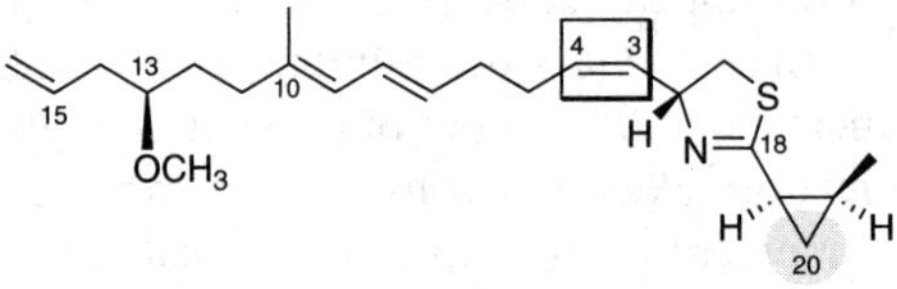

3 - Curacin A

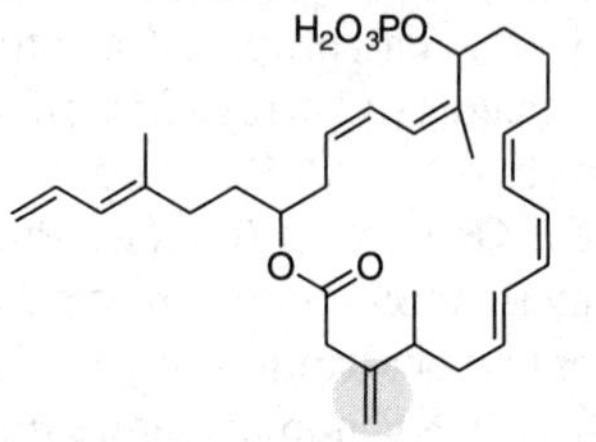

4 - Difficidin

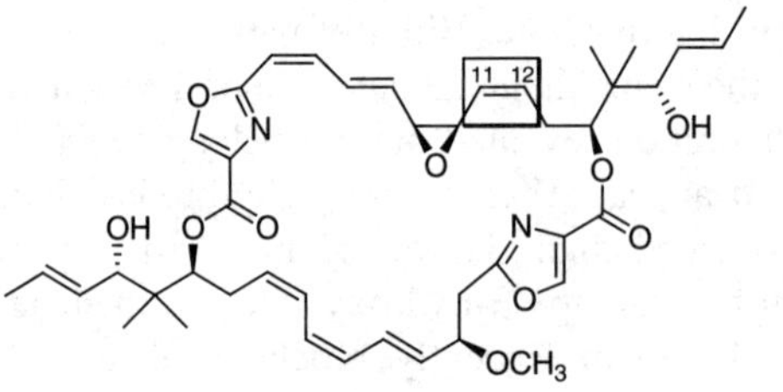

5 - Disorazole A_1

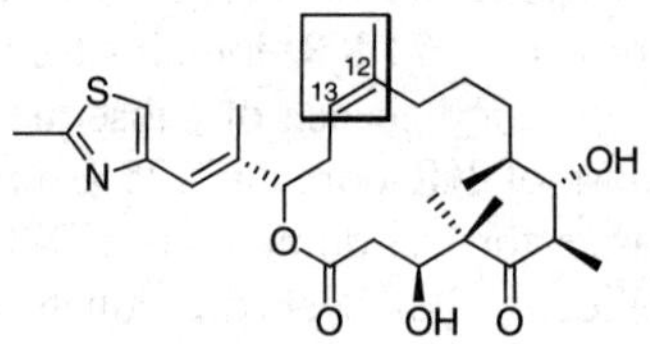

6 - Epothilone D

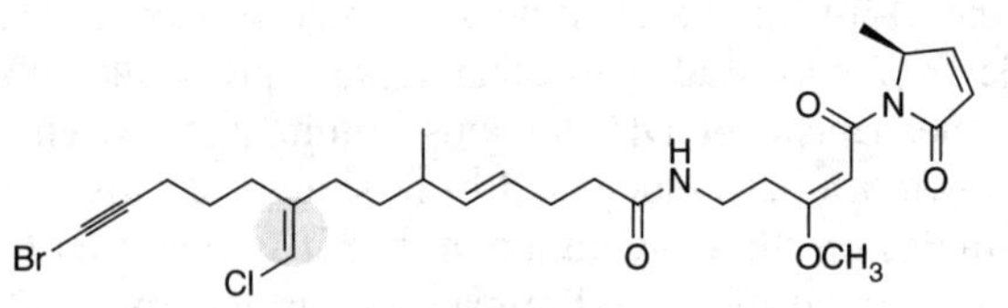

7 - Jamaicamide

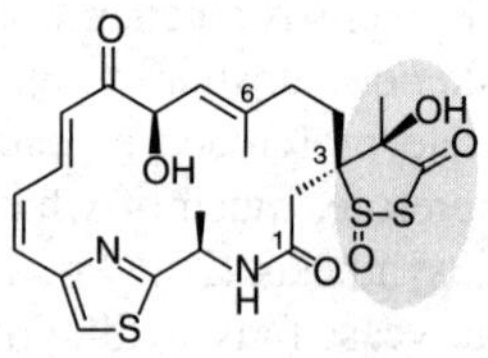

8 - Leinamycin

Figure 7.4 Double bond containing compounds discussed in this review. Cis-double bonds discussed in the text are boxed and modifications produced by the HMGS cassettes are shaded.

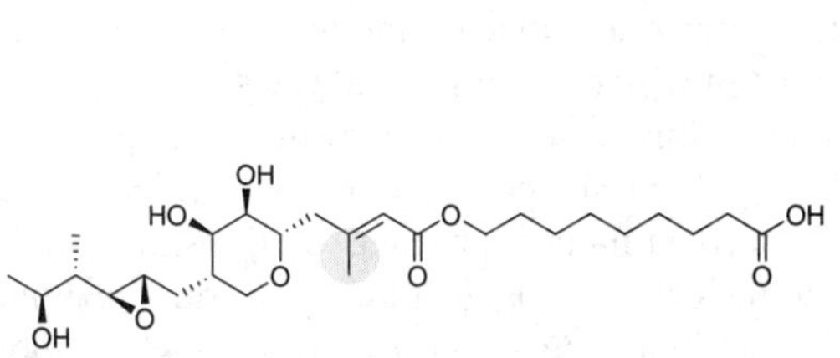

9 - Mupirocin

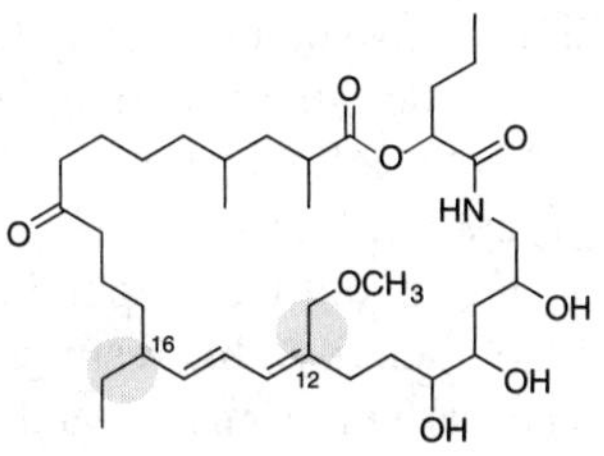

10 - Myxovirescin A

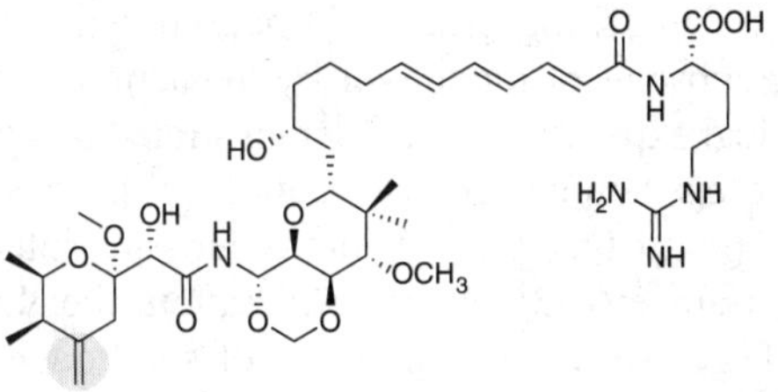

11 - Onnamide

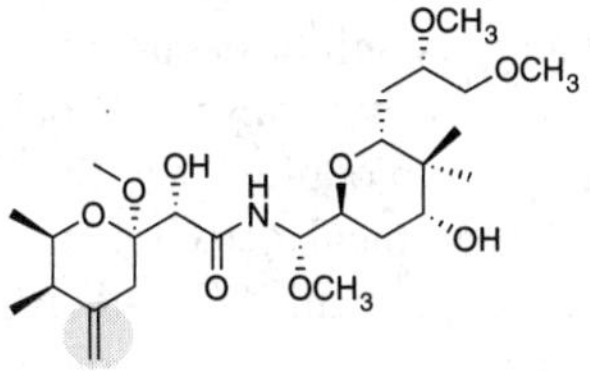

12 - Pederin

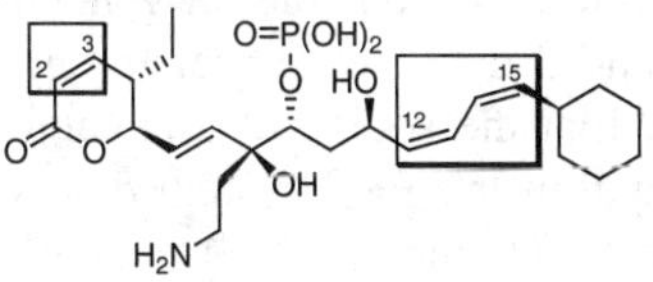

13 - Phoslactomycin

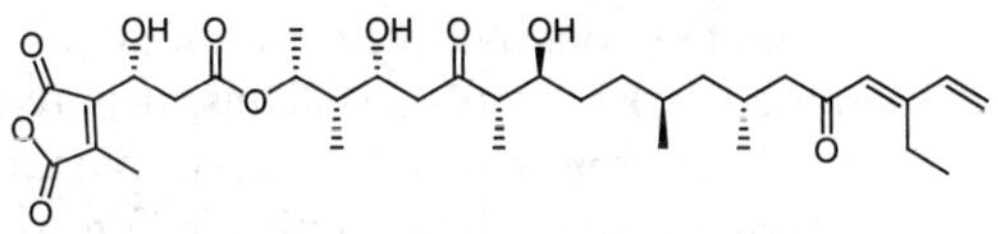

14 - Tautomycetin

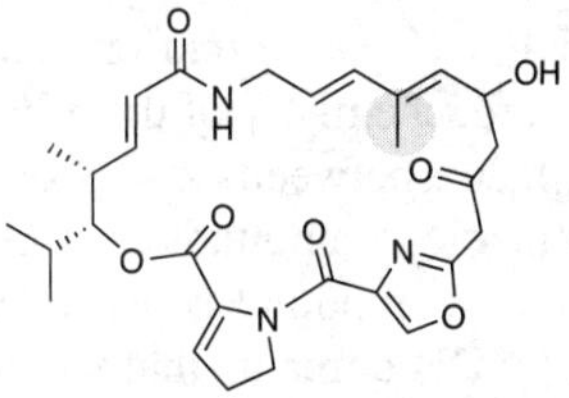

15 - Virginiamycin M

Figure 7.4 (*Continued*)

account for the infrequent occurrence of this double bond configuration. One explanation is that an isomerization event occurs that converts a *trans* double bond into a *cis* double bond. This isomerization activity could be specified by the PKS elongation module, much like previously identified epimerization activities that are known to exist in some PKS and NRPS modules. Alternatively, the combined activity of KR-DH domains within certain modules could directly establish the *cis* double bond. Finally, it is possible that after reduction of the β-keto functionality, a *trans* acting DH could catalyze dehydration to form a (*Z*)-*cis* double bond. This *trans* activity might derive from a discrete enzyme encoded within the biosynthetic gene cluster or from an adjacent module within the PKS pathway. Additionally, examples exist in which some general rules may not hold true. Chivosazol, which is a potential antitumor agent, is one such example (30).

The epothilones **6** (and Fig. 7.1a), which are produced by *Sorangium cellulosum*, are mixed NRPS-polyketide derived natural products that possess potent antitumor activity. Interestingly, some compounds feature a *cis* double bond between carbon atoms 12 and 13 that should be generated by PKS elongation module 4; however, sequence analysis of the epothilone biosynthetic gene cluster indicates that module 4 does not contain a DH domain requisite for the formation of the unsaturated bond (31). Thus, Tang et al. (31) hypothesized that the DH activity might occur from the subsequent module or by the action of a post-PKS modifying enzyme. Biochemical experiments later demonstrated that the DH domain of module 5 catalyzed the *cis* double bond formation (32). To account for this atypical activity, it is proposed that the 3-hydroxythioester intermediate undergoes an ACP_4-to-ACP_5 transfer (32). After dehydration, the thioester intermediate would then be transferred to the KS_5 domain for subsequent elongation. Similarly, the antitumor phoslactomycin compounds **13** feature three *cis* double bonds, two of which seem to be installed by a KR-DH pair (see below); however, the elongation module (Plm7) that should be responsible for generating the unsaturated bond between carbon atoms 2 and 3 does not appear to encode the required DH activity (33). Thus, it is likely that the source of this catalytic activity comes from either a different module within the PKS system or a separate enzyme that could act either before or after TE mediated termination of polyketide biosynthesis. Additional analysis is required to discriminate between these two possibilities.

Unlike the examples described above, most polyketide *cis* double bonds are installed through a successive KR-DH pair found embedded within the elongation module. In these cases, the stereochemistry of the 3-hydroxyacyl intermediate appears to be the discriminating factor between (*Z*)-*cis* or (*E*)-*trans* unsaturation. For example, the antitumor disorazole compounds **5** display up to three *cis* double bonds per monomer (note: final compound is a condensed dimer). Sequence analyses of KR domains preceding DH domains that would be responsible for *cis* double bond formation suggest that they all belong to the "A" class (34). Thus, they are predicted to generate a L-3-hydroxyacyl intermediate. It is expected that the subsequent DH domain preferentially recognizes the L-3-hydroxyl group to facilitate the generation of the *cis* double bond. Interestingly, the module responsible for incorporating the *cis* unsaturated bond between carbon atoms 11 and 12 is split between two polypeptide chains (34). It cannot be ruled out that this modular dissection may play a role in formation of this particular double bond, as the cleavage point occurs between the DH and KR domain. Furthermore, it is intriguing that the major product, disorazole A1, is composed of two nonidentical monomers that differ in saturation between carbon atoms 5 and 6. It has been suggested that the synthesis of the two different monomers is caused by poor activity of the DH domain (34); however, final proof requires additional experimental verification.

In addition to disorazole **5**, several other known examples of polyketide natural products exist that have *cis* double bonds installed by an embedded KR-DH domain pair. The potent antitumor compound curacin A **3** contains many interesting structural features. Among them is the presence of a *cis* double bond between carbon atoms 3 and 4. Sequence alignment of the KR domain encoded by *curG* (encoding the module responsible for generation of the *cis* double bond) suggests that it belongs to the "A" class of KR domains (35). Thus, this particular KR is predicted to generate a L-3-hydroxyacyl intermediate that is subsequently dehydrated to the *cis* double bond. Likewise, the KR domains that set up the 3 *cis* double bonds found in the linear mixed NRPS-polyketide natural product bacillaene **1** are predicted to produce L-3-hydroxyacyl intermediates (36). Interestingly, two of the elongation modules that incorporate *cis* olefins are split between two polypeptides (36). As in disorazole biosynthesis, it is possible that these modular dissections contribute to the configuration of the unsaturated bond that is introduced by these modules.

Finally, we consider the conjugated *cis* olefins that span carbon atoms 12-15 of the phoslactomycins **13**. Analysis of the KR sequences of elongation modules 1 and 2 could not clearly predict whether these reductive domains belonged to the "A" or "B" class. Therefore, Alhamadsheh et al. (37) employed a comprehensive biochemical study to elucidate the mechanism of *cis* olefin formation by the first elongation module. Two hypotheses were considered. First, the configuration of the double bond could develop directly from the combined activities of the embedded KR-DH domain pair. Alternatively, the KR-DH domains might establish a *trans* olefin that is isomerized subsequently to the observed *cis* configuration. To distinguish between these two possibilities, Alhamadsheh et al. (37) genetically inactivated both the loading module and elongation module of Plm1 and conducted feeding experiments with diketide analogs containing both *cis* and *trans* olefins. Results from this work indicated clearly that only the *cis* olefin containing diketide is accepted as a substrate for elongation module 2, suggesting that the product of module 1 must contain the *cis* double bond. Furthermore, this work demonstrated nicely that the phoslactomycin biosynthetic pathway cannot process *trans* diketide intermediates into mature products, which rules out the possibility of an isomerization domain in downstream modules.

7.2.3 Terminal Double Bonds

Termination of polyketide biosynthesis typically involves the TE mediated cleavage of the ACP-bound thioester, followed by cyclization to generate a macrolactone. Alternatively, the TE catalyzes the simple hydrolysis of the thioester to generate a linear free acid product. Here, we consider two of the relatively few known examples of polyketide natural products that are neither a macrocycle nor a free acid, but instead terminate with a double bond.

Aside from containing a *cis* double bond noted above, the antitumor polyketide compound curacin A **3** also features a terminal olefin. Previously reported feeding studies suggested that the formation of the terminal double bond develops from successive decarboxylation and dehydration events (35). The biosynthetic gene cluster responsible for curacin A biosynthesis has been identified and initially characterized, which enables the putative assignment of domains within the predicted elongation modules (35). Like most other known polyketide biosynthetic pathways, the final elongation module of the curacin pathway, CurM, contains a terminal thioesterase domain that presumably plays a role in formation of the terminal olefin; however, biochemical evidence for such a role is lacking. Interestingly, domain analysis of CurM also predicts the presence of a sulfotransferase (ST) domain immediately preceding the TE domain. Typically, ST domains are responsible for transferring a sulfuryl group from a donor molecule (such as 3′-phosphoadenosine-5′-phosphosulfate, PAPS) to a variety of acceptor carbohydrates, proteins and other low-molecular weight metabolites (38). Although STs have been characterized from both eubacterial and eukaryotic organisms, the presence of an ST domain within a PKS system is unprecedented.

Current efforts in our laboratory include elucidating the roles of the ST and TE domains in curacin A biosynthesis, and in particular, their potential functions in terminal olefin formation. One possible mechanism that is currently under consideration is shown in Fig. 7.3b. On reduction of the β-keto functionality of the ACP-bound thioester intermediate, the ST domain transfers a sulfuryl group from the donor molecule PAPS to the 3-hydroxyl group of the thioester chain. Consistent with this hypothesis, bioinformatic analysis suggests that the putative curacin ST domain contains the signature PAPS binding pocket (unpublished data); however, no experimental evidence suggests that this ST domain can catalyze the sulfuryl transfer or that the ST domain can bind the requisite PAPS donor molecule. Assuming our hypothesis is correct and that the ST domain functions as proposed, transfer of the sulfurylated intermediate to the TE domain would initiate hydrolytic termination of curacin A biosynthesis to produce the linear free acid. At this point, one of several chemical steps can be envisioned. Following hydrolysis, the TE may catalyze decarboxylation, after which the formation of the double bond would occur in a concerted process by displacement of the sulfate leaving group. Alternatively, a separately encoded enzyme might be responsible for decarboxylating the free acid generated by the TE domain. It is also conceivable that on TE catalyzed hydrolysis, the decarboxylation reaction occurs spontaneously due to the presence of the sulfate leaving group at carbon 3.

Similar to curacin, the polyketide metabolite tautomycetin **14** also possesses a terminal olefin. This polyketide metabolite has potential medicinal value because of its novel immunosuppressive activities (39). The tautomycetin biosynthetic gene cluster has been sequenced recently, which enables domain composition analysis of the terminal elongation module (40). Unlike the terminal module involved in curacin biosynthesis, the final tautomycetin elongation module does not contain the unusual ST domain. This finding may suggest that formation of the terminal olefin of tautomycetin **14** occurs by a different chemical mechanism. The final elongation module does contain a TE domain, which presumably terminates tautomycetin biosynthesis through generation of the free acid (Fig. 7.3c).

As described for curacin, it is possible that this TE domain can also catalyze the subsequent decarboxylation event; however, in lieu of an activated leaving group at carbon 3, it is reasonable to expect that dehydration to remove the hydroxyl at carbon 3 would also be a catalyzed event. Alternatively, the terminal olefin could be installed during the post-PKS maturation of the polyketide to the final tautomycetin product. DNA sequence analysis of open reading frames that are downstream of the tautomycetin PKS gene cluster reveals two potential candidates, *tmcJ* and *tmcM*, which may be involved in double bond formation (40). Bioinformatic analyses suggest that *tmcJ* might encode for a putative decarboxylase and that the gene product of *tmcM* is a potential dehydratase.

7.3 ATYPICAL PKS DOMAINS

The wide distribution of PKSs in the microbial world and the extreme chemical diversity of their products do in fact result from a varied use of the well-known catalytic domains described above for the canonical PKS systems. Taking a theoretic view of polyketide diversity, González-Lergier et al. (41) have suggested that even if the starter and extender units are fixed, over 100,000 linear heptaketide structures are possible using only the 5 common reductive outcomes at the β-carbon position (ketone, (*R*- or *S*-) alcohol, *trans*-double bond, or alkane). Recently, it has become apparent that even this does not represent the upper limit for polyketide diversification. To create chemical functionalities beyond those mentioned above, nature has recruited some enzymes from sources other than fatty acid synthesis (the mevalonate pathway in primary metabolism is one example) not typically thought of as type I PKS domains. Next, we explore the ways PKS-containing systems have modified these domains for the catalysis of some unique chemistries observed in natural products.

7.3.1 Methyl Groups at the α- and β-carbons

As described above for polyketide biosynthesis, the presence or absence of a methyl group on the α-carbon position of the growing polyketide chain is most often governed by the selection of the extender unit (malonyl-CoA versus methylmalonyl-CoA). However, in PKS systems that use *trans* acyltransferases (AT-less type I PKSs) (8, 42), the module by module control over extender unit selection is sometimes not possible. In most cases, malonyl-CoA is used as the extender unit, and a methyl group can be added to selected positions through the action of an embedded methyl transferase (MT) domain or discrete MT enzyme. For example, the C-6 methyl group of leinamycin **8** is thought to be installed by the MT embedded in LnmJ (43), the C-10 methyl of curacin A **3** is generated via the MT domain in CurJ (35), and the gem dimethyl groups on C-8 and C-18 of bryostatin **2**, most likely are the consequence of the MT domains in BryB and BryC (44).

In contrast to the α-carbon methylations, the incorporation of methyl or methylene groups (or functional groups derived from such groups) at the β-position represents the assimilation of a full cassette of enzymes into the typical PKS machinery. Recently, a subset of type I modular PKSs (and hybrid NRPS/PKS megasynthases) have been identified that contain multiple enzymes acting in *trans* during the traditional linear assembly-line process to accomplish β-branching. Termed HMG-CoA synthase (HMGS) cassettes, these enzyme systems provide a unique method of expanding the repertoire of the traditional reductive domains (KR, DH, ER). These enzymes work in conjunction with the PKS machinery to create unique functionalities observed at the branch points that include the pendant methyl groups of bacillaene **1** (36, 45, 46), mupirocin **9** (47), and virginiamycin M **15** (48), which are the methoxymethyl and ethyl groups of myxovirescin A **10** (49, 50); the exo methylene groups of difficidin **4** (45), onnamide A **11** (51), and pederin **12** (51–53); the cyclopropyl ring of curacin A **3** (35, 54); the vinyl chloride of jamaicamide **7** (55); the unique 1,3-dioxo-1,2-dithiolane moiety of leinamycin **8** (43); and the exocyclic olefins in bryostatin **2** (Fig. 7.4) (44).

7.3.2 HMG-CoA Synthase Cassettes

In primary metabolism, HMG-CoA synthase (HMGS) is responsible for the condensation of C-2 of acetyl-CoA onto the β-ketone of acetoacetyl-CoA to form 3-hydroxyl-3-methylglutaryl-CoA and free CoASH (Fig. 7.5a) (56). Several secondary metabolite pathways have been identified over the past five years that perform an analogous reaction, although they seem to use ACP-tethered acyl groups (Fig. 7.5b) as opposed to acyl-CoA substrates. After generation of the HMG-ACP analog on the growing polyketide chain, the product is usually dehydrated and decarboxylated to yield the branched intermediate. Found in 11 pathways to date (36, 45–55), included in the cassette are a discrete ACP, a decarboxylative KS (active site cysteine is replaced with a serine), an HMGS, and one or two enoyl CoA hydratase-like (ECH) domains. (Table 7.1, Figs. 7.5, 7.6).

7.3.3 HMGS Cassette Biochemistry

Three HMGS-containing cassettes (those in the curacin A, bacillaene, and myxovirescin pathways) have been validated biochemically and will serve as the basis for our analysis of the individual components in this complex (46, 54, 57, 58). The mechanistic and structural details for HMG-CoA synthase in primary metabolism have been elucidated for both bacterial and eukaryotic HMGSs (59–63). Although polyketide HMGSs share only 20–30% sequence identity with their primary metabolism homologs (in both prokaryotes and eukaryotes), multiple sequence alignment reveals that the key catalytic residues (Glu/Cys/His) are conserved. As shown in Fig. 7.5b, the first step in

TABLE 7.1 HMGS containing biosynthetic pathways and their producing organisms

Natural product	Producing organism	Discrete ACP	KS (Cys→Ser)	HMGS	ECH$_1$ (Dehydration)	ECH$_2$ (Decarboxylation)
Bacillaene	*Bacillus subtilis* 168/*B. amyloiquefaciens*	AcpK/ BaeF	PksF	PksG/ BaeG	PksH/ BaeH	PksI/BaeI
Bryostatin	*Candidatus* Endobugula sertula		BryQ	BryR		
Curacin	*Lyngbya majuscula*	CurB	CurC	CurD	CurE	CurF N-terminal domain
Difficidin	*Bacillus amyloiquefaciens*	DifC		DifN		DifO
Jamaicamide	*Lyngbya majuscula* JHB	JamF	JamG	JamH	JamI	JamJ N-terminal domain
Leinamycin	*Streptomyces atroolivaceus*	LnmL		LnmM	LnmF	
Mupirocin	*Pseudomonas fluorescens*	Macp14	MupG	MupH	MupJ	MupK
Myxovirescin A (antibiotic TA)	*Myxococcus xanthus* DK1622	TaB & TaE	TaK	TaC & TaF	TaX	TaY
Onnamide	Symbiont bacterium of *Theonella swinhoei*			OnnA		OnnB (embedded)
Pederin	Symbiont bacterium of *Paederus fuscipes*	PedN	PedM	PedP	PedL	PedI (embedded)
Virginiamycin M	*Streptomyces virginiae*		VirB	VirC	VirD	VirE

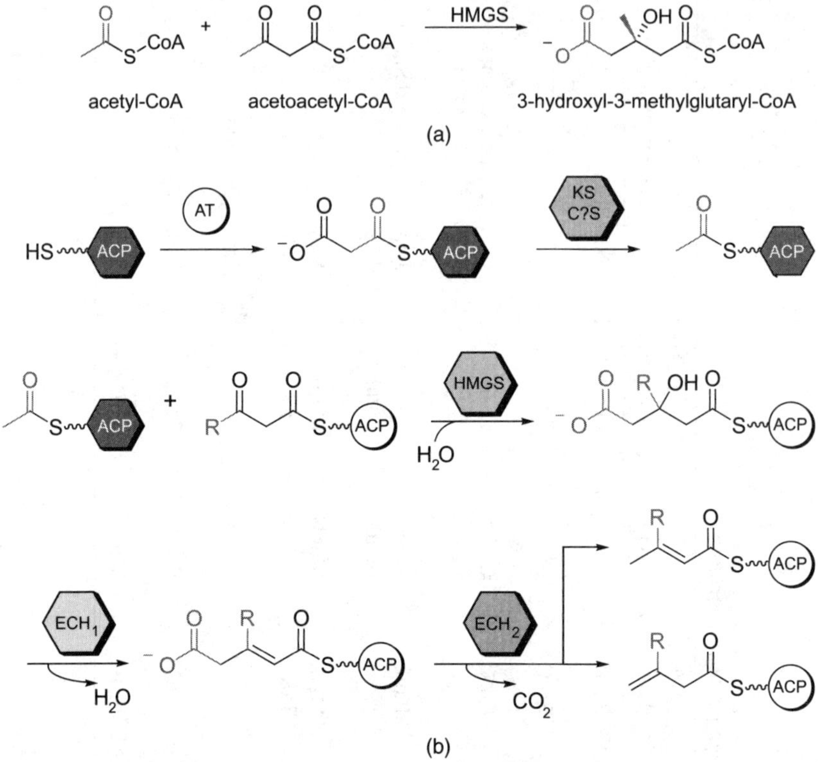

(a)

(b)

Figure 7.5 HMGS cassette reaction scheme. (a) HMG-CoA synthase (HMGS) reaction from primary metabolism. (b) An HMGS cassette can convert the β-ketone to an alkene (β,γ or γ,δ double bond) with a pendant methyl (or ethyl) group.

the formation of the HMG-intermediate is the generation of of acetyl-ACP. This step is accomplished through the loading via an AT of malonyl-CoA [or perhaps methylmalonyl-CoA in at the case of TaE (58)]. Then, the decarboxylative KS converts the malonyl-ACP into acetyl-ACP, after which the tethered acetyl group is condensed onto the β-ketone of the polyketide intermediate. Finally, formation of the HMG-analog is completed on addition of water.

Processing of the HMG-intermediate can vary considerably, but typically proceeds via dehydration and decarboxylation catalyzed by two enoyl-CoA hydratase-like domains (Fig. 7.5b). Based on sequence similarity, the members of the crotonase fold family observed in these HMGS cassettes can be subdivided into two groups, termed ECH_1 and ECH_2 (54). The successive dehydration and decarboxylation steps are catalyzed by the ECH_1 and ECH_2 enzymes/domains, respectively. Evidence for the specific function of the curacin ECH_1 and ECH_2 enzyme pair from the curacin pathway has been demonstrated using a coupled enzyme assay and ESI-FT-ICR MS (54). Using purified ECH_1 (CurE) and ECH_2 (the N-terminal domain of CurF) overexpressed in *E. coli*, (*S*)-HMG-ACP was

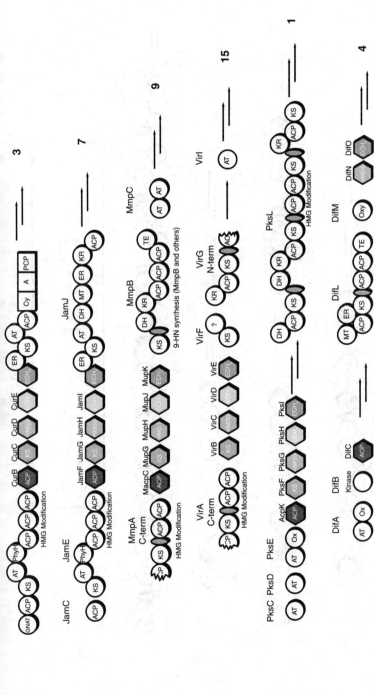

Figure 7.6 HMGS containing biosynthetic pathways. Portions of the PKS and PKS/NRPS pathways where the HMGS and related enzymes are located. *Abbreviations*: A—Adenylation, ACP—acyl carrier protein, AT—acyltransferase, Cy—cyclization, DH—dehydratase, ER—enoyl reductase, GNAT—GCN5-related *N*-acetyltransferase, KS—ketosynthase, KR—ketoreductase, MT—methyltransferase, Ox—Oxidase, Oxy—Oxygenase, PCP—peptide carrier protein, PhyH—phytanoyl-CoA dioxygenase, PS—pyrone synthase, TE—thioesterase, ?—unknown function, *—inactive domain.

179

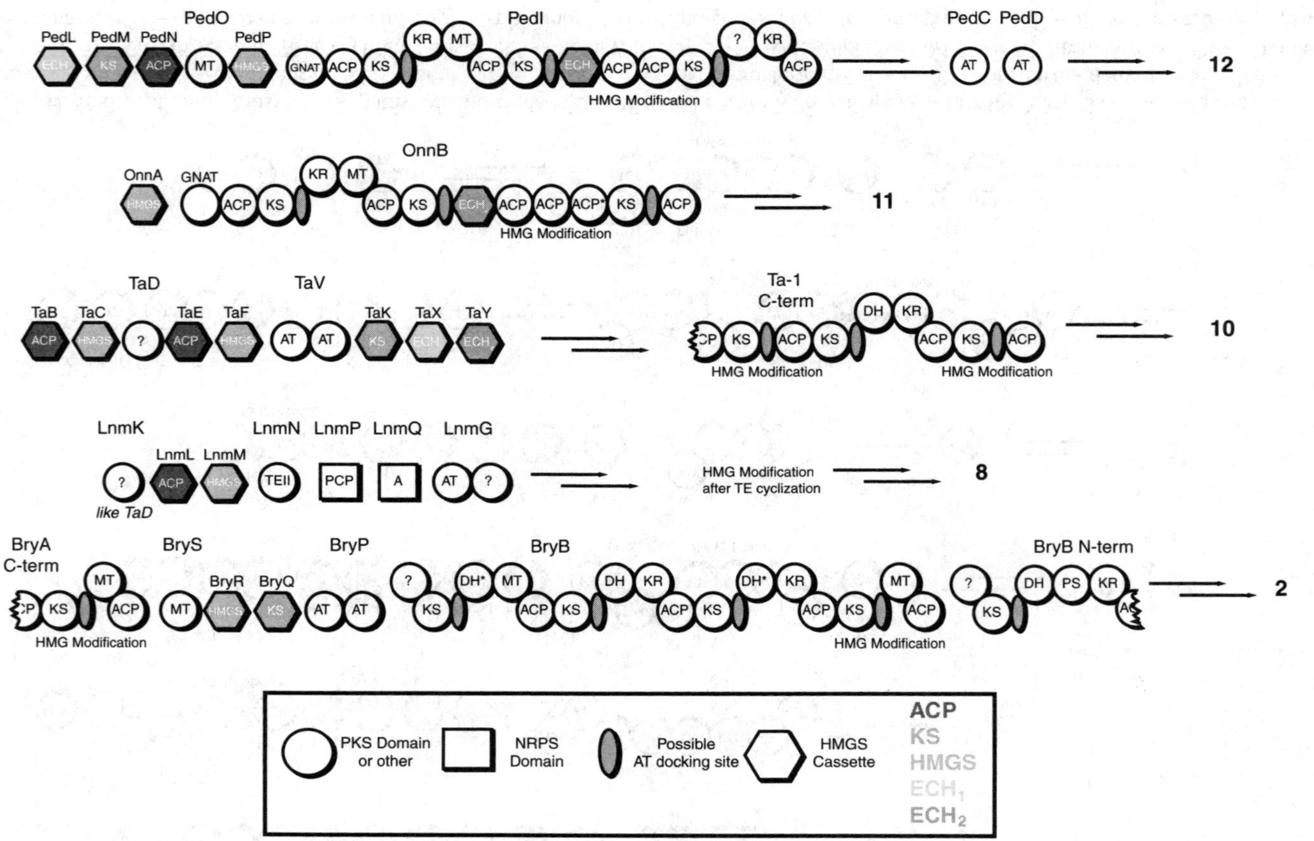

Figure 7.6 (Continued)

converted first to 3-methylglutaconyl-ACP then to 3-methylcrotonyl-ACP, which is the gained intermediate for subsequent formation of the cyclopropyl ring. Insights into the mechanism of the CurF ECH_2 decarboxylation have been gained based on the recent crystal structure of the curacin ECH_2 domain (64). Additional *in vitro* evidence for the function of these enzymes has been generated using proteins from the PksX pathway of *Bacillus subtilis* (46) and the myxovirescin pathway from *Myxococcus xanthus* (58). Prior to the identification of bacillaene as the product of the PksX pathway, Calderone et al. (46) and Dorrestein et al. (57) reported the function of several discrete enzymes. Using radioactive biochemical assays together with mass spectrometry, they assigned functional roles to AcpK, PksC, the tandem ACPs in PksL, PksF, PksG, PksH, and PksI. Using the model acceptor ACP, acetoacetyl-ACP, and malonyl-CoA in combination with the above proteins, a Δ^2-isoprenyl-S-carrier protein was generated (46). Most recently, a similar *in vitro* investigation was conducted using the homologous enzymes from the myxovirescin pathway (58). The HMGS cassette reaction sequence proposed above held fast for the myxovirescin pathway, although the generation of the propionyl- or methylmalonyl-S-ACP could not be demonstrated. The authors have suggested that perhaps additional enzymes are yet to be identified to fill these roles to complete the β-ethylation at C16 in **10**. Two variations on the HMGS cassette theme already have been identified. In the biosynthesis of bryostatin **2**, and leinamycin **8**, one or both of the ECH-mediated steps is likely omitted based on the final natural product structures. The details of these deviations have not yet been established.

Recently, *in vivo* evidence for the function of these HMGS cassettes has come from the Müller lab (49, 50, 65). To date, all HMGS cassette proteins for myxovirescin A (TaB/TaC, TaE/TaF, TaK, TaX, TaY) has been individually deleted, and the impact on the products of the engineered *Myxococcus xanthus* strains has been analyzed. In addition, analysis of products from ΔtaV (the trans-acting AT), ΔtaH (a cytochrome P450 that is thought to hydroxylate the HMGS-installed β-methyl group at C12 of **10**), and ΔtaQ (an O-methyl transferase necessary for completing the transformation to the final methoxy methyl functionality) strains have provided insights into this complex pathway. Production of myxovirescin A was abolished (or greatly reduced) in all above deletion strains. Appearance of novel myxovirescin analogs (β-methyl vs β-ethyl at C16) in the ΔtaE & ΔtaF *strains* appears to be a result of TaB or TaC complementation, which provides direct evidence for TaE/TaF in the formation of the ethyl branch point. However, independent biochemical verification of TaF function has been difficult to obtain (58).

7.3.4 HMGS Cassette Architecture

Analysis of the placement of the known HMGS cassettes identified to date into their biosynthetic clusters reveals a variety of possible architectures (Fig. 7.6). For example, the ECH_2 decarboxylase exists as a discrete enzyme downstream of the ECH_1 dehydratase (mupirocin and others), as an N-terminal domain of a large PKS (curacin and jamaicamide), and as an embedded domain (pederin and onnamide). Although most clusters published to date are mixed PKS/NRPS systems with *in trans* ATs and tandem ACPs at the site of HMGS modification, exceptions exist for each example (difficidin is PKS only, curacin and jamaicamide contain embedded ATs, and bryostatin and myxovirescin do not contain tandem ACPs at the site of HMGS modification).

As HMGS enzyme cassettes have been identified and functionally characterized only recently, many mechanistic details as well as the key protein–protein interactions needed to orchestrate communication among the polypeptide components remain unclear. Details on how the individual proteins are brought to the correct place in the pathway to perform their functions are still unknown for most pathways. In the case of the PksX/bacillaene pathway, some intriguing microscopy performed on *B. subtilis* suggests that the bacillaene proteins are clustered into a huge mega-enzyme factory inside the bacterial cell (66). Whether this organization extends (or is limited) to the other members of HMGS cassette containing pathways remains to be observed. Additionally, in some pathways, key enzymes have yet to be identified. Two lingering questions include, 1) Which AT domain loads the discrete ACP in the embedded AT systems typified by the curacin and jamaicamide pathways? and 2) where are the missing domains located in the incomplete cassettes? Despite these remaining issues, the stage is now set for these unique suites of enzymes to be included and applied in the growing metabolic engineering/combinatorial biosynthesis toolbox.

The goal of this review has been to highlight a series of novel systems for creating chemical diversity in polyketide natural product biosynthesis. This review includes the mechanistic basis for introduction of *trans* or *cis* double bonds within linear or macrocyclic compounds, or assembly of the rare terminal alkene in select secondary metabolites such as curacin and tautomycetin. Similarly, introduction of methyl groups to create branch points or gem dimethyl functionality can occur by several processes that have been dissected in several systems over the past few years. Finally, one of the most intriguing new methods for introduction of diverse branching functionality involves the HMGS-containing enzymes that are being identified in a growing number of PKS and mixed NRPS-PKS pathways. The rapidly increasing knowledge and mechanistic understanding of these complex metabolic systems will provide growing opportunities to engineer chemical diversity using rational approaches.

7.4 ACKNOWLEDGMENTS

Research on modular type I polyketide synthases in the Sherman laboratory is supported generously by grants from the National Institutes of Health (GM076477, CA108874, TW007404), the Hans and Ella McCollum Vahlteich Research Fund at the University of Michigan College of Pharmacy, and the H.W. Vahlteich Professorship in Medicinal Chemistry. J.D.K. is supported by an NRSA postdoctoral fellowship (GM075641) from the NIH.

REFERENCES

1. Koehn FE, Carter GT. The evolving role of natural products in drug discovery. Nat. Rev. Drug Discov. 2005;4:206–220.

2. Newman DJ, Cragg GM. Natural products as sources of new drugs over the last 25 years. J. Nat. Prod. 2007;70:461–477.

3. Newman DJ, Cragg GM, Snader KM. The influence of natural products upon drug discovery. Nat. Prod. Rep. 2000;17:215–234.

4. Stinear TP, Mve-Obian A, Small PLC, Frigui W, Pryor MJ, Brosch R, Jenkin GA, Johnson PDR, Davies JK, Lee RE, Adusumilli S, Garnier T, Haydock SF, Leadlay PF, Cole AT. Giant plasmid-encoded polyketide synthases produce the macrolide toxin of *Mycobacterium ulcerans*. Proc. Natl. Acad. Sci. U.S.A. 2004;101:1345–1349.

5. Kumar P, Schelle MW, Jain M, Lin FL, Petzold CJ, Leavell MD, Leary JA, Cox JS, Bertozzi CR. PapA1 and PapA2 are acyltransferases essential for the biosynthesis of the *Mycobacterium tuberculosis* virulence factor Sulfolipid-1. Proc. Natl. Acad. Sci. U.S.A. 2007;104:11221–11226.

6. Partida-Martinez LP, Hertweck C. A gene cluster encoding rhizoxin biosynthesis in "Burkholderia rhizoxina", the bacterial endosymbiont of the fungus Rhizopus microsporus. ChemBioChem. 2007;8:41–45.

7. Hill AM. The biosynthesis, molecular genetics and enzymology of the polyketide-derived metabolites. J. Nat. Prod. 2006;23:256–320.

8. Shen B. Polyketide biosynthesis beyond the type I, II and III polyketide synthase paradigms. Curr. Opin. Chem. Biol. 2003;7:285–295.

9. Staunton J, Weissman KJ. Polyketide biosynthesis: a millennium review. Nat. Prod. Rep. 2001;18:380–416.

10. Reeves CD. The enzymology of combinatorial biosynthesis. Crit. Rev. Biotechnol. 2003;23:95–147.

11. Sherman DH, Smith JL. Clearing the skies over modular polyketide synthases. ACS Chem. Biol. 2006;1:505–509.

12. Tang Y, Kim CY, Mathews II, Cane DE, Khosla C. The 2.7 Angstrom crystal structure of a 194-kDa homodimeric fragment of the 6-deoxyerythronolide B synthase. Proc. Natl. Acad. Sci. 2006;103:11124–11129.

13. Giraldes JW, Akey DL, Kittendorf JD, Sherman DH, Smith JL, Fecik RA. Structural and mechanistic insights into polyketide macrolactonization from polyketide-based affinity labels. Nat. Chem. Biol. 2006;2:531–536.

14. Akey DL, Kittendorf JD, Giraldes JW, Fecik RA, Sherman DH, Smith JL. Structural basis for macrolactonization by the pikromycin thioesterase. Nat. Chem. Biol. 2006;2:537–542.

15. Keatinge-Clay AT, Stroud RM. The structure of a ketoreductase determines the organization of the beta-carbon processing enzymes of modular synthases. Structure 2006;14:737–748.

16. Tsai SC, Lu H, Cane DE, Khosla C, Stroud RM. Insights into channel architecture and substrate specificity from crystal structures of two macrocycle-forming thioesterases of modular polyketide synthases. Biochemistry 2002;41:12598–12606.

17. Maier T, Jenni S, Ban N. Architecture of mammalian fatty acid synthase at 4.5 A resolution. Science 2006;311:1258–1262.

18. Keatinge-Clay AT. A tylosin ketoreductase reveals how chirality is determined in polyketides. Chem. Biol. 2007;14:898–908.

19. Khosla C, Tang Y, Chen AY, Schnarr NA, Cane DE. Structure and mechanism of the 6-deoxyerythronolide B synthase. Annu. Rev. Biochem. 2007;76:11.11–11.27.

20. Weissman KJ, Leadlay PF. Combinatorial biosynthesis of reduced polyketides. Nat. Rev. Microbiol. 2005;3:925–936.

21. Kittendorf JD, Sherman DH. Developing tools for engineering hybrid polyketide synthetic pathways. Curr. Opin. Biotechnol. 2006;17:597–605.

22. Baltz RH. Molecular engineering approaches to peptide, polyketide and other antibiotics. Nat. Biotechnol. 2006;24:1533–1540.

23. Rodriquez E, McDaniel R. Combinatorial biosynthesis of anti microbials and other natural products. Curr. Opin. Microbiol. 2001;4:526–534.

24. Reid R, Piagentini M, Rodriguez E, Ashley G, Viswanathan N, Carney J, Santi DV, Hutchinson CR, McDaniel R. A model of structure and catalysis for ketoreductase domains in modular polyketide synthases. Biochemistry 2003;42:72–79.

25. Caffrey P. Conserved amino acid residues correlating with ketoreductase stereospecificity in modular polyketicle synthases. ChemBioChem. 2003;4:654–657.

26. Xue Y, Zhao L, Liu HW, Sherman DH. A gene cluster for macrolide antibiotic biosynthesis in *Streptomyces venezuelae*: architecture of metabolic diversity. Proc. Natl. Acad. Sci. U.S.A 1998;95:12111–12116.

27. Wu JQ, Zaleski TJ, Valenzano C, Khosla C, Cane DE. Polyketide double bond biosynthesis. Mechanistic analysis of the dehydratase-containing module 2 of the picromycin/methymycin polyketide synthase. J. Am. Chem. Soc. 2005;127: 17393–17404.

28. Leesong M, Henderson BS, Gillig JR, Schwab JM, Smith JL. Structure of a dehydratase-isomerase from the bacterial pathway for biosynthesis of unsaturated fatty acids: two catalytic activities in one active site. Structure 1996;4:253–264.

29. Sedgwick B, Morris C, French SJ. Stereochemical course of dehydration catalyzed by yeast fatty-acid synthetase. J. Chem. Soc. Chem. Commun. 1978;193–194.

30. Perlova O, Klaus G, Kaiser O, Hans A, Müller R. Identification and analysis of the chivosazol biosynthetic gene cluster from the myxobacterial model strain *Sorangium cellulosum* So ce56. J. Biotechnol. 2006;121:174–191.

31. Tang L, Shah S, Chung L, Carney J, Katz L, Khosla C, Julien B. Cloning and heterologous expression of the epothilone gene cluster. Science 2000;287:640–642.

32. Tang L, Ward S, Chung L, Carney JR, Li Y, Reid R, Katz L. Elucidating the mechanism of cis double bond formation in epothilone biosynthesis. J. Am. Chem. Soc. 2004;126:46–47.

33. Palaniaappan N, Kim BS, Sekiyama Y, Osada H, Reynolds KA. Enhancement and selective production of phoslactomycin B, a protein phosphatase IIa inhibitor, through identification and engineering of the corresponding biosynthetic gene cluster. J. Biol. Chem. 2003;278:35552–35557.

34. Carvalho R, Reid R, Viswanathan N, Gramajo H, Julien B. The biosynthetic genes for disorazoles, potent cytotoxic compounds that disrupt microtubule formation. Gene 2005;359:91–98.

35. Chang Z, Sitachitta N, Rossi JV, Roberts MA, Flatt PM, Jia J, Sherman DH, Gerwick WH. Biosynthetic pathway and gene cluster analysis of curacin A, an antitubulin natural product from the tropical marine cyanobacterium *Lyngbya majuscula*. J. Nat. Prod. 2004;67:1356–1367.

36. Butcher RA, Schroeder FC, Fischbach MA, Straight PD, Kolter R, Walsh CT, Clardy J. The identification of bacillaene, the product of the PksX megacomplex in *Bacillus subtilis*. Proc. Natl. Acad. Sci. 2007;104:1506–1509.

37. Alhamadsheh MM, Palaniappan N, DasChouduri S, Reynolds KA. Modular polyketide synthases and cis double bond formation: establishment of activated cis-3-cyclohexylpropenoic acid as the diketide intermediate in phoslactomycin biosynthesis. J. Am. Chem. Soc. 2007;129:1910–1911.

38. Negishi M, Pedersen LG, Petrotchenko E, Shevtsov S, Gorokhov A, Kakuta Y, Pedersen LC. Structure and function of sulfotransferases. Arch. Biochem. Biophys. 2001;390:149–157.

39. Shim JH, Lee HK, Chang EJ, Chae WJ, Han JH, Han DJ, Morio T, Yang JJ, Bothwell A, Lee SK. Immunosuppressive effects of tautomycetin in vivo and in vitro via T cell-specific apoptosis induction. Proc. Natl. Acad. Sci. U.S.A. 2002;99:10617–10622.

40. Choi SS, Hur YA, Sherman DH, Kim ES. Isolation of the biosynthetic gene cluster for tautomycetin, a linear polyketide T cell-specific immunomodulator from *Streptomyces* sp. CK4412. Microbiology 2007;153:1095–1102.

41. González-Lergier J, Broadbelt LJ, Hatzimanikatis V. Theoretical considerations and computational analysis of the complexity in polyketide synthesis pathways. J. Am. Chem. Soc. 2005;127:9930–9938.

42. Cheng YQ, Tang GL, Shen B. Type I polyketide synthase requiring a discrete acyltransferase for polyketide biosynthesis. Proc. Natl. Acad. Sci. U.S.A. 2003;100: 3149–3154.

43. Tang GL, Cheng YQ, Shen B. Leinamycin biosynthesis revealing unprecedented architectural complexity for a hybrid polyketide synthase and nonribosomal peptide synthetase. Chem. Biol. 2004;11:33–45.

44. Sudek S, Lopanik NB, Waggoner LE, Hildebrand M, Anderson C, Liu H, Patel A, Sherman DH, Haygood MG. Identification of the putative bryostatin polyketide synthase gene cluster from "*Candidatus* Endobugula sertula", the uncultivated microbial symbiont of the marine bryozoan *Bugula neritina*. J. Nat. Prod. 2007;70:67–74.

45. Chen XH, Vater J, Piel J, Franke P, Scholz R, Schneider K, Koumoutsi A, Hitzeroth G, Grammel N, Strittmatter AW, Gottschalk G, Süssmuth RD, Borriss R. Structural and functional characterization of three polyketide synthase gene clusters in *Bacillus amyloliquefaciens* FZB 42. J. Bacteriol. 2006;188:4024–4036.

46. Calderone CT, Kowtoniuk WE, Kelleher NL, Walsh CT, Dorrestein PC. Convergence of isoprene and polyketide biosynthetic machinery: isoprenyl-S-carrier proteins in the *pksX* pathway of *Bacillus subtilis*. Proc. Natl. Acad. Sci. U.S.A. 2006;103: 8977–8982.

47. El-Sayed AK, Hothersall J, Cooper SM, Stephens ER, Simpson TJ, Thomas CM. Characterization of the mupirocin biosynthesis gene cluster from *Pseudomonas fluorescens* NCIMB 10586. Chem. Biol. 2003;10:419–430.

48. Pulsawat N, Kitani S, Nihira T. Characterization of biosynthetic gene cluster for the production of virginiamycin M, a streptogramin type A antibiotic, in *Streptomyces virginiae*. Gene 2007;393:31–42.

49. Simunovic V, Zapp J, Rachid S, Krug D, Meiser P, Müller R. Myxovirescin A biosynthesis is directed by hybrid polyketide synthases/nonribosomal peptide synthetase, 3-hydroxy-3-methylglutaryl-CoA synthases, and trans-acting acyltransferases. ChemBioChem. 2006;7:1206–1220.

50. Simunovic V, Müller R. 3-Hydroxy-3-methylglutaryl-CoA-like synthases direct the formation of methyl and ethyl side groups in the biosynthesis of the antibiotic myxovirescin A. ChemBioChem. 2007;8:497–500.

51. Piel J, Hui D, Wen G, Butzke D, Platzer M, Fusetani N, Matsunaga S. Antitumor polyketide biosynthesis by an uncultivated bacterial symbiont of the marine sponge *Theonella swinhoei*. Proc. Natl. Acad. Sci. U.S.A. 2004;101:16222–16227.

52. Piel J, Wen G, Platzer M, Hui D. Unprecedented diversity of catalytic domains in the first four modules of the putative pederin polyketide synthase. ChemBioChem. 2004;5:93–98.

53. Piel J. A polyketide synthase-peptide synthetase gene cluster from an uncultured bacterial symbiont of *Paederus* beetles. Proc. Natl. Acad. Sci. U.S.A. 2002;99: 14002–14007.

54. Gu L, Jia J, Liu H, Håkansson K, Gerwick WH, Sherman DH. Metabolic coupling of dehydration and decarboxylation in the curacin A pathway: functional identification of a mechanistically diverse enzyme pair. J. Am. Chem. Soc. 2006;128:9014–9015.

55. Edwards DJ, Marquez BL, Nogle LM, McPhail K, Goeger DE, Roberts MA, Gerwick WH. Structure and biosynthesis of the jamaicamides, new mixed polyketide-peptide neurotoxins from the marine cyanobacterium *Lyngbya majuscula*. Chem. Biol. 2004;11:817–833.

56. Lange BM, Rugan T, Martin W, Croteau R. Isoprenoid biosynthesis: The evolution of two ancient and distinct pathways across genomes. Proc. Natl. Acad. Sci. U.S.A. 2000;97:13172–13177.

57. Dorrestein PC, Bumpus SB, Calderone CT, Garneau-Tsodikova S, Aron ZD, Straight PD, Kolter R, Walsh CT, Kelleher NL. Facile detection of acyl and peptidyl intermediates on thiotemplate carrier domains via phosphopantetheinyl elimination reactions during tandem mass spectrometry. Biochemistry 2006;45:12756–12766.

58. Calderone CT, Iwig DF, Dorrestein PC, Kelleher NL, Walsh CT. Incorporation of nonmethyl branches by isoprenoid-like logic: multiple β-alkylation events in the biosynthesis of myxovirescin A1. Chem. Biol. 2007;14:835–846.

59. Campobasso N, Patel M, Wilding IE, Kallender H, Rosenberg M, Gwynn MN. *Staphylococcus aureus* 3-hydroxy-3-methylglutaryl-CoA synthase. J. Biol. Chem. 2004;279:44883–44888.

60. Steussy CN, Robison AD, Tetrick AM, Knight JT, Rodwell VW, Stauffacher CV, Sutherlin AL. A structural limitation on enzyme activity: the case of HMG-CoA synthase. Biochemistry 2006;45:14407–14414.

61. Theisen ML, Misra I, Saadat D, Campobasso N, Miziorko HM, Harrison DHT. 3-Hydroxy-3-methylglutaryl-CoA synthase intermediate complex observed in "real-time". Proc. Natl. Acad. Sci. U.S.A. 2004;101:16442–16447.

62. Steussy CN, Vartia AA, Burgner JW II, Sutherlin A, Rodwell VW, Stauffacher CV. X-ray crystal structures of HMG-CoA synthase from *Enterococcus faecalis* and a complex with its second substrate/inhibitor acetoacetyl-CoA. Biochemistry 2005;44:14256–14267.

63. Pojer F, Ferrer JL, Richard SB, Nagegowda DA, Chye ML, Bach TJ, Noel JP. Structural basis for the design of potent and species-specific inhibitors of 3-hydroxy-3-methylglutaryl-CoA synthases. Proc. Natl. Acad. Sci. U.S.A. 2006;103: 11491–11496.

64. Geders TW, Gu L, Mowers JC, Liu H, Gerwick WH, Håkansson K, Sherman DH, Smith JL. Crystal structure of the ECH_2 catalytic domain of CurF from *Lyngbya majuscula:* Insights inot a decarboxylase involved in polyketide chain β-branching. J. Biol. Chem. 2007;282:35954–35963.

65. Simunovic V, Müller R. Mutational analysis of the myxovirescin biosynthetic gene cluster reveals novel insights into the functional elaboration of polyketide backbones. ChemBioChem. 2007;8:1273–1280.

66. Straight PD, Fischbach MA, Walsh CT, Rudner DZ, Kolter R. A singular enzymatic megacomplex from *Bacillus subtilis*. Proc. Natl. Acad. Sci. U.S.A. 2007;104: 305–310.

8

POLYKETIDE POLYETHERS

ALISON M. HILL

School of Biosciences, University of Exeter, Stocker Road, Exeter, United Kingdom

This review covers the biosynthesis of terrestrial and marine polyethers and discusses their biologic properties and the molecular genetics and enzymology of the proteins responsible for their formation. The biosynthesis of monensin, nanchangmycin, nonactin, and the marine polyether ladders are discussed in detail. Novel enzymes found only in type I polyketide polyether gene clusters that are responsible for the epoxidation and cyclization of polyene biosynthetic intermediates are described. The macrotetrolide biosynthetic gene cluster, which is an ACP-less type II polyketide synthase that functions noniteratively is reviewed.

The first polyether antibiotic to be isolated was nigericin **1** in 1951. By 1983, more than 70 terrestrial polyether antibiotics had been reported and Cane et al. (1) had proposed a unified stereochemical model to account for their biosynthesis, including the polyene–polyepoxide model of polyether formation. It took almost 20 years for the first polyether gene cluster to be reported (2). In contrast, the first marine polyethers were reported in 1981, and a model to explain the biosynthesis of marine polyether ladder compounds was proposed in 2006.

8.1 BIOLOGICAL BACKGROUND

Interest in these compounds (Fig. 8.1) derives from their ability to transport ions across biologic membranes, and some terrestrial polyethers have been used widely in veterinary medicine. Marine polyethers are responsible for numerous cases of human food poisoning and toxic algal tides, which cause massive fish kills.

Natural Products in Chemical Biology, First Edition. Edited by Natanya Civjan.
© 2012 John Wiley & Sons, Inc. Published 2012 by John Wiley & Sons, Inc.

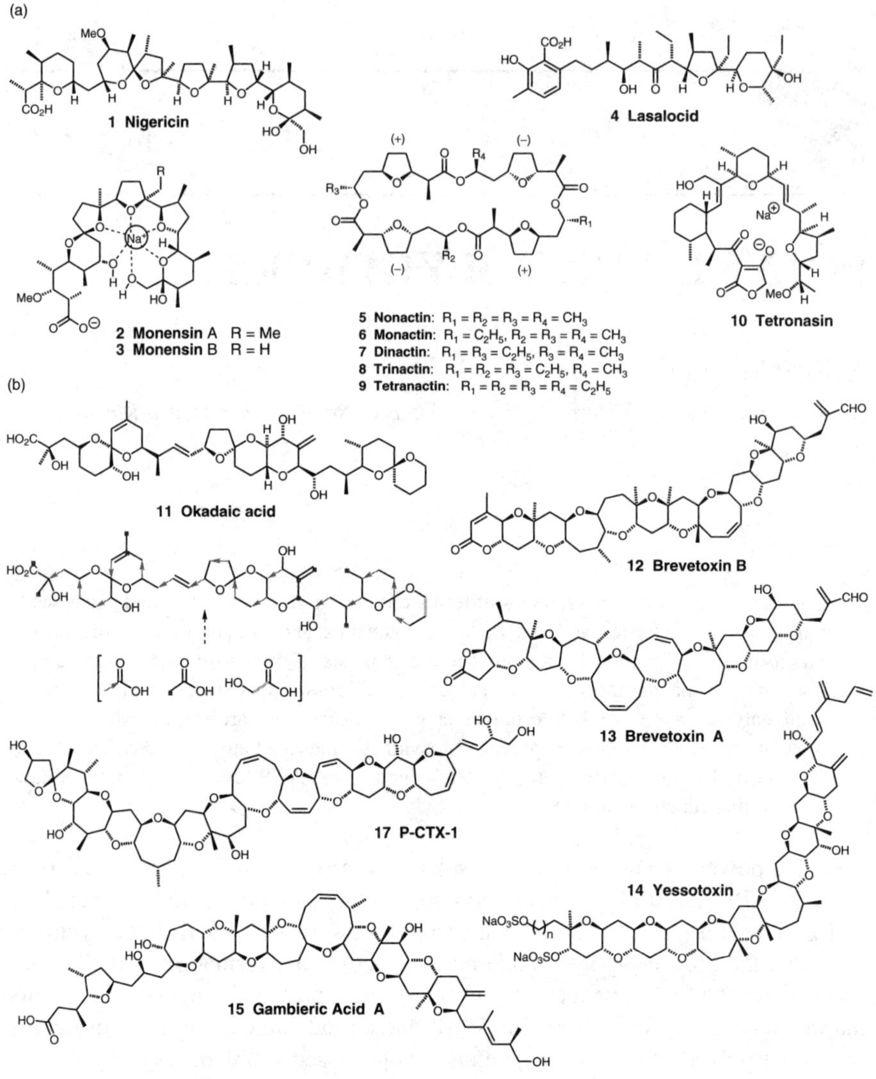

(a)

1 Nigericin

4 Lasalocid

2 Monensin A R = Me
3 Monensin B R = H

5 Nonactin: $R_1 = R_2 = R_3 = R_4 = CH_3$
6 Monactin: $R_1 = C_2H_5, R_2 = R_3 = R_4 = CH_3$
7 Dinactin: $R_1 = R_3 = C_2H_5, R_2 = R_4 = CH_3$
8 Trinactin: $R_1 = R_2 = R_3 = C_2H_5, R_4 = CH_3$
9 Tetranactin: $R_1 = R_2 = R_3 = R_4 = C_2H_5$

10 Tetronasin

(b)

11 Okadaic acid

12 Brevetoxin B

13 Brevetoxin A

17 P-CTX-1

14 Yessotoxin

15 Gambieric Acid A

Figure 8.1 (a) Terrestrial polyethers; (b) marine polyethers.

8.1.1   Terrestrial Polyethers

At physiological pH, the ionophores are ionized with the fat-soluble part of the molecule residing in the lipid bilayer of the membrane and the ionized moiety in the aqueous milieu (4). Binding of the metal ion takes place at the membrane surface: As successive ether oxygen atoms from the ionophore bind to the metal, it loses its solvated water molecules, thereby forming a neutral zwitterionic metal–ionophore complex. Transport across the membrane can now take place, and at the opposite surface, the process is reversed to leave the metal cation and the anionic ionophore on the other side of the membrane. The ionophore must bind another cation (usually a proton) to return to its original starting point.

The anionic ionophores are highly selective for particular metal cations, and both kinetic and thermodynamic terms determine which cation is selected. Note that the thermodynamic stability of the metal–ionophore complex does not always determine the transport rate. For example, nigericin forms a much more stable complex with potassium but transports sodium much more quickly. In the presence of both sodium and potassium, however, the extra stability of the nigericin–potassium complex results in the preferential transport of potassium over sodium (4).

Selectivities of ionophores are shown in Table 8.1 (5, 6). The carboxyl group may (e.g., nigericin) or may not (e.g., monensin 2) be involved in cation liganding via an ionic bond. Monensin has six liganding oxygen atoms but none for ionic bonds, and it is a weaker complexing agent than is nigericin. Lasalocid 4 can form a singly charged complex with doubly charged cations such as Ca^{2+}. The broad specificity of lasalocid results from the metals sitting on top rather than within the oxygen system of the ionophore. For nonactin 5, the ester carbonyl oxygen atoms and the ether oxygen atoms are all liganding and form the apices of a cube. Nonactin adopts a conformation resembling the seam of a tennis ball (5).

Commercial interest in the polyketide polyethers stems from their antibiotic activity. Many polyethers of the class are potent coccidiostats and can control coccidial infections in poultry when added at approximately 100 ppm to the feed (7). Cocciodosis is caused from an infection in the digestive tract by parasitic protozoa, most commonly in poultry from the genus *Eimeria*. Monensin was the first polyether antibiotic to be patented and was released onto the market in 1971. It is still widely used today and is marketed under the tradename of Coban® (Elanco, Toronto, Ontario, Canada). After monensin treatment, coccidia have been observed to literally explode at one stage in their lifecycle because of the increase in osmotic pressure generated from the exchange of sodium outside the organism for protons inside (4). Lasalocid was launched in 1976 under the tradename Avatec® (Alpharma, Inc., Bridgewater, NJ) (7), and unlike monensin, it cannot be used for laying hens and must be withheld from poultry for

TABLE 8.1 **Ionophore selectivities (5, 6)**

Ionophore	MW	Selectivity sequence
Nigericin (1)	724	K > Rb > Na > Cs > > Li
Monensin A (2)	670	Na ≫ K > Rb > Li > Cs
Lasalocid (4)	590	Cs > Rb ≈ K > Na > Li; Ba > Sr > Ca > Mg
Nonactin (5)	736	NH_4 > K ≈ Rb > Cs > Na
Tetronasin (10)	602	Ca > Mg > Na

three days before slaughter. The use of these compounds has led to some controversy as small amounts of the antibiotics have been detected in food for human consumption. Lasalocid residues have been detected in eggs, including organic eggs, and in 2004, the Soil Association recommended that maximum residue limits and acceptable daily intake levels should be set in line with Australian limits of 50 μg/kg in eggs and 1 μg/kg, respectively.

Ionophores can also be used in small doses to control bacteria in the rumen of cattle, which alters the ruminal fermentation to increase the amount of propionate (6). It results in an improvement in the feed efficiency and a reduction in methane output. Rumensin® (Elanco) is the tradename for monensin. Nigericin and the divalent antiporter tetronasin 10 also exert the same effect. Changes in ruminal fermentation patterns have been shown not to be cation specific, but the ionophore must have sodium or potassium and proton antiporter activity.

More recently, antimalarial activity has also been reported for monensin, nigericin, and lasalocid (8, 9). *Plasmodium falciparum* is sensitive to these compounds at all stages, but the schizont stages were most sensitive with 12 hours required for complete parasite clearance (8). In infected mice treated with 10 mg/kg of monensin, all mice were cured compared with one third of the mice given nigericin. A combination of the two drugs was shown to be even more effective allowing the dose to be decreased. The biological effect is thought to develop from the depletion of protons from the parasite's intracellular vesicles that results in acidification of the parasite cytosol, thereby inhibiting hemoglobin digestion, mitosis, and endocytic and exocytic pathways (9).

8.1.2 Marine Polyethers

Isolation of marine polyethers (Fig. 8.1b) is more recent than from terrestrial sources with structures for okadaic acid 11 and brevetoxin B 12 reported in 1981. Since that time, numerous polyether structures have been isolated from marine sources and several have had their biosynthetic pathways investigated through classical feeding experiments [e.g., okadaic acid (10)].

Considerable interest in these compounds exists as many are implicated in more than 60,000 cases of poisoning every year, of which 1.5% of these are fatal (11). These toxins are produced by marine microalgae, predominantly dinoflagellates, and are associated with red tides and lead to massive fish kills and mollusc contamination. Most dinoflagellate toxins are polyketides and include many polyethers with distinctive ring junctions resembling the rungs of a ladder. The dinoflagellate toxins are hazardous if inhaled or ingested. They are odorless, tasteless, and not destroyed by cooking or autoclaving. These toxins are extremely toxic in minute quantities, and no acceptable exposure limits have been established.

Okadaic acid 11 is a potent and specific inhibitor of protein phosphatases produced by the dinoflagellates *Prorocentrum lima*, *Dinophysis fortii*, and *Dinophysis accuminata*. It accumulates in bivalves and is one of the main toxins responsible for diarrhetic shellfish poisoning (DSP) (11). It has a highly unusual biosynthesis that has generated a lot of speculation because of the presence of "isolated" acetate chain methyl carbon atoms (10). Okadaic acid is a potent inhibitor of protein phosphatase A (11).

The brevetoxins (e.g., brevetoxin A **13** and brevetoxin B **12**) are a family of lipid soluble neurotoxins produced by *Karenia brevis* (formally known as *Gymnodinium breve*) that are structurally similar to yessotoxin. They are thought to exert their biological action through the depolarization of the sodium channels of the excitable membranes (11). Binding of brevetoxin to the voltage-gated sodium channels changes its function by shifting the activation voltage for channel opening to a more negative value and inhibiting the inactivation of opened channels (resulting in persistent activation). If ingested, they cause neurotoxic shellfish poisoning. Brevetoxin derivatives have been patented as a treatment for cystic fibrosis, mucociliary dysfunction, and pulmonary diseases.

Yessotoxins **14** are produced by dinoflagellates of the genera *Protoceratium* and *Gonyaulax*. They cause selective disruption of the E-cadherin-catenin system in epithelial cells (11). In common with the other ladder shaped marine polyethers, yessotoxin interacts with transmembrane helix domains.

The gambieric acids **15**, which are isolated from *Gambierdiscus toxicus*, are potent antifungal compounds that are particularly effective against filamentous fungi but inactive against yeasts. They are also cytotoxic, but they do not exhibit the neurotoxicity that is associated with other large marine polyethers such as yessotoxins, brevetoxins, and maitotoxins (11).

With a molecular weight of 3422 Da, maitotoxin **16** (see Fig. 8.4a) is the largest nonproteinaceous natural product isolated (3). It is a marine polyether ladder produced by the dinoflagellate *G. toxicus*. Maitotoxin is a powerful activator of voltage-insensitive Ca^{2+} channels and exerts its biological effect through the increase of intracellular Ca^{2+} concentration (11). It can be used as a tool for studies on cellular events associated with Ca^{2+} flux. It stimulates synthesis and secretion of the nerve growth factor. The minimum lethal dose of the toxin in mice is 0.17 μg/kg (intraperitoneally).

The ciguatoxins (CTXs) are responsible for the symptoms of ciguatera fish poisoning caused by ingesting certain tropical and semitropical fish from the Indo-Pacific Oceans and Caribbean Sea. Ciguatoxin poisoning is the most frequent food-borne illness related to fin-fish consumption although it is rarely fatal. Symptoms include neurological, cardiovascular, and gastrointestinal disorders, and ciguatoxins can be detected in all body fluids (12). Most people recover slowly over time, which can take from weeks to years. P-CTX1 **17** is the most toxic of all ciguatoxins with an LD_{50} of 0.25 μg/kg intraperitoneally. The toxins derive from the *Gambierdiscus* spp., which produce less-potent compounds that are biotransformed in the livers of the fin-fish to the more toxic ciguatoxins. The compounds activate the voltage-gated sodium channels that result in an increase of intracellular sodium. In high doses, ciguatoxins block the voltage-gated potassium channels that lead to membrane depolarization and that contribute to a lowering of action potential threshold (12).

8.2   CHEMISTRY

Classical feeding experiments with both stable and radioactive isotopic labels (7) enabled the biosynthetic origin of the polyethers to be elucidated and for a general stereochemical model to be proposed (1). More recent work on this class of compounds has focused on a genetic approach, and unusual and interesting genes specific to polyether biosynthesis have been isolated from these clusters.

8.2.1   Monensin

Monensins A **2** and B **3** are polyether ionophores produced by *Streptomyces cinnamonensis* that differ only in the sidechain at C16 (ethyl/methyl). Monensin acts as a specific ionophore to dissipate ionic gradients across cell membranes and is used widely in veterinary medicine and as a food additive in animal husbandry (7). Antimalarial activity has also been reported (8, 9). Monensin is the best studied of the polyether ionophore antibiotics, and it was the first to have its gene cluster sequenced (2).

Early feeding studies established that monensin A is biosynthesized from a classical polyketide pathway and is derived from five acetate, seven propionate, and one butyrate (for monensin B, an additional propionate unit replaces the butyrate) (7). Four of the nine oxygen atoms are derived from molecular oxygen, with the remaining five deriving from the corresponding carboxylic acid precursors (Fig. 8.2a). Based on these initial experiments, it was proposed that the monensin PKS produced a linear *E,E,E*-triene precursor **24** that was oxidized and cyclized to give the final structure (1). Alternative proposals using *Z,Z,Z*- and *E,Z,Z*-trienes have been made (2).

Publication of the gene cluster for monensin (Table 8.2) (2, 13–17) showed that the PKS comprised 12 modules in eight contiguous open reading frames consistent with the production of the linear triene premonensin **25**. The loading module contains an N-terminal KSQ domain that functions as a malonylCoA decarboxylase to generate starter units *in situ*. The monensin PKS does not contain an integrated C-terminal thioesterase domain at the end of module 12; instead an unusual 121 amino acid extension, rich in glycine, asparagines, and glutamine, was found (13). *MonCII*, which was originally assigned as an epoxide cyclase, has been shown to hydrolyze monensinyl NAC thioester **21** (Fig. 8.2c) as well as two other model substrates (14). Deletion of *monCII* gave a mutant that produced none or only trace amounts of monensin, which is consistent with its role as a chain-terminating TE. Complementation of *monCII* on a plasmid restored monensin production. Moreover, cell-free extracts from the *monCII* mutant treated with KOH gave significant amounts of monensin that resulted from the hydrolysis of a monensinyl ester or thioester present in the cell-free extract. Two ORFs, *monAIX* and *monAX*, have been shown to function as Type II thioesterases. Deletion of *monAIX* and/or *monAX* resulted in a modest drop in the monensin titer, which is consistent with their editing Type II TE role (14).

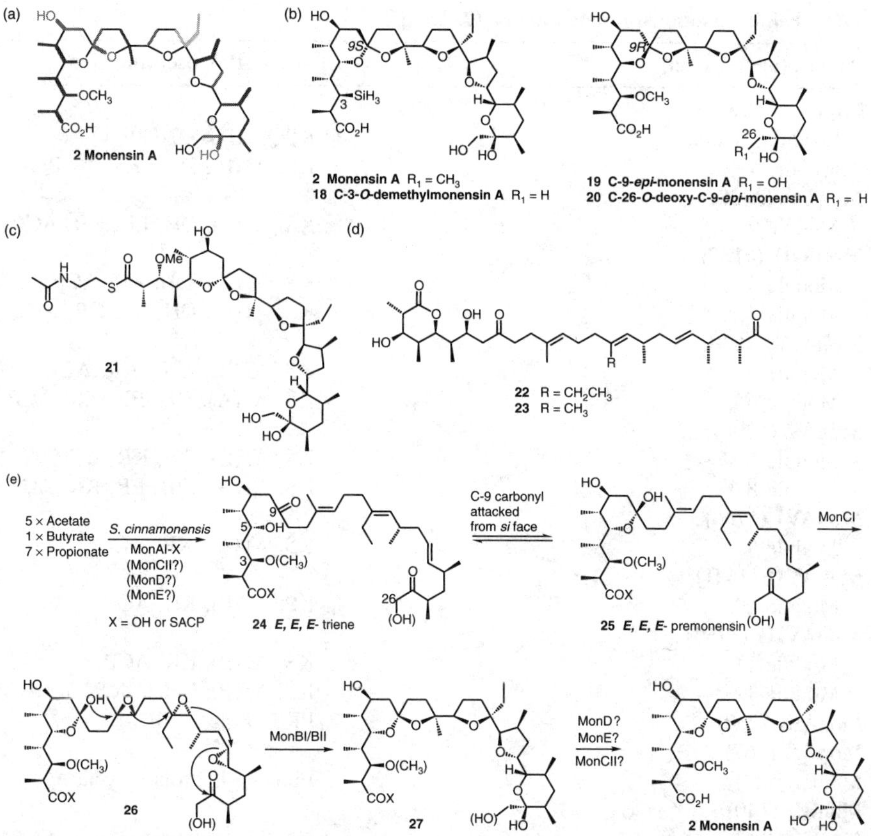

Figure 8.2 (a) Classic biosynthesis of monensin A (7); (b) novel monensins (16); (c) monensinyl *N*-acetylcysteamine thioester (14); (d) triene lactones produced by *monC1* mutant (15); (e) biosynthesis of monensin: The timing of the methylation and hydroxylation steps and release from the PKS are not known for certain (2, 13–16).

Genes for late steps such as methylation (*monE*) and hydroxylation (*monD*) and several regulatory genes (*monH, RI, RII*) were found in the cluster (2, 13). *MonT* is proposed to be involved in monensin export and consequently ensures the producing strain is self-resistant. The *monCI* gene product is a flavin-dependent epoxidase. Deletion of *monCI* results in complete loss of monensin production and the accumulation of the linear *E, E, E*-triene lactones (**22, 23**), (Fig. 8.2d), which differ from premonensin **25** only by the different cyclization pattern of the polyketide chain: δ-lactone (**22, 23**) compared with a hemi-ketal (**25**) (15). Strains in which the adjacent genes *monBI* and/or *monBII* are deleted in addition to *monCI* give the same results. No oxidized derivatives of **22** and **23** were found, which suggests that the epoxidase encoded by *monCII* is necessary and sufficient for epoxidation of all three double bonds. These experiments are

TABLE 8.2   Monensin gene cluster (2, 13–17)

Polypeptide (size aa)	Proposed function
MonAI (3025)	
Loading module	KSQ, AT(A), ACP
Module 1	KS, AT(P), DH,[a] KR, ACP
MonAII (2238)	
Module 2	KS, AT(P), DH, ER, KR, ACP
MonAIII (4132)	
Module 3	KS, AT(A), DH, KR, ACP
Module 4	KS, AT(P), DH, ER, KR, ACP
MonAIV (4038)	
Module 5	KS, AT(B),[b] DH, KR, ACP
Module 6	KS, AT(A), DH, ER, KR, ACP
MonAV (4106)	
Module 7	KS, AT(P), DH, KR, ACP
Module 8	KS, AT(A), DH, ER, KR, ACP
MonAVI (1700)	
Module 9	KS, AT(A), KR, ACP
MonAVII (1641)	
Module 10	KS, AT(P), KR, ACP
MonAVIII (3753)	
Module 11	KS, AT(P), KR, ACP
Module 12	KS, AT(P), DH,[a] ER,[c] KR, ACP
MonAIX (268)	PKS, Type II TE
MonAX (267)	
MonBI (144)	Epoxide hydrolase/cyclase[d]
MonBII (140)	
MonCI (496)	Non-haem epoxidase
MonCII (299)	Chain terminating TE[e]
MonD (458)	Hydroxylase
MonE (276)	Methyltranferase
MonH, MonRI, MonRII	Transcriptional regulators
MonT (511)	Efflux protein, resistance protein

Reproduced from Reference 17 with permission from the Royal Society of Chemistry.
[a]"Unnecessary" domain (sequence is indistinguishable from active Mon DH domains).
[b]The AT of module 5 should incorporate an ethylmalonate extender unit, and it contains a signature sequence very close to the propionate sequence.
[c]Predicted to be inactive.
[d]Originally assigned as double-bond isomerase.
[e]Originally assigned as epoxide cyclase.

consistent with an E,E,E-configured triene as originally proposed by Cane et al. (1). Heterologous expression of MonCI in *Streptomyces coelicolor* gave a strain that could convert linalool to linalool oxide with a 10–20-fold greater conversion activity than studies carried out previously with *S. cinnamonensis* (13).

MonBI and *monBII* are highly homologous to each other and significantly similar to a Δ^3-3-ketosteroid isomerase of *Comomonas testosteroni*. Originally, it was proposed that *monBI* and *monBII* interconverted *E* double bonds to *Z* via an extended enolate ion during polyketide biosynthesis in two modules (resulting in an *E,Z,Z*-triene intermediate) (2). Their assignment has been revised in light of new experimental evidence that has shown that they are involved in the epoxide ring opening and concomitant polyether ring formation (16). Deletion of *monBI* and/or *monBII* gave strains that produced no monensin. Instead, C-3-*O*-demethylmonensins **18**, C-9-*epi*-monensins **19**, and C-26-deoxy-*epi*-monensins **20** (Fig. 8.2b) were produced in addition to numerous minor components. The major products from these mutant strains were the C-3-*O*-demethyl analogs. Hence, in the absence of the C-3-methoxy group, nucleophilic attack by the C-5 hydroxyl group from the *si* face of the C-9 carbonyl giving the natural epimer of the spiroketal takes place (Fig. 8.2e). In contrast, the presence of the methoxy group at C-3 leads to the C-5 hydroxy attacking the C-9 carbonyl from the *re* face leading to the unnatural spiroketal epimer **19**. The C-26-deoxy-*epi*-monensins result from them being poor substrates for MonE. Treatment with acid converted the *epi*-monensins into the more thermodynamically stable corresponding monensins. Additional minor metabolites were detected in the cell culture but were produced in insufficient quantities to allow for structure determination. LC-MS analysis revealed that their molecular weights were identical to monensins A/B and C-3-*O*-demethylmonensins A/B, which suggests that they were intermediates with the correct oxidation state but cyclized incompletely. Treatment of these minor metabolites with acid led to their conversion to monensins A/B and C-3-*O*-demethylmonensins A/B. Taken together, these results show that the MonB enzymes are epoxide hydrolase/cyclase enzymes that accelerate, but do not change, the stereochemical course of polyether ring formation (16). Since the pattern of products in MonBI and MonBII mutants seem to be identical, it seems that their actions are somehow coordinated and could in principle exist as a heterodimer in *S. cinnamonensis*. Although the arrangement of the proteins is not known, it is clear that ring closure in monensin biosynthesis occurs stepwise in a precise order under enzymatic control from one or both MonB enzymes. The timing of the hydroxylation step at C-26 has also been shown to occur before either methylation or polyether ring formation (16). The biosynthetic pathway for monensin is shown in Fig. 8.2e.

8.2.2 Nanchangmycin

Streptomyces nanchangensis produces the polyether antibiotic nanchangmycin **28**, the 16-membered macrolide meilingmycin, and at least two other antibiotics of unknown structure (18).

The nanchangmycin gene cluster comprises 30 ORFs in a region of 132 kb of DNA (Table 8.3) (19, 20). The type I PKS, encoding 14 modules in 11 ORFs, is arranged in two distinct groups (*nanA1-A6, nanA7-A11*) with the sugar biosynthesis genes located in between. A KSQ domain is located in the loading domain in NanA1 and so decarboxylation of malonate provides the acetate starter unit. The type II ACP (NanA10) is proposed to act as an independent ACP

TABLE 8.3   Nanchangmycin gene cluster (17, 19, 20)

Polypeptide (size aa)	Function
NanA1 (2902)	
Loading Module	KSQ, AT(A), ACP
Module 1	KS, AT(P), *DH*, KR,[a] ACP
NanA2 (2223)	
Module 2	KS, AT(P), DH, ER, KR, ACP
NanA3 (4032)	
Module 3	KS, AT(A), DH, KR, ACP
Module 4	KS, AT(P), DH, ER, KR, ACP
NanA4 (3956)	
Module 5	KS, AT(P), *DH*, KR,[a] ACP
Module 6	KS, AT(A), DH,[a] *ER*, KR, ACP
NanA5 (3979)	
Module 7	KS, AT(P), DH, KR, ACP
Module 8	KS, AT(A), DH, ER, KR, ACP
NanA6 (1665)	
Module 9	KS, AT(A), KR,[a] ACP
NanA7 (1646)	
Module 10	KS, AT(P), KR, ACP
NanA8 (3455)	
Module 11	KS, AT(P), KR, ACP
Module 12	KS, AT(P), DH, KR, ACP
NanA9 (802)	
Module 13	KS, AT(P)
NanA10 (104)	ACP
NanA11 (2187)	
Module 14	KS, AT(P), DH, ER, KR, ACP, CR
NanE (290)	Type I TE[b]
NanG1-G5	4-*O*-methyl-L-rhodinose biosynthesis
NanI (313)	Epoxide hydrolase/cyclase[c]
NanM (305)	Methyl transferase
NanO (478)	Epoxidase
NanP (423)	Cytochrome P450
NanR1-R4	Regulation
NanT1-T5	Transporters, chemoreceptor, regulator

Reproduced from Reference 17 with permission from the Royal Society of Chemistry.

[a] Predicted to be inactive.
[b] Originally assigned as an epoxide hydrolase.
[c] Originally assigned as a ketosteroid isomerase.

for module 13 (*nanA9*), which lacks an ACP function. Either side of *nanA10* three unique polyether genes are located: *nanI*, *nanO*, and *nanE*, which are homologous to *monBI/BII*, *monCI*, and *monCII*. The epoxidase encoded by *nanO* oxidizes the triene precursor **30**, which is then cyclized by NanI (Fig. 8.3b). Originally, the chain-releasing (CR) domain located in module 14 was thought

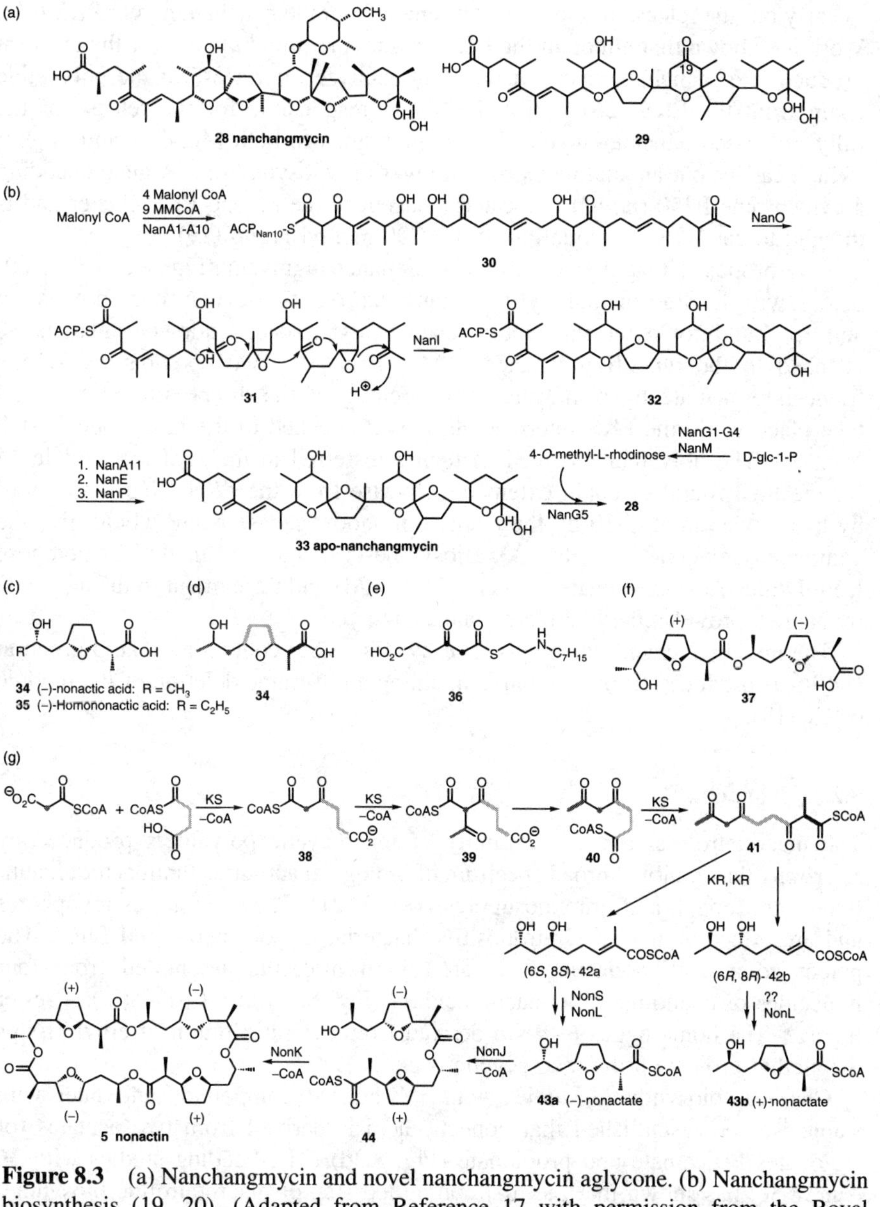

Figure 8.3 (a) Nanchangmycin and novel nanchangmycin aglycone. (b) Nanchangmycin biosynthesis (19, 20). (Adapted from Reference 17 with permission from the Royal Society of Chemistry.) (c) Structure of building blocks for the macrotetrolides. (d) Biosynthetic origin of carbon skeleton of nonactic acid (21). (e) (2,3-^{13}C$_2$)-3-ketoadipate N-caprylcysteamine thioester (22). (f) Homochiral nonactate dimmer (23). (g) Nonactin biosynthesis (21). (Adapted from Reference 17 with permission from the Royal Society of Chemistry.)

to carry out the release of the polyether chain from the nanchangmycin PKS (19). Work has shown that although the CR domain is required *in vivo* for the efficient production of nanchangmycin, it does not catalyze the release of the polyketide chain from the PKS (20). NanE, is in fact responsible for the release of the fully processed nanchangmycin. It is homologous to both MonC11 and NigCII (which carries out an analogous role in nigericin biosynthesis). A gene encoding a cytochrome P450 (*nanP*) is located adjacent to the second PKS cluster and is thought to catalyze the oxidation of the C30 methyl group (19).

It is proposed that the biosynthesis of nanchangmycin (Fig. 8.3b) (19, 20) begins with loading of malonyl CoA onto NanA1. Subsequent decarboxylation and 13 rounds of extension would result in a tetradecaketide intermediate **30** attached to the type II independent ACP NanA10. Two epoxidations of the triene intermediate by NanO and ring opening of the di-epoxide **31** by NanI take place while the PKS intermediate is still attached to the independent ACP NanA10. The polyether chain **32** is then transferred to the ACP of module 14 for the final round of chain extension. Release from the PKS (NanE) followed by hydroxylation at C-30 by the putative monooxygenase NanP would give the completed aglycone polyether **33**. Biosynthesis of the 4-*O*-methyl-L-rhodinose from D-glucose-1-phosphate (NanG1-G4, NanM) and attachment to the aglycone by NanG5 provides the completed nanchangmycin.

The novel nanchangmycin aglycone **29** (Fig. 8.3a) containing a keto group at C-19 was produced from a mutant containing an in-frame deletion of the module 6 KR (19).

8.2.3 Nonactin

The macrotetrolides **5-9** are a family of macrocyclic polyethers produced by *S. griseus* that exhibit a broad spectrum of biological activities (antibacterial, antitumor, antifungal, and immunosuppressive) (5, 21). They can act as ionophores and are effective against Gram-positive bacteria, mycobacteria, and fungi. The parent compound, nonactin **5**, is an achiral molecule assembled from four molecules of enantiomeric nonactic acid **34** (Fig. 8.3c) in a (+)(−)(+)(−)-ester linkage. The homologues **6–9** are derived from the substitution of ethyl groups for methyl groups on the macrocyclic backbone.

Classical biosynthetic studies with ^{14}C-labeled compounds and later with stable isotopes established that nonactic acid is derived from two acetates (or malonates), succinate and propionate (Fig. 8.3d) (21). Feeding studies with **36** (Fig. 8.3e) has shown that the first committed step of macrotetrolide biosynthesis is the coupling of the succinate unit with an acetate (or malonate) to give α-ketoadipate **40** (22). The late steps of nonactate biosynthesis (Fig. 8.3g) were shown to involve the cyclization of **42a** into (−)-nonactate and **42b** into (+)-nonactate, which demonstrates that a pair of enantiocomplementary pathways are operating in nonactate biosynthesis (21). (±)-Nonactic acid is incorporated efficiently into nonactin. Both (+) and (−)-nonactic acids as well as their dimers have been isolated from cultures of *S. griseus*.

Genetic characterization of the nonactin gene cluster has revealed that an ACP-less Type II PKS is used to assemble the macrotetrolides (21). The non-PKS is highly unusual, catalyzing both C–C and C–O bond-forming reactions, functioning noniteratively and acting on acyl CoA substrates (which is typical of Type III PKS). Twenty-three non genes were identified, which includes five discrete KS proteins, NonJKPQU, and four discrete KR proteins, NonEMNO. Heterologous expression of these 23 nongenes in *S. lividans* resulted in macrotetrolide production, thereby confirming that only these genes are required for macrotetrolide biosynthesis; sequestration of an ACP from elsewhere on the *S. griseus* genome does not occur.

The proposed biosynthetic pathway (Fig. 8.3g) requires three condensation and four reduction steps that would require at least three KS and four KR genes. It has not been established whether the five KS and four KR proteins function independently, but it has been demonstrated that all nine enzymes are required for macrotetrolide biosynthesis. Shen and Kwon (21) have suggested that the macrotetrolide PKS comprises noniteratively functioning subunits, which is unprecedented in all Type II PKSs known to date. NonJ and NonK have been assigned roles in dimerization (see below), which leaves NonPQU to carry out the three C–C bond-forming reactions. A mutant lacking the NonPQU genes was able to convert nonactic acid into nonactin, whereas deletion of NonJK resulted in no macrotetrolide formation. Hence, NonPQU represents a novel Type II minimal PKS that acts noniteratively, does not have an ACP, and uses acyl CoA substrates directly for polyketide biosynthesis.

NonJ has been shown to catalyze the first dimerization step to give (−)-non-actyl-(+)-non-actyl CoA **44**, and NonK is responsible for the stereo-specific cyclodimerization to afford nonactin (21). All C–C bond forming KSs are characterized by a Cys–His–His (Asn) catalytic triad (including NonPQU), whereas NonKJ are characterized by a mutated catalytic triad: Cys–Gly/Tyr–His. Replacement of the conserved cysteine residue in NonJ or NonK with glycine gave mutants that could not catalyze the transformation of (±)–nonactic acid into nonactin. Taken together, NonJK cannot function as a decarboxylase and instead catalyze the C–O bond-forming steps in nonactin biosynthesis using the same active site cysteine that other KSs use for C–C bond formation. In common with NonPQU, NonJK act noniteratively on CoA substrates.

The product of the minimal PKS, **41**, is the branch point of the pathway that diverges into a pair of enantio-specific pathways, each of which involves two KRs and affords (+)- or (−)-**42**. The nonactate synthase, NonS, catalyzes the intramolecular Michael addition of (−)-nonactic acid **43a** from (−)-**42a**. A nonS mutant supplemented with (±)-nonactic acid could produce nonactin, monactin, and dinactin, but not trinactin and tetranactin, which require one/two (−)-homononactate moieties, respectively. This process suggests that NonS can cyclize (−)-**42a** into (−)-nonactate **43a** and (−)-homononactate **35** only and that another nonactate synthase is required for the cyclization of (+)-**42b** to give (+)-nonactate **43b** and (+)-homononactate.

Based on its high sequence homology to CoA ligases, NonL was identified as a CoA ligase catalyzing the transformation of (±)-nonactic acid into (±)-nonactyl CoA. The sequence of reactions for the biosynthesis of nonactin and the other macrotetrolides in *S. griseus* is shown in Fig. 8.3g.

S. griseus is likely to be protected from nonactin by two mechanisms. The first mechanism seems to involve pumping excess antibiotic from the cell. Two genes, orf5 and orf6, which are clustered with the rest of the nonactin biosynthesis genes, show homology with ABC transporter genes. The second mechanism is the enzyme catalyzed hydrolysis of nonactin and homologues by NonR, which is homologous to serine protease and esterase enzymes and confers tetranactin resistance to *S. lividans* TK24. Overexpressed NonR has been shown to stereospecifically catalyze the breakage of the macrotetrolide ring into homochiral nonactate dimers **37** (Fig. 8.3f) in a two-step process (23). The macrotetrolides are converted initially into their seco-tetramer species and subsequently hydrolyzed to the dimer; in both cases, it is the bond between the alcohol of the (+)-nonactate and the carboxylate of (−)-nonactate that is cleaved. No trimer or monomers were detected.

8.2.4   Marine Polyether Ladders

All marine polyether ladders thus far characterized (e.g., maitotoxin **16** (Fig. 8.4a), the brevetoxins **12** and **13** and hemibrevetoxin B, the yessotoxins **14** (and the truncated adriatoxin), the Pacific and Caribbean ciguatoxins, the gambieric acids **15** and gamberiol (the gymnocins and brevenal) can be grouped into 14 backbone structures (3).

Retrobiosynthetic analysis of these structures has led to the development of a model for the biosynthesis of these complex structures from the cyclization of a polyepoxide precursor (3). The model accounts for the conserved stereochemistry of the numerous ring junctions in these polyethers, which are *syn/trans* (Fig. 8.4b). The required configuration of the rings can be derived from stereochemically identical all (*R,R*)- or (*S,S*)-*trans* epoxides, which are derived from the appropriate polyene (which may contain over 20 double bonds). Epoxidation of the polyene precursor takes place from the same face, and consequently, it is possible for a single monooygenase with broad specificity to produce all *trans* epoxides. Ring closure requires an *endo*-selective opening of each epoxide.

Independently, Prasad and Shimizu (24) as well as Lee et al. (25) proposed that brevetoxin A is biosynthesized from the cyclization of a polyepoxide precursor in a series of S_N2 (*R,R*)-*trans* epoxide openings (Fig. 8.4c). Gallimore and Spencer (3) argue that the nine disfavored *endo*-tet closures required for this mechanism makes it mechanistically unlikely and point out that an alternative cascade of S_N2 epoxide openings in the opposite direction from all (*S,S*)-*trans* epoxides yields the same structure.

Although it can be envisaged that a *trans*-polyene is converted by a monooxygenase to a polyepoxide intermediate, which is processed by an epoxide hydrolase

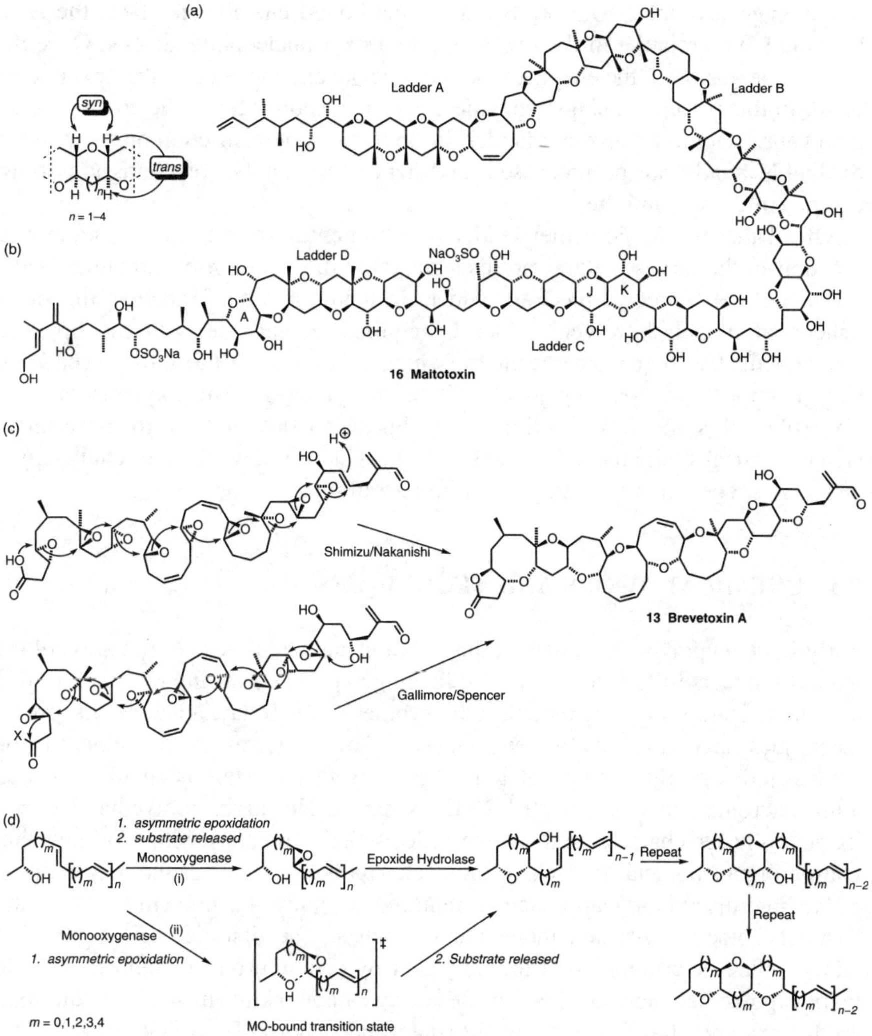

Figure 8.4 (a) Maitotoxin. (b) Common structure feature of marine polyether ring junctions (3). (c) Brevetoxin A biosynthesis as proposed by 1) Shimizu/Nakanishi (24, 25) and 2) Gallimore and Spencer from all (*S,S*)-*trans* epoxides (3). (d) Possible enzymatic routes to a fused polyether using 1) a mono-oxygenase and an epoxide hydrolase or 2) a monooxygenase only (3).

(cf. MonB in monensin biosynthesis) to the polycyclic ether, because of the reactivity of the polyepoxide intermediate (which may contain in excess of 20 reactive epoxide groups), this seems unlikely. An alternative mechanism is for the epoxidation and cyclization steps to be coupled in an iterative process whereby the production of an epoxide is followed by ring closure (Fig. 8.4d). Gallimore and Spencer (3) suggest that it may be possible for this process to be affected by the monooxygenase only: After epoxidation, the bound enzyme stabilizes the *endo* transition state relative to the *exo* as the hydroxyl nucleophile attacks. Once the ring has been closed, the enzyme would dissociate and move on to the next double bond. In this manner, a single enzyme could be responsible for the conversion of a polyene chain to a polyether ladder. In both cases, no restriction on the number of double bonds can be processed, and hence, the number of contiguous rings can be (in theory) infinite.

Of all the marine polyether ladder metabolites examined, only a single ring junction in the largest natural product, maitotoxin, could not be explained using the model (3). Three of the four ladders in maitotoxin conformed to the stereochemical model; however, ladder C requires an epoxide with the opposite stereochemistry to the other centers, which gives rise to the only exceptional ring junction ("the J-K ring junction") in any of the known polyether ladders. To explain this anomaly, Gallimore and Spencer (3) refer back to the original stereochemical assignment by Satake et al., which is described as challenging, and they suggest that the assignment be reexamined.

8.3   CHEMICAL TOOLS AND TECHNIQUES

Early work on polyethers relied on the isolation and structure determination of the natural products followed by classical feeding experiments using both radioactive and stable isotopes to elucidate their biosynthesis (7). Today, advances in genetic techniques have allowed the gene sequences of the terrestrial polyethers to be studied although the number of published polyether clusters is small compared with macrolides and mixed NRPS/PKS systems. No marine polyether has had its gene cluster characterized, which reflects the difficulties with working with marine organisms and the lack of molecular tools available for these systems.

An interdisciplinary approach is required to study the polyethers involving chemists, biochemists, and molecular biologists. The first step is the isolation and structure determination of the polyether by natural product chemists. For the marine polyether ladders, this can be a huge undertaking in itself. Maitotoxin **16**, for example, has four ladders, 32 rings, and a molecular weight of 3422 Da! Sophisticated NMR experiments were required to assign all stereocenters, and yet it was only detailed retrobiosynthetic analysis by Gallimore and Spencer that raised the possibility that one ring junction may have been assigned incorrectly (3).

Locating the polyether gene cluster can be done by using probes specific for polyketide genes to screen a cosmid library of total DNA [for both monensin **2** (2, 13) and nanchangmycin **28** (19), probes from the erythromycin PKS were used]. Alternatively, the polyether's resistance gene can be used to locate the gene cluster [e.g., nonactin **5** (26)]. Sequencing and mapping of the cosmids in the library allows the gene cluster size and composition to be determined. More than one PKS sequence can be revealed, so care must be taken to ensure that the gene cluster of interest is being sequenced. For example, eight clusters were identified from the nanchangmycin producer *S. nanchangensis*, and the nanchangmycin gene cluster was identified through gene disruption (18). The organization and assignment of function of each ORF is carried out by sequence comparison. Cloning and overexpression of individual enzymes in the gene cluster is an important step in confirming the function and mechanism of a particular ORF. For example, sequence comparisons of both the monensin and the nanchangmycin clusters revealed no TE function required for the release of the polyketide from the PKS. However, cloning, overexpression, and characterization of MonCII (14) and NanE (20) resulted in the reassignment from an epoxide cyclase to Type I TE.

The preparation of mutants is an important tool to investigate the role of individual enzymes, and it is used to ascertain which genes in the cluster are essential for the biosynthesis of the polyether and whether modified/truncated products are formed. The effected gene can then be supplied *in trans* on a plasmid to see whether activity can be restored. Chemists are required to analyze the fermentation mixtures for minor and truncated metabolites. In this manner, the triene lactones **22** and **23** were isolated from a 7L culture of a mutant strain of *S. cinnamonensis* lacking *MonCI*, and their structure was determined by NMR (15).

HPLC-MS is an essential technique for the separation and identification of these trace compounds, but unless sufficient quantities can be isolated for structure determination by NMR, chemists are also required to synthesize standards for comparison. Characterization of the gene cluster requires chemists to synthesize intermediates (often with stable isotopes at specific places) to be used as substrates for purified enzymes or to confirm the structure of products formed from modified genes.

REFERENCES

1. Cane DE, Celmer WD, Westley JW. Unified stereochemical model of polyether antibiotic structure and biogenesis. J. Am. Chem. Soc. 1983;105:3594–3600.

2. Leadlay PF, Staunton J, Oliynyk M, Bisang C, Cortés J, Frost E, Hughes-Thomas ZA, Jones MA, Kendrew SG, Lester JB, Long PF, McArthur HAI, McCormick EL, Oliynyk Z, Stark CBW, Wilkinson CJ. Engineering of complex polyketide biosynthesis – insights from sequencing of the monensin biosynthetic gene cluster. J. Indus. Microb. Biotechnol. 2001;27:360–367.

3. Gallimore AR, Spencer JB. Stereochemical uniformity in marine polyether ladders – implication for the biosynthesis and structure of maitotoxin. Angew. Chem. Int. Ed. 2006;45:4406–4413.

4. Riddell FG. Ionophoric antibiotics. Chem. Br. 1992;28:533–537.

5. Pressman BC. Biological applications of Ionophores. Ann. Rev. Biochem. 1976;45:501–530.

6. Fellner V, Sauer FD, Kramer JKG. Effect of nigericin, monensin, and tetronasin on biohydrogenation in continuous flow-through ruminal fermenters. J. Dairy Sci. 1997;80:921–928.

7. Westley JW. Polyether antibiotics–biosynthesis. Antibiotics IV. JW Corcoran, ed. 1981. Springer-Verlag, New York. pp. 41–73.

8. Gumila C, Ancelin M-L, Delort A-M, Jeminet G, Vial HJ. Characterisation of the potent in vitro and in vivo antimalarial activities of ionophore compounds. Antimicrob. Agents. Chemother. 1997;41:523–529.

9. Adovelande J, Schrével J. Carboxylic ionophores in malaria chemotherapy: the effects of monensin and nigericin on *plasmodium falciparium in vitro* and *plasmodium vinckei petteri in vivo*. Life Sci. 1996;59:309–315.

10. Needham J, Hu T, McLachlan JL, Walter JA, Wright JLC. Biosynthetic studies of the DSP toxin DTX-4 and an okadaic acid diol ester. J. Chem. Soc., Chem. Commun. 1995; 1623–1624.

11. Garcia Camacho F, Gallardo Rodríguez J, Sánchez Mirón A, Cerón García MC, Belarbi EH, Chisti Y, Molina Grima E. Biotechnological significance of toxic marine dinoflagellates. Biotech. Adv. 2007;25:176–194.

12. Nicholson GM, Lewis RJ. Ciguatoxins: Cyclic polyether modulators of voltage-gated ion channel function. Mar. Drugs 2006;4:82–118.

13. Oliynyk M., Stark CBW, Bhatt A, Jones MA, Hughes-Thomas ZA, Wilkinson, C, Oliynyk Z, Demydchuk, Y, Staunton J, Leadlay PF. Analysis of the biosynthetic gene cluster for the polyether antibiotic monensin in *Streptomyces cinnamonensis* and evidence for the role of *monB* and *monC* in oxidative cyclization. Mol. Microbiol. 2003;49:1179–1190.

14. Harvey BM, Hong H, Jones MA, Hughes-Thomas ZA, Goss RM, Heathcote ML, Bolanos-Garcia VM, Kroutil W, Staunton J, Leadlay PF, Spencer JB. Evidence that a novel thioesterase is responsible for polyketide chain release during biosynthesis of the polyether ionophore monensin. ChemBioChem 2006;7:1435–1442.

15. Bhatt A, Stark CBW, Harvey BM, Gallimore AR, Demydchuk YA, Spencer JB, Staunton J, Leadlay PF. Accumulation of an *E,E,E*-triene by the monensin-producing polyketide synthase when oxidative cyclization is blocked. Angew. Chem. Int. Ed. 2005;44:7075–7078.

16. Gallimore AR, Stark CBW, Bhatt A, Harvey BM, Demydchuk Y, Bolanos-Garcia V, Fowler DJ, Staunton J, Leadlay PF, Spencer JB. Evidence for the role of the *monB* genes in polyether ring formation during monensin biosynthesis. Chem. Biol. 2006;13:453–460.

17. Hill AM, The biosynthesis, molecular genetics and enzymology of the polyketide-derived metabolites. Nat. Prod. Rep. 2006;23:256–321.

18. Sun Y, Zhou X, Liu J, Bao K., Zhang G, Tu G, Kieser T, Deng Z. '*Streptomyces nanchangensis*', a producer of the insecticidal polyether antibiotic nanchangmycin and the antiparasitic macrolide meilingmycin, contains multiple polyketide gene clusters. Microbiology 2002;148:361–371.

19. Sun Y, Zhou X, Dong H, Tu G, Wang M, Wang B, Deng Z. A complete gene cluster from *Streptomyces nanchangensis* NS3226 encoding biosynthesis of the polyether ionophore nanchangmycin. Chem. Biol. 2003;10:431–441.

20. Liu T, You D, Valenzano C., Sun Y, Li J, Yu Q, Zhou X, Cane DE, Deng Z. Identification of NanE as the thioesterase for polyether chain release in nanchangmycin biosynthesis. Chem. Biol. 2006;13:945–955.

21. Shen B, Kwon H-J. Macrotetrolide biosynthesis: a novel type ii polyketide synthase. Chem. Rec. 2002;2:389–396.

22. Nelson ME, Priestley ND. Nonactin biosynthesis: the initial committed step is the condensation of acetate (malonate) and succinate. J. Am. Chem. Soc. 2002;124: 2894–2902.

23. Cox JE, Priestley ND. Nonactin biosynthesis: the product of the resistance gene degrades nonactin stereospecifically to form homochiral nonactate diners. J. Am. Chem. Soc. 2005;127:7976–7977.

24. Prasad AVK, Shimizu Y. The structure of hemibrevetoxin B: a new type of toxin in the gulf of mexico red tide organism. J. Am. Chem. Soc. 1989;111:6476–6477.

25. Lee MS, Qin GW, Nakanishi K, Zagoeski MG. Biosynthetic studies of brevetoxins, potent neurotoxins produced by the dinoflagellate *Gymnodinium breve*. J. Am. Chem. Soc. 1989;111:6234–6241.

26. Smith WC, Xiang L, Shen B. Genetic localization and molecular characterization of the *nonS* gene required for macrotetrolide biosynthesis in *Streptomyces griseus* DSM40695. Antimicrob. Agents Chemother. 2000;44:1809–1817.

FURTHER READING

For a review on polyether biosynthesis (as well as polyketides and mixed NRPS/PKS metabolites): Hill AM. The biosynthesis, molecular genetics and enzymology of the polyketide-derived metabolites. Nat. Prod. Rep. 2006;23:256–320.

For a review on polyether biosynthesis in dinoflagellates: Rein KS, Snyder RV. The biosynthesis of polyketide metabolites by dinoflagellates. Adv. Appl. Microbiol. 2006; 50:93–125.

For a review on the isolation and structure elucidation of the marine polyethers: Yasumoto T. The chemistry and biological function of natural marine toxins. Chem. Rec. 2001;1:228–242.

The first report of cloning polyketide genes from a dinoflagellate: Kubota T, Iinuma Y, Kobayashi J. Cloning of polyketide synthase genes from amphidinolide-producing dinoflagellate *Amphidinium* sp. Biol. Pharm. Bull. 2006;29:1314–1318.

For practical information on the manipulation of *Streptomyces* spp.: Kieser T, Bibb MJ, Buttner MJ, Chater KF, Hopwood DA. Practical Streptomyces Genetics. 2000. John Innes Foundation, Norwich.

Information on polyether residues in eggs: Too Hard to Crack- Eggs with Residues— Executive Summary. Information Sheet Soil Association. *http://www.soilassociation. org/library*.

9

ALKALOIDS

SARAH E. O'CONNOR

Department of Chemistry, Massachusetts Institute of Technology, Cambridge, Massachusetts

How nature synthesizes complex secondary metabolites, or natural products, can be studied only by working within the disciplines of both chemistry and biology. Alkaloids are a complex group of natural products with diverse mechanisms of biosynthesis. This chapter highlights the biosynthesis of four major classes of plant-derived alkaloids. Only plant alkaloids for which significant genetic information has been obtained were chosen for review. Isoquinoline alkaloid, terpenoid indole alkaloid, tropane alkaloid, and purine alkaloid biosynthesis are described here. The chapter is intended to provide an overview of the basic mechanism of biosynthesis for selected members of each pathway. Manipulation of these pathways by metabolic engineering is highlighted also.

Alkaloids are a highly diverse group of natural products related only by the presence of a basic nitrogen atom located at some position in the molecule. Even among biosynthetically related classes of alkaloids, the chemical structures are often highly divergent. Although some classes of natural products have a recognizable biochemical paradigm that is centrally applied throughout the pathway, for example, the "assembly line" logic of polyketide biosynthesis (1), the biosynthetic pathways of alkaloids are as diverse as the structures. It is difficult to predict the biochemistry of a given alkaloid based solely on precedent, which makes alkaloid biosynthesis a challenging, but rewarding, area of study.

Natural Products in Chemical Biology, First Edition. Edited by Natanya Civjan.
© 2012 John Wiley & Sons, Inc. Published 2012 by John Wiley & Sons, Inc.

9.1   BIOLOGIC BACKGROUND

Hundreds of alkaloid biosynthetic pathways have been studied by chemical strategies, such as isotopic labeling experiments (2, 3). However, modern molecular biology and genetic methodologies have facilitated the identification of alkaloid biosynthetic enzymes. This chapter focuses on pathways for which a significant amount of genetic and enzymatic information has been obtained. Although alkaloid natural products are produced by insects, plants, fungi, and bacteria, this chapter focuses on four major classes of plant alkaloids: the isoquinoline alkaloids, the terpenoid indole alkaloids, the tropane alkaloids, and the purine alkaloids.

In general, plant biosynthetic pathways are understood poorly when compared with prokaryotic and fungal metabolic pathways. A major reason for this poor understanding is that genes that express complete plant pathways typically are not clustered together on the genome. Therefore, each plant enzyme often is isolated individually and cloned independently. However, several enzymes involved in plant alkaloid biosynthesis have been cloned successfully, and many more enzymes have been purified from alkaloid-producing plants or cell lines (4–6). Identification and study of the biosynthetic enzymes has a significant impact on the understanding of the biochemistry of the pathway. Furthermore, genetic information also can be used to understand the complicated localization patterns and regulation of plant pathways. This chapter focuses on the biochemistry responsible for the construction of plant alkaloids and summarizes the biosynthetic genes that have been identified to date. Some of these pathways have been the subject of metabolic engineering studies; the results of these studies are mentioned here also. An excellent, more detailed review that covers the biochemistry and genetics of plant alkaloid biosynthesis up until the late 1990s is available also (7).

9.2   ISOQUINOLINE ALKALOIDS

The isoquinoline alkaloids include the analgesics morphine and codeine as well as the antibiotic berberine (Fig. 9.1a). Morphine and codeine are two of the most important analgesics used in medicine, and plants remain the main commercial source of the alkaloids (8). Development of plant cell cultures of *Eschscholzia californica*, *Papaver somniferum*, and *Coptis japonica* has aided in the isolation and cloning of many enzymes involved in the biosynthesis of isoquinoline alkaloids (9).

9.2.1   Early Steps of Isoquinoline Biosynthesis

Isoquinoline biosynthesis begins with the substrates dopamine and *p*-hydroxyphenylacetaldehyde (Fig. 9.1b). Dopamine is made from tyrosine by hydroxylation and decarboxylation. Enzymes that catalyze the hydroxylation and decarboxylation steps in either order exist in the plant, and the predominant

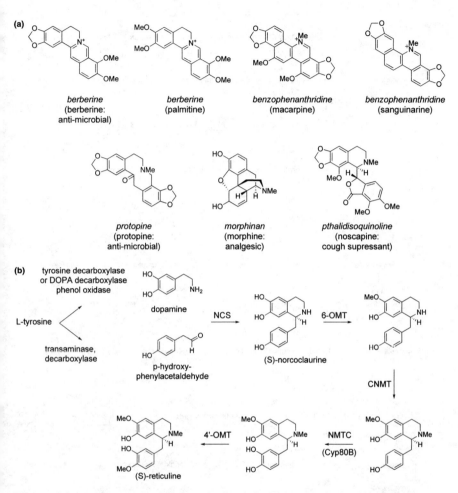

Figure 9.1 (a) Representative isoquinoline alkaloids. (b) Early biosynthetic steps of the isoquinoline pathway yield the biosynthetic intermediate (S)-reticuline, the central biosynthetic intermediate for all isoquinoline alkaloids. (c) Berberine and sanguinarine biosynthesis pathways. (d) Morphine biosynthesis. NCS, norcoclaurine synthase; 6-OMT, norcoclaurine 6-O-methyltransferase; CNMT, coclaurine N-methyltransferase (Cyp80B); NMTC, N-methylcoclaurine 3'-hydroxylase; 4'-OMT, 3'-hydroxy-N-methylcoclaurine 4'-O-methyltransferase; BBE, berberine bridge enzyme; SOMT, scoulerine 9-O-methyltransferase; CS, canadine synthase; TBO, tetrahydroprotoberberine oxidase; CHS, cheilanthifoline synthase; SYS, stylopine synthase; NMT, N-methyltransferase; NMSH, N-methylstylopine hydroxylase; P6H protopine 6-hydroxylase; DHPO, dihydrobenzophenanthridine oxidase; RO, reticuline oxidase; DHR, dihydroreticulinium ion reductase; STS, salutaridine synthase; SalR, salutaridine reductase; SalAT, salutaridinol acetyltransferase; COR, codeinone reductase.

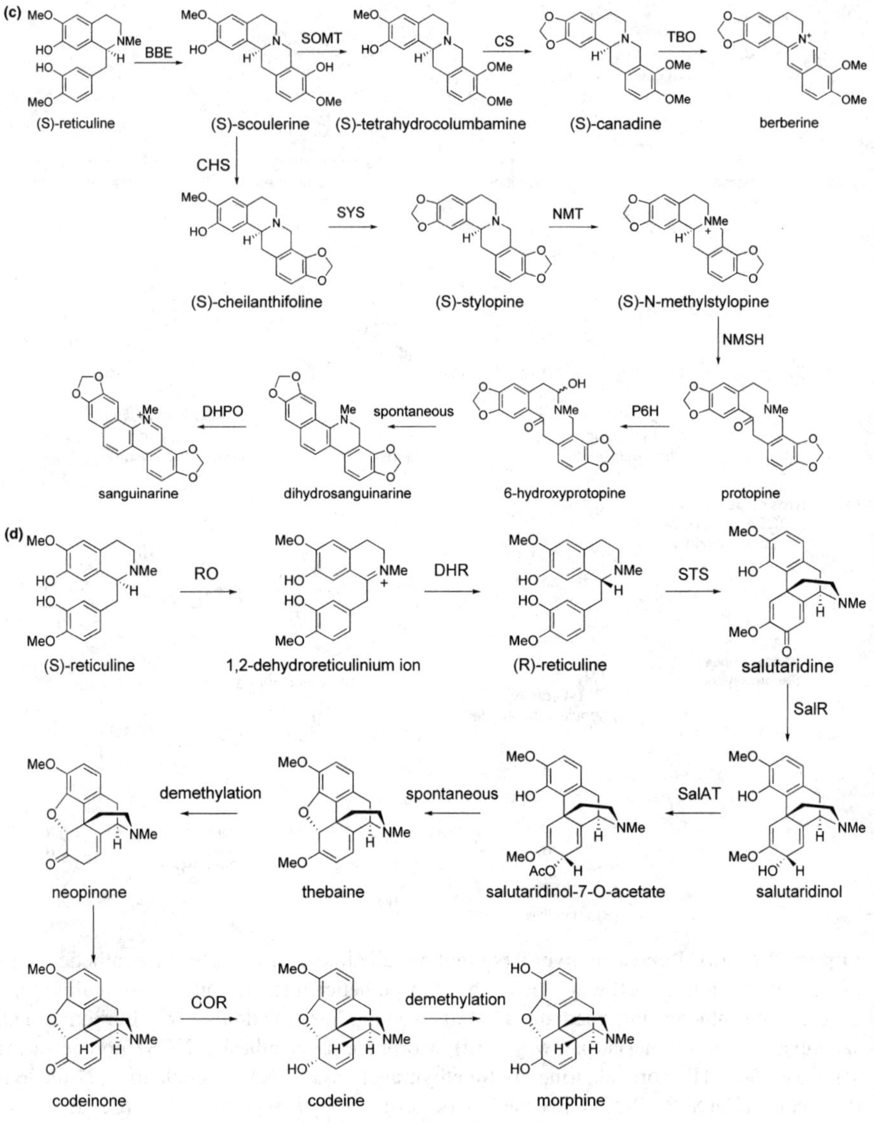

Figure 9.1 (*Continued*)

pathway for formation of dopamine from tyrosine is not clear. The second substrate, *p*-hydroxyphenylacetaldehyde, is generated by transamination and decarboxylation of tyrosine (10, 11).

Condensation of dopamine and *p*-hydroxyphenylacetaldehyde is catalyzed by norcoclaurine synthase to form (S)-norcoclaurine (Fig. 9.1b). Two norcoclaurine synthases with completely unrelated sequences were cloned (*Thalictrum flavum* and *C. japonica*) and heterologously expressed in *E. coli* (12–14). One is homologous to iron-dependent diooxygenases, whereas the other is homologous to a pathogenesis-related protein. Undoubtedly, future experiments will shed light on the mechanism of these enzymes and on how two such widely divergent sequences can catalyze the same reaction.

One of the hydroxyl groups of (S)-norcoclaurine is methylated by a S-adenosyl methionine-(SAM)-dependent O-methyl transferase to yield (S)-coclaurine. This enzyme has been cloned, and the heterologously expressed enzyme exhibited the expected activity (15–17). The resulting intermediate is then N-methylated to yield N-methylcoclaurine, an enzyme that has been cloned (18, 19). N-methylcoclaurine, in turn, is hydroxylated by a P450-dependent enzyme (CYP80B), N-methylcoclaurine 3′-hydroxylase, that has been cloned (20, 21). The 4′ hydroxyl group then is methylated by the enzyme 3′-hydroxy-N-methylcoclaurine 4′-O-methyltransferase (4′-OMT) to yield (S)-reticuline, the common biosynthetic intermediate for the berberine, benzo(c)phenanthridine, and morphinan alkaloids (Fig. 9.1b). The gene for this methyl transferase also has been identified (15, 22). These gene sequences also were used to identify the corresponding *T. flavum* genes that encode the biosynthetic enzymes for reticuline from a cDNA library (23). At this point, the biosynthetic pathway then branches to yield the different structural classes of isoquinoline alkaloids.

9.2.2 Berberine Biosynthesis

(S)-reticuline is converted to (S)-scoulerine by the action of a well-characterized flavin-dependent enzyme, berberine bridge enzyme (Fig. 9.1c). This enzyme has been cloned from several plant species, and the mechanism of this enzyme has been studied extensively (24–28). (S)-scolerine is then O-methylated by scoulerine 9-O-methyltransferase to yield (S)-tetrahydrocolumbamine. Heterologous expression of this gene in *E. coli* yielded an enzyme that had the expected substrate specificity (29). A variety of O-methyl transferases also have been cloned from *Thalictrum tuberosum* (30). The substrate-specific cytochrome P450 oxidase canadine synthase (31) that generates the methylene dioxy bridge of (S)-canadine has been cloned (32). The final step of berberine biosynthesis is catalyzed by a substrate-specific oxidase, tetrahydroprotoberberine oxidase, the sequence of which has not been identified yet (33).

Overproduction of berberine in *C. japonica* cell suspension cultures was achieved by selection of a high-producing cell line (34) with reported productivity of berberine reaching 7 g/L (35). This overproduction is one of the first demonstrations of production of a benzylisoquinoline alkaloid in cell culture at levels necessary for economic production. This cell line has facilitated greatly the identification of the biosynthetic enzymes.

9.2.3 Sanginarine Biosynthesis

The biosynthesis of the highly oxidized benzo(c)phenanthidine alkaloid sanguinarine is produced in a variety of plants and competes with morphine production in opium poppy. The pathway to sanguinarine has been elucidated at the enzymatic level (Fig. 9.1c) (36). Sanguinarine biosynthesis starts from (S)-scoulerine, as in berberine biosynthesis. Methylenedioxy bridge formation then is catalyzed by the P450 cheilanthifoline synthase to yield cheilanthifoline (37). A second P450 enzyme, stylopine synthase, catalyzes the formation of the second methyenedioxy bridge of stylopine (37). Stylopine synthase from *E. californica* has been cloned (38). Stylopine then is N-methylated by (S)-tetrahydroprotoberberine *cis*-N-methyltransferase to yield (S)-*cis*-N-methylstylopine, an enzyme that has been cloned from opium poppy (39). A third P450 enzyme, (S)-*cis*-N-methylstylopine hydroxylase, then forms protopine. Protopine is hydroxylated by a fourth P450 enzyme, protopine 6-hydroxylase, to yield an intermediate that rearranges to dihydrosanguinarine (40). This intermediate also serves as the precursor to the benzo(c)phenanthridine alkaloid macarpine (Fig. 9.1a). The copper-dependent oxidase dihydrobenzophenanthridine oxidase, which has been purified (41, 42), then catalyzes the formation of sanguinarine from dihydrosanguinarine.

Additional enzymes from other benzo(c)phenanthidine alkaloids have been cloned. For example, an O-methyl transferase implicated in palmitine biosynthesis has been cloned (43).

9.2.4 Morphine Biosynthesis

The later steps of morphine biosynthesis have been investigated in *P. somniferum* cells and tissue. Notably, in morphine biosynthesis, (S)-reticuline is converted to (R)-reticuline, thereby epimerizing the stereocenter generated by norcoclaurine synthase at the start of the pathway (Fig. 9.1d). (S)-reticuline is converted to (R)-reticuline through a 1,2-dehydroreticuline intermediate. Dehydroreticuline synthase catalyzes the oxidation of (S)-reticuline to 1,2-dehydroreticulinium ion (44). This enzyme has not been cloned but has been purified partially and shown to be membrane-associated. This intermediate then is reduced by dehydroreticuline reductase, an NADPH-dependent enzyme that stereoselectively transfers a hydride to dehydroreticulinium ion to yield (R)-reticuline. This enzyme has not been cloned yet but has been purified to homogeneity (45).

Next, the key carbon–carbon bond of the morphinan alkaloids is formed by the cytochrome P450 enzyme salutaridine synthase. Activity for this enzyme has been detected in microsomal preparations, but the sequence has not been identified (46). The keto moiety of the resulting product, salutaridine, then is stereoselectively reduced by the NADPH-dependent salutaridine reductase to form salutardinol. The enzyme has been purified (47), and a recent transcript analysis profile of *P. sominiferum* has resulted in the identification of the clone (48). Salutaridinol acetyltransferase, also cloned, then transfers an acyl group from acetyl-CoA to the newly formed hydroxyl group, which results in the formation of salutaridinol-7-O-acetate (49). This modification sets up the molecule to undergo a spontaneous reaction in which the acetate can act as a leaving group. The resulting product, thebaine, then is demethylated by an as yet uncharacterized enzyme to yield neopinione, which exists in equilibrium with its tautomer codeinone. The NADPH-dependent codeinone reductase catalyzes the reduction of codeinone to codeine and has been cloned (50, 51). Finally, codeine is demethylated by an uncharacterized enzyme to yield morphine.

The localization of isoquinoline biosynthesis has been investigated at the cellular level in intact poppy plants by using *in situ* RNA hybridization and immunoflouresence microscopy. The localization of 4'-OMT (reticuline biosynthesis), berberine bridge enzyme (saguinarine biosynthesis), salutaridinol acetyltransferase (morphine biosynthesis), and codeinone reductase (morphine biosynthesis) has been probed. 4'-OMT and salutaridinol acetyltransferase are localized to parenchyma cells, whereas codeinone reductase is localized to laticifer cells in sections of capsule (fruit) and stem from poppy plants. Berberine bridge enzyme is found in parenchyma cells in roots. Therefore, this study suggests that two cell types are involved in isoquinoline biosynthesis in poppy and that intercellular transport is required for isoquinoline alkaloid biosynthesis (52). Another study, however, implicates a single cell type (sieve elements and their companion cells) in isoquinoline alkaloid biosynthesis (53, 54). Therefore, it is not clear whether transport of pathway intermediates is required for alkaloid biosynthesis or whether the entire pathway can be performed in one cell type. Localization of enzymes in alkaloid biosynthesis is difficult, and, undoubtedly, future studies will provide more insight into the trafficking involved in plant secondary metabolism.

9.2.5 Metabolic Engineering of Morphine Biosynthesis

In attempts to accumulate thebaine and decrease production of morphine (a precursor to the recreational drug heroine), codeinone reductase in opium poppy plant was downregulated by using RNAi (8). Silencing of codeinone reductase results in the accumulation of (S)-reticuline but not the substrate codeinone or other compounds on the pathway from (S)-reticuline to codeine. However, the overexpression of codeinone reductase in opium poppy plants did result, in fact, in an increase in morphine and other morphinan alkaloids, such as morphine, codeine, and thebaine, compared with control plants (55). Gene expression levels in low morphine-producing poppy plants have been analyzed also (56). Silencing of berberine bridge enzyme in opium poppy plants also resulted in a change in alkaloid profile in the plant latex (57).

The cytochrome P450 responsible for the oxidation of (S)-N-methylcoclaurine to (S)-3'-hydroxy-N-methylcocluarine has been overexpressed in opium poppy plants, and morphinan alkaloid production in the latex is increased subsequently to 4.5 times the level in wild-type plants (58). Additionally, suppression of this enzyme resulted in a decrease in morphinan alkaloids to 16% of the wild-type level. Notably, analysis of a variety of biosynthetic gene transcript levels in these experiments supports the hypothesis that this P450 enzyme plays a regulatory role in the biosynthesis of benzylisoquinoline alkaloids. Collectively, these studies highlight that the complex metabolic networks found in plants are not redirected easily or predictably in all cases.

9.3 TERPENOID INDOLE ALKALOIDS

The terpenoid indole alkaloids have a variety of chemical structures and a wealth of biologic activities (Fig. 9.2a) (59, 60). Terpenoid indole alkaloids are used as anticancer, antimalarial, and antiarrhythmic agents. Although many biosynthetic genes from this pathway remain unidentified, studies have correlated terpenoid indole alkaloid production with the transcript profiles of *Catharanthus roseus* cell cultures (61).

9.3.1 Early Steps of Terpenoid Indole Alkaloid Biosynthesis

All terpenoid indole alkaloids are derived from tryptophan and the iridoid terpene secologanin (Fig. 9.2b). Tryptophan decarboxylase, a pyridoxal-dependent enzyme, converts tryptophan to tryptamine (62, 63). The enzyme strictosidine synthase catalyzes a stereoselective Pictet–Spengler condensation between tryptamine and secologanin to yield strictosidine. Strictosidine synthase (64) has been cloned from the plants *C. roseus* (65), *Rauwolfia serpentine* (66), and, *Ophiorrhiza pumila* (67). A crystal structure of strictosidine synthase from *R. serpentina* has been reported (68, 69), and the substrate specificity of the enzyme can be modulated (70).

Strictosidine then is deglycosylated by a dedicated β-glucosidase, which converts it to a reactive hemiacetal intermediate (71–73). This hemiacetal opens to form a dialdehyde intermediate, which then forms dehydrogeissoschizine. The enol form of dehydrogeissoschizine undergoes 1,4 conjugate addition to produce the heteroyohimbine cathenamine (74–76). A variety of rearrangements subsequently act on deglycosylated strictosidine to yield a diversity of indole alkaloid products (77).

9.3.2 Ajmaline Biosynthesis

The biosynthetic pathway for ajmaline in *R. serpentina* is one of the best-characterized terpenoid indole alkaloid pathways. Much of this progress has been detailed in an extensive review (78). Like all other terpenoid indole alkaloids, ajmaline, an antiarrhythmic drug with potent sodium channel-blocking properties (79), is derived from deglycosylated strictosidine (Fig. 9.2c).

A membrane–protein fraction of an *R. serpentina* extract transforms labeled strictosidine (80, 81) into sarpagan-type alkaloids. The enzyme activity is dependent on NADPH and molecular oxygen, which suggests that sarpagan bridge enzyme may be a cytochrome P450 enzyme. Polyneuridine aldehyde esterase hydrolyzes the polyneuridine aldehyde methyl ester, which generates an acid that decarboxylates to yield epi-vellosamine. This enzyme has been cloned from a *Rauwolfia* cDNA library, heterologously expressed in *E. coli*, and subjected to detailed mechanistic studies (82, 83).

In the next step of the ajmaline pathway, vinorine synthase transforms the sarpagan alkaloid epi-vellosamine to the ajmalan alkaloid vinorine (84).

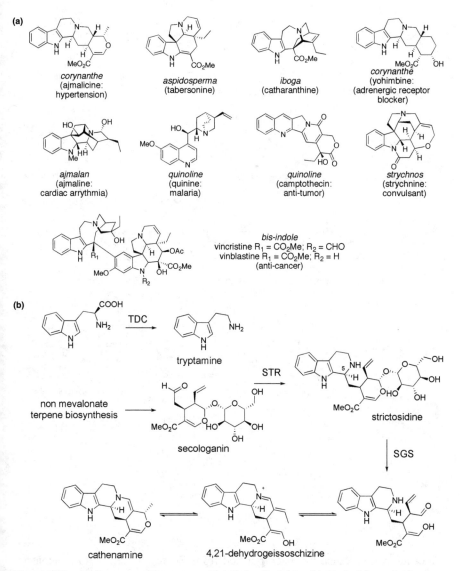

Figure 9.2 (a) Representative terpenoid indole alkaloids. (b) Early biosynthetic steps of the terpenoid indole alkaloid pathway yield the strictosidine, the central biosynthetic intermediate for all terpenoid indole alkaloids. (c) Ajmaline biosynthesis. (d) Ajmalicine and tetrahydroalstonine biosynthesis. (e) Vindoline biosynthesis from tabersonine. TDC, tryptophan decarboxylase; STR, strictosidine synthase; SGS, strictosidine glucosidase; SB, sarpagan bridge enzyme; PNAE, polyneuridine aldehyde reductase; VS, vinorine synthase; VH, vinorine hydroxylase; VR, vomilenine reductase; DHVR, dihydrovomilenine reductase; AAE, 17-O-acetyl-ajmalanesterase; NMT, norajmaline-N-methyltransferase; T16H, tabersonine-16-hydroxylase; HTOM, 16-hydroxytabersonine-16-O-methyltransferase; NMT, N-methyltransferase; D4H, desacetoxyvindoline-4-hydroxylase; DAT, desacetylvindoline O-acetyltransferase.

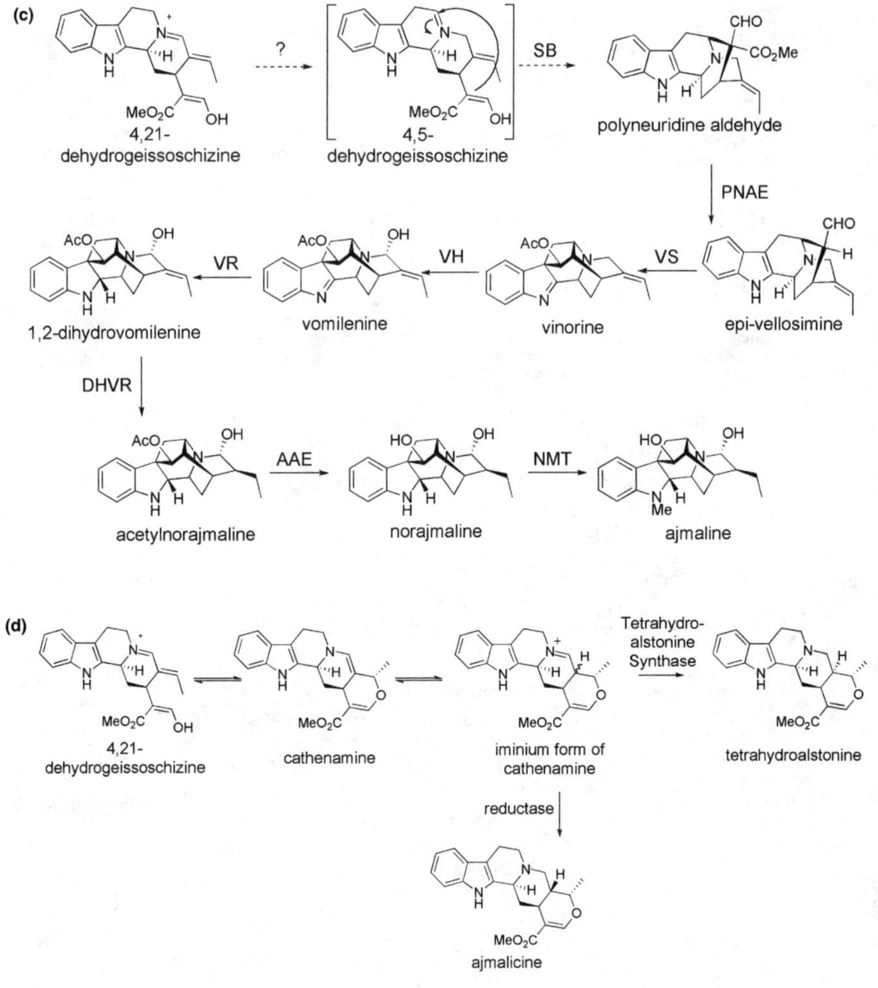

Figure 9.2   (*Continued*)

Vinorine synthase also has been purified from *Rauwolfia* cell culture, subjected to protein sequencing, and cloned from a cDNA library (85, 86). The enzyme, which seems to be an acetyl transferase homolog, has been expressed heterologously in *E. coli*. Crystallization and site-directed mutagenesis studies of this protein have led to a proposed mechanism (87).

Vinorine hydroxylase hydroxylates vinorine to form vomilene (88). Vinorine hydroxylase seems to be a P450 enzyme that requires an NADPH-dependent reductase. This enzyme is labile and has not been cloned yet. Next, the indolenine bond is reduced by an NADPH-dependent reductase to yield 1,2-dihydrovomilenene. A second enzyme, 1,2-dihydrovomilenene reductase, then reduces this product to acetylnorajmaline. Partial protein sequences have

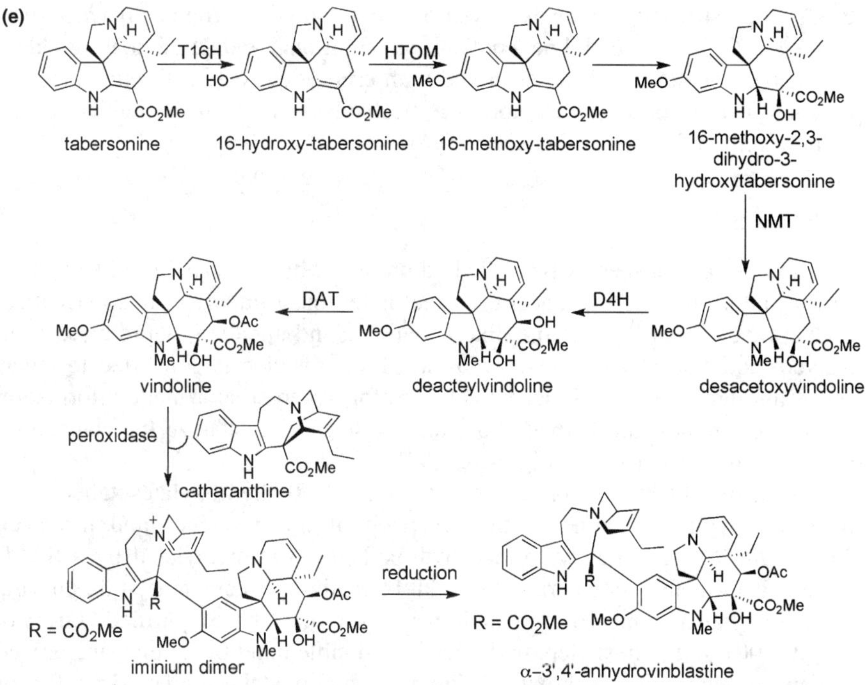

(e)

tabersonine — T16H → 16-hydroxy-tabersonine — HTOM → 16-methoxy-tabersonine → 16-methoxy-2,3-dihydro-3-hydroxytabersonine

NMT

vindoline ← DAT — deacteylvindoline ← D4H — desacetoxyvindoline

peroxidase → catharanthine

iminium dimer, R = CO₂Me — reduction → α-3',4'-anhydrovinblastine, R = CO₂Me

Figure 9.2 (*Continued*)

been obtained for both of the purified reductases. Although several putative clones that encode these proteins have been isolated, the activity of these clones has not been verified yet (89, 90).

An acetylesterase then hydrolyzes the acetyl link of acetylnorajmaline to yield norajmaline. This esterase has been purified from *R. serpentina* cell suspension cultures, and a full-length clone has been isolated from a cDNA library. Expression of the gene in tobacco leaves successfully yielded protein with the expected enzymatic activity (91). In the final step of ajmaline biosynthesis, an N-methyl transferase introduces a methyl group at the indole nitrogen of norajmaline. Although this enzymatic activity has been detected in crude cell extracts, the enzyme has not been characterized additionally (92).

9.3.3 Ajmalicine and Tetrahydroalstonine

Ajmalicine (raubasine) affects smooth muscle function and is used to help prevent strokes (93), and tetrahydroalstonine exhibits antipsychotic properties (Fig. 9.2d) (94). These compounds are found in a variety of plants, including *C. roseus* and *R. serpentina*. A partially purified NADPH-dependent reductase isolated from a tetrahydroalstonine that produces a *C. roseus* cell line was shown to catalyze the conversion of cathenamine, a spontaneous reaction product that results after strictosidine deglycosylation, to tetrahydroalstonine *in vitro* (95). A second *C. roseus* cell line contains an additional reductase that produces ajmalicine. Labeling studies performed with crude *C. roseus* cell extracts in the presence of D_2O or NADPD support a mechanism in which the reductase acts on the iminium form of cathenamine (96).

9.3.4 Vindoline

Vindoline, an aspidosperma-type alkaloid produced by *C. roseus*, is a key precursor for vinblastine, an anticancer drug that is the most important pharmaceutical product of *C. roseus*. Vindoline, like ajmalicine and ajmaline, is produced from deglycosylated strictosidine. Deglycosylated strictosidine is converted to tabersonine through a series of biochemical steps for which no enzymatic information exists. More details are known about the six steps that catalyze the elaboration of tabersonine to vindoline (Fig. 9.2e) (97).

Tabersonine-16-hydroxylase, a cytochrome P450, hydroxylates tabersonine to 16-hydroxy-tabsersonine in the first step of this sequence and has been cloned (98, 99). The newly formed hydroxyl group is methylated by a SAM-dependent O-methyl transferase to yield 16-methoxy-tabersonine; this enzyme (16-hydroxytabersonine-16-O-methyltransferase) has been purified but not cloned (100). In the next step, hydration of a double bond by an uncharacterized enzyme produces 16-methoxy-2,3-dihydro-3-hydroxytabersonine. Transfer of a methyl group to the indole nitrogen by an N-methyl transferase yields desacetoxyvindoline. This methyl transferase activity has been detected only in differentiated plants, not in plant cell cultures (101). The resulting intermediate, deacteylvindoline, is produced by the oxoglutatarate-dependent dioxygenase enzyme desacetylvindoline 4-hydroxylase. This enzyme has been cloned and also is absent from plant cell cultures (102). In the last step, desacteylvindoline is acetylated by desacteylvindoline O-acetyl transferase. This enzyme, also absent from nondifferentiated plant material, has been cloned successfully (103).

As in morphine biosynthesis, the knowledge of the enzyme sequences allows a more detailed understanding of the localization of the enzymes (104). Strictosidine synthase (Fig. 9.2b) seems to be localized to the vacuole (105), and strictosidine glucosidase is believed to be associated with the membrane of the endoplasmic reticulum (73, 106). Tabersonine-16-hydroxylase is associated with the endoplasmic reticulum membrane (98); N-methyl transferase activity is believed to be associated with the thykaloid, a structure located within the chloroplast (101, 107); and vindoline-4-hydroxylase and desacetylvindoline O-acetyltransferase are believed to be localized to the cytosol (Fig. 9.2e) (107, 108). Overall, extensive subcellular trafficking of biosynthetic intermediates is required for vindoline biosynthesis.

Aside from subcellular compartmentalization, specific cell types are required for the biosynthesis of some terpenoid alkaloids. Several enzymes involved in the early stages of secologanin biosynthesis seem to be localized to the phloem parenchyma, as evidenced by immunocytochemistry and *in situ* RNA hybridization studies (109). However, additional studies have suggested that these genes also are observed in the epidermis and laticifers (110). Studies of the localization of vindoline biosynthetic enzymes by using immunocytochemistry and *in situ* RNA hybridization strongly suggest that the mid-part of the vindoline pathway (tryptophan decarboxylase, strictosidine synthase, and tabersonine-16-hydroxylase) takes place in epidermal cells of leaves and stems. However, the later steps catalyzed by desacetylvindoline 4-hydroxylase and desacetylvindoline O-acetyltransferase take place in specialized cells, the laticifers, and idioblasts (109–112). As with isoquinoline alkaloid biosynthesis, deconvolution of the enzyme localization patterns remains a challenging endeavor.

9.3.5 Vinblastine

Vinblastine is a highly effective anticancer agent currently used clinically against leukemia, Hodgkin's lymphoma, and other cancers. (113, 114). Vinblastine is derived from dimerization of vindoline and another terpenoid indole alkaloid, catharanthine. The dimerization of catharanthine and vindoline is believed to proceed via the formation of an iminium intermediate with catharanthine (Fig. 9.2e). This iminium intermediate is reduced to form anhydrovinblastine, a naturally occurring compound in *C. roseus* plants (115). In support of this mechanism, anhydrovinblastine is incorporated into vinblastine and vincristine in feeding studies (116–119).

Peroxidase containing fractions of plant extracts were found to catalyze the formation of the bisindole dehydrovinblastine from catharanthine and vindoline (120, 121). A peroxidase from *C. roseus* leaves has been demonstrated to convert vindoline and catharanthine to anhydrovinblastine *in vitro* (122, 123). Because the dimerization of these *C. roseus* alkaloids also can be catalyzed by peroxidase from horseradish in reasonable yields (124), it is interesting to speculate that anhydrovinblastine may be a by-product of isolation; after lysis of the plant material, nonspecific peroxidases are released from the vacuole and may act on vindoline and catharanthine.

9.3.6 Metabolic Engineering of Terpenoid Indole Alkaloids

Strictosidine synthase and tryptophan decarboxylase have been overexpressed in *C. roseus* cell cultures (125, 126). Generally, overexpression of tryptophan decarboxylase does not seem to have a significant impact on alkaloid production, although overexpression of strictosidine synthase does seem to improve alkaloid yields. Overexpression of tryptophan and secologanin biosynthetic enzymes in *C. roseus* hairy root cultures resulted in modest increases in terpenoid indole alkaloid production (127, 128). Secologanin biosynthesis seems to be the rate-limiting factor in alkaloid production (129). Precursor-directed biosynthesis experiments with a variety of tryptamine analogs suggest that the biosynthetic pathway can be used to produce alkaloid derivatives (130). Strictosidine synthase and strictosidine glucosidase enzymes also have been expressed successfully heterologously in yeast (131); however, efforts to express heterologously terpenoid indole alkaloids currently are limited because the majority of the biosynthetic genes remain uncloned.

Transcription factors that upregulate strictosidine synthase (132), as well as a transcription factor that coordinately upregulates expression of several terpenoid indole alkaloid biosynthetic genes, have been found (133). Several zinc finger proteins that act as transcriptional repressors to tryptophan decarboxylase and strictosidine synthase also have been identified (134). Manipulation of these transcription factors may allow tight control of the regulation of terpenoid indole alkaloid production. Interestingly, expression of a transcription factor from *Arabidopsis thaliana* in *C. roseus* cell cultures results in an increase in alkaloid production (135).

9.4 TROPANE ALKALOIDS

The tropane alkaloids hyoscyamine and scopolamine (Fig. 9.3a) function as acetylcholine receptor antagonists and are used clinically as parasympatholytics. The illegal drug cocaine also is a tropane alkaloid. The tropane alkaloids are biosynthesized primarily in plants of the family *Solonaceae*, which includes *Hyoscyamus*, *Duboisia*, *Atropa*, and *Scopolia* (136, 137). Nicotine, although perhaps not apparent immediately from its structure, is related biosynthetically to the tropane alkaloids (Fig. 9.3b).

Tropane alkaloid biosynthesis has been studied at the biochemical level, and several enzymes from the biosynthetic pathway have been isolated and cloned, although the pathway has not been elucidated completely at the genetic level (Fig. 9.3b) (138). L-arginine is converted to the nonproteogenic amino acid L-ornithine by the urease enzyme arginase. Ornithine decarboxylase then decarboxylates ornithine to yield the diamine putrescine. In *Hyoscyamus*, *Duboisia*, and *Atropa*, putrescine serves as the common precursor for the tropane alkaloids.

Putrescine is N-methylated by a SAM-dependent methyl transferase that has been cloned to yield N-methylputrescine (139, 140). Putrescine N-methyl transferase now has been cloned from a variety of plant species (141–143), and site-directed mutagenesis and homology models have led to insights into the structure function relationships of this enzyme (143). N-methylputrescine then is oxidized by a diamine oxidase to form 4-methylaminobutanal, which then spontaneously cyclizes to form the N-methyl-D-pyrrolinium ion (144, 145). This enzyme, which has been cloned, seems to be a copper-dependent amine oxidase (146, 147). Immunoprecipitation experiments suggest that this enzyme associates with the enzyme S-adenosylhomocysteine hydrolase (148). The pyrrolinium ion then is converted to the tropanone skeleton by as yet uncharacterized enzymes (Fig. 9.3b). Although no enzymatic information is available, chemical labeling studies have indicated that an acetate-derived moiety condenses with the pyridollium ion; one possible mechanism is shown in Fig. 9.3b (136).

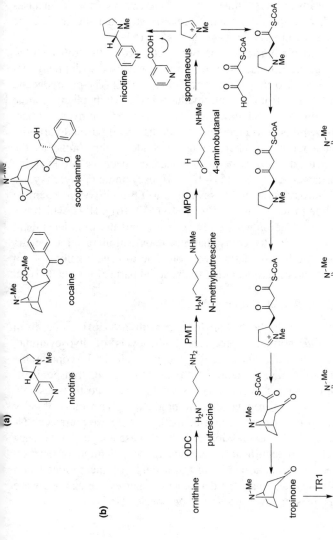

Figure 9.3 (a) Representative tropane and nicotine alkaloids. (b) Tropane biosynthesis. ODC, ornithine decarboxylase; PMT, putrescine N-methyltransferase; MPO, diamine oxidase; TR1, tropinone reductase 1; H6H, hyocyamine 6b-hydroxylase.

223

Tropanone then is reduced via an NADPH-dependent reductase to tropine that has been cloned from *Hyoscyamus niger* (149, 150). All tropane-producing plants seem to contain two tropinone reductases, which create a branch point in the pathway. Tropinone reductase I yields the tropane skeleton (Fig. 9.3b), whereas tropinone reductase II yields the opposite stereocenter, pseudotropine (151). Tropane is converted to scopolamine or hyoscyamine, whereas the TRII product pseudotropine leads to calystegines (152). These two tropinone reductases have been crystallized, and site-directed mutagenesis studies indicate that the stereoselectivity of the enzymes can be switched (153, 154).

The biosynthesis of scopolamine is the best characterized of the tropane alkaloids. After action by tropinone reductase I, tropine is condensed with phenyllactate through the action of a P450 enzyme to form littorine (155). The phenyllactate moiety is believed to derive from an intermediate involved in phenylalanine metabolism (136). Littorine then undergoes rearrangement to form hyoscyamine. The enzyme that catalyzes this rearrangement, which has been purified partially, seems to proceed via a radical mechanism using S-adenysylmethione as the source of an adenosyl radical (156). Labeling studies have been used to examine the mechanism of rearrangement (136)(157–159). Hyoscyamine 6β-hydroxylase (H6H) catalyzes the hydroxylation of hyoscyamine to 6β-hydroxyhyoscyamine as well as the epoxidation to scopolamine (Fig. 9.3b) (160, 161). H6H, which has been cloned and expressed heterologously (162), is a nonheme, iron-dependent, oxoglutarate-dependent protein. It seems that the epoxidation reaction occurs much more slowly than the hydroxylation reaction. The tropane alkaloids seem to be formed in the roots and then transported to the aerial parts of the plant (163).

9.4.1 Metabolic Engineering of Tropane Alkaloids

Atropa belladonna plants have been transformed with an H6H clone from *H. niger*. *A. belladonna* normally produces high levels of hyoscyamine, the precursor for the more pharmaceutically valuable alkaloid scopolamine (Fig. 9.3b). However, after transformation with the H6H gene, transgenic *A. belladonna* plants were shown to accumulate scopolamine almost exclusively (164). Additionally, the levels of tropane alkaloid production in a variety of hairy root cultures were altered by overexpression of methyltransferase putrescine-N-methyltransferase and H6H. Overexpression of both of these enzymes in a hairy root cell culture resulted in significant increases in scopolamine production (164, 165). Fluorinated phenyllactic acid substrates could be incorporated into the pathway (166), and several substrates derived from putrescine analogs were turned over by the enzymes of several *Solonaceae* species (167).

9.5 PURINE ALKALOIDS

9.5.1 Caffeine Biosynthesis

Caffeine, a purine alkaloid, is one of the most widely known natural products. Caffeine is ingested as a natural component of coffee, tea, and cocoa, and the

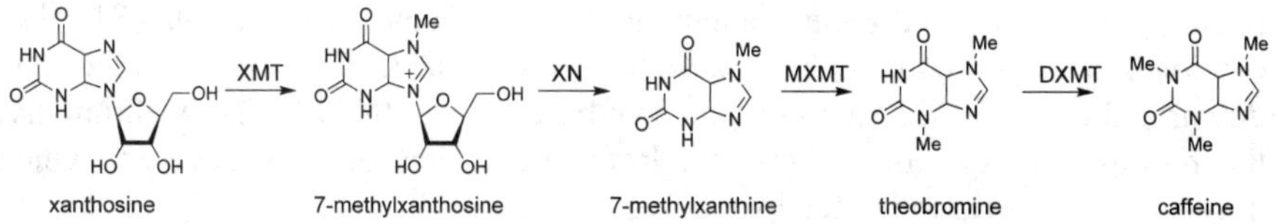

Figure 9.4 Caffeine biosynthesis. XMT, xanthosine N-methyltransferase (also called 7-methylxanthosine synthase); XN, methylxanthosine nucleotidase; MXMT, 7-methylxanthine-N-methyltransferase (also called theobromine synthase); DXMT, dimethylxanthine-N-methyltransferase (also called caffeine synthase).

impact of caffeine on human health has been studied extensively. The biosynthetic pathway of caffeine has been elucidated on the genetic level. Caffeine biosynthesis has been studied most widely in the plant species *Coffea* (coffee) and *Camellia* (tea) (168, 169).

Xanthosine, which is derived from purine metabolites, is the first committed intermediate in caffeine biosynthesis (Fig. 9.4). Xanthosine can be formed from *de novo* purine biosynthesis, S-adenosylmethione (SAM) cofactor, the adenylate pool, and the guanylate pool (169). *De novo* purine biosynthesis and the adenosine from SAM are believed to be the most important sources of xanthosine (168, 170).

The biosynthesis of caffeine begins with the methylation of xanthosine to yield N-methylxanthosine by the enzyme xanthosine N-methyltransferase (XMT) (also called 7-methylxanthosine synthase) (171–173). N-methylxanthosine is converted to N-methylxanthine by methylxanthine nucleosidase, an enzyme that has not been cloned yet (174). N-methylxanthine is converted to theobromine by 7-methylxanthine-N-methyltransferase (MXMT) (also called theobromine synthase), a second N-methyltransferase (171, 175). Theobromine is converted to caffeine by a final N-methyltransferase, dimethylxanthine-N-methyltransferase (DXMT) (also called caffeine synthase) (171).

Coffee and tea plants seem to contain a variety of N-methyltransferase enzymes that have varying substrate specificity (168, 169). For example, a caffeine synthase enzyme isolated from tea leaves catalyzes both the N-methylation of N-methylxanthine and theobromine (176). The substrate specificity of the methyltransferases can be changed by site-directed mutagenesis (177), and the crystal structure of two of the N-methyltransferases has been reported (178).

9.5.2 Metabolic Engineering of Caffeine Biosynthesis

Caffeine may act as a natural insecticide in plants. When the three N-methyltransferase genes were overexpressed in tobacco, the resulting increase in caffeine production improved the tolerance of the plants to certain pests (179). Conversely, coffee beans with low caffeine levels would be valuable commercially, given the demand for decaffeinated coffee. Because of the discovery of these N-methyltransferase genes, genetically engineered coffee plants with reduced caffeine content now can be constructed (180, 181). For example, a 70% reduction in caffeine content in *Coffea* was obtained by downregulating MXMT (theobromine synthase) using RNAi (182). Additionally, the promoter of one of the N-methyltransferases has been discovered, which may allow transcriptional gene silencing (183).

REFERENCES

1. Fischbach MA, Walsh CT. Assembly-line enzymology for polyketide and nonribosomal Peptide antibiotics: logic, machinery, and mechanisms. Chem. Rev. 2006; 106:3468–3496.

2. Herbert RB. The Biosynthesis of Secondary Metabolites. 2nd edition. 1989. Chapman and Hall, London.

3. Dewick PM. Medicinal Natural Products: A Biosynthetic Approach. 2nd edition. 2002. John Wiley and Sons, Ltd., New York.

4. Hashimoto T, Yamada Y. New genes in alkaloid metabolism and transport. Curr. Opin. Biotech. 2003;14:163–168.

5. de Luca V, Laflamme PT. The expanding universe of alkaloid biosynthesis. Curr. Opin. Plant Biol. 2001;4:225–233.

6. Facchini PJ. Alkaloid biosynthesis in plants: biochemistry, cell biology, molecular regulation and metabolic engineering applications. Annu. Rev. Plant Physiol. Plant Mol. Biol. 2001;52:29–66.

7. Kutchan TM. Molecular genetics of plant alkaloid biosynthesis. In: The Alkaloids: Chemistry and Biology, vol.50. Cordell GA, ed. 1998. Academic Press, San Diego, CA. pp.257–316.

8. Allen RS, Millgate AG, Chitty JA, Thisleton J, Miller JAC, Fist AJ, Gerlach WL, Larkin PJ. RNAi-mediated replacement of morphine with the nonnarcotic alkaloid reticuline in opium poppy. Nat. Biotechnol. 2004;22:1559–1566.

9. Facchini PJ, Bird DA, Bourgault R, Hagel JM, Liscombe DK, MacLeod BP, Zulak KG. Opium poppy: a model system to investigate alkaloid biosynthesis in plants. Can. J. Bot. 2005;83:1189–1206.

10. Battersby AR, Binks R, Francis RJ, McCaldin DJ, Ramuz H. Alkaloid biosynthesis. IV. 1-Benzylisoquinolines as precursors of thebaine, codeine, and morphine. J. Chem. Soc. 1964:3600–3610.

11. Rueffer M, Zenk MH. Distant precursors of benzylisoquinoline alkaloids and their enzymatic formation. Naturforsch. 1987;42c:319–332.

12. Samanani N, Facchini PJ. Purification and characterization of norcoclaurine synthase. J. Biol. Chem. 2002;277:33878–33883.

13. Samanani N, Liscombe DK, Facchini PJ. Molecular cloning and characterization of norcoclaurine synthase, an enzyme catalyzing benzylisoquinoline alkaloid biosynthesis. Plant J. 2004;40:302–313.

14. Minami H, Dubouzet E, Iwasa K, Sato F. Functional analysis of norcoclaurine synthase in Coptis japonica. J. Biol. Chem. 2007;282:6274–6282.

15. Morishige T, Tsujita T, Yamada Y, Sato F. Molecular characterization of the S-adenosyl-L-methionine:3′-hydroxy-N-methylcoclaurine   4′-O-methyltransferase involved in isoquinoline alkaloid biosynthesis in Coptis japonica. J. Biol. Chem. 2000;275:23398–23405.

16. Ounaroon A, Decker G, Schmidt J, Lottspeich F, Kutchan TM. (R,S)-reticuline 7-O-methyltransferase and (R,S)-norcoclaurine 6-O-methyltransferase of Papaver somniferum-cDNA cloning and characterization of methyl transfer enzymes of alkaloid biosynthesis in opium poppy. Plant J. 2003;36:808–819.

17. Sato F, Tsujita T, Katagiri Y, Yoshida S, Yamada Y. Purification and characterization of S-adenosyl-L-methionine: norcoclaurine 6-O-methyltransferase from cultured Coptis japonica cells. Eur. J. Biochem. 1994;225:125–131.

18. Choi K-B, Morishige T, Shitan N, Yazaki K, Sato F. Molecular cloning and characterization of coclaurine N-methyltransferase from cultured cells of Coptis japonica. J. Biol. Chem. 2002;277:830–835.

19. Choi K-B, Morishige T, Sato F. Purification and characterization of coclaurine N-methyltransferase from cultured Coptis japonica cells. Phytochemistry 2001;56:649–655.

20. Huang F-C, Kutchan TM. Distribution of morphinan and benzo(c)phenanthridine alkaloid gene transcript accumulation in Papaver somniferum. Phytochemistry 2000; 53:555–564.

21. Pauli HH, Kutchan TM. Molecular cloning and functional heterologous expression of two alleles encoding (S)-N-methylcoclaurine 3′-hydroxylase (CYP80B1), a new methyl jasmonate-inducible cytochrome P-450-dependent mono-oxygenase of benzylisoquinoline alkaloid biosynthesis. Plant J. 1998;13:793–801.

22. Ziegler J, Diaz-Chavez ML, Kramell R, Ammer C, Kutchan TM. Comparative macroarray analysis of morphine containing Papaver somniferum and eight morphine free Papaver species identifies an O-methyltransferase involved in benzylisoquinoline biosynthesis. Planta 2005;222:458–471.

23. Samanani N, Park S-U, Facchini PJ. Cell type-specific localization of transcripts encoding nine consecutive enzymes involved in protoberberine alkaloid biosynthesis. Plant Cell. 2005;17:915–926.

24. Dittrich H, Kutchan TM. Molecular cloning, expression, and induction of berberine bridge enzyme, an enzyme essential to the formation of benzophenanthridine alkaloids in the response of plants to pathogenic attack. Proc. Natl. Acad. Sci. 1991; 88:9969–9973.

25. Facchini PJ, Penzes C, Johnson AG, Bull D. Molecular characterization of berberine bridge enzyme genes from opium poppy. Plant Physiol. 1996;112:1669–1677.

26. Hauschild K, Pauli HH, Kutchan TM. Isolation and analysis of a gene bbe1 encoding the berberine bridge enzyme from the California poppy Eschscholzia californica. Plant Mol. Biol. 1998;36:473–478.

27. Kutchan TM, Dittrich H. Characterization and mechanism of the berberine bridge enzyme, a covalently flavinylated oxidase of benzophenanthridine alkaloid biosynthesis in plants. J. Biol. Chem. 1995;270:24475–24481.

28. Winkler A, Hartner F, Kutchan TM, Glieder A, Macheroux P. Biochemical evidence that berberine bridge enzyme belongs to a novel family of flavoproteins containing a bi-covalently attached FAD cofactor. J. Biol. Chem. 2006;281:21276–21285.

29. Takeshita N, Fujiwara H, Mimura H, Fitchen JH, Yamada Y, Sato F. Molecular cloning and characterization of S-adenosyl-L-methionine:scoulerine-9-O-methyltransferase from cultured cells of Coptis japonica. Plant Cell Physiol. 1995;36: 29–36.

30. Frick S, Kutchan TM. Molecular cloning and functional expression of O-methyltransferasese common to isoquinoline alkaloid and phenylpropanoid biosynthesis. Plant J. 1999;17:329–339.

31. Galneder E, Rueffer M, Wanner G, Tabata M, Zenk MH. Alternative final steps in Coptis japonica cell cultures. Plant Cell Rep. 1988;7:1–4.

32. Ikezawa N, Tanaka M, Nagayoshi M, Shinkyo R, Sakaki T, Inouye K, Sato F. Molecular cloning and characterization of CYP719, a methylenedioxy bridge-forming enzyme that belongs to a novel P450 family, from cultured Coptis japonica cells. J. Biol. Chem. 2003;278:38557–38565.

33. Amann M, Nagakura N, Zenk MH. Purification and properties of (S)-tetrahydroprotoberberine oxidase from suspension cultures of Berberis wilsoniae. Eur. J. Biochem. 1988;175:17–25.

34. Sato F, Yamada Y. High berberine-producing cultures of Coptis-japonica cells. Phytochemistry 1984;23:281–285.

35. Matsubara K, Kitani S, Yoshioka T, Morimoto T, Fujita Y. High-density culture of Coptis-japonica cells increases berberine production. J. Chem. Technol. Biotechnol. 1989;46:61–69.

36. Zenk MH. The formation of benzophenanthridine alkaloids. Pure Appl. Chem. 1994;66:2023–2028.

37. Bauer W, Zenk MH. Two methylenedioxy bridge forming cytochrome P-450 dependent enzymes are involved in (S)-stylopine biosynthesis. Phytochemistry 1991;30:2953–2961.

38. Ikezawa N, Iwasa K, Sato F. Molecular cloning and characterization of methylene-dioxy bridge-forming enzymes involved in stylopine biosynthesis in E. californica. FEBS J. 2007;274:1019–1035.

39. Liscombe DK, Facchini PJ. Molecular cloning and characterization of tetrahydropro-toberberine cis-N-methyltransferase, an enzyme involved in alkaloid biosynthesis in opium poppy. J. Biol. Chem. 2007;282:14741–14751.

40. Tanahashi T, Zenk MH. Elicitor induction and characterization of microsomal pro-topine 6-hydroxylase, the central enzyme in benzophenanthridine alkaloid biosynthesis. Phytochemistry 1990;29:1113–1122.

41. Schumacher HM, Zenk MH. Partial purification and characterization of dihy-drophenanthridine oxidase from Eschscholtzia californica cell suspension cultures. Plant Cell Rep. 1988;7:43–46.

42. Arakawa H, Clark WG, Psenak M, Coscia CJ. Purification and characterization of dihydrobenzophenanthridine oxidase from elicited Sanguinaria canadensis cell cultures. Arch. Biochem. Biophys. 1992;299:1–7.

43. Morishige T, Dubouzet E, Choi K-B, Yazaki K, Sato F. Molecular cloning of colum-bamine O-methyltransferase from cultured Coptis japonica cells. Eur. J. Biochem. 2002;269:5659–5667.

44. Hirata K, Poeaknapo C, Schmidt J, Zenk MH. 1,2-Dehydroreticuline synthase, the branch point enzyme opening the morphinan biosynthetic pathway. Phytochemistry 2004;65:1039–1046.

45. De-Eknamkul W, Zenk MH. Purification and properties of 1,2-dehydroreticuline reductase from Papaver somniferum seedlings. Phytochemistry 1992;31:813–821.

46. Gerardy R, Zenk MH. Formation of salutaridine from (R)-reticuline by a membrane-bound cytochrome P-450 enzyme from Papaver somniferum. Phytochemistry 1992;32:79–86.

47. Gerardy R, Zenk MH. Purification and characterization of salutaridine:NADPH 7-oxidoreductase from Papaver somniferum. Phytochemistry 1993;34:125–132.

48. Ziegler J, Voigtlaender S, Schmidt J, Kramell R, Miersch O, Ammer C, Gesell A, Kutchan TM. Comparative transcript and alkaloid profiling in Papaver species identifies a short chain dehydrogenase/ reductase involved in morphine biosynthesis. Plant J. 2006;48:177–192.

49. Grothe T, Lenz R, Kutchan TM. Molecular characterization of the salutaridinol 7-O-acetyltransferase involved in morphine biosynthesis in opium poppy Papaver somniferum. J. Biol. Chem. 2001;276:30717–30723.

50. Lenz R, Zenk MH. Stereoselective reduction of codeinone, the penultimate enzymic step during morphine biosynthesis in Papaver somniferum. Tetrahedron Lett. 1995;36:2449–2452.

51. Unterlinner B, Lenz R, Kutchan TM. Molecular cloning and functional expression of codeinone reductase. Plant J. 1999;18:465–475.

52. Weid M, Ziegler J, Kutchan TM. The roles of latex and the vascular bundle in morphine biosynthesis in the opium poppy, Papaver somniferum. Proc. Natl. Acad. Sci. U.S.A. 2004;101:13957–13962.

53. Samanani N, Alcantara J, Bourgault R, Zulak KG, Facchini PJ. The role of phloem sieve elements and laticifers in the biosynthesis and accumulation of alkaloids in opium poppy. Plant J. 2006;47:547–563.

54. Bird DA, Franceschi VR, Facchini PJ. A tale of three cell types: alkaloid biosynthesis is localized to sieve elements in opium poppy. Plant Cell 2003;15:2626–2635.

55. Larkin PJ, Miller JAC, Allen RS, Chitty JA, Gerlach WL, Frick S, Kutchan TM, Fist AJ. Increasing morphinan alkaloid production by overexpressing codeinone reductase in transgenic Papaver somniferum. Plant Biotechnol. J. 2007;5:26–37.

56. Millgate AG, Pogson G, Wilson IW, Kutchan TM, Zenk MH, Gerlach WL, Fist AJ, Larkin PJ. Analgesia: morphine-pathway block in top1 poppies. Nature 2004;431:413–414.

57. Frick S, Chitty J, Kramell R, Schmidt J, Allen R, Larkin P, Kutchan T. Transformation of opium poppy with antisense berberine bridge enzyme gene via somatic embryogenesis results in an altered ratio of alkaloids in latex but not in roots. Transgenic Res. 2004;13:607–613.

58. Frick S, Kramell R, Kutchan TM. Metabolic engineering with a morphine biosynthetic P450 in opium poppy surpasses breeding. Met. Eng. 2007;9:169–176.

59. Leonard J. Recent progress in the chemistry of monoterpenoid indole alkaloids derived from secologanin. Nat. Prod. Rep. 1999;16:319–338.

60. Van der Heijden R, Jacobs DI, Snoeijer W, Hallard DVR. The Catharanthus alkaloids: pharmacognosy and biotechnology. Curr. Med. Chem. 2004;11:607–628.

61. Rischer H, Oresic M, Seppanen-Laakso T, Katajamaa M, Lammertyn F, Ardiles-Diaz W, Van Montagu MCE, Inze D, Oksman-Caldentey K-M, Goossens A. Gene-to-metabolite networks for terpenoid indole alkaloid biosynthesis in Catharanthus roseus cells. Proc. Natl. Acad. Sci. U.S.A. 2006;103:5614–5619.

62. de Luca V, Marineau C, Brisson N. Molecular cloning and analysis of a cDNA encoding a plant tryptophan decarboxylase. Proc. Natl. Acad. Sci. U.S.A. 1989;86: 2582–2586.

63. Facchini PJ, Huber-Allanach KL, Tari LW. Plant aromatic L-amino acid decarboxylases: evolution, biochemistry, regulation, and metabolic engineering applications. Phytochemistry 2000;54:121–138.

64. Kutchan TM. Strictosidine: from alkaloid to enzyme to gene. Phytochemistry 1993;32:493–506.

65. McKnight TD, Roessner CA, Devagupta R, Scott AI, Nessler C. Nucleotide sequence of a cDNA encoding the vacuolar protein strictosidine synthase from Catharanthus roseus. Nuc. Acids Res. 1990;18:4939.

66. Kutchan TM, Hampp N, Lottspeich F, Beyreuther K, Zenk MH. The cDNA clone for strictosidine synthase from Rauvolfia serpentina. DNA sequence determination and expression in Escherichia coli. FEBS Lett. 1988;237:40–44.

67. Yamazaki Y, Sudo H, Yamazaki M, Aimi N, Saito K. Camptothecin biosynthetic genes in hairy roots of Ophiorrhiza pumila: cloning, characterization and differential expression in tissues and by stress compounds. Plant Cell Physiol. 2003;44:395–403.

68. Ma X, Panjikar S, Koepke J, Loris E, Stockigt J. The structure of *Rauvolfia serpentina* strictosidine synthase is a novel six-bladed beta-propeller fold in plant proteins. Plant Cell 2006;18:907–920.

69. Koepke J, Ma X, Fritzsch G, Michel H, Stoeckigt J. Crystallization and preliminary X-ray analysis of strictosidine synthase and its complex with the substrate tryptamine. Acta Crystallographica D. 2005;D61:690–693.

70. Chen S, Galan MC, Coltharp C, O'Connor SE. Redesign of a central enzyme in alkaloid biosynthesis. Chem. Biol. 2006;13:1137–1141.

71. Gerasimenko I, Sheludko Y, Ma X, Stockigt J. Heterologous expression of a Rauvolfia cDNA encoding strictosidine glucosidase, a biosynthetic key to over 2000 monoterpenoid indole alkaloids. Eur. J. Biochem. 2002;269:2204–2213.

72. Brandt V, Geerlings A, Tits M, Delaude C, Van der Heijden R, Verpoorte R, Angenot L. New strictosidine b-glucosidase from Strychnos mellodora. Plant Physiol. Biochem. 2000;38:187–192.

73. Geerlings A, Ibanez MM-L, Memelink J, Van der Heijden R, Verpoorte R. Molecular cloning and analysis of strictosidine b-D-glucosidase, an enzyme in terpenoid indole alkaloid biosynthesis in Catharanthus roseus. J. Biol. Chem. 2000;275:3051–3056.

74. Kan-Fan C, Husson HP. Isolation and biomimetic conversion of 4,21-dehydrogeissoschizine. J. Chem. Soc. Chem. Comm. 1979:1015–1018.

75. El-Sayed M, Choi YH, Frederich M, Roytrakul S, Verpoorte R. Alkaloid accumulation in Catharanthus roseus cell suspension cultures fed with stemmadenine. Biotech. Lett. 2004;26:793–798.

76. Heinstein P, Hofle G, Stockigt J. Involvement of cathenamine in the formation of N-analogues of indole akaloids. Planta Med. 1979;37:349–357.

77. Lounasmaa M, Hanhinen P. Biomimetic formation and interconversion in the heteroyohimbine series. Heterocycles 1998;48:1483–1492.

78. Ruppert M, Ma X, Stoeckigt J. Alkaloid biosynthesis in Rauvolfia-cDNA cloning of major enzymes of the ajmaline pathway. Curr. Org. Chem. 2005;9:1431–1444.

79. Brugada J, Brugada P, Brugada R. The ajmaline challenge in Brugada syndrome. A useful tool or misleading information? European Heart J. 2003;24:1085–1086.

80. Stockigt J. Enzymatic biosynthesis of monoterpenoid indole alkaloids: ajmaline, sarpagine and vindoline. Studies Organ. Chem. 1986;26:497–511.

81. Schmidt D, Stockigt J. Enzymatic formation of the sarpagan bridge: a key step in the biosynthesis of sarpagine and ajmaline type alkaloids. Planta Med. 1995;61: 254–258.

82. Mattern-Dogru E, Ma X, Hartmann J, Decker H, Stockigt J. Potential active site residues in polyneuridine aldehyde esterase a central enzyme of indole alkaloid biosynthesis by modelling and site directed mutagenesis. Eur. J. Biochem. 2002;269:2889–2896.

83. Dogru E, Warzecha H, Seibel F, Haebel S, Lottspeich F, Stockigt J. The gene encoding polyneuridine aldehyde esterase of monoterpenoid indole alkaloid biosynthesis in plants is an ortholog of the a/b hydrolase super family. Eur. J. Biochem. 2000;267:1397–1406.

84. Pfitzner A, Stoeckigt J. Biogenetic link between sarpagine and ajmaline type alkaloids. Tetrahedron Lett. 1983;24:5197–5200.

85. Gerasimenko I, Ma X, Sheludko Y, Mentele R, Lottspeich F, Stockigt J. Purification and partial amino acid sequences of the enzyme vinorine synthase involved in a crucial step of ajmaline biosynthesis. Bioorg. Med. Chem. 2004;12:2781–2786.

86. Bayer A, Ma X, Stockigt J. Acetyltransfer in natural product biosynthesis functional cloning and molecular analysis of vinorine synthase. Bioorg. Med. Chem. 2004;12:2787–2795.

87. Ma X, Koepke J, Panjikar S, Fritzsch G, Stoeckigt J. Crystal estructure of vinorine synthase, the first representative of the BAHD superfamily. J. Biol. Chem. 2005;280:13576–13583.

88. Falkenhagen H, Polz L, Takayama H, Kitajima M, Sakai S, Aimi N, Stockigt J. Substrate specificity of vinorine hydroxylase, a novel membrane bound key enzyme of Rauwolfia indole alkaloid biosynthesis. Heterocycles 1995;41:2683–2690.

89. Gao S, von Schumann G, Stockigt J. A newly detectedreductase from Rauvolfia closes a gap in the biosynthesis of the antiarrhythmic alkaloid ajmaline. Planta Med. 2002;68:906–911.

90. von Schumann G, Gao S, Stockigt J. Vomilenine reductase: a novel enzyme catalyzing a crucial step in the biosynthesis of the therapeutically applied antiarrhythmic alkaloid ajmaline. Bioorg. Med. Chem. 2002;10:1913–1918.

91. Ruppert M, Woll J, Giritch A, Genady E, Ma X, Stoeckigt J. Functional expression of an ajmaline pathway-specific esterase from Rauvolfia in a novel plant-virus expression system. Planta 2005;222:888–898.

92. Stockigt J, Pfitzner A, Keller PI. Enzymic formation of ajmaline. Tetrahedron Lett. 1983;244:2485–2486.

93. Li S, Long J, Ma Z, Xu Z, Li J, Zhang Z. Assessment of the therapeutic activity of a combination of almitrine and raubasine on functional rehabilitation following ischaemic stroke. Curr. Med. Res. Opin. 2004;20:409–415.

94. Costa-Campos L, Iwu M, Elisabetsky E. Lack of pro-convulsant activity of the antipsychotic alkaloid alstonine. J. Ethnopharmacol. 2004;93:307–310.

95. Hemscheidt T, Zenk MH. Partial purification and characterization of a NADPH dependent tetrahydroalstonine synthase from Catharanthus rosues cell suspension cultures. Plant Cell Rep. 1985;4:216–219.

96. Stockigt J, Hemscheidt T, Hofle G, Heinstein P, Formacek V. Steric cours of hydrogen transfer during enzymatic formation of 3a-heteroyohimbine alkaloids. Biochemistry 1983;22:3448–3452.

97. de Luca V. Biochemistry and molecular biology of indole alkaloid biosynthesis: the implication of recent discoveries. Rec. Adv. Phytochem. 2003;37:181–202.

98. St. Pierre B, de Luca V. A cytochrome P-450 monooxygenase catalyzes the first step in the conversion of tabersonine to vindoline in Catharanthus roseus. Plant Physiol. 1995;109:131–139.

99. Schroder G, Unterbusch E, Kaltenbach M, Schmidt J, Strack D, de Luca V, Schroder J. Light induced cytochrome P450 dependent enzyme in indole alkaoid biosynthesis: tabersonine 16-hydroxylase. FEBS Lett. 1999;458:97–102.

100. Cacace S, Schroder G, Wehinger E, Strack D, Schmidt J, Schroder J. A flavonol O-methyltransferase from Catharanthus roseus performing two sequential methylations. Phytochemistry 2003;62:127–137.

101. Dethier M, de Luca V. Partial purification of an N-methyltransferase involved in vindoline biosynthesis in Catharanthus roseus. Phytochemistry 1993;32:673–678.

102. Vazquez-Flota F, de Carolis E, Alarco A, de Luca V. Molecular cloning and characterization of desacetoxyvindoline-4-hydroxylase, a 2-oxoglutarate dependent-dioxygenase involved in the biosynthesis of vindoline in Catharanthus roseus (L.) G. Don.. Plant Mol. Biol. 1997;34:935–948.

103. St. Pierre B, Laflamme P, Alarco A, de Luca V. The terminal O-acetyltransferase involved in vindoline biosynthesis defines a new class of proteins responsible for coenzyme A-dependent acyl transfer. Plant J. 1998;14:703–713.

104. de Luca V, St. Pierre B. The cell and developmental biology of alkaloid biosynthesis. Trends Plant Sci. 2000;5:168–173.

105. McKnight TD, Bergey DR, Burnett RJ, Nessler C. Expression of enzymatically active and correctly targeted strictosidine synthase in transgeneic tobacco plants. Planta 1991;185:148–152.

106. Stevens LH, Blom TJM, Verpoorte R. Subcellular localization of tryptophan decarboxylase, strictosidine synthase and strictosidine glucosidase in suspension cultured cells of Catharanthus roseus and Tabernamontana divaricata. Plant Cell Rep. 1993;12:573–576.

107. de Luca V, Cutler AJ. Subcellular localization of enzymes involved in indole alkaloid biosynthesis in Catharanthus roseus. Plant Physiol. 1987;85:1099–1102.

108. de Carolis E, Chan F, Balsevich J, de Luca V. Isolation and characterization of a 2-oxoglutarate dependent dioxygenase involved in the second to last step in vindoline biosynthesis. Plant Physiol. 1990;94:1323–1329.

109. Burlat V, Oudin A, Courtois M, Rideau M, St. Pierre B. Coexpression of three MEP pathway genes and geraniol 10 hydroxylase in internal phloem parenchyma of Catharanthus roseus implicates multicellular translocation of intermediates during the biosynthesis of monoterpene indole alkaloids and isoprenoid derived primary metabolites. Plant J. 2004;38:131–141.

110. Murata J, de Luca V. Localization of tabersonine 16-hydroxylase and 16-OH tabersonine 16-O-methyl transferase to leaf epidermal cells defines them as a major site of precursor biosynthesis in the vindoline pathway in Catharanthus roseus. Plant J. 2005;44:581–594.

111. Irmler S, Schroder G, St. Pierre B, Crouch NP, Hotze M, Schmidt J, Strack D, Matern U, Schroder J. Indole alkaloid biosynthesis in Catharanthus roseus: new enzyme activities and identification of cytochrome P450 CYP72A1 as secologanin synthase. Plant J. 2000;24:797–804.

112. St. Pierre B, Vazquez-Flota F, de Luca V. Multicellular compartmentation of Catharanthus roseus alkaloid biosynthesis predicts intercellular translocation of a pathway intermediate. Plant Cell 1999;11:887–900.

113. Islam MN, Iskander MN. Microtubulin binding sites as target for developing anticancer agents. Mini-Rev. Med. Chem. 2004;4:1077–1104.

114. Beckers T, Mahboobi S. Natural, semisynthetic and synthetic microtubule inhibitors for cancer therapy. Drugs Future 2003;28:767–785.

115. Scott AI, Gueritte F, Lee SL. Role of anhydrovinblastine in the biosynthesis of the antitumor dimeric indole alkaloids. J. Am. Chem. Soc. 1978;100:6253–6255.

116. Stuart KL, Kutney JP, Honda T, Worth BR. Intermediacy of 3′-4′-dehydrovinblastine in the biosynthesis of vinblastine type alkaloids. Heterocycles 1978;9:1419–1426.

117. Baxster RI, Dorschel CA, Lee SL, Scott AI. Biosynthesis of the antitumor catharanthus alkaloids. Conversion of anhydrovinblastine into vinblastine. J. Chem. Soc. Chem. Comm. 1979:257–259.

118. Gueritte F, Bac NV, Langlois Y, Potier P. Biosynthesis of antitumour alkaloids from Catharanthus roseus. Conversion of 20′deoxyleurosidine into vinblastine. J. Chem. Soc. Chem. Comm. 1980:452–453.

119. Sottomayor M, Cardoso IL, Pereira LG, Barcel AR. Peroxidase and the biosynthesis of terpenoid indole alkaloids in the medicinal plant Catharanthus roseus (L.) G. Don.. Phytochem. Rev. 2004;3:159–171.

120. Endo T, Goodbody A, Vukovic J, Misawa M. Enzymes from Catharanthus roseus cell suspension cultures that couple vindoline and catharanthine to form 3′,4′-anhydrovinbastine. Phytochemistry 1988;27:2147–2149.

121. Smith JL, Amouzou E, Yamagushi A, McLean S, DiCosmo F. Peroxidase from bioreactor cultivated Catharanthus roseus cell cultures mediate biosynthesis of a-3′,4′-anhydrovinblastine. Biotechnol. Appl. Bioeng. 1988;10:568–575.

122. Hillou F, Costa M, Almeida I, Lopes Cardoso I, Leech M, Ros Barcelo A, Sottomayor M. Cloning of a peroxidase enzyme involved in the biosynthesis of pharmaceutically aciveterpenoid indole alkaloids in Catharanthus roseus. In: Proceedings of the VI International Plant Peroxidase Symposium. Acosta M, Rodriguez-Lopez JN, Pedreno MA, eds. 2002. pp.152–158.

123. Sottomayor M, Lopez-Serrano M, DiCosmo F, Ros Barcelo A. Purification and characterization of anhydrovinblastine synthase (peroxidase like) from Catheranthus roseus. FEBS Lett. 1998;428:299–303.

124. Goodbody AE, Endo T, Vukovic J, Kutney JP, Choi LSL, Misawa M. Enzymic coupling of catharanthine and vindoline to form 3′,4′-anhydrovinblastine by horseradish peroxidase. Planta Med. 1988;54:136–140.

125. Canel C, Lopes-Cardoso MI, Whitmer S, Van der Fits L, Pasquali G, Van der Heijden R, Hoge JHC, Verpoorte R. Effects of over-expression of strictosidine synthase and tryptophan decarboxylase on alkaloid production by cell cultures of Catharanthus roseus. Planta 1998;205:414–419.

126. Di Fiore S, Hoppmann V, Fischer R, Schillberg S. Transient gene expression of recombinant terpenoid indole alkaloid enzymes in Catharanthus roseus leaves. Plant Mol. Biol. Rep. 2004;22:15–22.

127. Ayora-Talavera T, Chappell J, Lozoya-Gloria E, Loyola-Vargas VM. Overexpression in Catharanthus roseus hairy roots of a truncated hamster 3-hydroxy-3-methylglutaryl-CoA reductase gene. Appl. Biochem. Biotechnol. 2002;97:135–145.

128. Peebles CAM, Hong S-B, Gibson SI, Shanks JV, San K-Y. Transient effects of overexpressing anthranilate synthase a and b subunits in Catharanthus roseus hairy roots. Biotech. Prog. 2005;21:1572–1576.

129. Hedhili S, Courdavault V, Gigioli-Guivarc'h N, Gantet P. Regulation of the terpene moiety biosynthesis of Catharanthus roseus alkaloids. Phytochem. Rev. 2007;6:341–351.

130. McCoy E, O'Connor SE. Directed biosynthesis of alkaloid analogues in the medicinal plant periwinkle. J. Am. Chem. Soc. 2006;128:14276–14277.

131. Geerlings A, Redondo FJ, Contin A, Memelink J, Van der Heijden R, Verpoorte R. Biotransformation of tryptamine and secologanin into plant terpenoid indole alkaloids by transgenic yeast. Appl. Microbiol. Biotechnol. 2001;56:420–424.

132. Menke FLH, Champion A, Kijne JW, Memelink J. A novel jasmonate- and elicitor-responsive element in the periwinkle secondary metabolite biosynthetic gene Str interacts with a jasmonate- and elicitor-inducible AP2-domain transcription factor, ORCA2. EMBO J. 1999;18:4455–4463.

133. Van der Fits L, Memelink J. ORCA3, a Jasmonate-responsive transcriptional regulator of plant primary and secondary metabolism. Science 2000;289:295–297.

134. Pauw B, Hilliou FAO, Martin VS, Chatel G, de Wolf CJF, Champion A, Pre M, Van Duijn B, Kijne JW, Van der Fits L, Memelink J. Zinc finger proteins act as transcriptional repressors of alkaloid biosynthesis genes in Catharanthus roseus. J. Biol. Chem. 2004;279:52940–52948.

135. Montiel G, Breton C, Thiersault M, Burlat V, Jay-Allemand C, Gantet P. Transcription factor Agamous-like 12 from Arabidopsis promotes tissue-like organization and alkaloid biosynthesis in Catharanthus roseus suspension cells. Met. Eng. 2007;9:125–132.

136. Humphrey AJ, O'Hagan D. Tropane alkaloid biosynthesis: a century old problem unresolved. Nat. Prod. Rep. 2001;18:494–502.

137. Hemscheidt T. Tropane and related alkaloids. Top. Curr. Chem. 2000;209:175–206.

138. Oksman-Caldenty KM, Arroo R. Regulation of tropane alkaloid metabolism in plants and plant cell cultures. In: Metabolic Engineering of Plant Secondary Metabolism. Verpoorte R, Alfermann AW, eds. 2000. Kluwer Academic Publishers, Dordrecht, the Netherlands. pp.253–281.

139. Hibi N, Higashiguchi S, Hashimoto T, Yamada Y. Gene expression in tobacco low-nicotine mutants. Plant Cell 1994;6:723–735.

140. Hibi N, Fujita T, Hatano M, Hashimoto T, Yamada Y. Putrescine N-methyltransferase in cultured roots of Hyoscyamus albus. Plant Physiol. 1992;100:826–835.

141. Hashimoto T, Shoji T, Mihara T, Oguri H, Tamaki K, Suzuki KI, Yamada Y. Intraspecific variability of the tandem repeats in Nicotiana putrescine N-methyltransferases. Plant Mol. Biol. 1998;37:25–37.

142. Suzuki K, Yamada Y, Hashimoto T. Expression of Atropa belladonna putrescine N-methyltransferase gene in root pericycle. Plant Cell Physiol. 1999;40:289–297.

143. Teuber M, Azemi ME, Namjoyan F, Meier A-C, Wodak A, Brandt W, Drager B. Putrescine N-methyltransferases—a structure–function analysis. Plant Mol. Biol. 2007;63:787–801.

144. McLauchlan WR, McKee RA, Evans DM. The purification and immunocharacterization of N-methylputrescine oxidase from transformed root cultures of Nicotinia tabacum. Planta 1993;191:440–445.

145. Haslam SC, Young TW. Purification of N-methylputrescine oxidase from Nicotiana rustica. Phytochemistry 1992;31:4075–4079.

146. Katoh A, Shoji T, Hashimoto T. Molecular cloning of N-methylputrescine oxidase from tobacco. Plant Cell Physiol. 2007;48:550–554.

147. Heim WG, Sykes KA, Hildreth SB, Sun J, Lu RH, Jelesko JG. Cloning and characterization of a Nicotiana tabacum methylputrescine oxidase transcript. Phytochemistry 2007;68:454–463.

148. Heim WG, Jelesko JG. Association of diamine oxidase and S-adenosylhomocysteine hydrolase in Nicotiana tabacum extracts. Plant Mol. Biol. 2004;56:299–308.

149. Hashimoto T, Nakajima K, Ongena G, Yamada Y. Two tropinone reductases with distinct stereospecificities from cultured roots of Hyoscyamus niger. Plant Physiol. 1992;100:836–845.

150. Rocha P, Stenzel O, Parr A, Walton NJ, Christou P, Drager B, Leech MJ. Functional expression of tropinone reductase I and hyoscyaine-6b-hydroxylase from Hyoscyamus niger in Nicotiana tabacum. Plant Sci. 2002;162:905–913.

151. Nakajima K, Yamashita A, Akama H, Nakatsu T, Kato H, Hashimoto T, Oda J, Yamada Y. Crystal structures of two tropinone reductases: different reaction stereospecificities in the same protein fold. Proc. Natl. Acad. Sci. 1998;95:4876–4881.

152. Dräger B. Chemistry and biology of calystegines. Nat. Prod. Rep. 2004;21:211–223.

153. Nakajima K, Kato H, Oda J, Yamada Y, Hashimoto T. Site directed mutagenesis of putative substrate binding residues reveals a mechanism controlling different substrate specificities of two tropinone reductases. J. Biol. Chem. 1999;274: 16563–16568.

154. Yamashita A, Kato H, Wakatsuki S, Tomizaki T, Nakatsu T, Nakajima K, Hashimoto T, Yamada Y, Oda J. Structure of tropinone reductase-II with NADP + and pseudotropine at 1.9A resolution: implication for stereospecific substrate binding and catalysis. Biochemistry 1999;38:7630–7637.

155. Li R, Reed DW, Liu E, Nowak J, Pelcher LE, Page JE, Covello PS. Functional genomic analysis of alkaloid biosynthesis in Hyoscyamus niger reveals a cytochrome P450 involved in littorine rearrangement. Chem. Biol. 2006;13:513–520.

156. Oliagnier S, Kervio E, Retey J. The role and source of 5′-deoxyadenosylradical in a carbon skeleton rearrangement catalyzed by a plant enzyme. FEBS Lett. 1998;437: 309–312.

157. Lanoue A, Boitel-Conti M, Portais JC, Laberche JC, Barbotin JN, Christen P, Sangwan-Norreel B. Kinetic study of littorine rearrangement in Datura innoxia hairy roots by 13C NMR spectroscopy. J. Nat. Prod. 2002;65:1131–1135.

158. Patterson S, O'Hagan D. Biosynthetic studies on the tropane alkaloid hyoscyamine in Datura stramonium; hyoscyamine is stable to in vivooxidtion and is not derived from littorine via a vicinal interchange process. Phytochemistry 2002;61:323–329.

159. Duran-Patron R, O'Hagan D, Hamilton JTG, Wong CW. Biosynthetic studies on the tropane ring system of the tropane alakloids from Datura stramonium. Phytochemistry 2000;53:777–784.

160. Yamada Y, Hashimoto T. Substrate specificity of the hyoscyamine 6 β-hydroxylase from cultured roots of Hyoscyamus niger. Proc. Japan Acad. B 1989;65:156–159.

161. Hashimoto T, Kohno J, Yamada Y. 6b-Hydroxyhyoscyamine epoxidase from cultured roots of Hyoscyamus niger. Phytochemistry 1989;28:1077–1082.

162. Matsuda J, Okabe S, Hashimoto T, Yamada Y. Molecular cloning of hyoscyamine 6β-hydroxylase, a 2-oxoglutarate-dependent dioxygenase, from cultured roots of Hyoscyamus niger. J. Biol. Chem. 1991;266:9460–9464.

163. Hashimoto T, Yamada Y. Scopolamine production in suspension-cultures and redifferentiated roots of Hyoscyamus niger. Planta Med. 1983;47:195–199.

164. Yun DJ, Hashimoto T, Yamada Y. Metabolic engineering of medicinal plants: transgenic Atropa belladonna with an improved alkaloid composition. Proc. Natl. Acad. Sci. 1992;89:11799–11803.

165. Sato F, Hashimoto T, Hachiya A, Tamura K, Choi K-B, Morishige T, Fujimoto H, Yamada Y. Metabolic engineering of plant alkaloid biosynthesis. Proc. Natl. Acad. Sci. U.S.A. 2001;98:367–372.

166. O'Hagan D, Robins RJ, Wilson M, Wong CW, Berry M. Fluorinated tropane alkaloids generated by directed biosynthesis in transformed root cultures of Datura stramonium. J. Chem. Soc. Perkins Trans. 1. 1999:2117–2120.

167. Boswell HD, Drager B, McLauchlan WR, Portseffen A, Robins DJ, Robins RJ, Walton NJ. Specificities of the enzymes of N-alkyltropane biosynthesis in Brugmansia and Datura. Phytochemistry 1999;52:871–878.

168. Ashihara H, Crozier A. Caffeine: a well known but little mentioned compound in plant science. Trends Plant Sci. 2001;6:407–413.

169. Crozier A, Ashihara H. The cup that cheers. Biochemist 2006;28:23–26.

170. Koshiishi C, Kato A, Yama S, Crozier A, Ashihara H. A new caffeine biosynthetic pathway in tea leaves: utilisation of adenosine released from the S-adenosyl-L-methionine cycle. FEBS Lett. 2001;499:50–54.

171. Uefuji H, Ogita S, Yamaguchi Y, Koizumi N, Sano H. Molecular cloning and characterization of three distinct N-methyltransferases involved in the caffeine biosynthetic pathway in coffee plants. Plant Physiol. 2003;132:372–380.

172. Mizuno K, Kato M, Irino F, Yoneyama N, Fujimura T, Ashihara H. The first committed step reaction of caffeine biosynthesis: 7-methylxanthosine synthase is closely homologous to caffeine synthase in coffee. FEBS Lett. 2003;547:56–60.

173. Waldhauser SSM, Gillies FM, Crozier A, Baumann TW. Separation of the N-methyltransferase the key enzyme in caffeine biosynthesis. Phytochemistry 1997;45:1407–1414.

174. Stoychev G, Kierdaszuk B, Shugar D. Xanthosine and xanthine: substrate properties with purine nucleoside phosphorylases and relevance to other enzyme systems. Eur. J. Biochem. 2002;269:4048–4057.

175. Ogawa M, Herai Y, Koizumi N, Kusano T, Sano H. 7-methylxanthine methyltransferase of coffee plants. Gene isolation and enzymatic properties. J. Biol. Chem. 2001;276:8213–8218.

176. Kato M, Mizuno K, Crozier A, Fumjimura T, Ashihara H. Plant biotechnology: caffeine synthase gene from tea leaves. Nature 2000;406:956–957.

177. Yoneyama N, Morimoto H, Ye CX, Ashira H, Mizuno K, Kato M. Substrate specificity of N-methyltransferase involved in purine alkaloid synthesis is dependent on one amino acid residue of the enzyme. Mol. Gen. Genomics 2006;275:125–135.

178. McCarthy AA, McCarthy JG. The structure of two N-methyltransferases from the caffeine biosynthetic pathway. Plant Physiol. 2007;144:879–889.

179. Kim YS, Uefuji H, Ogita S, Sano H. Transgenic tobacco plants producing caffeine: a potential new strategy for insect pest control. Transgenic Res. 2006;15:667–672.

180. Ashira H, Zheng XQ, Katahira R, Morimoto M, Ogita S, Sano H. Caffeine biosynthesis and adenine metabolism in transgenic Coffea canephora plants with reduced expression of N-methyltransferase genes. Phytochemistry 2006;67:882–886.

181. Ogita S, Uefuji H, Morimoto H, Sano H. Application of RNAi to confirm theobromine as the major intermediate for caffeine biosynthesis in coffee plants with potential for construction of decaffeinated varieties. Plant Mol. Biol. 2004;54:931–941.

182. Ogita S, Uefuji H, Yamaguchi Y, Koizumi N, Sano H. Producing decaffeinated coffee plants. Nature 2003;423:823.

183. Satyanarayana KV, Kumar V, Chandrasheker A, Ravishankar GA. Isolation of promoter for N-methyltransferase gene associated with caffeine biosynthesis in Coffea canephora. J. Biotechnol. 2005;119:20–25.

FURTHER READING

Ashihara H, Crozier A., Caffeine: a well known but little mentioned compound in plant science. Trends Plant Sci. 2001;6:407–413.

de Luca V, Laflamme PT. The expanding universe of alkaloid biosynthesis. Curr. Opin. Plant Biol. 2001;4:225–233.

Hemscheidt T. Tropane and related alkaloids. Top. Curr. Chem. 2000;209:175–206.

Humphrey AJ, O'Hagan D. Tropane alkaloid biosynthesis: a century old problem unresolved. Nat. Prod. Rep. 2001;18:494–502.

Kutchan TM. Molecular genetics of plant alkaloid biosynthesis. In: The Alkaloids: Chemistry and Biology, vol.50. Cordell GA, ed. 1998. Academic Press: San Diego, CA. pp.261–263.

O'Connor SE, Maresh JM. Chemistry and biology of terpene indole alkaloid biosynthesis. Nat. Prod. Rep. 2006;23:532–547.

10

COFACTORS

ILKA HAASE AND MARKUS FISCHER
Institute of Food Chemistry, University of Hamburg, Hamburg, Germany

ADELBERT BACHER, WOLFGANG EISENREICH, AND FELIX ROHDICH
Institute of Biochemistry, Department Chemie, Technische Universität München, München, Germany

Whereas plants and certain microorganisms can generate all required coenzymes from CO_2 or simple organic precursors, animals must obtain precursors (designated as vitamins) for a major fraction of their coenzymes from nutritional sources. Still, most vitamins must be converted into the actual coenzymes by reactions catalyzed by animal enzymes. The structures and biosynthetic pathways of some coenzymes are characterized by extraordinary complexity. Enzymes for coenzyme biosynthesis have frequently low catalytic rates, and some of them catalyze reactions with highly unusual mechanisms.

Many coenzymes (cofactors) involved in human and animal metabolism were discovered in the first half of the twentieth century, and their isolation and structure elucidation were hailed as milestones as shown by the impressive number of Nobel prizes awarded for research in that area. Studies on coenzyme biosynthesis were typically initiated in the second half of the twentieth century and have generated a massive body of literature that continues to grow rapidly because the area still involves many incompletely resolved problems. In parallel, numerous novel coenzymes were discovered relatively recently by studies of microorganisms. In this chapter, the terms "cofactor" and "coenzyme" are used as synonyms.

Cofactor biosynthesis is a very broad and multifaceted topic. This chapter summarizes basic concepts of major cofactors. Detailed status reports on the

---

Natural Products in Chemical Biology, First Edition. Edited by Natanya Civjan.
© 2012 John Wiley & Sons, Inc. Published 2012 by John Wiley & Sons, Inc.

biosynthesis of individual coenzymes can be found in excellent reviews that are quoted (1).

10.1 GENERAL ASPECTS

10.1.1 Biosynthetic Origin of Coenzymes in Autotrophic and Heterotrophic Organisms

Plants, algae, and certain microorganisms can generate all their organic components, including their entire coenzyme repertoire, from CO_2. Certain microorganisms can also generate their entire biomass from single organic nutrients, such as carbohydrates or carboxylic acids. Animals, on the other hand, depend on plants and bacteria for the supply of many coenzymes or coenzyme precursors (designated vitamins) but synthesize others *de novo* from basic precursors derived from central intermediary metabolism (Table 10.1). The distribution of coenzyme biosynthetic pathways in eubacteria and fungi is very complex; whereas certain species are auxotrophic with regard to coenzymes, the range is from complete

TABLE 10.1   Origin of coenzymes in humans

Coenzyme	Partial Synthesis From	De novo Synthesis From	Nutritional
FMN, FAD	Riboflavin (Vitamin B$_2$), ATP		
Tetrahydrofolate	Folate, Dihydrofolate		
Tetrahydrobiopterin		GTP	
Molybdopterin		GTP	
Thiamine pyrophosphate	Thiamine (Vitamin B$_1$)		
Pyridoxal phosphate	Pyridoxine, Pyridoxal (Vitamin B$_6$)		
Coenzyme B$_{12}$	Vitamin B$_{12}$, ATP		
Coenzyme A, Pantetheine	Pantothenate, ATP		
Porphyrins		Succinyl-CoA, Glycine	
Fe/S Clusters		Cysteine	
Coenzyme Q10 (ubiquinone)		Acetyl-CoA, Tyrosine	
Vitamin K (menaquinone)			Nutritional
Vitamin C			Nutritional
NADH, NADPH	Nicotinate, Nicotinamide, Nicotinamide riboside, ATP		
Biotin			Nutritional
Retinal	Carotenoids		Nutritional

self-sufficiency to virtually complete dependence on exogenous coenzymes or coenzyme precursors. However, even organisms that depend on exogenous vitamins must convert those into the actual coenzymes by more or less complex enzymatic transformations.

10.1.2 Basic Precursors for Coenzyme Biosynthesis

Precursors for cofactor biosynthesis are drawn from all major pools of central intermediary metabolism, including carbohydrates, amino acids, purine nucleotides, and carboxylic acids.

10.1.3 Convergent Biosynthetic Pathways

Numerous cofactors are generated via convergent pathways starting from two or more different precursors (e.g., the biosynthesis of vitamins B_1, B_2, and tetrahydrofolate). Occasionally, a single precursor is sufficient for biosynthesis, for example, in case of vitamin C and tetrahydrobiopterin.

10.1.4 Alternative Pathways Exist for Several Coenzymes

Whereas the biosyntheses of some coenzymes (e.g., tetrahydrobiopterin) seem to proceed via unique pathways, several different pathways have been described for the biosynthesis of others. In certain cases, the different pathways are variations on a common theme; for example, in the biosynthesis of vitamin B_2, a sequence of a deamination and reduction can proceed in different order. In other cases, entirely different reactions afford a given intermediate, for example in the case of nicotinic acid biosynthesis. Another example is the biosynthesis of isoprenoid building blocks and their downstream products, in which the existence of a second pathway besides the classic mevalonate pathway had been ignored until recently. Notably, higher plants use both isoprenoid pathways but for different final products.

10.1.5 Some Coenzymes Develop by very Complex Pathways and/or Reactions

Whereas tetrahydrobiopterin is biosynthesized from GTP via just three enzyme-catalyzed steps (2), some coenzyme biosynthetic pathways are characterized by enormous complexity. Thus, the biosynthesis of vitamin B_{12} requires five enzymes for the biosynthesis of the precursor uroporhyrinogen III (**16**) from succinyl-CoA (**10**) and glycine (**11**) that is then converted into vitamin B_{12} via the sequential action of about 20 enzymes (3). Additional enzymes are involved in the synthesis of the building blocks aminopropanol and dimethylbenzimidazole (4, 5). Vitamin B_{12} from nutritional sources must then be converted to coenzyme B_{12} by mammalian enzymes. Ultimately, however, coenzyme B_{12} is used in humans by only two enzymes, albeit of vital importance, which are involved in fatty acid and amino acid metabolism (6). Notably, because plants do not generate corrinoids, animals depend on bacteria for their supply of vitamin B_{12} (which may be obtained in recycled form via nutrients such as milk and meat) (7).

However, the sheer complexity of a pathway is not an indicator for the vitamin status of a given class of compounds, as opposed to endogenous biosynthesis in mammals. Thus, animals biosynthesize molybdopterin, which is a cofactor involved in certain redox reactions, from basic building blocks using at least eight enzymes, whereas folic acid has vitamin status and must be obtained by animals from nutritional sources (8).

It should also be noted that some pathways to be discussed below depend on highly unusual chemical reactions. To give just one example, the formation of the pyridine ring system of vitamin B_6 depends on a protein that catalyzes a complex series of reactions, including carbohydrate isomerization, imine formation, ammonia addition, aldol-type condensation, cyclization, and aromatization (9).

10.2 SPECIFIC BIOSYNTHETIC PATHWAYS

10.2.1 Biosynthesis of Iron/Sulfur Clusters

Iron/sulfur clusters are inorganic cofactors that are used in all cells (10). They comprise S^{2-} ions and iron ions in the $+2$ or $+3$ state (Fig. 10.1). Iron/sulfur clusters are essential cofactors for numerous redox and nonredox enzymes, alone or in tandem with organic cofactors such as flavocoenzymes and/or pyridine nucleotides. The simplest structural type is the rhombic [2Fe-2 S] cluster (3). [3Fe-4 S] (4) and [4Fe-4 S] (5) clusters are characterized by distorted cubic

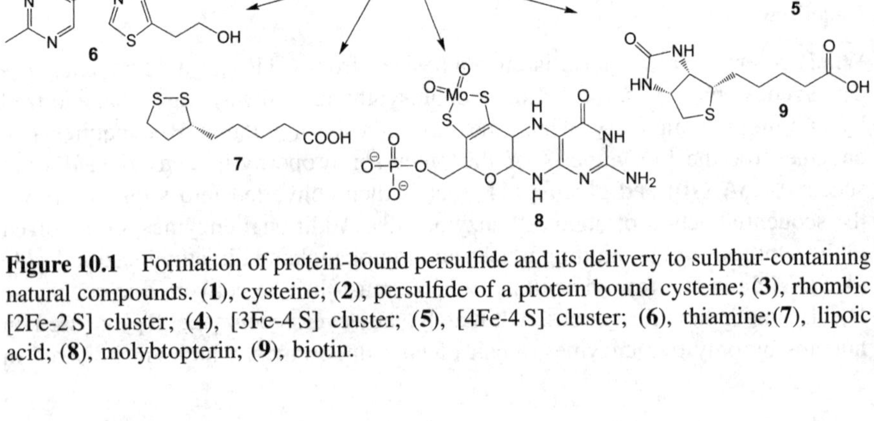

Figure 10.1 Formation of protein-bound persulfide and its delivery to sulphur-containing natural compounds. (**1**), cysteine; (**2**), persulfide of a protein bound cysteine; (**3**), rhombic [2Fe-2 S] cluster; (**4**), [3Fe-4 S] cluster; (**5**), [4Fe-4 S] cluster; (**6**), thiamine;(**7**), lipoic acid; (**8**), molybtopterin; (**9**), biotin.

symmetry (10, 11). Clusters can form aggregates, and other metal ions can replace iron ions or can be present additionally.

Whereas many cognate apoenzymes can be reconstituted with iron/sulfur clusters by simple and essentially alchemistic procedures using Fe^{2+} and sulfide ions under anaerobic conditions, a highly complex enzymatic machinery is used *in vivo* for the synthesis of iron/sulfur clusters and their transfer to the target enzymes. Sulfide ions required for cluster synthesis are obtained from cysteine (**1**) via a persulfide of a protein-bound cysteine residue (**2**); pyridoxal phosphate is required for the formation of the persulfide intermediate (Fig. 10.1) (12).

In eukaryotes, the formation of iron/sulfur clusters proceeds inside mitochondria (13). The mitochondrial enzymes are orthologs of the eubacterial *isc* proteins and are characterized by very slow rates of evolution. Iron/sulfur clusters are initially assembled on IscU protein (prokaryotic) or Isu protein (eukaryotic) that serves as a scaffold. They can be exported to the cytoplasm in which they can become part of cytoplasmic enzymes by the assistance of proteins that serve as iron chaperones.

The persulfide intermediate (**2**) can also serve as a sulfur source for the biosynthesis of thiamine (**6**), lipoic acid (**7**), molybdopterin (**8**), and biotin (**9**) (Fig. 10.1).

10.2.2 Tetrapyrroles

A large and structurally complex family of coenzymes, including various hemes and chlorophylls, and corrinoids, including coenzyme B_{12} and the archaeal coenzyme F_{430}, are characterized by their macrocyclic tetrapyrrole structure (14, 15). These coenzymes contain a metal ion (Fe, Mg, Co, or Ni) at the center of the tetrapyrrole macrocycle, which is specifically introduced by enzyme catalysis. These compounds are all derived from δ-aminolevulinic acid (**12**) that can be biosynthesized by two independent pathways, that is, from glycine (**11**) and succinyl-CoA (**10**) in animals and some bacteria (e.g., *Rhodobacter*) or from glutamyl-tRNA (**13**) in plants, many eubacteria, and archaea (Fig. 10.2) (16, 17). Two molecules of δ-aminolevulinic acid (**12**) are condensed under formation of porphobilinogen (**14**). Oligomerization of porphobilinogen affords hydroxymethylbilane (**15**), in which all pyrrole rings share the same orientation of their substituents. Ring D is then inverted by a rearrangement that affords uroporphyrinogen III (**16**) (3). Side-chain modification and the incorporation of iron by ferrochelatase (18) afford the various heme cofactors, including heme a (**18**), that carries an isoprenoid side chain. Starting from (**16**), a sequence of partial reduction, side-chain modification and incorporation of Mg^{2+} affords chlorophyll (**17**), which also carries an isoprenoid side chain. Partial reduction, ring contraction, and incorporation of Co^{2+} afford vitamin B_{12} (**20**) as well as several analogs that are found in archaea. Moreover, archaea incorporate nickel into the corrinoid coenzyme F_{430} (**19**) that plays a central role in the biosynthesis of methane (19).

Because plants are devoid of vitamin B_{12}, the supply of humans and animals is ultimately of bacterial origin (although humans can obtain vitamin B_{12} via

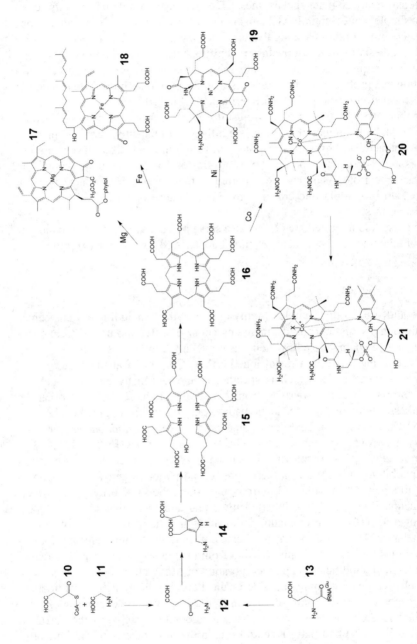

Figure 10.2 Biosynthesis of tetrapyrroles. (**10**), succinyl-CoA; (**11**), glycine; (**12**), δ-aminolevulinic acid; (**13**), glutamyl-tRNA; (**14**), porphobilinogen; (**15**), hydroxymethylbilane; (**16**), uroporphyrinogen III; (**17**), chlorophyll a; (**18**), heme a; (**19**), coenzyme F_{430}; (**20**), vitamin B_{12}; (**21**), coenzyme B_{12}.

animal products such as meat, milk, and milk products). A glycoprotein designated intrinsic factor that is secreted by the human gastric mucosa enables the take-up of the vitamin, which is then converted into coenzyme B_{12} by a series of enzyme reactions that occur in the human host (7).

10.2.3 A Family of Coenzymes Biosynthesized From GTP: Tetrahydrofolate, Tetrahydrobiopterin, Flavocoenzymes, Molybdopterin

Several coenzymes comprising a pyrimidine ring motif are derived from GTP (22) (Fig. 10.3). Specifically, this group comprises two members of the B vitamin group, riboflavin (vitamin B_2) (24) and folic acid/tetrahydrofolate (33). Two other members of the group, tetrahydrobiopterin (31) and molybdopterin (8), are biosynthesized *de novo* in animals and do not have vitamin status (20, 21).

The first committed step in the biosyntheses of these compounds is the hydrolytic opening of the imidazole ring of GTP, which affords a diaminopyrimidine-type intermediate. In the biosynthetic pathways of folate, tetrahydrobiopterin, and methanopterin (34), the respective diaminopyrimidine intermediate undergoes ring closure by means of an intramolecular condensation that involves parts of the ribose side chain of GTP, which affords a 2-amino-4-pteridinone compound (29).

The 4-aminobenzoate moiety of tetrahydrofolic acid is obtained from the shikimate pathway of aromatic amino acid biosynthesis via chorismate. Interestingly, apicomplexan protozoa may have conserved the complex shikimate pathway for the single purpose to generate 4-aminobenzoate as a tetrahydrofolate precursor, whereas aromatic amino acids are obtained from external sources.

The formation of intermediate (27) (compound Z) in the biosynthesis of molybdopterin (8) proceeds a rearrangement that involves the ribose side chain as well as C-8 of GTP for the formation of the tetracyclic ring system.

The carbocyclic moiety of vitamin B_2 is assembled from two molecules of a deoxytetrulose phosphate (22). The carbocyclic moiety of the deazaflavin-type coenzymes (36) is provided by the tyrosine precursor, 4-hydroxyphenylpyruvate.

Notably, the GTP cyclohydrolases that catalyze the first committed step in the pathways of tetrahydrofolate biosynthesis in plants and microorganisms and of tetrahydrobiopterin in animals are orthologs.

10.2.4 Thiamine Pyrophosphate

The biosynthesis of thiamine pyrophosphate (46) in microorganisms and plants is characterized by extraordinary complexity (21, 23). Animals are dependent on nutritional sources but can convert unphosphorylated thiamine (that is not an intermediate of the bacterial biosynthetic pathway) into thiamine pyrophosphate in two steps.

In bacteria, the pyrimidine precursor (38) is derived from 5-aminoimidazole ribotide (37), an intermediate of the basic branch of purine biosynthesis, which supplies all carbon atoms for (38) by a complex rearrangement reaction (the fate

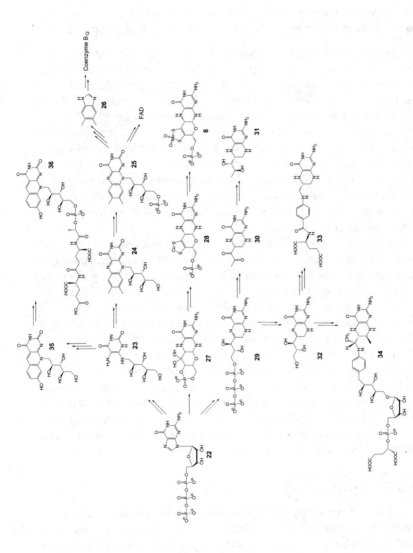

Figure 10.3 Coenzymes biosynthesized from GTP. (8), molybdopterin; (22), GTP; (23), 5-amino-6-ribitylamino-2,4(1*H*,3*H*)-pyrimidinedione; (24), 5,6-dimethylbenzimidazole; (25), FMN; (26), riboflavin; (27), precursor Z; (28), metal containing pterin; (29), dihydroneopterin triphosphate; (30), 6-pyruvoyl-tetrahydropterin; (31), 6(R)-5,6,7,8-tetrahydrobiopterin; (32), dihydroneopterin; (33), 6(S)-5,6,7,8-tetrahydrofolate; (34), 5,6,7,8-tetrahydromethanopterin; (35), 5-deaza-7,8-didemethyl-8-hydroxyribo-flavin; (36), coenzyme F_{420}.

of the individual carbon atoms is indicated by Greek letters in Fig. 10.4). In yeasts, a totally unrelated reaction sequence uses carbon atoms from vitamin B_6 (**39**) that are indicated by roman letters in Fig. 10.4 for the assembly of the thiamine precursor (**38**), whereas the nitrogen atoms and one additional carbon atom are introduced from histidine (**40**).

In bacteria, the thiazole moiety (**42**) of thiamine is derived from 1-deoxy-D-xylulose 5-phosphate (**43**) that can also serve as a precursor for pyridoxal in many eubacteria (Fig. 10.5) and for isoprenoids via the nonmevalonate pathway (cf. isoprenoid cofactors). The sulfur atom is derived from the persulfide that also serves as precursor for iron/sulfur clusters and for biotin (**6**) and thiooctanoate (**7**) (Fig. 10.1). C2 and N3 of the thiazole moiety of thiamine have been reported to stem from tyrosine in *Escherichia coli* and from glycine in *Bacillus subtilis*, respectively. Yeasts use ADP-ribulose (**44**) derived from NAD as precursor (24).

In plants, little is known about the basic building blocks and the reactions involved in thiamine biosynthesis. An early study with chloroplasts of spinach indicated that 1-deoxy-D-xylulose 5-phosphate, tyrosine, and cysteine act as precursors of the thiazole moiety in analogy to the pathway in *E. coli*. More recently, it has been shown that a homolog of the THIC protein that converts 5-aminoimidazole ribotide into (**38**) is essential (25). These results suggest that the plant pathway is similar to the pathway in prokaryotes but not to that in yeast.

10.2.5 Pyridoxal Phosphate

In many eubacteria, 1-deoxy-D-xylulose phosphate (**43**) serves as a common precursor for the biosynthesis of vitamins B_1 and B_6 and for the biosynthesis of isoprenoids via the nonmevalonate pathway. Condensation of (**43**) with 3-amino-1-hydroxyacetone phosphate (**47**) (biosynthesized from D-erythrose 4-phosphate) affords pyridoxine 5′-phosphate (**48**), (Fig. 10.5a). A sequence of elimination, tautomerization, and water addition precedes cyclization via an aldol condensation (26, 27). Then, pyridoxine 5′-phosphate can be converted into pyridoxal 5′-phosphate (**39**) by oxidation with molecular oxygen.

A more recently discovered second pathway starts from ribulose 5-phosphate (**49**) that is condensed with ammonia and glyceraldehyde phosphate or its isomerization product dihydroxyacetone phosphate (**50**), which affords pyridoxal 5′-phosphate in a single enzyme-catalyzed reaction step (Fig. 10.5b) (28, 29). This pathway seems to be widely distributed; it is used in plants (30) and has also been shown to proceed in fungi, archaea, and most eubacteria.

In mammals, dietary vitamin forms, including pyridoxal, pyridoxol, and pyridoxamine, can all be converted to the respective coenzyme forms by phosphorylation.

10.2.6 Pyridine Nucleotides

Animals and yeasts can synthesize nicotinamide from tryptophan via hydroxyanthranilic acid (**52**) and quinolinic acid (**53**), (Fig. 10.6a) (31), but the biosynthetic

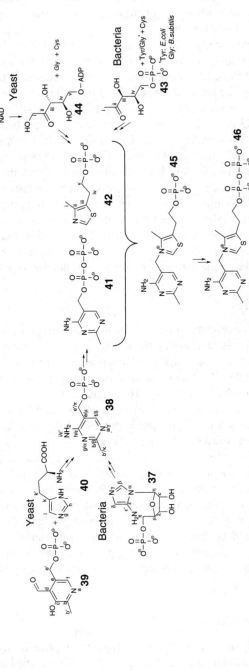

Figure 10.4 Biosynthesis of thiamine (vitamin B₁). (**37**), aminoimidazole ribotide; (**38**), 2-methyl-4-amino-5-hydroxymethyl-pyrimidine phosphate; (**39**), pyridoxal 5′-phosphate; (**40**), histidine; (**41**), 2-methyl-4-amino-5-hydroxymethyl-pyrimidine pyrophosphate; (**42**), 4-methyl-5-β-hydroxyethylthiazole phosphate; (**43**), 1-deoxy-D-xylulose 5-phosphate; (**44**), 5-ADP-D-ribulose; (**45**), thiamine phosphate; (**46**), thiamine pyrophosphate.

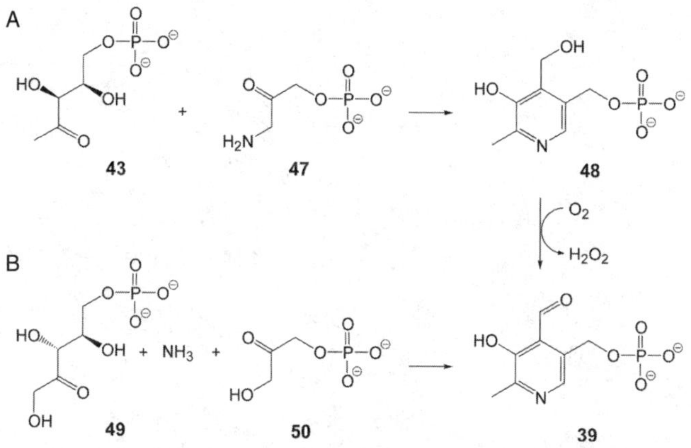

Figure 10.5 Formation of the pyridoxine ring in vitamin B$_6$. (a) deoxyxylulose phosphate-dependent pathway; (b) deoxyxylulose phosphate-independent pathway. (**43**), 1-deoxy-D-xylulose 5-phosphate; (**47**), 3-amino-1-hydroxyacetone 1-phosphate; (**48**), pyridoxine 5′-phosphate; (**49**), ribulose 5-phosphate; (**50**), dihydroxyacetone phosphate; (**39**), pyridoxal 5′-phosphate.

capacity of humans is limited. On a diet that is low in tryptophan, the combined contributions of endogenous synthesis and nutritional supply of precursors, such as nicotinic acid, nicotinamide, and nicotinamide riboside, may be insufficient, which results in cutaneous manifestation of niacin deficiency under the clinical picture of pellagra. Exogenous supply of nicotinamide riboside was shown to promote NAD$^+$-dependent Sir2-function and to extend life span in yeast without calorie restriction (32).

Bacteria and plants use aspartate (**54**) and dihydroxyacetone phosphate (**50**) as precursors for the biosynthesis of nicotinamide via quinolinic acid (**53**), (Fig. 10.6b) (33).

The transformation of precursors into NAD (**56**) and NADP (**57**) follow the same pathway in all organisms. A ribosyl phosphate residue can be transferred to biosynthetic quinolinic acid or to preformed nicotinamide affording (**55**) (or its amide) that is converted to NAD by adenylation. Subsequent phosphorylation yields NADP.

10.2.7 Pantothenate

Pantothenate (also designated vitamin B$_5$, Fig. 10.7) (**64**), is biosynthesized *de novo* in plants and many microorganisms but must be obtained from nutritional sources by animals (20, 34). The branched carboxylic acid (**63**) is obtained from α-ketoisovalerate (**61**), which is an intermediate of valine biosynthesis, via (**62**). β-Alanine (**60**) is obtained by decarboxylation of aspartate (**54**) in microorganisms. Plants and yeasts can biosynthesize β-alanine from spermine (**58**) (35). An additional pathway to (**60**) starting from uracil (**59**) has been reported in

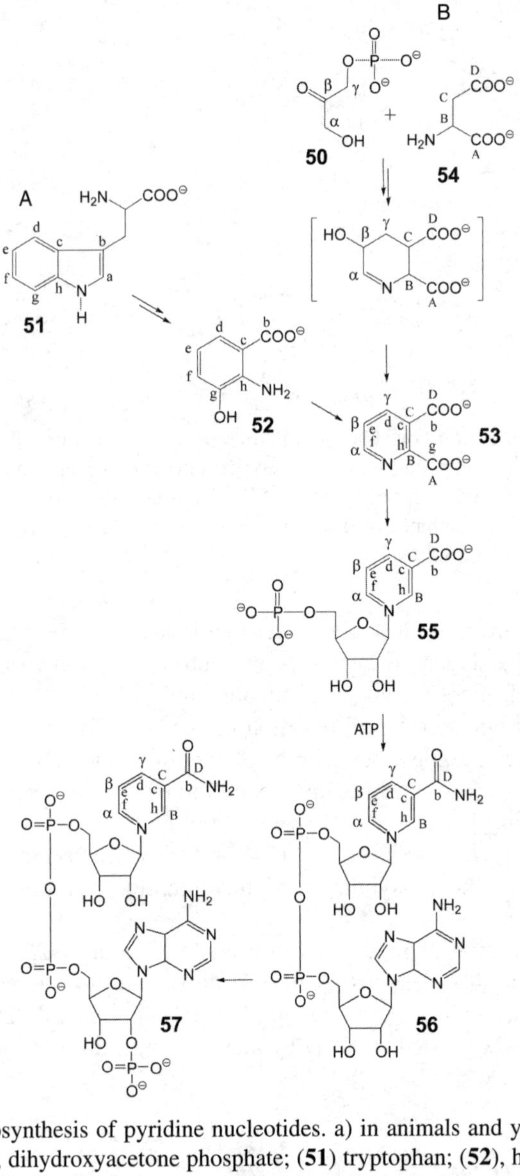

Figure 10.6 Biosynthesis of pyridine nucleotides. a) in animals and yeasts; b) in plants and bacteria. (**50**), dihydroxyacetone phosphate; (**51**) tryptophan; (**52**), hydroxyanthranilic acid; (**53**), quinolinic acid; (**54**), aspartate; (**55**), nicotinic acid mononucleotide; (**56**), NAD, (**57**), NADP.

plants, where the downstream steps that lead to pantothenate are understood incompletely.

Organisms of all biological kingdoms convert (**64**) into the cysteamine deriva-tive phosphopantetheine (**65**) using L-cysteine as substrate. (**65**) is converted to coenzyme A (**66**) by attachment of an adenosine moiety via a pyrophosphate

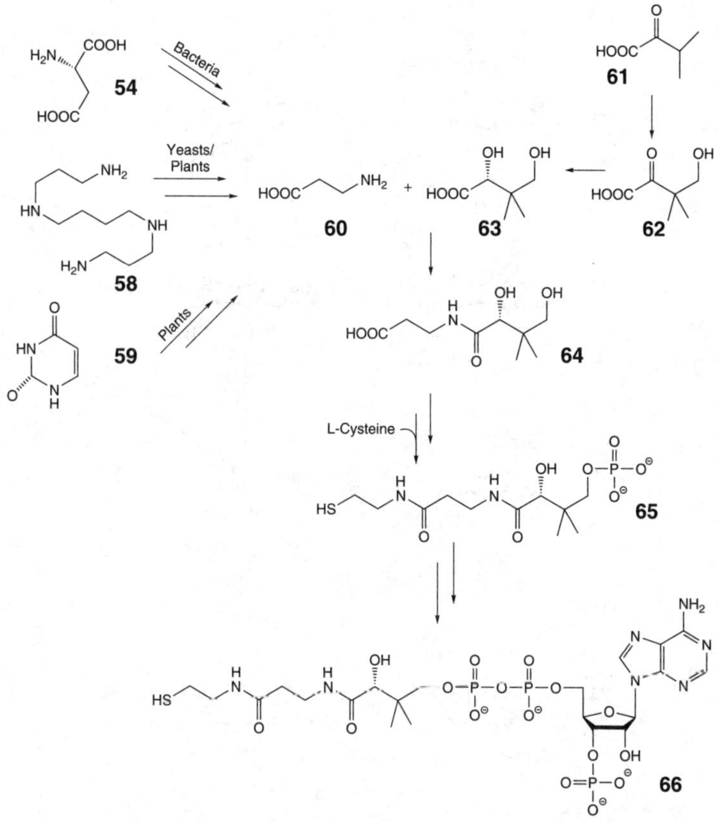

Figure 10.7 Biosynthesis of pantothenate. (**54**), aspartate; (**58**), spermine; (**59**), uracil; (**60**), β-alanine; (**61**), α-ketoisovalerate; (**62**), ketopantoate; (**63**), pantoate; (**64**), pantothenate; (**65**), pantetheine 4-phosphate; (**66**), coenzyme A.

linker and phosphorylation of the ribose moiety. Phosphopantetheine can be attached covalently to serine residues of acyl carrier proteins that are parts of fatty acid synthases and polyketide synthases.

10.2.8 Vitamin C

Whereas most mammals can synthesize ascorbic acid (vitamin C), (**75**) from D-glucose 1-phosphate (**67**) via the pathway shown in Fig. 10.8b, humans and guinea pigs lack the last enzyme of that pathway and are therefore dependent on nutritional sources (36, 37). Plants use the pathway shown in Fig. 10.8a that has been elucidated relatively recently (38, 39). Yeasts produce and use a five-carbon analog, which is called erythroascorbic acid, instead of ascorbate. The biosynthetic pathway of erythroascorbic acid involves the oxidation of D-arabinose to D-arabino-1,4-lactone, which is then oxidized to erythroascorbic acid.

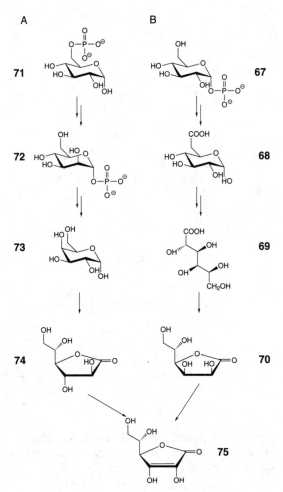

Figure 10.8 Biosynthesis of ascorbic acid (vitamin C). (a) in plants; (b) in mammals. (**67**), D-glucose 1-phosphate; (**68**), D-glucuronate; (**69**), L-gulonate; (**70**), L-gulono-1,4-lactone; (**71**), D-glucose 6-phosphate; (**72**), D-mannose 1-phosphate; (**73**), L-galactose; (**74**), L-galactono-1,4-lactone; (**75**), ascorbic acid. The enzyme converting (**70**) into (**75**) is missing in primates and guinea pigs.

10.2.9 Pyrroloquinoline Quinone

Pyrroloquinoline quinone (**77**, Fig. 10.9; PQQ) serves as cofactor of bacterial oxidoreductases (40). The heated debate whether PQQ has vitamin character for animals is still inconclusive (41–43).

PQQ is derived from a peptide precursor that contains conserved glutamate (**13**) and tyrosine (**76**) residues. All carbon and nitrogen atoms of the precursor amino acids are incorporated into the product (44). Gene clusters involved in

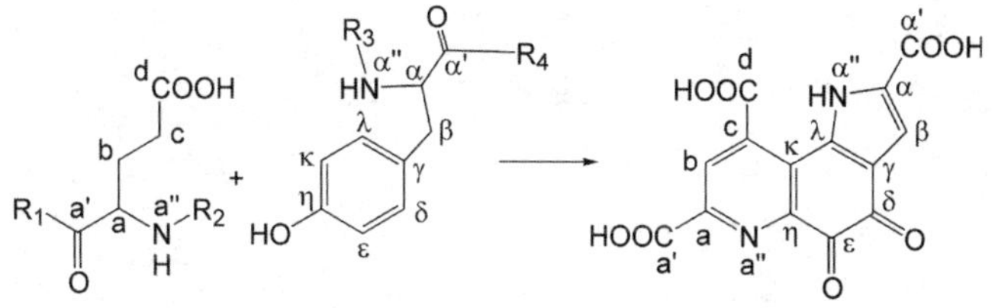

Figure 10.9 Biosynthesis of PQQ. (**13**), glutamate (protein bound); (**76**), tyrosine (protein bound); (**77**), PQQ.

this pathway have been studied in considerable detail. The X-ray structure of the enzyme that catalyzes the final reaction step has been determined, and reaction mechanisms have been proposed on that basis (45). However, details of the biosynthetic pathway are still incompletely understood.

10.2.10 Biotin

Biotin (vitamin H, (**6**), Figs. 10.1 and 10.10) acts a cofactor of carboxylases. It can be produced in bacteria, plants, and some fungi (46). The biosynthetic pathway involves four steps that start from alanine (**78**) and pimeoyl-CoA (**79**). Carboxylation and cyclization of (**81**) affords dethiobiotin (**82**), which is then converted into biotin (**6**) by the iron/sulfur protein, biotin synthase, in an unusual radical mechanism (47).

10.2.11 Isoprenoid Cofactors

Isoprenoids are one of the largest classes of natural products that comprise at least 35,000 reported members (48). Many of these compounds play crucial roles in human metabolism as hormones, vitamins (vitamins A, D, E and K), quinine-type cofactors of respiratory chain enzymes (ubiquinone), membrane constituents, and functionally important side chains of signal cascade proteins (Fig. 10.11). Chlorophyll (**17**, Fig. 10.2) and heme a (**18**) have isoprenoid side chains.

Whereas vitamin E (**99**) and vitamin A or its biosynthetic precursor, β-carotene, must be obtained by animals from dietary sources, many other isoprenoids, including the quinone type coenzyme Q family (where individual representatives differ by the length of their side chains), can be synthesized *de novo* by vertebrates. 3-Hydroxy-3-methylglutaryl-CoA reductase, the enzyme that catalyzes the conversion of (*S*)-3-hydroxy-3-methylglutaryl-CoA (**84**) to mevalonate (**85**), is one of the most important drug targets for the prevention of cardiovascular disease (49, 50).

All isoprenoids are biosynthesized from two isomeric 5-carbon compounds, isopentenyl diphosphate (IPP), (**86**) and dimethylallyl diphosphate (DMAPP), (**87**) (Fig. 10.6). The mammalian pathway for the biosynthesis of these key

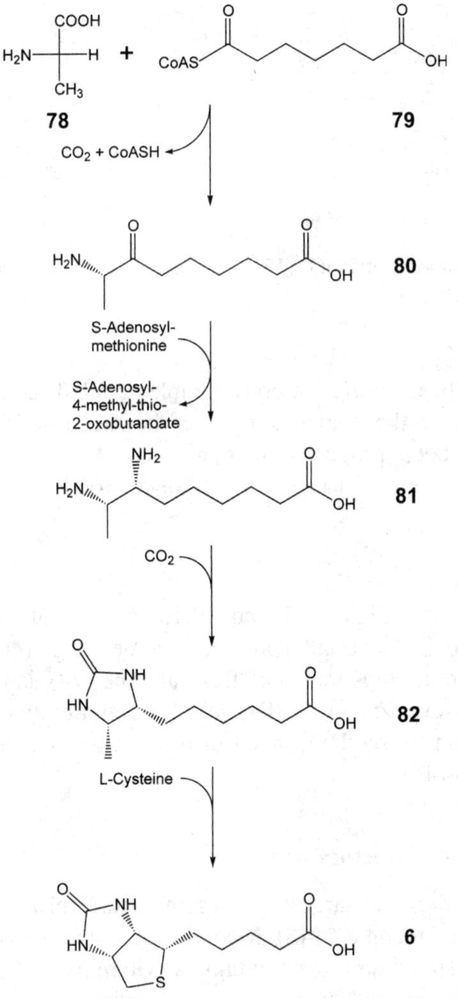

Figure 10.10 Biosynthesis of biotin. (**78**), alanine; (**79**), pimeloyl-CoA; (**80**), 7-keto-8-amino-pelargonic acid; (**81**), 7,8-diamino-pelargonic acid; (**82**), dethiobiotin; (**6**), biotin.

biosynthetic precursors from three acetyl-CoA units (**83**) via mevalonate (**85**) had been elucidated in the 1950s (51). In the wake of that pioneering work, it became established dogma that all terpenoids are invariably of mevalonate origin, even in the face of significant aberrant findings.

The existence of a second pathway that affords IPP and DMAPP was discovered in the 1990s. The details of that nonmevalonate pathway were then established in rapid sequence by the combination of isotope studies, comparative genomics, and enzymology (52). The delayed discovery of the nonmevalonate pathway can serve as paradigm for pitfalls in the elucidation of biosynthetic pathways.

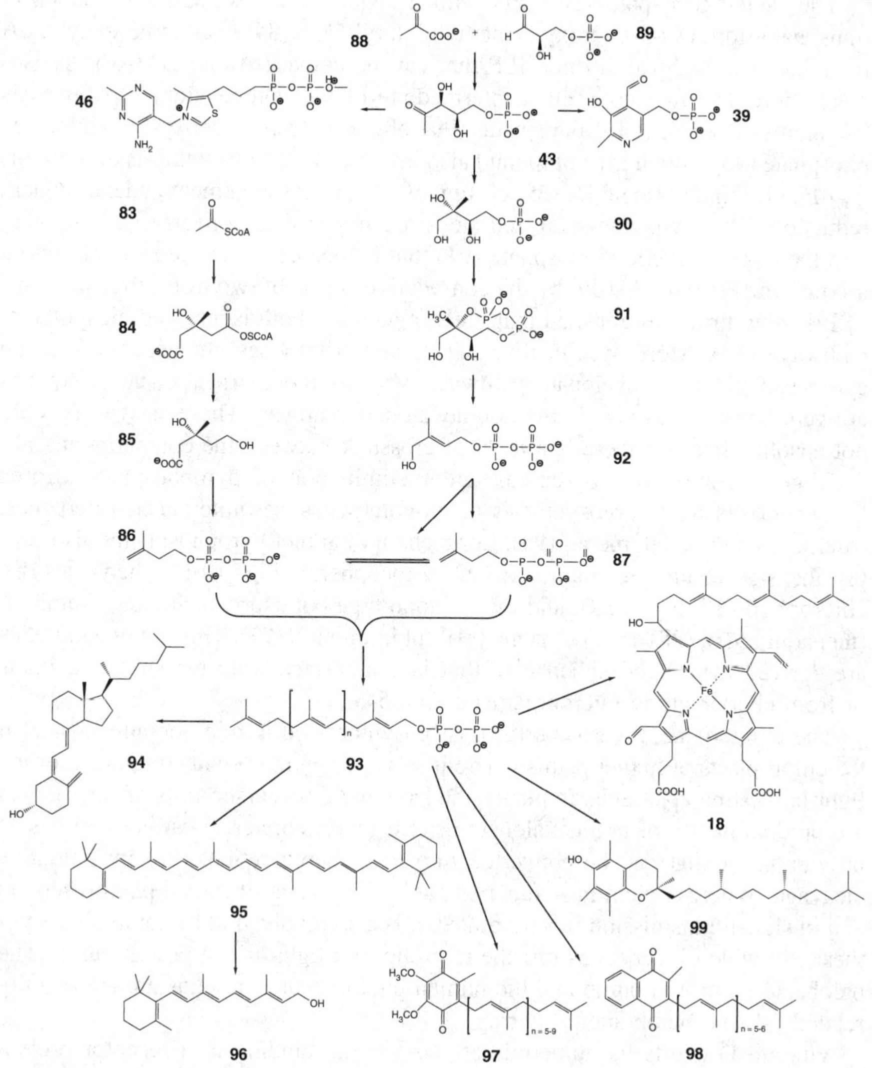

Figure 10.11 Biosynthesis of isoprenoid type cofactors. (**18**), Heme a; (**39**), pyridoxal 5'-phosphate; (**43**), 1-deoxy-D-xylulose 5-phosphate; (**46**), thiamine pyrophosphate; (**83**), acetyl-CoA; (**84**), (S)-3-hydroxy-3-methylglutaryl-CoA; (**85**), mevalonate; (**86**), isopentenyl diphosphate (IPP); (**87**), dimethylallyl diphosphate (DMAPP); (**88**), pyruvate; (**89**), D-glyceraldehyde 3-phosphate; (**90**), 2C-methyl-D-erythritol 4-phosphate; (**91**), 2C-methyl-erythritol 2,4-cyclodiphosphate; (**92**), 1-hydroxy-2-methyl-2-(E)-butenyl 4-diphosphate; (**93**), polyprenyl diphosphate; (**94**), cholecalciferol; (**95**), β-carotene; (**96**), retinol; (**97**), ubiquinone; (**98**), menaquinone; (**99**), α-tocopherol.

The mevalonate pathway starts with a sequence of two Claisen condensations that afford (*S*)-3-hydroxy-3-methylglutaryl-CoA (**84**) from three acetyl-CoA moieties. The pathway affords IPP that can be converted into DMAPP by isomerization. The first committed intermediate of the nonmevalonate pathway is 2*C*-methyl-D-erythritol 4-phosphate (**90**) obtained from 1-deoxy-D-xylulose 5-phosphate (**43**), which is a compound also involved in the biosynthesis of vitamins B$_1$ (**46**, cf. Fig. 10.4) and B$_6$ (**39**, cf. Fig. 10.5), by rearrangement and subsequent reduction. Three enzyme-catalyzed steps are required to convert the compound into the cognate cyclic diphosphate (**91**) that is then converted reductively into a mixture of IPP and DMAPP by the consecutive action of two iron/sulfur proteins.

It is now firmly established that green plants use both isoprenoid biosynthesis pathways (52). More specifically, sterols and triterpenes are generated in the cytoplasm via the mevalonate pathway, whereas monoterpenes and diterpenes are generated in plastids via the nonmevalonate pathway. These assignments are not absolute because there is a level of crosstalk between the compartments.

Oligomerization of isoprenoids under elimination of pyrophosphate affords the precursors for the biosynthesis of monoterpenes, sesquiterpenes, diterpenes, triterpenes, and tetraterpenes (**93**). Long-chain oligomer pyrophosphates also supply the side chains of vitamin E (**99**, α-tocopherol, Fig. 10.11), heme a (**18**), chlorophyll (**17**, Fig. 10.2), and the quinone type cofactors, including vitamin K (menaquinone), (**98**) and coenzyme Q10 (ubiquinone), (**97**). The quinone moieties are derived from hydroxybenzoate that is synthesized from tyrosine in animals or from chorismate in microorganisms (53, 54).

The tetraterpene, β-carotene (**95**), is biosynthesized in microorganisms and in the chloroplasts of higher plants where it serves as an important component of the light-harvesting apparatus. In plants, the isoprenoid precursor units of carotenoids are predominantly of nonmevalonate origin. In vertebrates, β-carotene serves as provitamin A that can be converted into the vitamin retinol (**96**) by oxidative cleavage. Whereas vitamin A functions as a component of retinal photoreceptors and in signal transmission in vertebrates, it is also involved in bacterial photosynthesis, in which it serves as prosthetic group in a light-driven proton pump. The halobacterial proton pump and the human photoreceptor proteins are structurally related 7-helix membrane proteins.

Vitamin D exerts its numerous effects via the binding to a receptor protein that serves as a transcription factor. It is included in this review in light of its essential status for human health.

Humans can generate vitamin D$_3$ (cholecalciferol), (**94**), (Fig. 10.11) by endogenous biosynthesis but require dietary sources under certain environmental conditions. More specifically, (**94**) that can be obtained via the endogenous mevalonate pathway can be photochemically converted into provitamin D in light-exposed skin areas. The transformation requires ultraviolet light. At higher geographical latitudes, light exposure of the skin can be a limiting factor. The fact that skin pigmentation retards the photochemical formation of vitamin D may have acted as a selective factor that favored pale skin when modern humans migrated to areas with higher geographic latitude.

10.2.12 Biosynthesis of Archaeal Coenzymes

Archaebacteria that were discovered only during recent decades are now recognized as a third kingdom of life besides eubacteria and eukaryotes. An important subgroup of archaea can generate energy by conversion of carbon dioxide or low molecular weight organic acids into methane. The pathway of methanogenesis has been shown to implicate several unique coenzymes (55). Specifically, 5-deaza-8-hydroxy-10-ribitylisoalloxazine (factor F_0), (35), which is an analog of riboflavin, serves as the business end of coenzyme F_{420} (36, Fig. 10.3), whose designation is based on its characteristic absorption maximum at 420 nm. Factor F_0 is biosynthesized from the pyrimidine type intermediate (23) of the riboflavin biosynthetic pathway, which affords the pyrimidine ring and the ribityl side chain, whereas the carbocyclic moiety is derived from the shikimate pathway via 4-hydroxyphenylpyruvate (56, 57). In contrast to the coenzymes described below, deazaflavin-type coenzymes are not strictly limited to methanogenic bacteria and are also found in streptomycetes and mycobacteria.

The tetrapyrrole-type coenzyme F_{430} (19) was named on basis of its absorption maximum at 430 nm. The nickel-chelating factor is biosynthesized via the porphyrin biosynthetic pathway (Fig. 10.2) (19). For the handling of one-carbon fragments that play a central role in their metabolism, methanogenic bacteria use methanopterin (34, Fig. 10.3). The tetrahydropterine system that serves as the business end of the methanopterin coenzyme family is structurally similar to tetrahydrofolate, and the biosynthetic pathway starting from GTP is similar to that of tetrahydrofolate (Fig. 10.3). The ribitylaniline moiety is derived from ribose and from the shikimate pathway via 4-aminobenzoate (55).

10.2.13 Coenzymes as Biosynthetic Precursors

Some coenzymes serve as biosynthetic precursors that afford structural parts of other coenzymes. Thus, the benzenoid moiety of the flavocoenzyme FMN serves as a precursor for the lower ligand (26) of the central cobalt ion in vitamin B_{12} (20) (Fig. 10.3) (5). Pyridoxal and NAD are used as precursors for the biosynthesis of thiamine in yeast (Fig. 10.4) (23, 24).

10.2.14 Branched Coenzyme Biosynthesis Pathways

Branching of pathways is relevant in several cases. Thus, intermediates of the porphyrin biosynthetic pathway serve as precursors for chlorophyll (17, Fig. 10.2) and for the corrinoid ring systems of vitamin B_{12} (20, Fig. 10.2) (17). 1-Deoxy-D-xylulose 5-phosphate (43) serves as an intermediate for the biosynthesis of pyridoxal 5'-phosphate (39, Fig. 10.5), for the terpenoid precursor IPP (86) via the nonmevalonate pathway (Fig. 10.11), and for the thiazole moiety of thiamine pyrophosphate (46, Fig. 10.4). 7,8-Dihydroneopterin triphosphate (29, Fig. 10.3) serves as intermediate in the biosynthetic pathways of tetrahydrofolate (33) and tetrahydrobiopterin (31). The closely related compound 7,8-dihydroneopterin $2',3'$-cyclic phosphate is the precursor of the archaeal cofactor, tetrahydromethanopterin (**34**) (58). A common pyrimidine-type intermediate (**23**) serves as precursor for flavin and deazaflavin coenzymes. Various sulfur-containing coenzymes [thiamine (**9**), lipoic acid (**7**), biotin (**6**), Fig. 10.1] use a pyrosulfide protein precursor that is also used for the biosynthesis of inorganic sulfide as a precursor for iron/sulfur clusters (12).

10.2.15 Requirement of Coenzymes for Their Own Biosynthesis

Several coenzymes are involved in the biosynthesis of their own precursors. Thus, thiamine is the cofactor of the enzyme that converts 1-deoxy-D-xylulose 5-phosphate (**43**) (the precursor of thiamine pyrophosphate, pyridoxal $5'$-phosphate and of isoprenoids via the nonmevalonate pathway) into 2 C-methyl-D-erythritol 4-phosphate (**90**, Fig. 10.11). Similarly, two enzymes required for the biosynthesis of GTP, which is the precursor of tetrahydrofolate, require tetrahydrofolate derivatives as cofactors (Fig. 10.3). When a given coenzyme is involved in its own biosynthesis, we are faced with a "hen and egg" problem, namely how the biosynthesis could have evolved in the absence of the crucially required final product. The answers to that question must remain speculative. The final product may have been formed via an alternative biosynthetic pathway that has been abandoned in later phases of evolution or that may persist in certain organisms but remains to be discovered. Alternatively, the coenzyme under study may have been accessible by a prebiotic sequence of spontaneous reactions. An interesting example in this respect is the biosynthesis of flavin coenzymes, in which several reaction steps can proceed without enzyme catalysis despite their mechanistic complexity.

In terms of coenzyme evolution, it is also noteworthy that the biosynthesis of a given coenzyme frequently requires the cooperation of other coenzymes. For example, the biosynthesis of riboflavin (**24**) requires tetrahydrofolate (**33**) for the biosynthesis of GTP serving as precursor (Fig. 10.3). Pyridoxal $5'$-phosphate is required for the biosynthesis of the activated pyrosulfide type protein (**2**) that serves as the common precursor for iron/sulfur clusters and various sulfur-containing organic coenzymes (Fig. 10.1).

10.2.16 Covalent Coenzymes

Whereas many coenzymes form noncovalent complexes with their respective apoenzymes, various flavoenzymes are characterized by covalently bound FMN (**25**) or FAD (Fig. 10.3). Covalent linkage involves the position 8α methyl group or the benzenoid carbon atom 6 of the flavin and a cysteine or histidine residue of the protein. The covalent CN or CS bond can be formed by autoxidation of the noncovalent apoenzyme/coenzyme precursor complex as shown in detail for nicotine oxidase (59).

Biotin (**6**, Fig. 10.10) and lipoic acid (**7**, Fig. 10.1) are attached enzymatically to apoenzymes via carboxamide linkage to specific lysine residues (60, 61). The pantothenyl moiety (**64**, Fig. 10.7) can also be linked covalently to proteins via amide linkage (62). Covalently bound heme is involved in heme M (63) and heme L-catalyzed reactions (64, 65).

Several covalently bound coenzymes, including pyruvoyl, methylidine imidazolone, topaquinone, and tryptophan tryptophyl quinine-type prosthetic groups are generated by posttranslational modification (66).

10.2.17 Cellular Topology of Coenzyme Biosynthesis in Eukaryotes

In bacteria, coenzyme biosynthesis is located in the cytoplasm. In eukaryotic cells, organelles play important roles in the biosynthesis of certain cofactors. For example, certain steps of iron/sulfur cluster biosynthesis proceed in mitochondria (10, 13). In plants, some steps of the biosynthesis of tetrahydrofolate, biotin, and lipoate proceed in mitochondria (7, 15), whereas the biosynthesis of vitamin B_2 is operative in plastids (67). In apicomplexan protozoa, enzymes in mitochondria, the apicoplast (an organelle that is believed to have a common evolutionary origin with chloroplasts), and the cytoplasmic compartment must cooperate for the biosynthesis of tetrapyrrole cofactors, thiamine, and isoprenoid cofactors (68, 69). The biosynthesis of carotenoids proceeds via the nonmevalonate pathway in the chloroplasts of plants (52). The same holds true for the biosynthesis of thiamine, pyridoxal, and chlorophyll (70).

10.2.18 Three-Dimensional Structures of Coenzyme Biosynthesis Enzymes

The rapid technological progress in X-ray crystallography has enabled the structural analysis of numerous enzymes involved in coenzyme biosynthesis. Complete sets of structures that cover all enzymes of a given pathway are available in certain cases such as riboflavin, tetrahydrobiopterin, and folic acid biosynthesis. Structures of orthologs from different taxonomic groups have been reported in certain cases. X-ray structures of enzymes in complex with substrates, products, and analogs of substrates, products, or intermediates have been essential for the elucidation of the reaction mechanisms. Structures of some coenzyme biosynthesis enzymes have been obtained by NMR-structure analysis.

Enzymes that are addressed by major drugs have been studied in particular detail. Thus, well above one hundred structures have been reported for dihydrofolate reductases from a variety of organisms, including major pathogens such as *Mycobacterium tuberculosis*, which is the causative agent of tuberculosis, and of *Plasmodium falciparum*, which is the most important of the *Plasmodium* spp. that causes malaria. The interaction of mammalian dihydrofolate reductases with inhibitors that are used as cytostatic agents and/or immunosuppressants is also documented extensively by X-ray structures.

The rapidly growing number of three-dimensional coenzyme biosynthesis enzyme structures in the public domain and the cognate publications are best addressed via the internet server of Brookhaven Protein Data Bank (*http://www.rcsb.org/pdb/home/home.do*). Queries can be targeted to individual enzymes or to entire pathways.

10.3 MEDICAL ASPECTS

10.3.1 Genetic Deficiency of Coenzyme Biosynthesis

Genetic defects have been reported for the biosynthesis of several coenzymes in humans. Typically, these rare anomalies cause severe neurological deficits that become apparent at birth or in early childhood.

Specifically, the deficiency of certain enzymes of tetrahydrobiopterin biosynthesis (GTP cyclohydrolase I, pyruvoyltetrahydrobiopterin synthase, Fig. 10.3) result in severe neurological and developmental deficit designated as atypical phenylketonuria caused by the ensuing deficiency in catecholamine type neurotransmitter biosynthesis. The condition can be treated with some success by the oral application of synthetic tetrahydrobiopterin in large amounts. Tetrahydrobiopterin therapy has also been advocated for certain patients with classic phenylketonuria that results from mutations of phenylalanine hydroxylase (71). This therapeutic approach is based on the concept that the function of certain defective phenylalanine hydroxylases can be bolstered by increased amounts of the cognate coenzyme tetrahydrobiopterin. In fact, the relatively large number of patients with classic phenylketonuria may provide an economic incentive for the development of a biotechnological process for the bulk production of the coenzyme.

Deficiencies of enzymes involved in the transformation of the vitamin pantothenic acid (**64**) into the cognate coenzyme forms (**66**, Fig. 10.7) result in severe developmental and neurological deficits that affect few human patients (34, 72). Therapy with megadoses of pantothenic acid has been advocated, but their efficiency has yet to be demonstrated by stringent clinical studies (73).

Genetic defects of molybdopterin biosynthesis (Fig. 10.3) also result in severe neurologic and developmental deficits (74). Genetic defects in the biosynthesis of the quinine-type coenzyme Q10 (**97**, Fig. 10.11) can result in encephalopathy, myopathy, and renal disease (53).

Inherited defects or porphyrine biosynthetic enzymes can cause the accumulation of pathway intermediates that cannot be converted anymore with sufficient velocity. Various genetic forms of porphyria have been reported and result in liver toxicity, neurological damage and photosensitivity (75). Acquired forms of porphyria can be caused by a variety of toxic and pharmacologic agents.

The absorption of vitamin B_{12} (**20**) requires a glycoprotein-designated intrinsic factor that is secreted by the gastric mucosa (7, 76). The factor binds the vitamin and enables its subsequent transport across the ileal mucosa. Acquired failure to produce the intrinsic factor results in a complex disease that can present with hematological (macrocytotic anemia and pernicious anemia), neurological, or psychiatric symptoms or a combination thereof. Prior to the discovery of vitamin B_{12}, pernicious anemia was lethal. Initial treatment was based on the consumption of large amounts of uncooked liver. If diagnosed timely, the disease can now be cured easily by the parenteral administration of vitamin B_{12}. Notably, the liver can store large amounts of the vitamin.

10.3.2 Coenzyme Biosynthesis Enzymes as Anti-infective Drug Targets

Enzymes involved in coenzyme biosynthesis represent targets for anti-infective agents (77). The sulfonamides that were discovered in the 1930s were the first group of synthetic agents with a broad spectrum of activity against pathogenic bacteria and protozoa. Their mode of action, via inhibition of dihydropteroate synthetase in the biosynthetic pathway of tetrahydrofolate biosynthesis (Fig. 10.3), was elucidated only much later. Subsequent studies on compounds with antifolate activity afforded inhibitors of dihydrofolate reductase, which is the enzyme in that pathway that catalyzes the formation of tetrahydrofolate from dihydrofolate in organisms that synthesize the coenzyme *de novo* and from folate in organisms that rely on dietary sources. Trimethoprim, which is an inhibitor of dihydrofolate reductase (that is required for the use of nutritional folate and dihydrofolate as well as for the metabolic recycling of tetrahydrofolate coenzymes) became a widely used antimicrobial agent that is typically applied in combination with a sulfonamide.

Fosmidomycin, initially discovered as a product of *Streptomyces lavendulae* with antibacterial and herbicide activity, was shown to act via the inhibition of IspC protein that catalyzes the first committed step in the nonmevalonate pathway of isoprenoid biosynthesis that is absent in humans (**43** → **90**, Fig. 10.11). Based on these findings, the compound is now under clinical evaluation as an antimalarial drug (78).

In principle, other coenzyme biosynthetic pathways that occur in pathogenic bacteria but not in humans should qualify as anti-infective drug targets with a favorable toxicity profile. Novel anti-infective principles would be highly desirable in light of the rapid spread of resistant pathogens.

10.3.3 Coenzyme Biosynthesis as a Target for Cytostatic Agents

The development of the sulfonamides as antibacterial and antiprotozoan agents had preceded the discovery of its metabolic target in the biosynthesis of tetrahydrofolate (that was *per se* unknown in the 1930s) (Fig. 10.3). The discovery of the vitamin in the 1940s triggered a wave of research directed at additional inhibitors of its biosynthesis. This work resulted in the discovery of methotrexate that is widely used as a cytostatic agent predominantly for hematological malignancy, and also as an immunosuppressive agent used in the therapy of autoimmune disease such as Crohn's disease (79).

10.3.4 Drug Interaction with Coenzyme Biosynthesis Pathways

Mevastatin is an inhibitor of 3-hydroxy-3-methylglutaryl-CoA reductase (Fig. 10.11), (**84** → **85**) that was isolated from *Penicillium citrinum* in 1971. A group of structurally related compounds designated as statins is now widely used for the prevention and treatment of cardiovascular disease (80). Statins exert their desired effects via the lowering of low-density lipoprotein and probably also via reduced prenylation of small G-proteins of the Ras protein family that are involved in proinflammatory signaling. The highly pleiotrophic mevalonate pathway is the source of numerous other highly important metabolites including coenzyme Q10 (**97**). However, a review concludes that the suppression of coenzyme Q10 biosynthesis, which can be expected as a side result of statin therapy, is not a significant cause of rhabdomyolysis, a dreaded side effect of statin therapy (81).

10.3.5 Modulation of Vitamin D Biosynthesis by Environmental Factors

Fair-skinned humans face a dilemma because ultraviolet light exposure carries the risks of carcinogenesis and skin aging, whereas insufficient ultraviolet light carries the risk of vitamin D_3 (**94**, Fig. 10.11) insufficiency. The case is unique in so far as an endogenous biosynthetic pathway is subject to regulation by external lifestyle factors such as ultraviolet exposure and the use of chemical and physical sunscreens. Recent studies indicate that vitamin D deficiency is wide spread in the human population and is a risk factor for a wide variety of conditions, including cancer and autoimmune disease (82). The dilemma of ultraviolet protection and vitamin D sufficiency can be addressed easily by vitamin D supplements.

10.4 BIOTECHNOLOGY

10.4.1 Harnessing Biosynthetic Pathways for Vitamin Production

Certain vitamins that serve as coenzymes (vitamin C) or as precursors of coenzymes (all B group vitamins and certain carotenoids) are commercially produced in bulk amounts, which are used for human nutrition and animal husbandry, as antioxidants (vitamins C and E) and food colorants (vitamin B_2, carotenoids). Only a fraction of technically manufactured vitamins is used for inclusion in drugs.

Vitamin B_{12} is produced exclusively by bacterial fermentation technology. The chemical synthesis of vitamin B_2 has been superseded during the past two decades by fermentation processes using bacteria and yeasts. Carotene, various other carotenoids and vitamin A are produced by chemical synthesis, but a variety of biotechnological processes have been also been explored for their production. Notably, vitamin A deficiency continues to be a major cause for acquired blindness in developmental countries, although vitamin A and β-carotene, which serves a provitamin, can be produced at modest cost and in virtually unlimited quantity by existing technology. At least certain steps of the various technical vitamin C processes are also conducted by fermentation. For various other vitamins, biotechnological production may become competitive in the future.

REFERENCES

1. Leeper FJ, Smith AG. Editorial: vitamins and cofactors-chemistry, biochemistry and biology. Nat. Prod. Rep. 2007;24:923–926.

2. Thöny B, Auerbach G, Blau N. Tetrahydrobiopterin biosynthesis, regeneration and functions. Biochem. J. 2000;347:1–16.

3. Scott IA. Discovering nature's diverse pathways to vitamin B_{12}: a 35-year odyssey. J. Org. Chem. 2003;68:2529–2539.

4. Roth JR, Lawrence JG, Bobik TA. Cobalamin (coenzyme B_{12}): synthesis and biological significance. Annu. Rev. Microbiol. 1996;50:137–181.

5. Taga ME, Larsen NA, Howard-Jones AR, Walsh CT, Walker GC. BluB cannibalizes flavin to form the lower ligand of vitamin B_{12}. Nature 2007;446:449–453.

6. Brown KL. Chemistry and enzymology of vitamin B_{12}. Chem. Rev. 2005;105: 2075–2149.

7. Rebeille F, Ravanel S, Marquet A, Mendel RR, Smith AG, Warrene MJ. Roles of vitamins B_5, B_8, B_9, B_{12} and molybdenum cofactor at cellular and organismal levels. Nat. Prod. Rep. 2007;24:949–962.

8. Mendel RR. Biology of the molybdenum cofactor. J. Exp. Bot. 2007;58:2289–2296

9. Raschle T, Arigoni D, Brunisholz R, Rechsteiner H, Amrhein N, Fitzpatrick TB. Reaction mechanism of pyridoxal $5'$-phosphate synthase. Detection of an enzyme-bound chromophoric intermediate. J. Biol. Chem. 2007;282:6098–6105.

10. Johnson DC, Dean DR, Smith AD, Johnson MK. Structure, function and formation of biological iron-sulfur clusters. Annu. Rev. Biochem. 2005;74:247–281.

11. Imlay JA. Iron-sulphur clusters and the problem with oxygen. Mol. Microbiol. 2006;59:1073–1082.

12. Kessler D. Enzymatic activation of sulfur for incorporation into biomolecules in prokaryotes. FEMS Microbiol. Rev. 2006;30:825–840.

13. Lill R, Müehlenhoff U. Iron-sulfur protein biogenesis in eukaryotes: components and mechanisms. Annu. Rev. Cell Dev. Biol. 2006;22:457–486.

14. Leeper FJ. The biosynthesis of porphyrins, chlorophylls, and vitamin B_{12}. Nat. Prod. Rep. 1989;6:171–203.

15. Holliday GL, Thornton JM, Marquet A, Smith AG, Rebeillé F, Mendel RR, Schubert HL, Lawrence AD, Warren MJ. Evolution of enzymes and pathways for the biosynthesis of cofactors. Nat. Prod. Rep. 2007;24:972–987.

16. Panek H, O'Brian MR. A whole genome view of prokaryotic haem biosynthesis. Microbiology 2002;148(Pt 8): 2273–2282.

17. Tanaka R, Tanaka A. Tetrapyrrole biosynthesis in higher plants. Annu. Rev. Plant Biol. 2007;58:321–346.

18. Mendel RR, Smith AG, Marquet A, Warren MJ. Metal and cofactor insertion. Nat. Prod. Rep. 2007;24:963–971.

19. Warren MJ, Scott AI. Tetrapyrrole assembly and modification into the ligands of biologically functional cofactors. Trends Biochem. Sci. 1990;15:486–491.

20. Webb ME, Marquet A, Mendel RR, Rebeille F, Smith AG. Elucidating biosynthetic pathways for vitamins and cofactors. Nat. Prod. Rep. 2007;24:988–1008.

21. Roje S. Vitamin B biosynthesis in plants. Phytochem 2007;68:1904–1921.

22. Bacher A, Fischer M. Biosynthesis of flavocoenzymes. Nat. Prod. Rep. 2005;22: 324–350.

23. Begley TP. Cofactor biosynthesis: an organic chemist's treasure trove. Nat. Prod. Rev. 2006;23:15–25.

24. Chatterjee A, Jurgenson CT, Schroeder FC, Ealick SE, Begley TP. Biosynthesis of thiamin thiazole in eukaryotes: conversion of NAD to an advanced intermediate. J. Am. Chem. Soc. 2007;129:2914–2922.

25. Raschke M, Bürkle L, Müller N, Nunes-Nesi A, Fernie AR, Arigoni D, Amrhein N, Fitzpatrick TB. Vitamin B_1 biosynthesis in plants requires the essential iron sulfur cluster protein, THIC. Proc. Natl. Acad. Sci. U.S.A. 2007;104:19637–19642.

26. Fitzpatrick TB, Amrhein N, Kappes B, Macheroux P, Tews I. Two independent routes of de novo vitamin B_6 biosynthesis: not that different after all. Biochem. J. 2007;407:1–13.

27. Scott DE, Ciulli, Abell C. Coenzyme biosynthesis: enzyme mechanism, structure and inhibition. Nat. Prod. Rep. 2007;24:1009–1026.

28. Ehrenshaft M, Bilski P, Li MY, Chignell CF, Daub ME. A highly conserved sequence is a novel gene involved in de novo vitamin B_6 biosynthesis. Proc. Natl. Acad. Sci. U.S.A. 1999;96:9374–9378.

29. Osmani AH, May GS, Osmani SA. The extremely conserved pyroA gene of Aspergillus nidulans is required for pyridoxine synthesis and is required indirectly for resistance to photosensitizers. J. Biol. Chem. 1999;274:23565–23569.

30. Tambasco-Studart M, Titiz O, Raschle T, Forster G, Amrhein N, Fitzpatrick TB. Vitamin B_6 biosynthesis in higher plants. Proc. Natl. Acad. Sci. U.S.A. 2005;102: 13687–13692.

31. Denu JM. Vitamins and aging: pathways to NAD+ synthesis. Cell. 2007;129: 453–454.

32. Belenky P, Racette FG, Bogan KL, McClure JM, Smith JS, Brenner C. Nicotinamide riboside promotes Sir2 silencing and extends lifespan via Nrk and Urh1/Pnp1/Meu1 pathways to NAD^+. Cell. 2007;129:473–484.

33. Begley TP, Kinsland C, Mehl RA, Osterman A, Dorrestein P. The biosynthesis of nicotinamide adenine dinucleotides in bacteria. Vitam. Horm. 2001;61:103–119.

34. Webb ME, Smith AG, Abell C. Biosynthesis of pantothenate. Nat. Prod. Rep. 2004;21:695–721.

35. Chakauya E, Coxona KM, Whitney HM, Ashurst JL, Abell C, Smith AG. Pantothenate biosynthesis in higher plants: advances and challenges. Physiol. Plant. 2006;126:319–329.

36. Linster CL, Van Schaftingen E. Vitamin C: Biosynthesis, recycling and degradation in mammals. FEBS J. 2007;274:1–22.

37. Smirnoff N. L-ascorbic acid biosynthesis. Vitam. Horm. 2001;61:241–266.

38. Wheeler GL, Jones MA, Smirnoff, N. The biosynthetic pathway of vitamin C in higher plants. Nature 1998;393:365–369.

39. Linster CL, Gomez TA, Christensen KC, Young BD, Brenner C. Arabidopsis VTC2 encodes a GDP-L-galactose phosphorylase, the last unknown enzyme in the Smirnoff-Wheeler pathway to ascorbic acid in plants. J. Biol. Chem. 2007;282:18879–18885.

40. Davidson VL. Electron transfer in quinoproteins. Arch. Biochem. Biophys. 2004; 428:32–40.

41. Felton LM, Anthony C. Role of PQQ as a mammalian enzyme cofactor? Nature 2005;433:E10.

42. Kasahara T, Kato T. A new redox-cofactor vitamin for mammals. Nature 2003;422:832.

43. Rucker, R, Storms, D, Sheets, A, Tchaparian, E and Fascetti, A. Is pyrroloquinoline quinone a vitamin? Nature. 2005;433:E10–E11.

44. Magnusson OT, Toyama H, Saeki M, Schwarzenbacher R, Klinman JP. The structure of a biosynthetic intermediate of pyrroloquinoline quinone (PQQ) and elucidation of the final step of PQQ biosynthesis. J. Am. Chem. Soc. 2004;126:5342–5343.

45. Magnusson OT, Toyama H, Saeki M, Rojas A, Reed JC, Liddington RC, Klinman JP, Schwarzenbacher R. Quinone biogenesis: Structure and mechanism of PqqC, the final catalyst in the production of pyrroloquinoline quinone. Proc. Natl. Acad. Sci. U.S.A. 2004;101:7913–7918

46. Marquet A, Bui BT. Biosynthesis of biotin and lipoic acid. Vitam. Horm. 2001; 61:51–101.

47. Marquet A, Tse Sum Bui, B, Smith AG, Warren MJ. Iron–sulfur proteins as initiators of radical chemistry. Nat. Prod. Rep. 2007;24:1027–1040.

48. Sacchettini JC, Poulter CD. Creating isoprenoid diversity. Science 1997;277: 1788–1789.

49. Slater, EE and MacDonald, JS. Mechanism of action and biological profile of HMG CoA reductase inhibitors. A new therapeutic alternative. Drugs 1988;36 Suppl 3: 72–82.

50. Stancu C, Sima A. Statins: mechanism of action and effects. J. Cell. Mol. Med. 2001;5:378–387.

51. Qureshi N, Porter JW. Conversion of acetyl-coenzyme A to isopentenyl pyrophosphate. In: Biosynthesis of Isoprenoid Compounds. J. W. Porter and S. L. Spurgeon, eds. 1981. John Wiley, New York. pp. 47–94.

52. Eisenreich W, Bacher A, Arigoni D, Rohdich F. Biosynthesis of isoprenoids via the non-mevalonate pathway. Cell. Mol. Life Sci. 2004;61:1401–1426.

53. Siemieniuk E, Skrzydlewska E. Coenzyme Q10: its biosynthesis and biological significance in animal organisms and in humans. Postepy. Hig. Med. Dosw. 2005;59:150–159.

54. Szkopińska A, Ubiquinone. Biosynthesis of quinone ring and its isoprenoid side chain. Intracellular localization. Acta Biochim. Polonica. 2000;47:469–480.

55. Graham DE, White RH. Elucidation of methanogenic coenzyme biosyntheses: from spectroscopy to genomics. Nat. Prod. Rep. 2002;19:133–147.

56. Eisenreich W, Schwarzkopf B, Bacher A. Biosynthesis of nucleotides, flavins, and deazaflavins in *Methanobacterium thermoautotrophicum*. J. Biol. Chem. 1991;266:9622–9631.

57. Reuke B, Korn S, Eisenreich W, Bacher A. Biosynthetic precursors of deazaflavins. J. Bacteriol. 1992;174:4042–4049.

58. Grochowski LL, Xu H, Leung K, White RH. Characterization of an Fe^{2+}-dependent archaeal-specific GTP cyclohydrolase, MptA, from *Methanocaldococcus jannaschii*. Biochemistry 2007;46:6658–6667.

59. Koetter JW, Schulz GE. Crystal structure of 6-hydroxy-D-nicotine oxidase from *Arthrobacter nicotinovorans*. J. Mol. Biol. 2005;352:418–428.

60. Jitrapakdee S, Wallace JC. The biotin enzyme family: conserved structural motifs and domain rearrangements. Curr. Protein Pept. Sci. 2003;4:217–229.

61. Rucker RB, Wold F. Cofactors in and as posttranslational protein modifications. FASEB J. 1988;2:2252–2261.

62. Vagelos PR, Majerus PW, Alberts AW, Larrabee AR, Ailhaud GP. Structure and function of the acyl carrier protein. Fed. Proc. 1966;25:1485–1494.

63. Blair-Johnson M, Fiedler T, Fenna R. Human myeloperoxidase: structure of a cyanide complex and its interaction with bromide and thiocyanate substrates at 1.9 A resolution. Biochemistry 2001;40:13990–13997.

64. Singh AK, Singh N, Sharma S, Singh SB, Kaur P, Bhushan A, Srinivasan A, Singh TP. Crystal structure of Lactoperoxidase at 2.4 A resolution. J. Mol. Biol. 2007;376:1060–1075.

65. Furtmüller PG, Zederbauer M, Jantschko W, Helm J, Bogner M, Jakopitsch C, Obinger C. Active site structure and catalytic mechanisms of human peroxidases. Arch. Biochem. Biophys. 2006;445:199–213.

66. Davidson VL. Protein-derived cofactors. Expanding the scope of post-translational modifications. Biochemistry 2007;46:5283–5292.

67. Fischer M, Bacher A. Biosynthesis of vitamin B_2 in plants. Physiol. Plant. 2006;126: 304–318.

68. Ralph SA, van Dooren GG, Waller RF, Crawford MJ, Fraunholz, MJ, Foth, BJ, Tonkin, CJ, Roos, DS and McFadden, GI. Tropical infectious diseases: metabolic maps and functions of the *Plasmodium falciparum* apicoplast. Nat. Rev. Microbiol. 2004;2:203–216.

69. van Dooren GG, Stimmler LM, McFadden GI. Metabolic maps and functions of the *Plasmodium* mitochondrion. FEMS Microbiol. Rev. 2006;30:596–630.

70. Smith AG, Croft MT, Moulin M, Webb ME. Plants need their vitamins too. Curr. Op. Plant Biol. 2007;10:266–275.

71. Muntau AC, Röschinger W, Habich M, Demmelmair H, Hoffmann B, Sommerhoff CP, Roscher AA. Tetrahydrobiopterin as an alternative treatment for mild phenylketonuria. N. Engl. J. Med. 2002;347:2122–3212.

72. Tahiliani AG, Beinlich CJ. Pantothenic acid in health and disease. Vitam. Horm. 1991;46:165–228.

73. Hayflick SJ, Westaway SK, Levinson B, Zhou B, Johnson MA, Ching KH, Gitschier J. Genetic, clinical, and radiographic delineation of Hallervorden-Spatz syndrome. N. Engl. J. Med. 2003;348:33–40.

74. Schwarz G. Molybdenum cofactor biosynthesis and deficiency. Cell Mol. Life Sci. 2005;62:2792–2810.

75. Sarkany RP. Porphyria. From Sir Walter Raleigh to molecular biology. Adv. Exp. Med. Biol. 1999;455:235–241.

76. Toh BH, van Driel IR, Gleeson PA. Pernicious anemia. N. Engl. J. Med. 1997;337: 1441–1448.

77. Lange RP, Locher HH, Wyss PC, Then RL. The targets of currently used antibacterial agents: lessons for drug discovery. Curr. Pharm. Des. 2007;13:3140–3154.

78. Wiesner J, Ortmann R, Jomaa H, Schlitzer M. New antimalarial drugs. Angew. Chem. Int. Ed. Engl. 2003;42:5274–5293.

79. Gangjee A, Jain HD. Antifolates—past, present and future. Curr. Med. Chem. Anticancer Agents. 2004;4:405–410.

80. Mazighi M, Lavallee PC, Labreuche J, Amarenco P. Statin therapy and stroke prevention: what was known, what is new and what is next? Curr. Opin. Lipidol. 2007;18:622–625.

81. Harper CR, Jacobson TA. The broad spectrum of statin myopathy: from myalgia to rhabdomyolysis. Curr. Opin. Lipidol. 2007;18:401–408.

82. Holick MF. Vitamin D deficiency. N. Engl. J. Med. 2007;357:266–281.

11

ANTIBIOTICS

SERGEY B. ZOTCHEV

Department of Biotechnology, Norwegian University of Science and Technology, Norway

Antibiotics are synthesized by many bacterial, fungal, plant, and animal species as secondary metabolites not required for normal growth of the producing organisms. Medical usefulness of some antibiotics in treatment of infections and cancer prompted investigations into their biosynthesis, which have revealed great complexity and variability of enzymes and reactions involved. Although antibiotics are represented by compounds that belong to diverse chemical classes, some common themes are observed in their biosynthesis. Those themes comprise assembly of antibiotic scaffolds from activated precursors that originate from primary metabolism, followed by modification of the scaffolds with different chemical moieties. In this review, generalized schemes of antibiotic biosynthesis are discussed, along with the common enzymology, genetics, and methods used for studying antibiotic biosynthesis pathways.

Compounds with antibiotic activity, that is, those which inhibit the growth or kill other living organisms, are synthesized by many microbial, plant, and some animal species. The discovery of penicillin by Alexander Fleming in 1928 has opened a new era in human history. This antibacterial compound synthesized by a fungus *Penicillium notatum, Penicillium chrysogenum*, and several other fungi has been found to be active against a wide range of bacteria, and it was introduced into medical practice in the early 1940s (1). Since then, thousands of antibiotics that possess antimicrobial, antitumor, and insecticidal activities and that belong to different chemical classes have been isolated from natural sources. Some of them were introduced into medical practice and agriculture. Studies on biosynthesis of antibiotics proved to be important both for fundamental and

Natural Products in Chemical Biology, First Edition. Edited by Natanya Civjan.
© 2012 John Wiley & Sons, Inc. Published 2012 by John Wiley & Sons, Inc.

applied sciences, unraveling novel enzymatic mechanisms and providing tools for changing chemical structures of some antibiotics to improve their properties.

Because most antibiotics are natural products, their synthesis by living organisms relies on enzymes, proteins with specific catalytic activities. Thus, biosynthesis of antibiotics can generally be described as *enzyme-catalyzed synthesis of compounds with antibiotic activity in living organisms*. Several enzymes with different catalytic activities are usually involved in antibiotic biosynthesis, acting in a sequential manner in the synthesis of the antibiotic molecule and form a biosynthetic pathway. Our current understanding of antibiotic biosynthesis is mostly based on the studies of biosynthetic pathways in bacteria and fungi, which are the most prominent antibiotic producers among living organisms. Such studies have ensured considerable progress in understanding of genetics, enzymology, and chemistry of antibiotic biosynthesis. Several hundred antibiotic biosynthetic pathways have been elucidated, unraveling many new and unusual enzymatic mechanisms. Recent advances in the genome sequencing of antibiotic-producing organisms and analytical chemistry, along with the development of bioinformatics and chemoinformatics tools, proved to be very useful for elucidation of new biosynthetic pathways. In this chapter, general mechanisms behind antibiotic biosynthesis in bacteria and fungi are reviewed.

11.1 BIOLOGY OF ANTIBIOTIC BIOSYNTHESIS

The bacteria and fungi are especially profound in synthesizing chemically diverse antibiotics, whereas the plants and animals mostly produce ribosomally synthesized peptides and terpenes with antibiotic activity. Among the bacteria, species that belong to the order *Actinomycetales* produce over 60% of all known compounds with antibiotic activity. Some antibiotics of medical and agricultural importance, along with the information on their chemical class, biological activity, and producing organisms are presented in Table 11.1. Antibiotics are synthesized as secondary metabolites, which are not required for growth and maintenance of the producing organism as do the primary metabolites (amino acids, lipids, sugars, etc.). The debate on biological function of antibiotics in terms of their significance for the producing organisms is still ongoing. One obvious function of antibiotics would be inhibition of growth of competing organisms to ensure access to nutritional sources in the surrounding environment. However, in their natural environment, antibiotic-producing organisms may not synthesize antibiotics in significant amounts, which suggest an alternative biological role(s). Some antibiotics have been shown to modulate gene expression in bacteria in subinhibitory concentrations, which implies that antibiotics may play a role of signaling molecules providing means for communication between different species (2). Whatever the true biological function(s) of antibiotics, it must be important for the producing organism, considering the complexity of their biosynthetic pathways and metabolic burden imposed by the process of biosynthesis. As exemplified below, many enzymes

TABLE 11.1   Some medically and agriculturally important antibiotics produced by bacteria and fungi

Name	Chemical class	Biological activity	Producing organism
Amphotericin B	Polyene macrolide	Antifungal	*Streptomyces nodosus*
Bialaphos	Peptide	Herbicidal	*Streptomyces viridochromogenes*
Cephalosporin C[a]	β-Lactam	Antibacterial	*Acremonium chrysogenum*
Daptomycin	Lipopeptide	Antibacterial	*Streptomyces roseosporus*
Echinocandin B[a]	Lipopeptide	Antifungal	*Aspergillus nidulans var. echinulatus*
Erythromycin	Macrolide	Antibacterial	*Saccharopolyspora erythraea*
Ivermectin	Macrolide	Antihelminthic	*Streptomyces avermitilis*
Kanamycin	Aminoglycoside	Antibacterial	*Streptomyces kanamyceticus*
Lincomycin	Lincosamide	Antibacterial	*Streptomyces lincolnensis*
Monensin	Polyether	Anticoccidial	*Streptomyces cinnamonensis*
Penicillin G[a]	β-Lactam	Antibacterial	*Penicillium chrysogenum*
Pleuromutilin[a]	Diterpene	Antibacterial	*Clitopilus scyphoides*
Rifamycin	Ansamycin	Antibacterial	*Amycolatopsis mediterranei*
Spinosins	Macrolide	Insecticidal	*Saccharopolyspora spinosa*
Tetracycline	Aromatic polyketide	Antibacterial	*Streptomyces aureofaciens*
Vancomycin	Glycopeptide	Antibacterial	*Amycolatopsis orientalis*

[a] Antibiotics of fungal origin.

involved in antibiotic biosynthesis have apparently evolved from those dedicated to primary metabolism, and use the same precursors.

11.1.1   Antibiotic Biosynthesis: From Primary to Secondary Metabolism

With the exception of lantibiotics and bacteriocins, which are synthesized by ribosomes as short peptides and then posttranslationally modified, antibiotic molecules are assembled by specific enzymes from precursors mainly supplied by the primary metabolism. A generalized scheme for antibiotic biosynthesis process showing precursors that originated from primary metabolism is

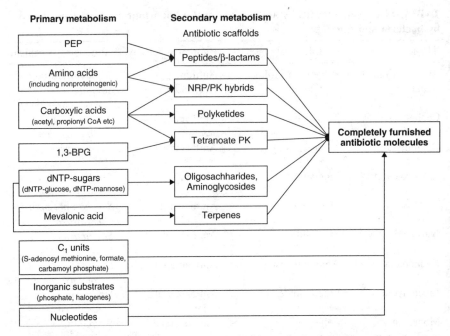

Figure 11.1 Generalized scheme of antibiotic biosynthesis that shows links between primary and secondary metabolites. *1,3-BPG*, 1,3-bisphosphoglycerate; *dNTP*, deoxynucleotidetriphosphate.

presented in Fig. 11.1. As a rule, biosynthesis of antibiotics comprises two stages: 1) assembly of antibiotic molecule skeletons, or scaffolds, which, in themselves, may possess little or no antibiotic activity; and 2) modification of the scaffolds by specific enzymes that result in appearance of additional chemical groups on the scaffold and yield completely furnished and fully active antibiotics. For example, in the first stage of biosynthesis of the macrolide antibiotic erythromycin, the macrolactone scaffold 6-deoxyerythronolide B that has no antibacterial activity is assembled from small carboxylic acid CoA esters. During the second stage, erythronolide is modified by means of glycosylation, hydroxylation, and methylation, which yields erythromycin A molecule with potent antibacterial activity (3).

In general, all polyketide antibiotics (e.g., erythromycin and tetracycline) scaffolds are assembled from coenzyme A esters of carboxylic acids, but they may also use other precursors for starter and extender building blocks. Those precursors include, for example, 1,3-bisphosphoglycerate that originates from glycolysis in the biosynthesis of tetranoate polyketides (e.g., tetronomycin) (4), or monosachharides and phosphoenolpyruvate (PEP) in biosynthesis of ansamycin polyketides (e.g., rifamycin) (5). Precursors for assembly of scaffolds of the nonribosomally synthesized peptide antibiotics (e.g., vancomycin and daptomycin) may include PEP and nonproteinogenic amino acids, such as β-hydroxytyrosine,

p-hydroxyphenylglycine, and β-hydroxyasparagine. Monosaccharides activated by attachment of deoxynucleotidetriphosphates are used as building blocks in biosynthesis of aminoglycoside and oligosaccharide antibiotics, and also for modification of many chemically diverse antibiotic scaffolds (6, 7). In the biosynthesis of terpene antibiotics, which are relatively rare in bacteria, mevalonic acid is used as a precursor. Mevalonate is synthesized from two acetyl-CoA molecules via sequential action of acetyl-CoA acetyltransferase, 3-hydroxy-3-methyl-glutaryl(HMG)-CoA synthase, and HMG-CoA reductase (8).

Primary metabolites, such as S-adenosyl methionine, formate, carbamoyl phosphate, nucleotides, as well as inorganic substrates, are also used as precursors for modification of certain antibiotic scaffolds.

The use of primary metabolites as precursors for antibiotic biosynthesis implies that their supply may be a limiting factor for the latter process in some cases, especially when the organism is actively growing. Early studies have indeed confirmed that antibiotic biosynthesis starts when the growth of the organism is slowing down, and it reaches its maximum with the cessation of growth (9). Understanding of the antibiotic biosynthetic pathways, especially the origins of precursors, is therefore important from the practical point of view in terms of increasing the yields during industrial production of antibiotics. Engineering of primary metabolic pathways that result in redirection of precursor flow toward antibiotic biosynthesis have been described for several antibiotic-producing organisms (10–12).

11.1.2 Regulation of Antibiotic Biosynthesis and Mechanisms of Self-Protection

It has been now firmly established that biosynthesis of antibiotics is subject to control by multiple regulatory systems. Several biological reasons may explain this complex regulation. Active synthesis of antibiotics represents a metabolic burden for the producing organisms, which draws heavily on the precursor, ATP, and electron carrier pools from primary metabolism (13). The latter suggests that, in natural environment, antibiotics are synthesized only when they provide a survival benefit. Antibiotic biosynthesis in most organisms is indeed activated in response to certain environmental stimuli. Such factors as nutrient limitation, change of temperature and pH in the environment, phage infection, presence of other organisms, or organic solvents are known to trigger antibiotic biosynthesis (14). Many antibiotic-producing organisms have a genetic capacity to synthesize several antibiotics with diverse chemical structures and biological activities. Consequently, in some cases, a competition for precursor supply may occur between the antibiotic biosynthesis pathways present in one organism, and a decision must be made on which pathway shall be activated in current circumstances. Although the global regulatory network involved in regulation of antibiotic biosynthesis remains poorly understood, considerable progress has been made toward unraveling parts of it. The latter can be exemplified by the regulatory cascades that involve γ-butyrolactones as signaling molecules discovered in *Streptomyces* bacteria (15). These small molecules bind to receptor proteins usually represented by repressors, which renders them inactive. The latter allows expression of certain activator proteins, which affect expression of antibiotic biosynthesis genes directly or indirectly. In most cases where the genetics of antibiotic biosynthesis have been studied in detail, biosynthetic genes and pathway-specific regulatory genes that control their expression were found in close proximity.

Another issue to be considered with respect to antibiotic biosynthesis is the resistance of the producing organism to the endogenous antibiotic. The most frequently used strategy for self-protection is active efflux of antibiotic molecules outside the producing cells, which is often driven by ATP hydrolysis (16). Besides the efflux, antibiotic-producing organisms may employ additional or alternative defense mechanisms, such as target modification (e.g., methylation of own ribosomal RNA), degradation (e.g., by β-lactamases), or modification of intracellularly accumulated antibiotic (17).

11.2 CHEMISTRY, ENZYMOLOGY, AND GENETICS OF ANTIBIOTIC BIOSYNTHESIS

11.2.1 Assembly of Antibiotic Scaffolds

The common theme in the biosynthesis of all antibiotics is use of activated precursors ("building blocks") for assembly of antibiotic scaffolds. Enzymes involved in this process either use already activated precursors from primary metabolism (such as acyl-CoAs), or they ensure their activation through adenylation, phosphorylation, attachment of nucleotide moieties, and so on. Common reactions involved in initiation of antibiotic biosynthesis and/or activation of "building blocks" are presented on Fig. 11.2. Biosynthesis of polyketide antibiotics, such as erythromycin, tetracycline etc proceeds through Claisen-type decarboxylative condensation of acyl-CoAs by polyketide synthases. Malonyl-CoA and methylmalonyl-CoA are the most common precursors in polyketide antibiotic scaffold biosynthesis, which yield acetate and propionate building blocks, respectively, incorporated into polyketides after decarboxylation. However, several examples of alternative building blocks are available in polyketide antibiotics, especially those used for initiation of the scaffold biosynthesis. These examples include glycerate (4), methoxy- or ethylmalonate (3, 18), 3-amino-5-hydroxybenzoic acid (19), and amino acids in the mixed polyketide synthase (PKS)-nonribosomal peptides synthetase (NRPS) pathways (20). Peptide antibiotics (e.g., vancomycin) scaffolds are synthesized by NRPS via condensation of amino acids. Specific domains within NRPS enzymes activate amino acids by adenylation; optionally modify aminoacyl-adenylate intermediates via epimerization, oxidation, and N-methylation, and they finally form peptide bonds to assemble peptide antibiotic scaffolds (21). In the case of lipopeptide antibiotics (e.g., daptomycin and friulimicin), scaffold synthesis starts with activation of a long chain fatty acid and its coupling to an amion acid by a specific domain of

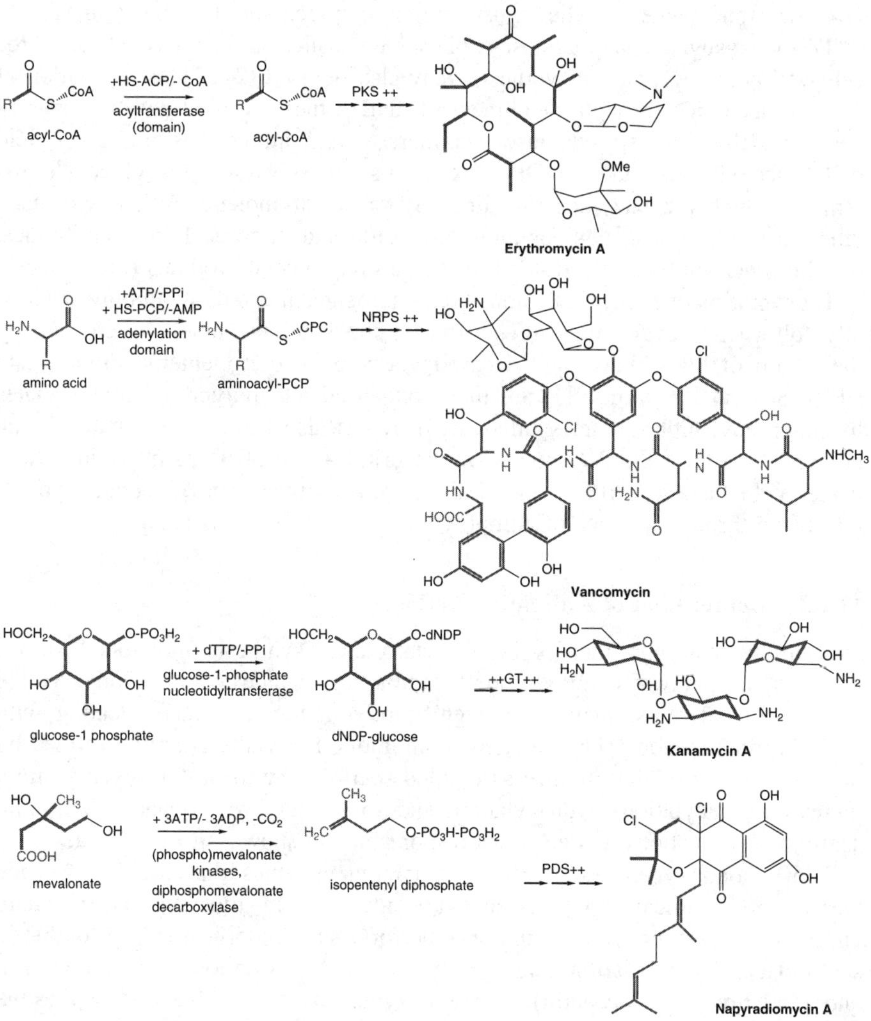

Figure 11.2 Examples on activation and incorporation of "building blocks" (shown in bold lines) into the scaffolds of polyketide, nonribosomal peptide, aminoglycoside, and terpene antibiotics. Common enzymes responsible for scaffolds' assembly: *PKS*; *NRPS*; *GT*, glycosyltransferase; *PDS*, polyprenyl diphosphate synthase. (++) indicate that additional enzymatic reactions required to synthesize the antibiotic molecule.

NRPS (22). Most nonribosomally synthesized peptide antibiotics contain nonproteinogenic amino acids, which are synthesized by dedicated enzymes (23).

"Building blocks" in the biosynthesis of aminoglycoside and oligosaccharide antibiotics scaffolds (e.g., kanamycin and avilamycin) are derived from monosaccharide-1-phosphates, which are converted to deoxysugar nucleotidyldiphosphates (dNDP-deoxysugars) via a series of enzymatic steps.

Glucose-1-phosphate is the most common precursor for biosynthesis of dNDP-deoxysugars, but mannose-6-phosphate, which is an intermediate in the cell wall peptidoglycan biosynthesis, may also be used (24). After activation of a monosaccharide through attachment of dNDP, they are modified by enzymes such as dehydratases, reductases, epimerase, and aminotransferases to yield dNDP-deoxysugars (25). dNDP-deoxysugars are substrates for glycosyltransferases, which are enzymes that link several sugar moieties either with each other to produce scaffolds for aminoglycoside and oligosaccharide antibiotics, or with other antibiotic scaffolds to provide glycosylated products (26).

Biosynthesis of terpene antibiotics (e.g., terpentecin and napyradiomycin) usually follows a mevalonate pathway, where the latter precursor is activated by the action of both kinase and decarboxylase to yield isopentenyl diphosphate (IPP). Several IPP molecules are then condensed into polyprenyl diphosphates by polyprenyl diphosphate synthases (8). It has also been shown that IPP can be synthesized via the 2-C-methyl-D-erythritol 4-phosphate pathway in certain bacteria. In this case, IPP is synthesized from pyruvate and glyceraldehyde-3-phosphate through a series of complex biochemical reactions (27).

11.2.2 Modification of Antibiotic Scaffolds

Despite the common themes in the scaffold biosynthesis mentioned above, biosynthetic pathways for different antibiotics, which even belong to the same chemical class, may differ significantly in the final steps dealing with scaffold modification. The late steps in antibiotic biosynthesis are performed by dedicated enzymes that modify assembled scaffolds by means of glycosylation, methylation, acylation, hydroxylation, and so on. Different types of antibiotic scaffold modifications and enzymes responsible are shown on Fig. 11.3a.

P450 monooxygenases are heme-contaning enzymes that catalyze a wide variety of chemical reactions and are involved in biosynthesis of many antibiotics. These enzymes commonly perform such modifications of antibiotic scaffolds as hydroxylation (e.g., erythromycin), epoxidation (e.g., pimaricin), and oxidation (e.g., nystatin). These enzymes have also been shown to be involved in oxidative cyclization of phenolic side chains during biosynthesis of certain antibiotics (e.g., vancomycin) (28). FMN-dependent monooxygenases and dioxygenases are also known to be involved in antibiotic biosynthesis, which modify antibiotic scaffolds via hydroxylation, as in the biosynthesis of actinorhodin and tetracenomycin (29, 30).

Several halogen-containing antibiotics have been described, for example, medically important antibacterial agents such as chlortetracycline and vancomycin, and chloro-indolocarnazoles with antitumor activity. Attachment of halogen atoms to antibiotic scaffolds is performed by several types of halogenases, which include haloperoxidases, and flavin- and a-ketoglutarate-dependent halogenases (31).

Methyltransferases involved in antibiotic biosynthesis commonly use S-adenosylmethionine as a methyl group donor, and they are mostly responsible

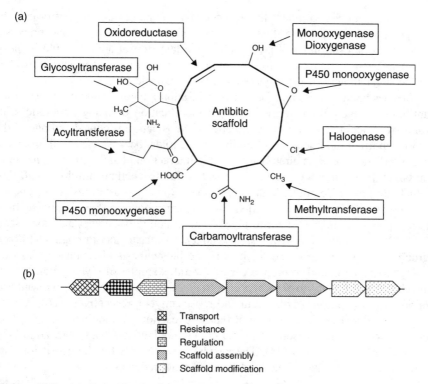

Figure 11.3 (a) Generalized scheme of the most common modifications of antibiotic scaffolds showing enzymes responsible. (b) Organization of a typical antibiotic biosynthesis gene cluster.

for methylation of deoxysugar moieties of antibiotics after their attachment to scaffolds by glycosyltransferases (32). However, methyltransferases modify antibiotic scaffolds directly, as it was shown for the biosynthesis of antibacterial agents clorobiocin and fosfomycin (33, 34).

Carbamoylation is a relatively rare modification step in antibiotic biosynthesis, whereas carbamoyl moiety seems to be important for biological activity. Carbamoyltransferases, which are the enzymes responsible for this modification, participate in biosynthesis of certain b-lactams (35), antibacterial antibiotic novobiocin (36), and antitumor agent geldanamycin (37). As a carbamoyl group donor, these enzymes use carbamoyl phosphate synthesized during amino acid metabolism to capture ammonia released during oxidative deamination of glutamate.

Scaffolds of antibiotics, such as aminocoumarins and some b-lactams, are modified with acyl moieties by acyltransferases. In the biosynthesis of the aminocoumarine antibacterial antibiotic clorobiocin, a specific acyltransferase attaches pyrrole-2 carboxyl moiety that originates from L-proline to the antibiotic scaffold (38). The last step in the biosynthesis of the b-lactam antibiotic cephalosporin C in *Acremonium chrysogenum* is performed by a specific acyltransferase that attaches an acetyl moiety to the precursor deacetylcephalosporin C (39). Acylation as a late step of antibiotic biosynthesis has also been demonstrated for chromomycin, where an acyl group is attached to the sugar moiety of glycosylated antibiotic scaffold (40).

Besides being responsible for assembly of aminoglycoside and oligosaccharide antibiotics, glycosyltransferases perform modification of many antibiotic scaffolds by attaching specific sugar moieties. Glycosylation is considered one of the most important modifications of antibiotic scaffolds, because chemical groups of sugar moieties are often directly involved in interaction of antibiotic with its cellular target. In addition, glycosylation helps to solubilize certain antibiotic scaffolds, which are otherwise highly hydrophobic (41). Some glycosyltransferases display relaxed specificity toward both dNDP-deoxysugar donor and antibiotic scaffold acceptor, which enables them to glycosylate structurally distinct antibiotic scaffolds (26). Usually, glycosyltransferase attaches sugar moiety through a hydroxyl group on antibiotic aglycone, such as in the biosynthesis of erythromycin, vancomycin, nystatin, and so on. However, several examples of glycosyltransferases attach a sugar moiety to a carbon or nitrogen atom, such as in the biosynthesis of antitumor antibiotics urdamycin and rebeccamycin, respectively (26).

Oxidoreductases catalyze several types of reactions in modification of antibiotic scaffolds, which perform dehydration, keto reduction (42), oxygenation (43), and oxidative cyclization (44). The enzymes responsible for these reactions are usually flavin- or a-ketoglutarate-dependent oxidoreductases.

11.2.3 Genetics of Antibiotic Biosynthesis

Biosynthesis of antibiotics requires concerted action of enzymes involved in both assembly of the scaffold and its modification. The latter is achieved through coordinated expression of genes that encode these enzymes ensured by the following: 1) colocalization (clustering) of biosynthetic genes in the genomes of producing organisms and 2) regulation of gene expression by pathway-specific regulatory genes. A schematic presentation of a typical gene cluster that governs biosynthesis of antibiotic is shown on Fig. 11.3b. The "core" of the antibiotic biosynthesis gene cluster is represented by the genes encoding enzymes responsible for scaffold biosynthesis (such as PKS, NRPS, glycosyltransferase, etc.). Genes for scaffold modification enzymes (hydroxylases, methyltransferases, acyltransferases, halogenases, glycosyltransferases, etc.) are usually found in the vicinity of the "core". One or more pathway-specific regulatory genes are typically present in the cluster, which regulate expression of biosynthetic genes. Resistance to its own antibiotic is ensured by the presence of genes encoding enzymes that inactivate endogenously accumulated antibiotic or modify antibiotic target in the producing organism. Often, genes encoding efflux pumps that export antibiotic molecules outside the cells are also found with the cluster. The latter enzymes not only protect the producing organisms from harmful action of their own antibiotics but also ensure delivery of the antibiotics to the surrounding environment.

Significant advances in understanding antibiotic biosynthesis have been made in the recent years mainly because of cloning and analysis of antibiotic biosynthesis gene clusters. Modern bioinformatics tools allow deduction of entire antibiotic biosynthetic pathways or parts thereof during analysis of biosynthetic genes (45). Moreover, access to these genes makes it possible to study their function in the producing organisms by mutational analysis, obtaining biosynthetic enzymes for kinetic and mechanistic studies, and so on (see next section). Besides fundamental scientific value, antibiotic biosynthesis genes can be used for generation of novel antibiotics by means of combinatorial biosynthesis and biosynthetic engineering. Both technologies require understanding of the functions of the genes, good knowledge on the entire antibiotic biosynthetic pathway, and tools for genetic manipulation of the producing organism or expression of the pathway in a heterologous host. Combinatorial biosynthesis is based on "mix-and-match" approach, where genes for biosynthesis of different antibiotics or parts thereof are combined in one host in an attempt to obtain "hybrid" antibiotics (46). In the recent years, this technology has mainly been used for expression of specific genes or parts of exogenous biosynthetic pathways in antibiotic-producing organisms to obtain new functionalities on antibiotic molecules (32). Biosynthetic engineering is focusing on manipulation of the intrinsic antibiotic biosynthesis genes to alter the structure of antibiotic in a predictable manner (47). Coupled with the knowledge on antibiotic's structure–activity relationship, combinatorial biosynthesis and biosynthetic engineering become powerful tools for generation of new antibiotics with improved properties.

11.3 TOOLS AND TECHNIQUES USED TO STUDY ANTIBIOTIC BIOSYNTHESIS

Antibiotic biosynthesis pathways may involve dozens of biochemical reactions; some of them represent unusual enzymology and chemistry. Because biosynthesis of antibiotics occurs in living cells, interference from other metabolic pathways certainly exists. The latter, as well as relaxed specificity of certain enzymes toward their substrates and occasional failure to perform dedicated reaction, usually leads to biosynthesis of a mixture of antibiotic congeners by the same pathway. If the congeners are many and their structures are complex, then the analysis of biosynthetic pathways may be complicated. However, identification of some congeners as intermediates in antibiotic biosynthesis can be of help in establishing the order of events in the pathway. The most reliable and complete information on antibiotic biosynthesis is obtained by using a combination of different methods described below and merging the information obtained into a complete picture.

11.3.1 Feeding Studies

Many antibiotics have a relatively complex chemical structure, and identification of the precursors that serve as "starting units" for antibiotic biosynthesis and/or for some unusual "building blocks" might be difficult to deduce from the structure. In such cases, a series of putative precursors labeled with either ^{13}C or ^{15}N are used in so-called "feeding" experiments. In essence, the labeled precursors are added to the growing culture of antibiotic-producing organism, which allows for their full or partial incorporation into the antibiotic molecule in the course of the biosynthesis. Antibiotic is then extracted, purified, and investigated with regard to incorporation of the labeled precursors (or parts thereof) by means of LC-MS/MS or NMR (48, 49). In some cases, stereochemically labeled precursors shall be used to determine the biosynthetic origin of certain "building blocks" (50). Besides determining the origin of the building blocks of antibiotic molecules, feeding studies using unnatural substrates for enzymes that participate in the biosynthesis can be performed. In such studies, the antibiotic-producing organism is fed a series of chemically synthesized substrates for a particular antibiotic biosynthesis enzyme, and the secondary metabolites produced are analyzed (51). Results of these experiments can reveal the flexibility of certain antibiotic biosynthesis enzymes and complete pathways in terms of unnatural substrates and their incorporation into antibiotic molecules. Such studies are important for engineered biosynthesis that may provide novel antibiotics with improved properties.

11.3.2 Mutational Analysis and Enzymatic Assays

A method frequently used for determination or confirmation of antibiotic biosynthetic pathways takes advantage of the availability of antibiotic biosynthesis genes. If these genes can be manipulated in either an original or heterologous host, then in many cases the exact function of the genes can be established through their inactivation and subsequent analysis of metabolite profiles in the resulting mutants. This method in itself is mostly used for experimental confirmation of a certain enzyme activity in antibiotic biosynthetic pathway that has been suggested from bioinformatics analysis. For example, involvement of specific glycosyltransferase or P450 monooxygenase in antibiotic biosynthesis can be confirmed through inactivation of respective genes in the producing organisms and identification of the resulting deglycosyl- or deoxy-intermediates (52, 53). However, this method relies heavily on analytical methods (LC-MS/MS, LC-MS-TOF, NMR) for identification of antibiotic biosynthesis intermediate(s) produced as a result of pathway interruption. In certain cases, these intermediates are either chemically unstable, scarcely available, or undergo extensive conversion by noncognate enzymes, which makes exact placement of a step performed by inactivated enzyme on a biosynthetic pathway rather difficult.

Providing that antibiotic biosynthesis intermediate can be purified, a more comprehensive analysis of a biochemical reaction performed by an enzyme responsible for the next biosynthetic step may be achieved. This result can be performed by expression of the enzyme either in the original host or heterologously, and by performing *in vitro* enzyme assays (54). Individually expressed and purified antibiotic biosynthesis enzymes can also be used in coupled *in vitro* assays to determine the order of events in the biosynthetic pathway (42).

11.3.3 Studies on Enzyme-Bound Substrates and Intermediates

Another method used extensively for studying molecular details of antibiotic biosynthesis involves investigations into particular biosynthetic enzymes with bound substrates or intermediates. The method allows investigation of substrate tolerance and specificity, timing and order of reactions, transfer of substrates and biosynthetic intermediates from one enzyme to another. Early examples of this method relied completely on radioactively labeled substrates, which were mixed with purified antibiotic biosynthesis enzymes, and substrate/intermediate protein complexes were separated by chromatographic methods. Purified complexes were then studied with respect to the amount of the substrate/intermediate bound, as well as the identity of the intermediates. The latter was achieved after denaturation of the complexes and identification of released intermediates by analytical methods (55, 56). Recent development of efficient protein expression systems and new analytical methods greatly enhanced this technique, which allows deeper mechanistic insights into the mode of action of enzymes involved in antibiotic biosynthesis. One new method involves electrospray ionization Fourier-Transform mass spectrometry (ESI-FTMS), which is particularly suitable for studies on modular biosynthetic enzyme systems, such as PKS and NRPS (57). ESI-FTMS offers high-resolution power in determining masses of proteins with bound intermediates in complex mixtures, which are produced as a result of limited proteolysis of the enzyme-substrate complexes. Using this method, it is possible to determine relative ratios of intermediates at specific sites on antibiotic biosynthesis enzymes and to identify particular intermediates directly.

Another recently introduced method involves crystallographic trapping of potential substrate in purified antibiotic biosynthesis enzyme that allows determination of enzymatic mechanism (58). In this method, crystals of a biosynthetic enzyme are soaked in potential substrate mimics, and subsequent crystallographic analysis allows identification of a true substrate, even in the mixture of apparently unstable intermediates.

REFERENCES

1. Chain E, Florey HW, Adelaide MB, Gardner AD, Heatley NG, Jennings MA, Orr-Ewing J, Sanders AG. Penicillin as a chemotherapeutic agent. Lancet 1940;236: 226–228.

2. Yim G, Wang HH, Davies J. Antibiotics as signalling molecules. Philos. Trans. R. Soc. Lond. B. Biol. Sci. 2007;362:1195–1200.

3. McDaniel R, Welch M, Hutchinson CR. Genetic approaches to polyketide antibiotics. 1. Chem. Rev. 2005;105:543–558.

4. Sun Y, Hong H, Gillies F, Spencer JB, Leadlay PF. Glyceryl-S-acyl carrier prtein as an intermediate in the biosynthesis of tetronate antibiotics. Chembiochem. 2008;9: 150–156.

5. Yu TW, Muller R, Muller M, Zhang X, Draeger G, Kim CG, Leistner E, Floss HG. Mutational analysis and reconstituted expression of the biosynthetic genes involved in the formation of 3-amino-5-hydroxybenzoic acid, the starter unit of rifamycin biosynthesis in amycolatopsis Mediterranei S699. J. Biol. Chem. 2001;276:12546–12555.

6. Llewellyn NM, Spencer JB. Biosynthesis of 2-deoxystreptamine-containing aminoglycoside antibiotics. Nat. Prod. Rep. 2006;23:864–874.

7. Luzhetskyy A, Vente A, Bechthold A. Glycosyltransferases involved in the biosynthesis of biologically active natural products that contain oligosaccharides. Mol. Biosyst. 2005;1:117–126.

8. Dairi T. Studies on biosynthetic genes and enzymes of isoprenoids produced by actinomycetes. J. Antibiot. 2005;58:227–243.

9. Hopwood DA. The Leeuwenhoek lecture, 1987. Towards an understanding of gene switching in Streptomyces, the basis of sporulation and antibiotic production. Proc. R. Soc. Lond. B. Biol. Sci. 1988;235:121–138.

10. Thykaer J, Nielsen J. Metabolic engineering of beta-lactam production. Metab. Eng. 2003;5:56–69.

11. Freitag A, Méndez C, Salas JA, Kammerer B, Li SM, Heide L. Metabolic engineering of the heterologous production of clorobiocin derivatives and elloramycin in Streptomyces coelicolor M512. Metab. Eng. 2006;8:653–661.

12. Reeves AR, Brikun IA, Cernota WH, Leach BI, Gonzalez MC, Weber JM. Engineering of the methylmalonyl-CoA metabolite node of Saccharopolyspora erythraea for increased erythromycin production. Metab. Eng. 2007;9:293–303.

13. Kleijn RJ, Liu F, van Winden WA, van Gulik WM, Ras C, Heijnen JJ. Cytosolic NADPH metabolism in penicillin-G producing and non-producing chemostat cultures of Penicillium chrysogenum. Metab. Eng. 2007;9:112–123.

14. Bibb MJ. Regulation of secondary metabolism in streptomycetes. Curr. Opin. Microbiol. 2005;8:208–215.

15. Horinouchi S. A microbial hormone, A-factor, as a master switch for morphological differentiation and secondary metabolism in Streptomyces griseus. Front Biosci. 2002;7:d2045–2057.

16. Martín JF, Casqueiro J, Liras P. Secretion systems for secondary metabolites: how producer cells send out messages of intercellular communication. Curr. Opin. Microbiol. 2005;8:282–293.

17. Hopwood DA. How do antibiotic-producing bacteria ensure their self-resistance before antibiotic biosynthesis incapacitates them? Mol. Microbiol. 2007;63:937–940.

18. Yu TW, Bai L, Clade D, Hoffmann D, Toelzer S, Trinh KQ, Xu J, Moss SJ, Leistner E, Floss HG. The biosynthetic gene cluster of the maytansinoid antitumor agent ansamitocin from Actinosynnema pretiosum. Proc. Natl. Acad. Sci. USA. 2002;99: 7968–7973.

19. Floss HG. Natural products derived from unusual variants of the shikimate pathway. Nat. Prod. Rep. 1997;14:433–452.

20. Du L, Shen B. Biosynthesis of hybrid peptide-polyketide natural products. Curr. Opin. Drug Discov. Devel. 2001;4:215–228.

21. Sieber SA, Marahiel MA. Molecular mechanisms underlying nonribosomal peptide synthesis: approaches to new antibiotics. Chem. Rev. 2005;105:715–738.

22. Baltz RH. Biosynthesis and genetic engineering of lipopeptide antibiotics related to daptomycin. Curr. Top Med. Chem. 2008;8:618–638.

23. Chen H, Tseng CC, Hubbard BK, Walsh CT. Glycopeptide antibiotic biosynthesis: enzymatic assembly of the dedicated amino acid monomer (S)-3,5-dihydroxyphenyl-glycine. Proc. Natl. Acad. Sci. USA. 2001;98:14901–14906.

24. Trefzer A, Salas JA, Bechthold A. Genes and enzymes involved in deoxysugar biosynthesis in bacteria. Nat. Prod. Rep. 1999;16:283–299.

25. Rupprath C, Schumacher T, Elling L. Nucleotide deoxysugars: essential tools for the glycosylation engineering of novel bioactive compounds. Curr. Med. Chem. 2005;12:1637–1675.

26. Luzhetskyy A, Méndez C, Salas JA, Bechthold A. Glycosyltransferases, important tools for drug design. Curr. Top Med. Chem. 2008;8:680–709.

27. Kuzuyama T, Seto H. Diversity of the biosynthesis of the isoprene units. Nat. Prod. Rep. 2003;20:171–183.

28. Hubbard BK, Walsh CT. Vancomycin assembly: nature's way. Angew. Chem. Int. Ed. Engl. 2003;42:730–765.

29. Valton J, Mathevon C, Fontecave M, Nivière V, Ballou DP. Mechanism and regulation of the Two-component FMN-dependent monooxygenase ActVA-ActVB from Streptomyces coelicolor. J. Biol. Chem. 2008;283:10287–10296.

30. Rafanan ER Jr, Hutchinson CR, Shen B. Triple hydroxylation of tetracenomycin A2 to tetracenomycin C involving two molecules of O(2) and one molecule of H(2)O. Org. Lett. 2000;2:3225–3227.

31. Neumann CS, Fujimori DG, Walsh CT. Halogenation strategies in natural product biosynthesis. Chem. Biol. 2008;15:99–109.

32. Méndez C, Luzhetskyy A, Bechthold A, Salas JA. Deoxysugars in bioactive natural products: development of novel derivatives by altering the sugar pattern. Curr. Top Med. Chem. 2008;8:710–724.

33. Westrich L, Heide L, Li SM. CloN6, a novel methyltransferase catalysing the methylation of the pyrrole-2-carboxyl moiety of clorobiocin. Chembiochem. 2003;4:768–773.

34. Woodyer RD, Li G, Zhao H, van der Donk WA. New insight into the mechanism of methyl transfer during the biosynthesis of fosfomycin. Chem. Commun. 2007;359–361.

35. Martín JF. New aspects of genes and enzymes for beta-lactam antibiotic biosynthesis. Appl. Microbiol. Biotechnol. 1998;50:1–15.

36. Xu H, Heide L, Li SM. New aminocoumarin antibiotics formed by a combined mutational and chemoenzymatic approach utilizing the carbamoyltransferase NovN. Chem. Biol. 2004;11:655–662.

37. Hong YS, Lee D, Kim W, Jeong JK, Kim CG, Sohng JK, Lee JH, Paik SG, Lee JJ. Inactivation of the carbamoyltransferase gene refines post-polyketide synthase modification steps in the biosynthesis of the antitumor agent geldanamycin. J. Am. Chem. Soc. 2004;126:11142–11143.

38. LinksXu H, Kahlich R, Kammerer B, Heide L, Li SM. CloN2, a novel acyltransferase involved in the attachment of the pyrrole-2-carboxyl moiety to the deoxysugar of clorobiocin. Microbiology 2003;149:2183–2191.

39. Lejon S, Ellis J, Valegård K. The last step in cephalosporin C formation revealed: crystal structures of deacetylcephalosporin C acetyltransferase from Acremonium chrysogenum in complexes with reaction intermediates. J. Mol. Biol. 2008;377:935–944.

40. Menéndez N, Nur-E-Alam M, Braña AF, Rohr J, Salas JA, Méndez C. Tailoring modification of deoxysugars during biosynthesis of the antitumour drug chromomycin A by Streptomyces griseus ssp. griseus. Mol. Microbiol. 2004;53:903–915.

41. Walsh C, Freel Meyers CL, Losey HC. Antibiotic glycosyltransferases: antibiotic maturation and prospects for reprogramming. J. Med. Chem. 2003;46:3425–3436.

42. Kallio P, Liu Z, Mäntsälä P, Niemi J, Metsä-Ketelä M. Sequential action of two flavoenzymes, PgaE and PgaM, in angucycline biosynthesis: chemoenzymatic synthesis of gaudimycin C. Chem. Biol. 2008;15:157–166.

43. Widboom PF, Fielding EN, Liu Y, Bruner SD. Structural basis for cofactor-independent dioxygenation in vancomycin biosynthesis. Nature 2007;447:342–345.

44. Arakawa K, Sugino F, Kodama K, Ishii T, Kinashi H. Cyclization mechanism for the synthesis of macrocyclic antibiotic lankacidin in Streptomyces rochei. Chem. Biol. 2005;12:249–256.

45. de Bruijn I, de Kock MJ, Yang M, de Waard P, van Beek TA, Raaijmakers JM. Genome-based discovery, structure prediction and functional analysis of cyclic lipopeptide antibiotics in Pseudomonas species. Mol. Microbiol. 2007;63:417–428.

46. Hopwood DA, Malpartida F, Kieser HM, Ikeda H, Duncan J, Fujii I, Rudd BA, Floss HG, Omura S. Production of 'hybrid' antibiotics by genetic engineering. Nature 1985;314:642–644.

47. Baltz RH. Molecular engineering approaches to peptide, polyketide and other antibiotics. Nat. Biotechnol. 2006;24:1533–1540.

48. Herold K, Xu Z, Gollmick FA, Grafe U, Hertweck C. Biosynthesis of cervimycin C, an aromatic polyketide antibiotic bearing an unusual dimethylmalonyl moiety. Org. Biomol. Chem. 2004;2:2411–2414.

49. Mahmud T, Flatt PM, Wu X. Biosynthesis of unusual aminocyclitol-containing natural products. J. Nat. Prod. 2007;70:1384–1391.

50. Schuhmann T, Vollmar D, Grond S. Biosynthetic origin of the methoxyl extender unit in bafilomycin and concanamycin using stereospecifically labeled precursors. J. Antibiot. 2007;60:52–60.

51. Jacobsen JR, Keatinge-Clay AT, Cane DE, Khosla C. Precursor-directed biosynthesis of 12-ethyl erythromycin. Bioorg. Med. Chem. 1998;6:1171–1177.

52. Luzhetskyy A, Fedoryshyn M, Dürr C, Taguchi T, Novikov V, Bechthold A. Iteratively acting glycosyltransferases involved in the hexasaccharide biosynthesis of landomycin A. Chem. Biol. 2005;12:725–729.

53. Mendes MV, Recio E, Fouces R, Luiten R, Martín JF, Aparicio JF. Engineered biosynthesis of novel polyenes: a pimaricin derivative produced by targeted gene disruption in Streptomyces natalensis. Chem. Biol. 2000;8:635–644.

54. Volokhan O, Sletta H, Ellingsen TE, Zotchev SB. Characterization of the P450 monooxygenase NysL, responsible for C-10 hydroxylation during biosynthesis of the polyene macrolide antibiotic nystatin in Streptomyces noursei. Appl. Environ. Microbiol. 2006;72:2514–2519.

55. Gevers W, Kleinkauf H, Lipmann F. Peptidyl transfers in gramicidin S biosynthesis from enzyme-bound thioester intermediates. Proc. Natl. Acad. Sci. USA. 1969;63:1335–1342.

56. Jackson MD, Gould SJ, Zabriskie TM. Studies on the formation and incorporation of streptolidine in the biosynthesis of the peptidyl nucleoside antibiotic streptothricin F. J. Org. Chem. 2002;67:2934–2941.

57. Dorrestein PC, Kelleher NL. Dissecting non-ribosomal and polyketide biosynthetic machineries using electrospray ionization Fourier-Transform mass spectrometry. Nat. Prod. Rep. 2006;23:893–918.

58. Ryan KS, Howard-Jones AR, Hamill MJ, Elliott SJ, Walsh CT, Drennan CL. Crystallographic trapping in the rebeccamycin biosynthetic enzyme RebC. Proc. Natl. Acad. Sci. USA. 2007;104:15311–11536.

FURTHER READING

Hopwood DA. Streptomyces in Nature and Medicine: the Antibiotic Makers. 2007, New York: Oxford University Press.

Wilkinson B, Micklefield J. Mining and engineering natural-product biosynthetic pathways. Nat. Chem. Biol. 2007;3:379–386.

Clardy J, Fischbach MA, Walsh CT. New antibiotics from bacterial natural products. Nat. Biotechnol. 2006;24:1541–1550.

Fischbach MA, Walsh CT. Assembly-line enzymology for polyketide and nonribosomal peptide antibiotics: logic, machinery, and mechanisms. Chem. Rev. 2006;106: 3468–3496.

PART III

NATURAL PRODUCTS IN MEDICINE

12

PHARMACEUTICALS: NATURAL PRODUCTS AND NATURAL PRODUCT MODELS

Sнео B. Singh

Merck Research Laboratories, Rahway, New Jersey

Natural products have played a vital role in the treatment of human ailments for thousands of years and continue to play a big role in the modern discovery of new agents for the treatment of diseases today. In certain therapeutic areas, natural products account for almost all key modern medicine used today. Many drugs are formulated and used directly as they are found in nature, some are derived directly from natural products by semisynthesis, and others are modeled after natural products. In this overview, examples of the pharmaceutically important natural products have been summarized.

12.1 INTRODUCTION

Natural product preparations have played a vital role in empiric treatment of ailments for thousands of years in many advanced civilizations and continue to play a significant role even today in various parts of the world. The use of plants and preparations derived from plants has been the basis for the sophisticated medical treatments in Chinese, Indian, and Egyptian civilizations for many thousands of years. These medical applications have been documented in the Chinese *Materia Medica* (1100 BC), in the Indian *Ayurveda* (1000 BC), and in Egyptian medicine as early as 2900 BC. Plant preparations continued to be the basis of medical treatments in the ancient Western world as well. This knowledge migrated through

Natural Products in Chemical Biology, First Edition. Edited by Natanya Civjan.
© 2012 John Wiley & Sons, Inc. Published 2012 by John Wiley & Sons, Inc.

Greece to Western Europe, including England, during the ancient period and led to its formal codification in the United Kingdom and to the publication of the *London Pharmacopoeia* in 1618.

The isolation of strychnine (1), morphine (2), atropine (3), colchicine (4), and quinine (5) in the early 1800s from the commonly used plants and their use for the treatment of certain ailments might constitute the early idea of "pure" compounds as drugs. E. Merck isolated and commercialized morphine (2) as the first pure natural product for the treatment of pain (1–3). Preparations of the Willow tree have been used as a painkiller for a long period in traditional medicine. Isolation of salicylic acid (6) as the active component followed by acetylation produced the semisynthetic product called "Aspirin" (7) that was commercialized by Bayer in 1899 for the treatment of arthritis and pain (4).

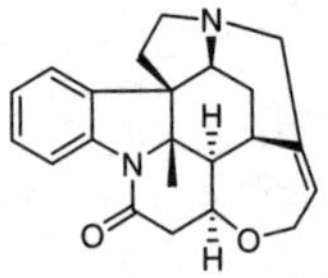

Strychnine (1) 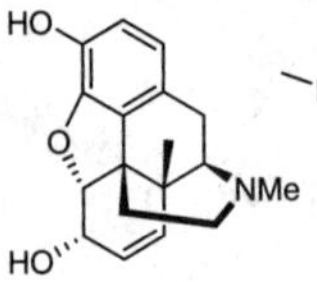   Morphine (2) 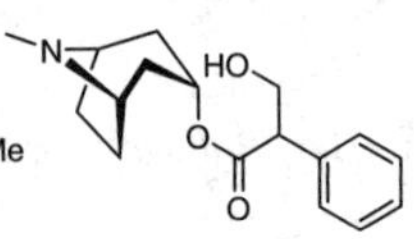   Atropine (3)

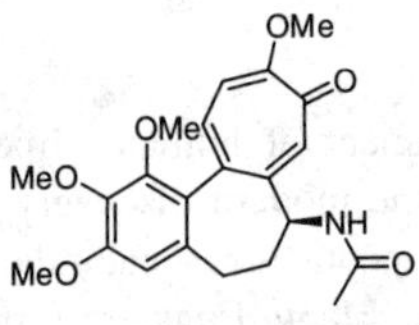

Colchicine (4) 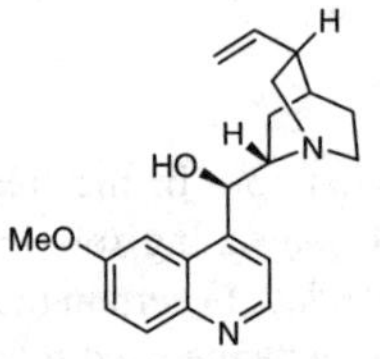   Quinine (5) 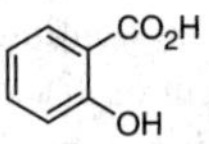   Salicylic acid (6) 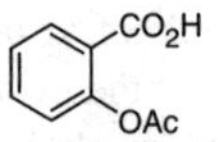   Aspirin (7)

The World Health Organization estimates that herbal and traditional medicines, derived mostly from plants, constitute primary health care for ~80% of the world population even today. The compounds produced by plants play significant roles in the treatment of diseases for the rest of the 20% of populations that are fortunate to use modern medicine. About 50% of the most prescribed drugs in the United States consist of natural products or their semisynthetic derivatives, or they were modeled after natural products. "Curare," the crude extract from the South American plant, *Chondodendron tomentosum*, and the derived purified compound tubocurarine, has been used as anesthetic in surgery until recently. Purified digitoxin, as well as the crude extracts that contain digitalis glycosides from foxglove plant, Digitalis lanata, is used as cardiotonic even today.

Drugs derived from microbial fermentations have played perhaps a bigger role in the modern drug discovery and have revolutionized the practice of medicine, which leads to saving human lives. Although the contribution of purified natural products as single agent drugs is significant in almost all therapies, their contribution in the treatment of bacterial infection is perhaps most critical (5). Natural products constitute drugs or leads to all but three classes of antibiotics. The

discovery of microbial natural products-based antibiotics began with the serendip-
itous observation by Fleming in 1929 that bacterial growth was prevented by the
growth of *Penicillium notatum*. Although this discovery was highly publicized
and very important, it took over 10 years before the active material, penicillin,
was purified and structurally elucidated by Florey and Chain in early 1940s. Subsequent commercialization was very quick, driven largely by the medical needs
of World War II. Penicillin was one of the first broad-spectrum antibiotics that
treated bacterial infection and saved millions of lives. Fleming, Florey, and Chain
were awarded the Nobel Prize in 1945 for their efforts on penicillin. The success of penicillin led to unparalleled efforts by government, academia, and the
pharmaceutical industry to focus drug-discovery efforts based on the newfound
"microbial" sources for the discovery of natural products beyond plants. However, initial efforts were mostly focused on the discovery of antibiotic compounds
from fermentations of a variety of microorganisms of not only fungal origin by
also soil-dwelling prokaryotes (e.g., *Streptomyces* spp.), which led to the discovery by 1962 of almost all novel classes of antibiotic scaffolds that are being used
today. The antibiotic discovery effort was performed largely by Fleming's method
of detection of antibacterial activity on petri plates. Zones of inhibition of bacterial strains on agar plates were measured after applying whole broth or extracts
obtained from microbial ferments (5). As newer biological assays and screening
techniques became available in the 1960s, microbial sources, along with plant and
marine sources, started to be used for screening against other therapeutic targets,
which led to the discovery of leads and drugs in those areas. Examples of these
discoveries will be discussed with the target areas. As time progressed, improved
technologies in biology and chemistry helped with the popularization of natural
products; natural product extracts became part of the screening resource in most
large pharmaceutical houses from 1960 through the 1980s until their de-emphasis
in the early 1990s. Therefore, natural product extracts became popular sources
for the screening against purified enzymes and receptors, an occurrence that led
to the identification of many nonantibiotic natural products that have revolutionized the practice of medicine, saved countless human lives, improved quality of
life, and perhaps helped increase life expectancy for humans.

12.2 ANTIBACTERIAL AGENTS

Natural products contribute to over 80% of all antibiotics that are in clinical practice today. Natural products contribute to all but three classes of antibiotics (3).

Penicillin was the first β-lactam and the first broad-spectrum antibiotic discovered that started the "Golden age" (1940–1962) of antibiotics. The structure
of penicillin contains a thiazolidine ring that is fused to a β-lactam ring. The
existence and stability of the β-lactam ring was highly controversial at the time
despite the availability of a single crystal X-ray structure of one of the penicillins. Penicillin G (**8**) was the first penicillin that was clinically used. Penicillin

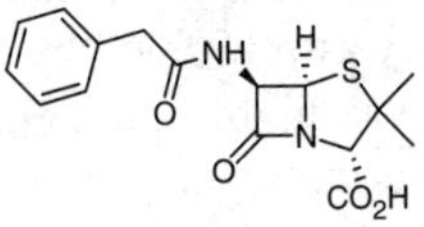

Penicillin G (8) 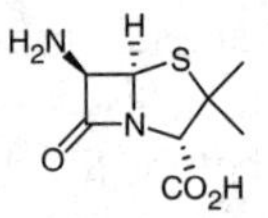 6-Amino-penicillin acid (9) 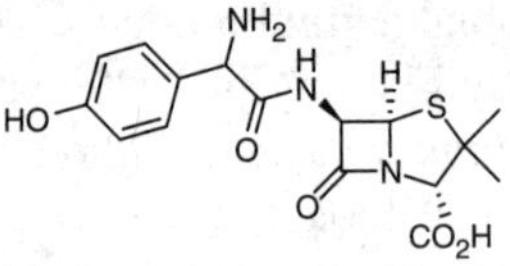 Amoxicillin (10) 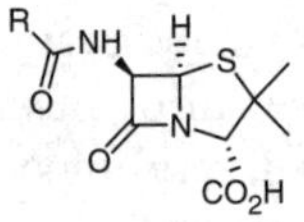 Penicillins (11)

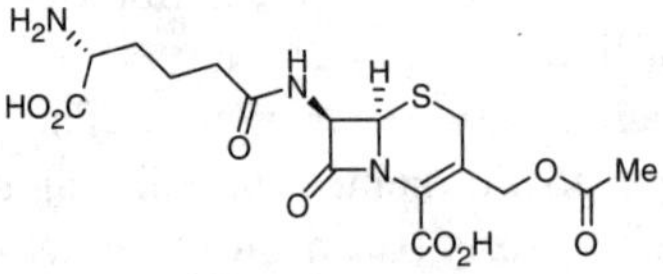

Cephalosporin C (12) 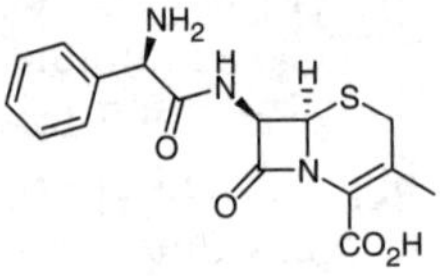 Cephalexin (13) 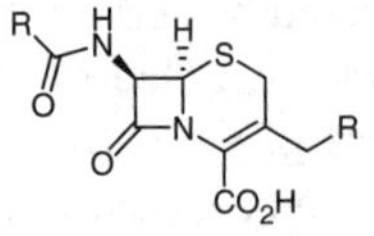 Cephalosporins (14)

G was converted easily by either chemical or biochemical means to 6-amino-penicillanic acid (9), which became the lead for the semisynthetic modifications that led to the synthesis of various penicillin derivatives. Some early derivatives (e.g., amoxicillin 10) are still in clinical use. Penicillins (general structure 11) bind to penicillin-binding proteins and inhibit the bacterial cell wall. Penicillins became targets of β-lactamases that opened the β-lactam ring and abolished the antibacterial effectiveness of these compounds (5, 6).

Cephalosporin C (12), a second class of β-lactam antibiotics, was first discovered from *Cephalosporium acremonium*, isolated from a sewer outfall of Sardinia, Italy in 1948. Although Cephalosporin C was less active than penicillin G, it was less prone to β-lactamase action and therefore attracted a lot of attention that led to the development of five generations of orally active clinical agents (e.g., cephalexin, 13 and general structure, 14) (5, 6).

Continued search for even better antibiotics led to the discovery of the highly potent and broadest-spectrum antibiotic thienamycin (15), the third class of the β-lactams, called carbapenems, in which the sulfur atom of the thiazolidine ring was replaced by a methylene group. Thienamycin was produced by *Streptomyces cattleya* (7). The primary amine group of thienamycin self-catalyzes the opening of the β-lactam ring, which leads to the concentration-dependent instability that poses a serious challenge for the fermentation-based production of the compound. The Merck group stabilized the compound by replacing the primary amine with an aminomethylidineamino group and synthesized imipenem (16) (8). They developed a highly efficient total synthesis that remains in commercial use today. Imipenem was approved for clinical use 23 years ago in 1985, but it remains one of the most important broad-spectrum hospital antibiotics in the market today. Like other β-lactams, several generations of carbapenems (general structure 17) have been approved for clinical use in recent years (9).

As resistance to β-lactam antibiotics increased because of the expression of a variety of β-lactamases, many groups focused their efforts on discovering compounds that could be more reactive to β-lactamases without having significant intrinsic antibiotic activities of their own and pharmacokinetic properties

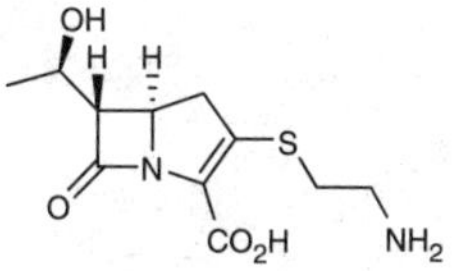

Thienamycin (15)

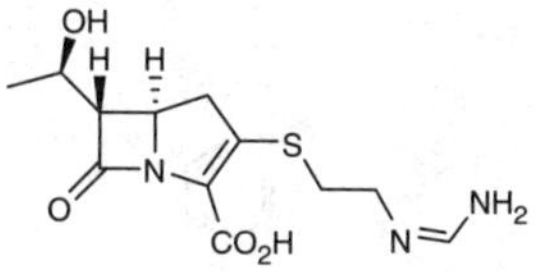

Imipenem (16)

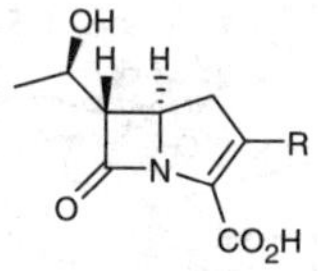

Carbapenems (17)

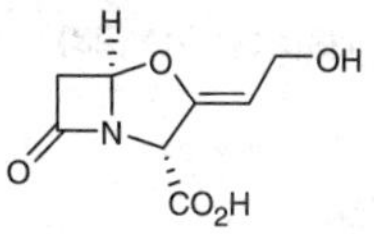

Clavulanic acid (18)

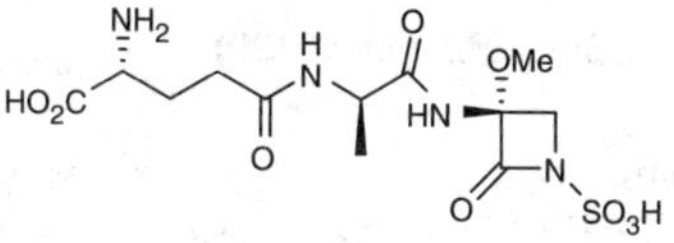

sulfazecin (19)

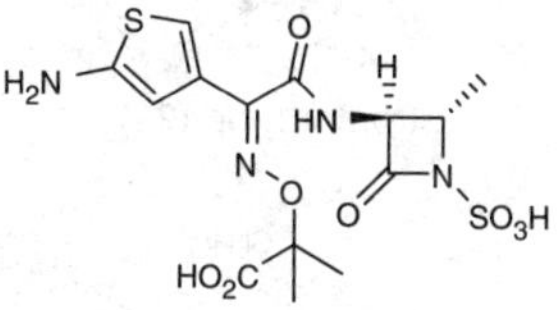

Aztreonam (20)

that would be similar to β-lactam antibiotics. This focus led to the discoveries of clavulanic acid (18) and monobactam sulfazecins (19). Nature effectively stabilized the latter monobactam structure by the addition of a *N*-sulfamic acid. The β-lactamase inhibitor clavulanic acid was combined with amoxicillin, which led to the development of a potent and successful antibacterial agent, Augmentin® (GSK, Surrey, UK) (10). Chemical modifications of the monobactam produced aztreonam (20), a clinical agent with a narrow spectrum but significantly improved activity against Gram-negative pathogens, particularly *Pseudomonas aeriginosa* (11).

Immediately after the discovery of penicillin, Waksman started efforts on soil-dwelling bacteria and discovered the first of the aminoglycosides, streptomycin (21) from *Streptomyces griseus*, in 1943 (6). Subsequently, a series of aminoglycosides was isolated. These aminoglycosides are potent broad-spectrum antibiotics and are potent inhibitors of protein synthesis. Unfortunately, nephrotoxicity limited their wider use, and they are used mainly for treatment of infections caused by Gram-negative bacteria. Continued efforts to screen prokaryotic organisms led to the discovery of the phenyl propanoids (chloramphenicol, 22) and tetracyclines. The latter is a major class of tetracyclic polyketides that were discovered from various species of *Streptomyces* spp. Although the parent tetracycline (23) was not used as an antibiotic to a great extent, the chloro derivative (Clortetracycline 24), oxytetracycline (25), and minocycline (26) are clinical agents. This class of compound suffered from the selection for rapid resistance via efflux mechanism that limited their use (6). Recently, however, chemical modifications of the A-ring yielded compounds that overcame the efflux pump and lead to the development of tigecycline (27) as an effective broad spectrum antibiotic (12).

Another large class of orally active protein syntheis inhibitor antibiotics that were produced by *Streptomyces* spp. is represented by 14-membered lactones generically callsed macrolides, exemplified by the first member, erythromycin (28) (6). Chemical modifications of this class of compounds led to many clinical

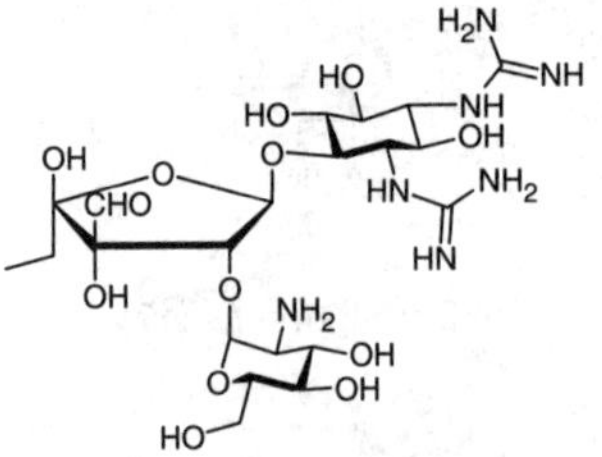

Streptomycin (21) 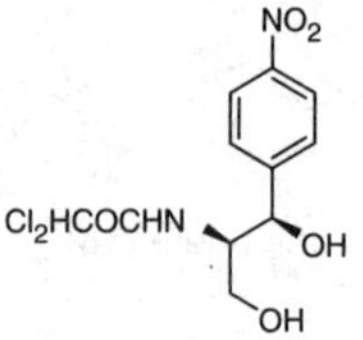 Chloramphenicol (22) 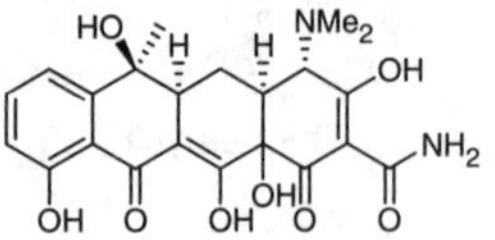 Tetracycline (23)

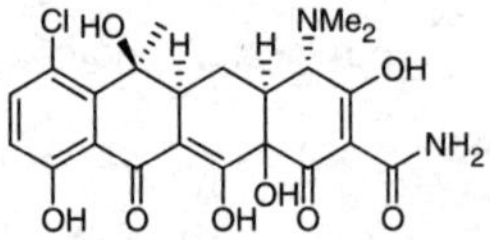

Chlortetracycline (24) 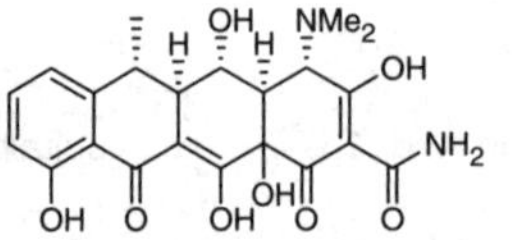 Doxycycline (25)

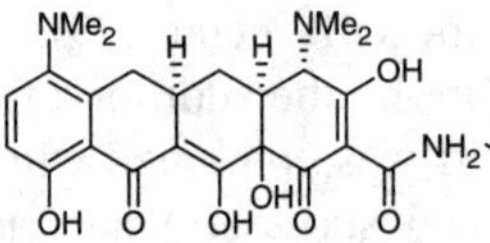

Minocycline (26) 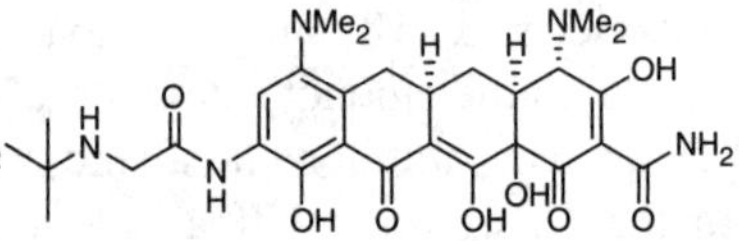 Tigecycline (27)

agents such as the aza derivatives, azithromycin (29) and ketolide (telithromycin, 30) (13, 14). Mupirocin (pseudomonic acid, 31) is another protein synthesis inhibitor that was isolated from *Pseudomonas fluorescens* and is used only as a topical agent (6).

Vancomycin (32), a glycopeptide produced by *Streptomyces orientalis*, is a key Gram-positive antibiotic, originally discovered in 1954, and remains a critical antibiotic in clinical practice even today for the treatment of Gram-positive bacterial infections (6). Teicoplanin (33), a related glycopeptide produced by *Streptomyces teicomyceticus*, is a newer antibiotic that complements vancomycin in the clinic but is not effective against vancomycin-resistant bacteria. Ramoplanin (34) represents another glycopeptide that is larger in molecular size and structurally different from vancomycin and teicoplanin; it is in the late stages of clinical development for treatment of Gram-positive bacterial infections. Glycopeptides inhibit the bacterial cell wall. Daptomycin (35), a cyclic lipopeptide produced by *Streptomyces roseosporus*, is one of the newest members of antibiotics approved for the clinical practice as a broad-spectrum Gram-positive agent. It works by depolariztion of the bacterial cell membrane (14). Streptogramins were discovered in the early 1960s but were used for humans only recently when a 70/30 mixture of dalfopristin (36) and quinupristin (37) with the trade name Synercid® King Pharmaceuticals, Bristol, NJ; was developed for the treatment of drug-resistant Gram-positive bacterial infections (15).

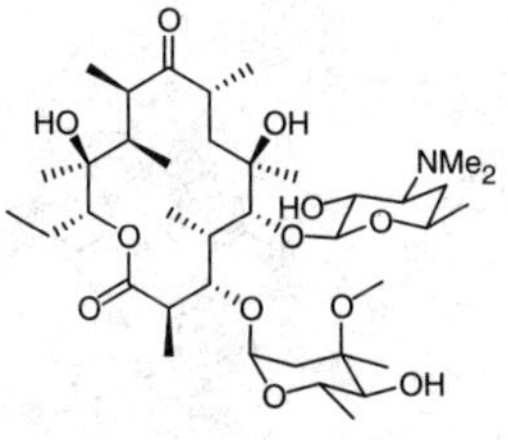

Erythromycin (**28**)

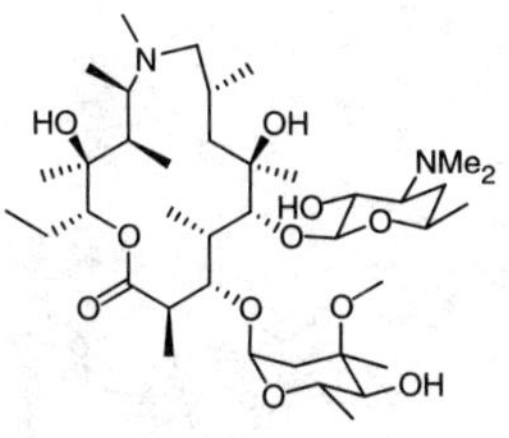

Azithromycin (**29**)

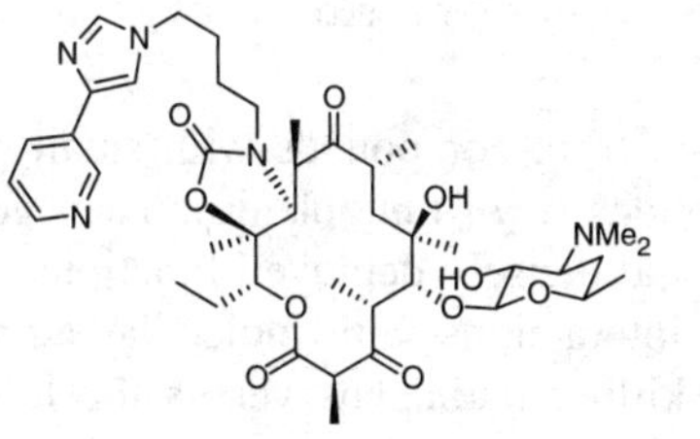

Telithromycin (**30**)

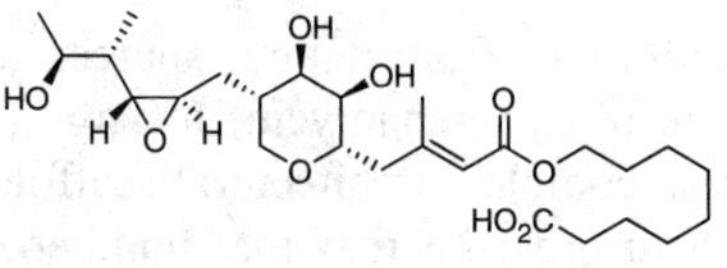

Mupirocin (**31**)

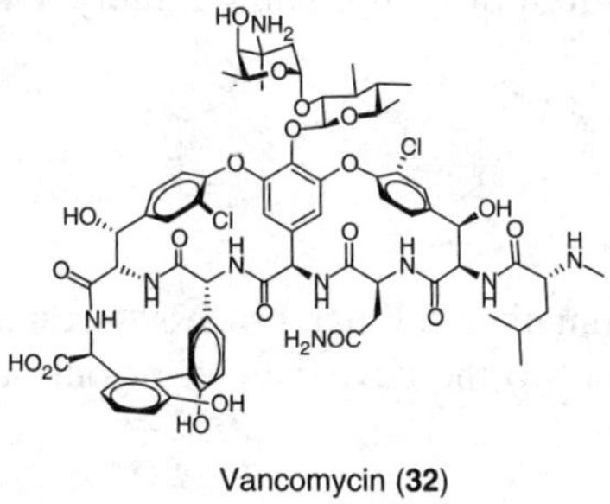

Vancomycin (**32**)

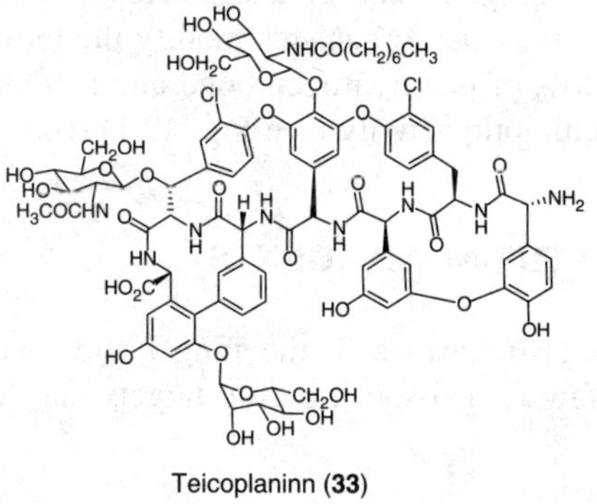

Teicoplaninn (**33**)

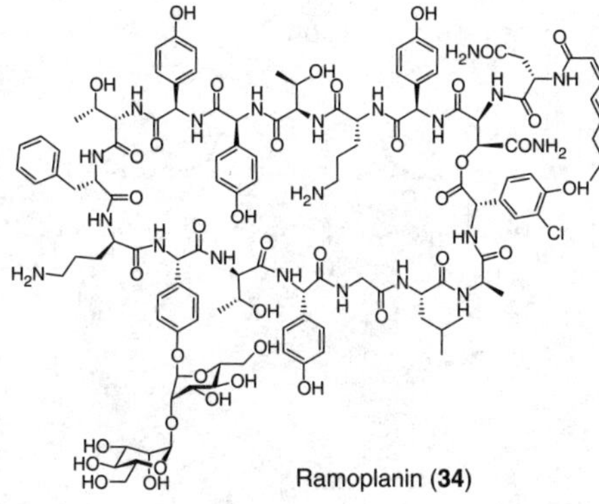

Ramoplanin (**34**)

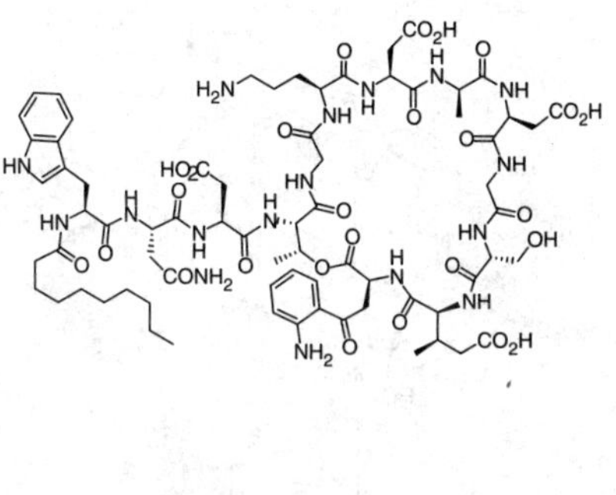

Daptomycin (**35**)

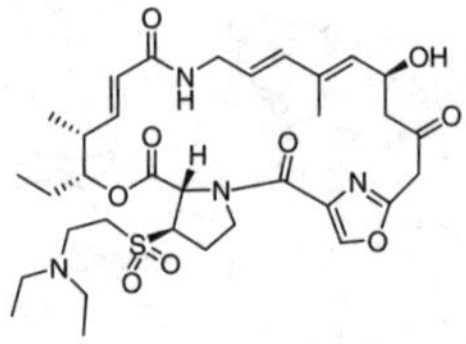

Dalfopristin (36) 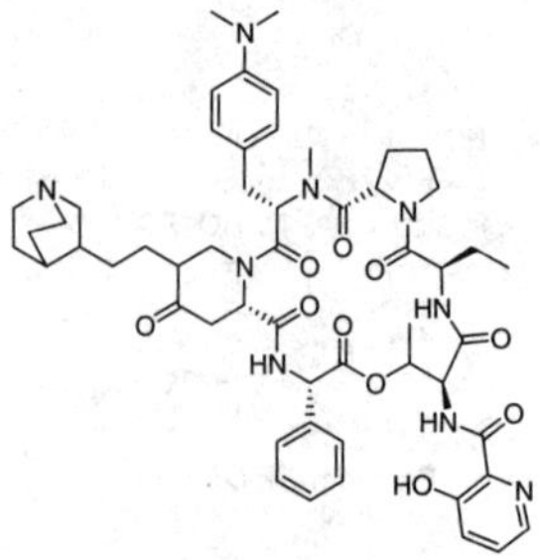   Quinupristin (37)

Antibiotics from natural sources range from compounds with small molecular size (e.g., thienamycin) to large peptides (e.g., ramoplanin). They generally possess complex architectural scaffolds and densely deployed functional groups, which affords the maximal number of interactions with molecular targets and often leads to exquisite selectivity for killing pathogens versus the host. This function is nicely illustrated by vancomycin binding to its target. Vancomycin has five hydrogen bond contacts with the D-Ala-D-Ala terminal end of peptidoglycan. Resistant organisms modify the terminal D-Ala with *D*-lactate, which leads to loss of one hydrogen bond and a 1000-fold drop in binding affinity and loss of antibiotic activity (see Fig. 12.1) (6).

12.3 ANTIFUNGAL AGENTS

Significant similarities in the fungal and mammalian cellular processes result in very few fungal-specific drug targets that lead to the development of only a

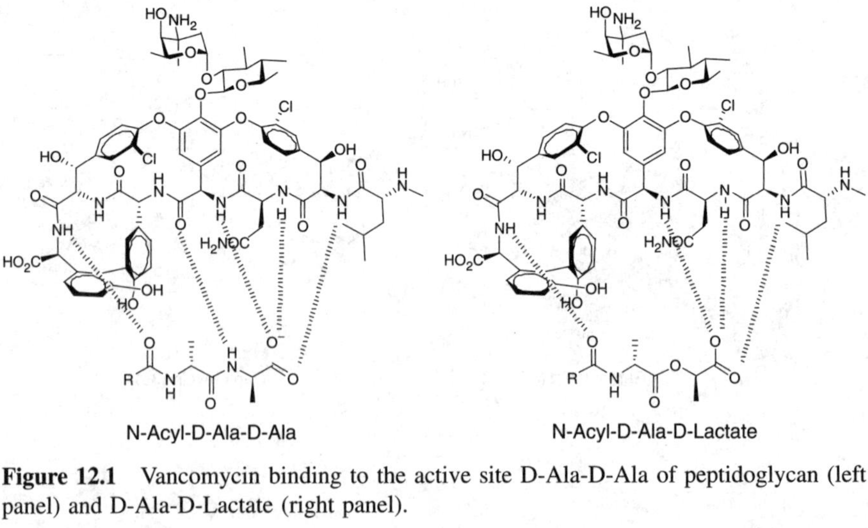

N-Acyl-D-Ala-D-Ala   N-Acyl-D-Ala-D-Lactate

Figure 12.1   Vancomycin binding to the active site D-Ala-D-Ala of peptidoglycan (left panel) and D-Ala-D-Lactate (right panel).

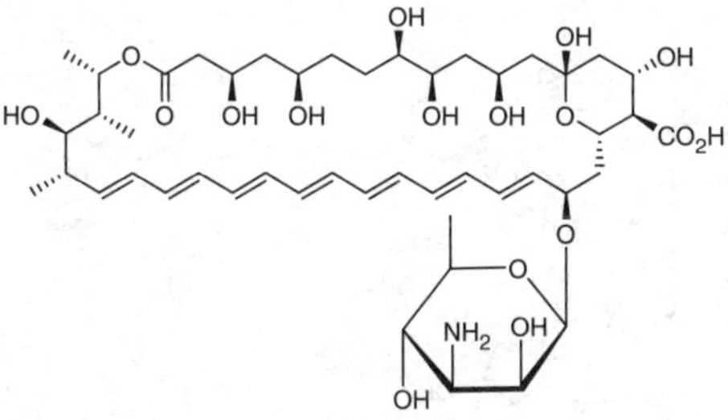

Amphotericin B (**38**)

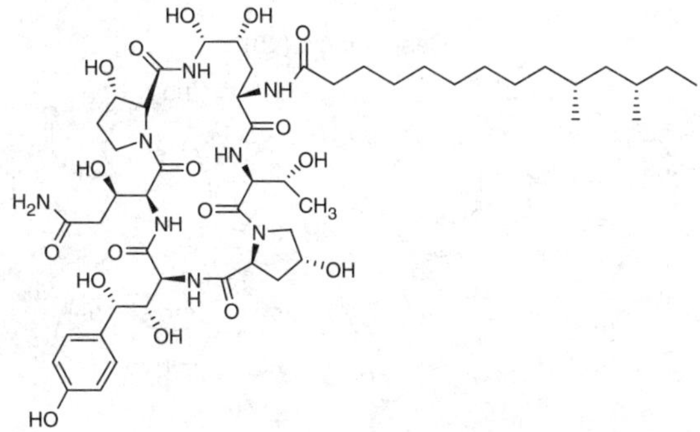

Pneumocandin Bo (**39**)

few quality and safe antifungal agents. Amphotericin B (**38**), a natural product that consists of a polyene lactone, is a highly effective broad-spectrum anti-fungal agent unfortunately with a very limited safety margin. Glucan synthesis was identified as a fungal-specific target that could be inhibited by a series of cyclic peptides called echinocandins, which were identified in the 1970s as hav-ing potent antifungal activities (16, 17). This identification provided impetus for the discovery and development of new related lipopeptides that led to the identi-fication of pneumocandins from *Glarea lozoyensis* (e.g., pneumocandin Bo, **39**). Chemical modifications at two sites of pneumocandin Bo led to the synthesis of caspofungin (**40**), which was the first in the class of glucan synthesis inhbitors; it is a "potent," highly effective, and safe antifungal agent approved for serious fungal infections in hospitals. Side chain replacements of the related cyclic pep-tide FR901379 (isolated from the fungus *Coleophoma empetri*) and echinocandin B (isolated from *Aspergillus nidulans*) led to two additional clinical agents of this class, micafungin (**41**) and anidulafungin (**42**), respectively. (16, 17)

12.4   ANTIMALARIALS

Quinine (**5**) isolated from Cinchona bark was one of the first antimalarials dis-covered, and it became a model for the discovery and development of some of

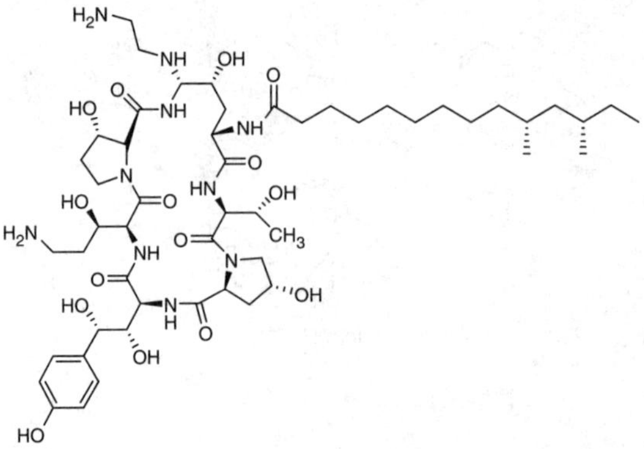

Caspofungin (**40**)

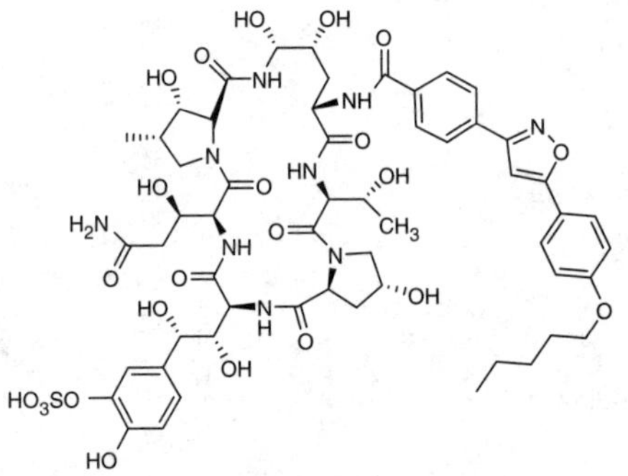

Micafungin (**41**)

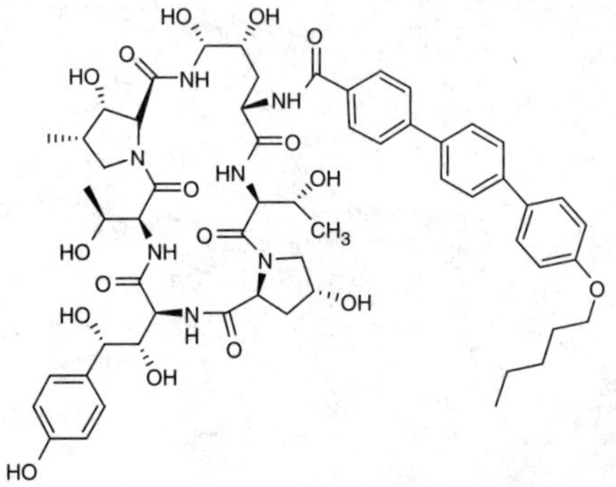

Anidulafungin (**42**)

the most successful antimalarial agents, chloroquine and its successors. However, the development of resistance by the malarial parasite *Plasmodium falciparum* for these drugs has rendered them ineffective. Artemisinin (**43**), a sesquiterpene peroxide originally isolated from a Chinese herb *Artemisia annua* in 1972 as an antimalarial agent, was chemically modified to a derivative, artemether (**44**), which is a very effective and widely used antimalarial agent (18). Unfortunately, limited supply of this plant-derived compound rendered it inaccessible for wider use. Biosynthetic genes of artimisinin have been identified and successfully transfected to an heterologous host, *Escherichia coli*. This method has allowed the production of an intermediate, amorphadiene (**45**) and artemisinic acid (**46**), which could be transformed chemically to artmether and potentially could relieve the strain of supply and could provide wider availability (19–21).

12.5   ANTIVIRALS

Most antiviral agents are based on nucleoside structures and have their origin from spongouridine (**47**) and spongothymidine (**48**) that were isolated from marine sponges in the 1950s by Bergmann and his coworkers (22–24). These natural

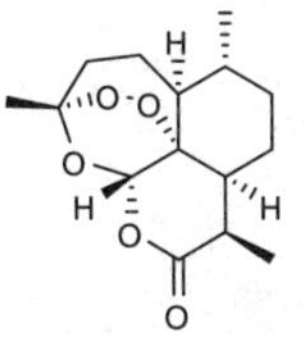

Artemisinin (43) 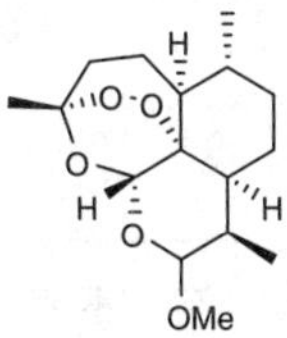   Artemether (44) 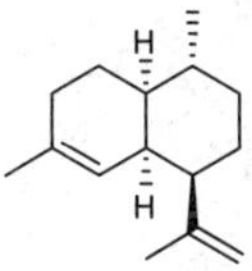   Amorphadiene (45) 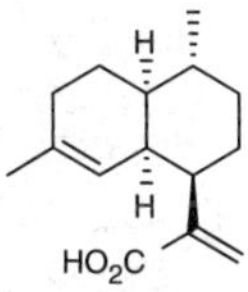   Artemisinic acid (46)

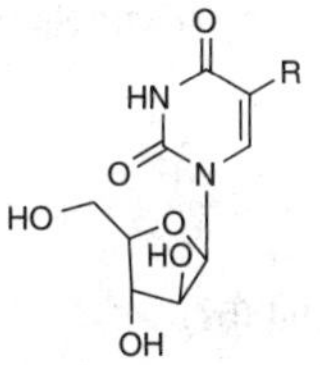

Spongouridine, R = H (47)   Ara-A (49) 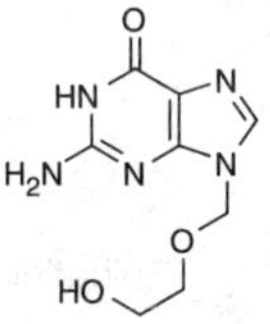   Acyclovir (50)
Spongothymidine, R = Me (48)

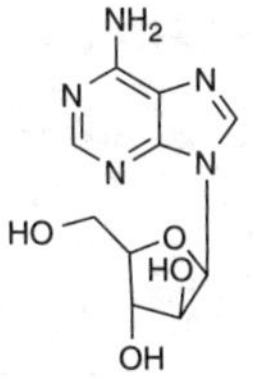

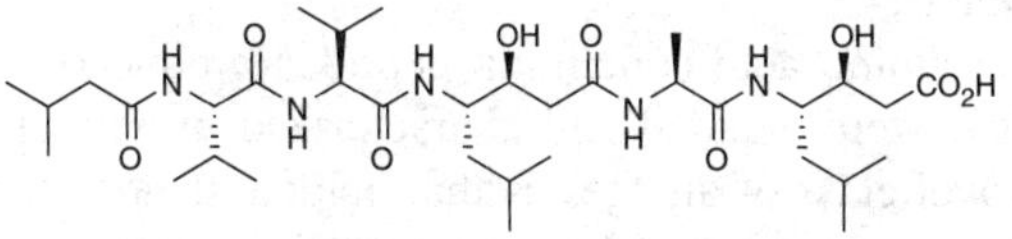

Pepstatin (51) 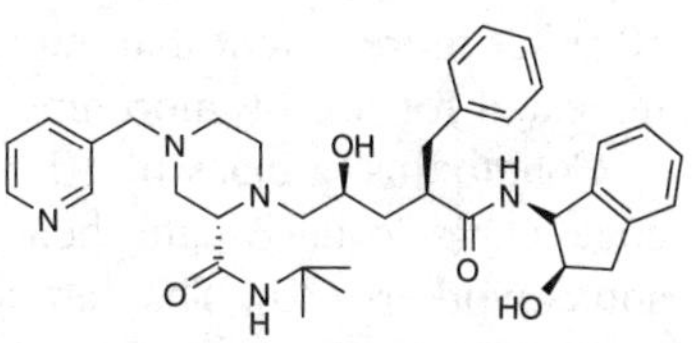   Indinavir (52)

nucleosides possessed sugars other than ribose and deoxyribose and provided rationale for the substitution of the sugars in the antiviral nucleosides with various sugar mimics, including linear polar groups that led to the synthesis of Ara-A (49) and acyclovir (50). HIV protease inhibitors were developed from pepstatin (51), a pepsin inhibitor produced by various fungal species. Pepstatin possesses as the structural component statine a β-hydroxy-γ-amino acid that mimics the transition-state intermediate of the hydrolytic reaction catalyzed by the proteases (25). This structure became the foundation for the rational peptidomimetics and design of all HIV protease inhibitors, for example, indinavir (Crixivan®, Merck & Co., Inc., Whitehouse Station, NJ; 52) and others (26).

12.6   ANTIPAIN AGENTS

Use of the opium poppy (*Papaver somniferum*) to ameliorate pain dates back thousands of years, and the active metabolite morphine (2) was isolated first from its extracts in 1806 followed by codeine (53) in 1832 (27, 28). Morphine and its derivatives are agonists of opiate receptors in the central nervous system and are some of the most effective pain relievers known and prescribed for postoperative pain. Morphine and codeine differ by substitution by methyl ether. Unfortunately, addictive properties of these compounds limit their use. Efforts have been made to reduce the addictive properties of morphine, which resulted in a semisynthetic derivative buprenorphine (54) (29). This compound is 25 to

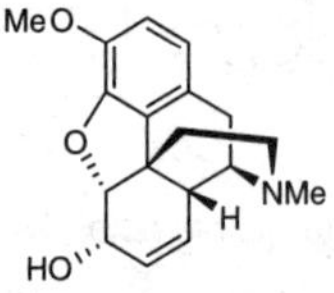

Codeine (53) 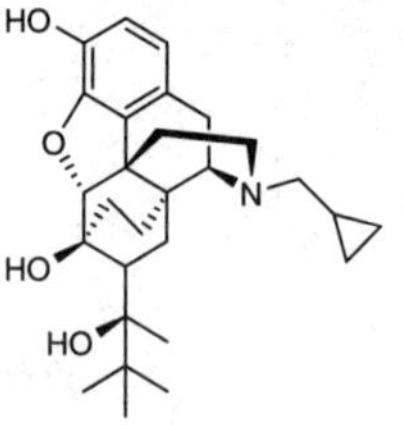 Buprenorphine (54) 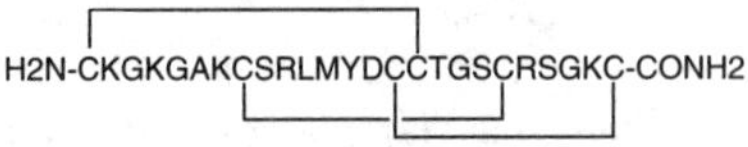 Ziconotide (55)

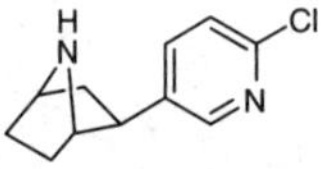

H2N-CKGKGAKCSRLMYDCCTGSCRSGKC-CONH2

Epibatidine (56) 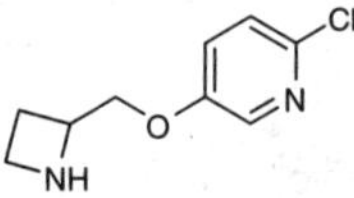 ABT594 (57)

50 times more potent than morphine with lower addictive potential and has been indicated for use by morphine addicts.

Conotoxins, a class of 10 to 35 amino acid-containing peptides produced by cone snails to intoxicate their prey, were isolated and characterized by Olivera and coworkers (30). They are a novel class of analgesics that helped identify the target and blocking of N-type Ca^{+2} channels. One compound, Ziconotide (55), was synthesized and developed as a treatment for severe chronic pain (31, 32).

The alkaloid epibatidine (56) was discovered from the skin of an Ecuadorian poison frog (*Epipedobates tricolor*), and its potent analgesic activity was demonstrated as early as in 1974 by Daly and coworkers (33–35). The paucity of the material delayed the structure elucidation and was only accomplished after the invention of newer and more sensitive NMR techniques in 1980s. Once the structure was elucidated as chloronicotine derivative, it was synthesized and was shown to antagonize nicotinic receptors in neurons. It did not show any specificity with similar receptors in other tissues and lacked a therapeutic index of any clinical value. However it served as a model for designing compounds with desired specificity and resulted in the synthesis of ABT594 (57), which is an agonist of the nicotinic acetylcholine receptor, is about 50-fold more potent than morphine without addictive properties, and was under advanced clinical development until its discontinuance in 2003 (36–39).

12.7 ANTIOBESITY

Lipstatin (58), comprised of a 3,5-dihydroxy-2-hexyl hexadeca-7,9-dienoic acid with cyclization of the hydroxy group at C-3 with the carboxylic acid to form a β-lactone and esterification of the hydroxy group at C-5 with a N-formyl leucine, was isolated from *Streptomyces toxytricini*. It inhibits gastric and pancreatic lipase and blocks intestinal absorption of lipids (40, 41). Reduction of the olefins led to the synthesis of tetrahydrolipstatin, which was approved in 1999 as Xenical (59) for the treatment of obesity (42–44).

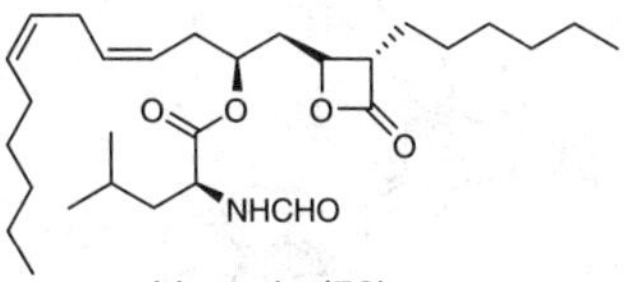

Lipstatin (58) 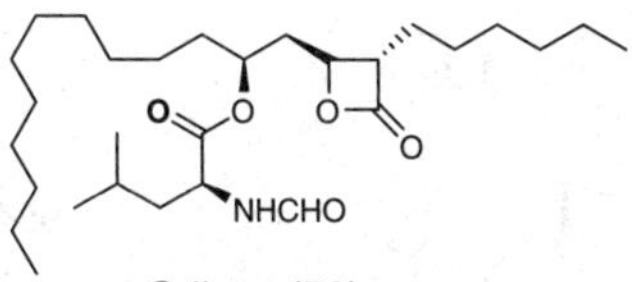 Orlistat (59) 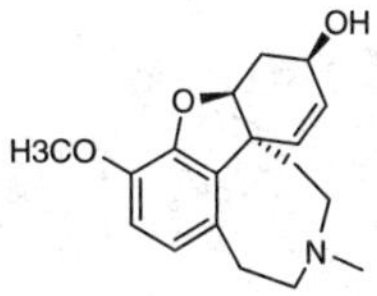 Galantamine (60)

12.8 ALZHEIMER'S AGENTS

Galantamine (60) is a tetracyclic alkaloid that was isolated originally from *Galanthus nivalis* and subsequently from *Narcissus* spp. (45). It was approved by trade name Reminyl® Johnson & Johnson, New Brunswick, NJ for the treatment of Alzheimer's disease. That galantamine, a selective inhibitor of acetylcholinesterase, was confirmed by X-ray structural characterization of galantamine bound to a plant acetylcholinesterase (46, 47).

12.9 ANTINEOPLASTIC AGENTS

Natural products have played a much bigger role, perhaps second only to antibacterial agents, in the discovery and development of anticancer agents either directly as drugs or leads to drugs (48). In fact, they contribute 64% of all approved cancer drugs (3). Taxol® BMS, NY (61, paclitaxel) is unarguably the most successful anticancer drug in clinical use. It is a taxane diterpenoid that was isolated from the Pacific yew *Taxus brevifolia* in 1967 as a cytotoxic agent but not pursued as a development candidate until its novel mode of action was determined in 1979 as a stabilizer of microtubule assembly (49–52). The discovery of the mechanism of action led to the United States National Cancer Institute (NCI) committing significant resources to the large-scale production, eventual clinical development, and approval of Taxol® (paclitaxel) by U.S. Food and Drug Administration in 1992 for treatment of breast, lung, and ovarian cancer. Another important plant product, camptothecin (62), an alkaloid from *Camptotheca acuminata*, was discovered in 1966, also by the Wall and Wani group (53). The development of this compound was also hampered by the lack of knowledge of the mechanism of action and most importantly by its extremely poor water solubility. Determination of the mechanism of action as an inhibitor of topoisomerases-I led to significant efforts both by NCI and the pharmaceutical industry, which resulted in chemical modification of the structure, introduction of water-solubilizing groups (e.g., amino), and development of several derivatives as anticancer agents exemplified by topotecan (63) and irinotecan (64) (54, 55).

The Vinca alkaloids, vinblastine (65) and vincristine (66), isolated from *Catharanthus roseus*, have contributed significantly to the understanding and treatment of cancer (56, 57). These compounds bind to tubulin and inhibit cell division by inhibiting mitosis; they were perhaps the best-known anticancer agents before

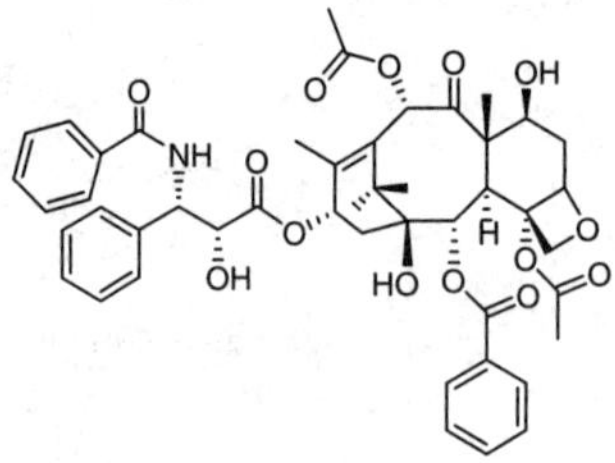

Taxol (**61**) 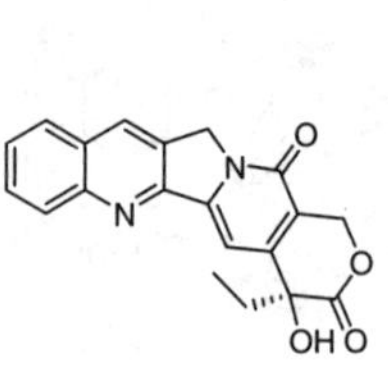   Camptothecin (**62**) 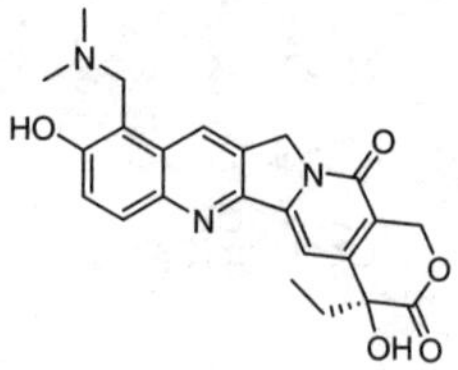   Topotecan (**63**)

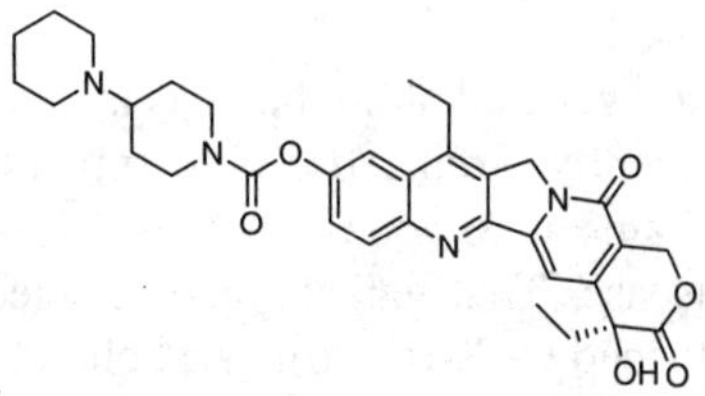

Irinotecan (**64**)

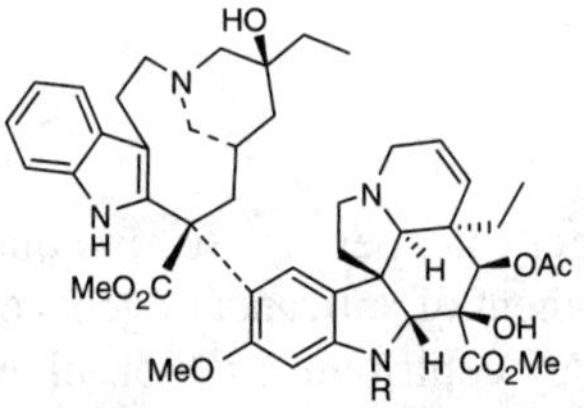

Vinblastine, R = Me(**65**)
Vincristine, R = CHO(**66**)

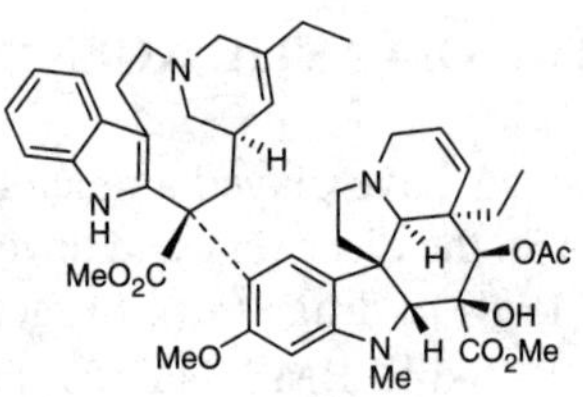

Vinorelbine (**67**)

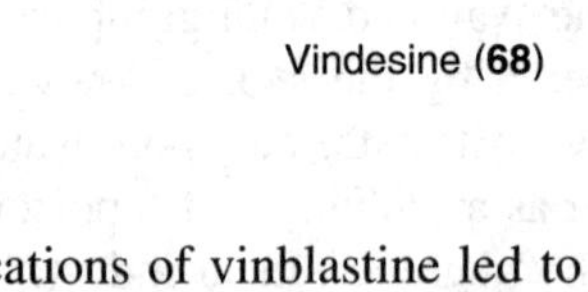

Vindesine (**68**)

Taxol®. Chemical modifications of vinblastine led to the clinical agents vinorelbine (**67**) (58) and vindesine (**68**) (48, 59–61). Podophyllotoxin (**69**) was isolated from various species of the genus *Podophyllum* spp. as an anticancer agent. Chemical modifications of the naturally occurring epimer, epipodophyllotoxin (**70**), led to the synthesis and development of etoposide (**71**) and teniposide (**72**) as clinical agents (48, 54, 62–65).

Combretastatin A4 phosphate (**73**) is a phosphate prodrug of combretastatin A4, a *cis*-stilbene, isolated from *Combretum caffrum* (66). Combretastatin A4

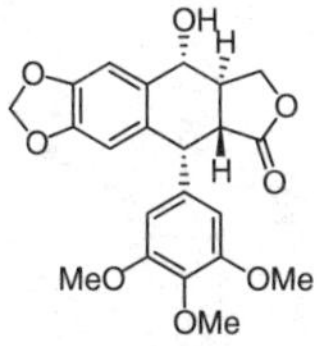

Podophyllotoxin (**69**) 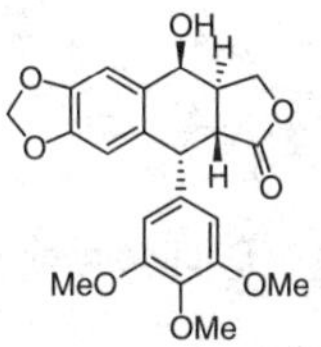   Epipodophyllotoxin (**70**) 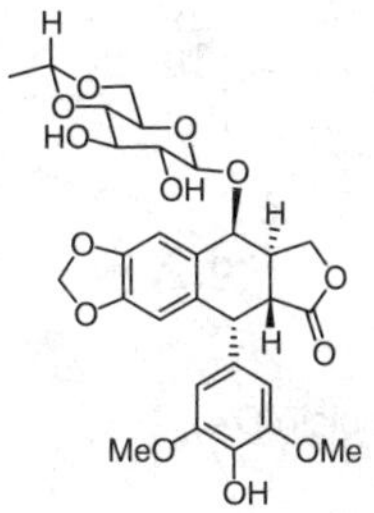   Etoposide(**71**) 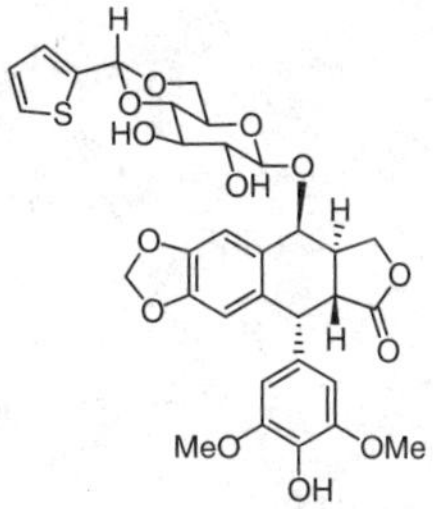   Teniposide (**72**)

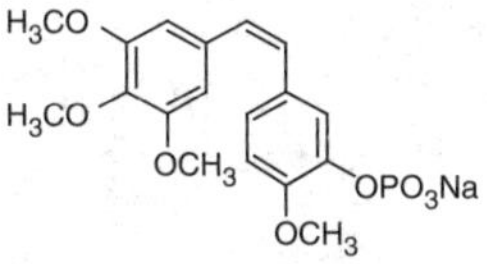

Combretastatin A4
phosphate (**73**)

is one of the many combretastatins that inhibits tubulin polymerization (67), shows efficacy against solid tumor, is a vascular targeting agent that blocks the blood supply to solid tumors, and is in Phase II/III clinical development for the treatment of various types of tumors as a vascular targeting agent (68–71).

Microbial sources have been a very rich source for cancer chemotherapeutic agents. Of particular note is the *Streptomyces* spp., which has been responsible for the production of many approved anticancer agents that are in clinical practice. These agents are represented by highly diverse structural classes exemplified by the anthracycline family (e.g., doxorubicin, **74**) (72–74), actinomycin family (e.g., dactinomycin, **75**), glycopeptides family (e.g., bleomycins A2 and B2, **76** and **77**) (75), and mitomycin family (e.g., mitomycin C, **78**) (72, 76). All these compounds specifically interact with DNA for their mode of action.

Staurosporine (**79**) produced by *Streptomyces* spp. is a potent inhibitor of protein kinase C (77–79). This compound inhibits many other kinases with almost equal potency and has become a great tool for the study of kinases. Lack of selectivity for protein kinase C has significantly hampered the development of this compound. Recently, however, several compounds derived from this lead have entered in the clinic for potential treatment of cancer. These include 7-deoxystaurosporine (**80**) and CGP41251 (**81**) (80, 81). CGP41251 shows multiple modes of action including inhibition of angiogenesis *in vivo*.

Microbial sources other than *Streptomyces* spp. have also provided highly interesting and structurally diverse compounds. Discovery of epothilones from myxobacterial strains by a German group (82) and the Merck group (83, 84) constitute a breakthrough discovery. The Merck group used an assay that mimicked Taxol® at the active site for the screening of natural products that led to the isolation of epothilones A (**82**) and B (**83**). The discovery of a unique structural class,

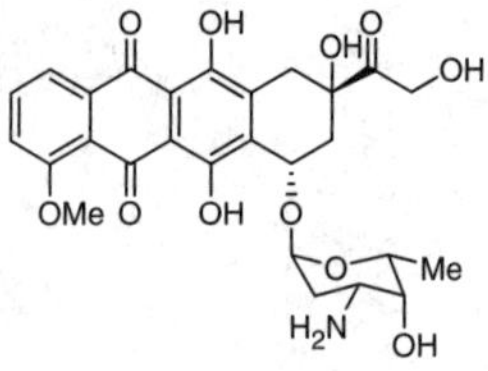

Doxorubicin (74)

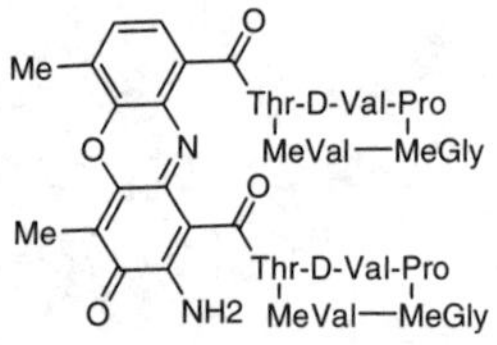

Dactinomycin (75)

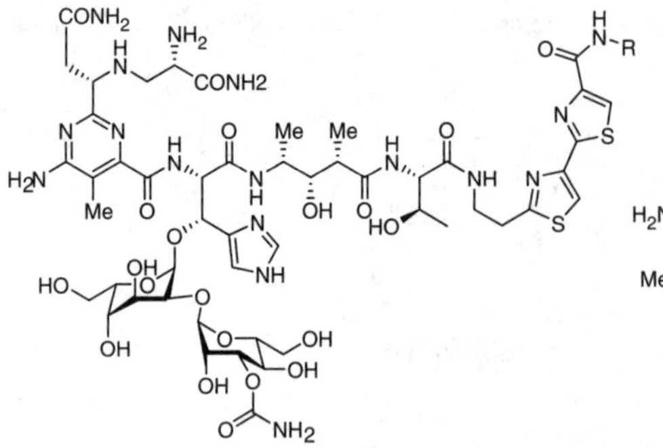

Bleomycin A$_2$, R=CH$_2$CH$_2$CH$_2$S+(CH$_3$)$_2$ (76)
Bleomycin B$_2$, R=CH$_2$CH$_2$CH$_2$CH$_2$NHC(NH)NH$_2$, (77)

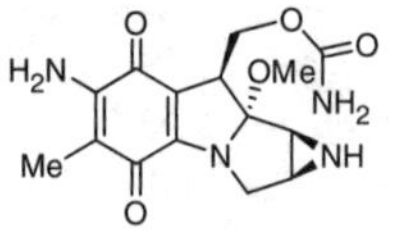

Mitomycin C (78)

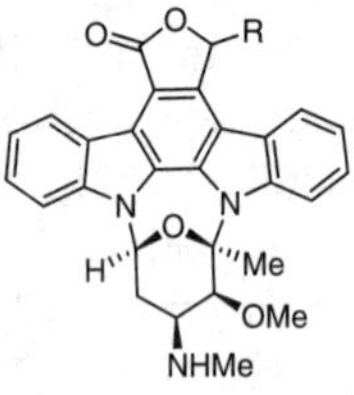

Staurosporine,R = H (79)
7-Hydrxystaurosporine, R =OH (80)

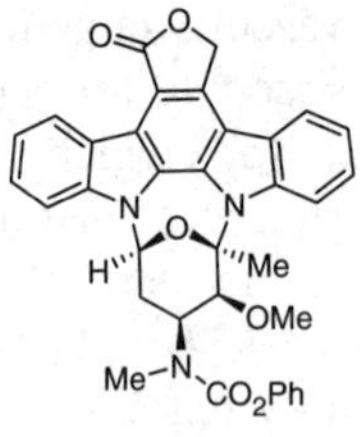

CGP41251 (81)

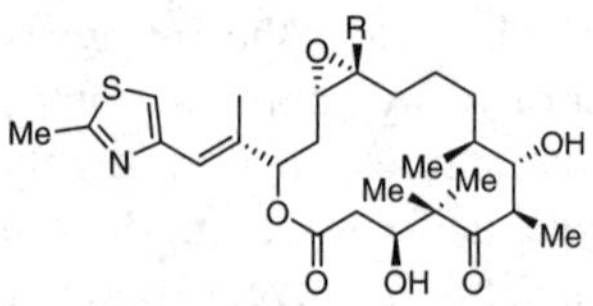

Epothilone A, R = H (82)
Epothilone B, R = Me (83)

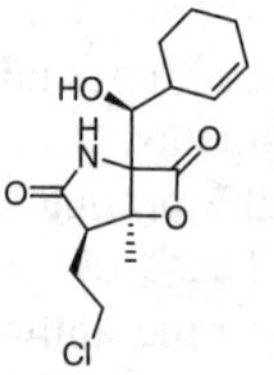

Salinosporamide A (84)

interesting biological activity, and clinically proven mode of action drew significant attention from the scientific community and led to a variety of approaches, including combinatorial biosynthesis, chemical modifications, and total synthesis, that permitted preparation of many derivatives with improved potency and drug-like properties. A series of these compounds have entered human clinical trials, and many are in the late stages of development. Epothilone discovery and development has been reviewed (85).

Recent pursuit of marine microbial sources led to the isolation of salinosporamide A (84). It is a β-lactone produced by the marine bacteria *Salinispora tropica* and is a proteasome inhibitor (86). Mechanistically, it works by specific covalent modification of the target. This compound has entered human clinical development for treatment of multiple myeloma (87–89).

Although use of marine microbial sources for the discovery of natural products is a somewhat recent phenomenon, marine natural products from higher species have contributed tremendously to the discovery of novel architecturally complex compounds as anticancer agent leads with one, Ecteinascidin-743, now approved in the European Union for treatment of sarcoma. The discovery of natural products derived from marine sources exploded in the 1970s not only because of increased level of NCI funding but also because of technological advancements in the techniques for collection of specimens, chemical isolation, and structural elucidation of low amounts of compounds initially isolated. Because of the fear of a limited supply of marine sources for large-scale production, marine natural products have remained the exclusive purview of academia except for a small Spanish pharmaceutical company, PharmaMar, which collaborates closely with academia and governments. Bryostatins are among the most interesting marine natural products known. They were isolated from the bryozoan *Bugula neritina*. They are a series of polyketide macro lactones represented here by the major congener bryostatin I (**85**), which is a modulator of protein kinase C and has been subjected to several human clinical trials (90, 91). Recently, combination studies have been recommended for Phase I and Phase II trials, mainly under the auspices of the NCI. Modeling studies along with a diligent chemical design approach has led to the synthesis of a simplified analog (**86**) that has been shown to be equally active as bryostatin I in most *in vitro* studies (92, 93). This compound stands a better chance of being produced at larger scale by total synthesis.

Dolastatins are a class of peptides comprised of mostly nonribosomal amino acids. They were isolated from a sea hare *Dolabella auricularia* (94). Dolastatin-10 (**87**) is one of the most potent and the best-studied members (95, 96). It exerts it antitumor effect by inhibiting tubulin polymerization and binds at the vinca alkaloid binding site (97, 98). Dolastatin-10 has been studied in Phase II human clinical trials but was discontinued because of lack of efficacy (99). Auristatin PE (**88**), a synthetic analog, seems to be more promising and is being studied in Phase II human trials (100).

Discodermolide (**89**) was isolated from *Discodermia dissoluta* by using a P388 cell line toxicity bioassay; later it was determined that it stabilized microtubule assembly better than Taxol[®], and it drew a lot of attention as an anticancer agent (101, 102). Its development, like that of many other complex marine natural

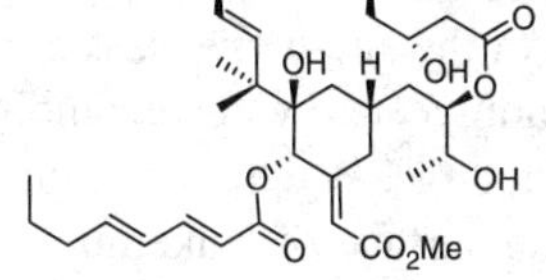

Bryostatin 1(**85**) 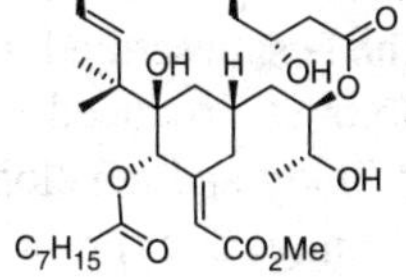   (**86**)

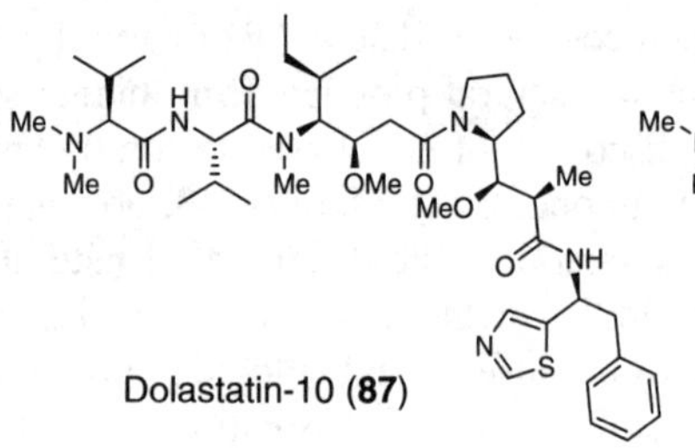

Dolastatin-10 (**87**)

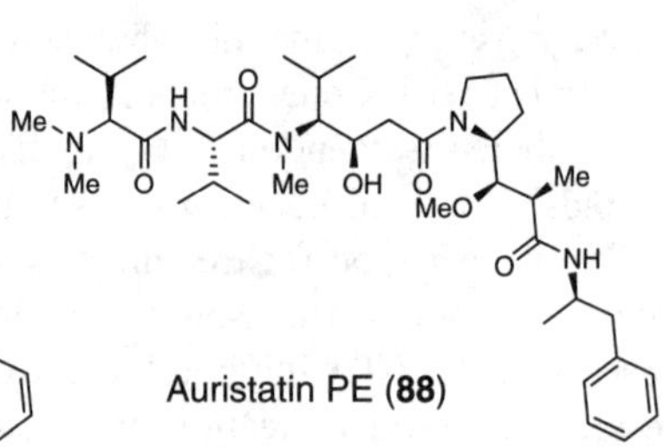

Auristatin PE (**88**)

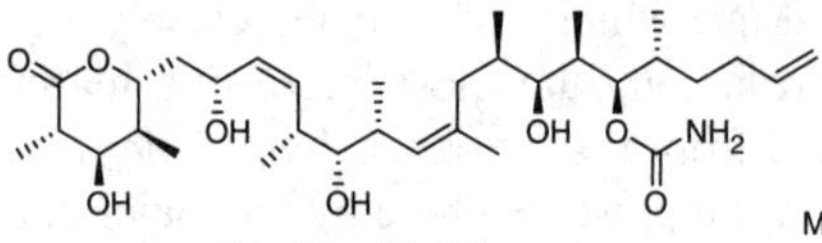

Discodermolide (**89**)

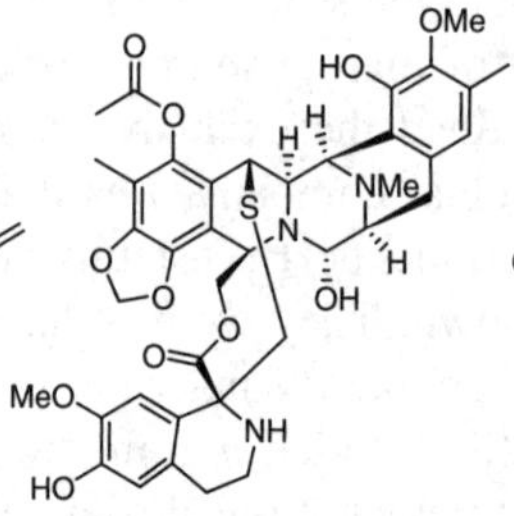

Ecteinascidin 743 (**90**)

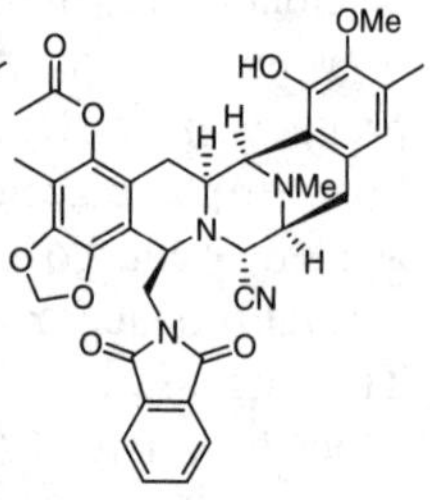

Phthalascidin (**91**)

products, was hampered because of the lack of ample supply of the material required for the clinical studies. In this case, the supply problem was overcome by the synthetic efforts of the Novartis process group. They synthesized it on a large enough scale to allow clinical studies. Unfortunately, its development seems to have been halted because of toxicity at Phase I (103, 104).

Several other novel, structurally and mechanistically diverse marine natural products have entered various preclinical and clinical studies. One of these products is ecteinascidin-743 (**90** ET-743), isolated from the tunicate *Ecteinscidea turbinata* (105, 106) and recently approved for the treatment of sarcoma in the European Union as the first "direct-from-the-sea" drug. Total synthesis and methods developed during the total synthesis allowed the preparation of a simpler analog phthalascidin (**91**) with comparable activities (107–109).

Hemiasterlin (**92**), a tripeptide isolated from a sponge that was chemically modified to HTI-286 (**93**), which binds to the vinca binding site of tubulin, depolarizes microtubules; it entered into clinical development but was apparently dropped (110, 111).

One difficulty with the cytotoxic agents that are used for the treatment of cancer is the differentiation of cytotoxicity between target tumor cells and normal cells. In an innovative approach, the Wyeth group took advantage of a tumor cell-specific drug delivery mechanism of antibodies. They conjugated calicheamicin (**94**), perhaps the best described member of the ene-diyne class of highly cytotoxic antitumor antibiotics produced by Actinomycetes, with recombinant humanized IgG$_4$ kappa antibody and developed Mylotarg® Wyeth, Madison, NJ (**95**), which binds to CD33 antigens expressed on the surface of leukemia blasts. Mylotarg® is an effective and less toxic treatment of myeloid leukemia (112–115).

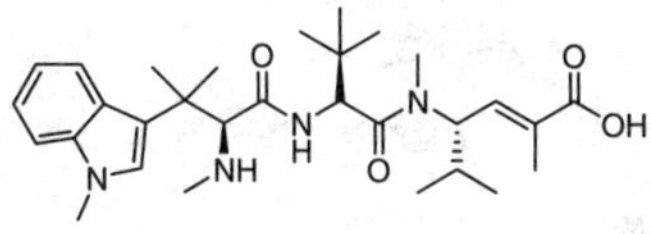

Hemiasterlin (**92**)

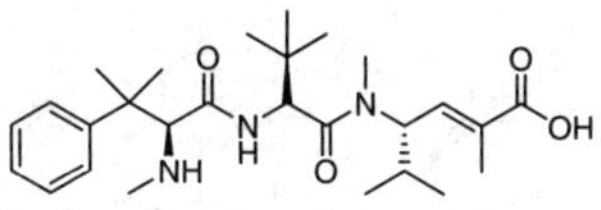

HTI-286 (**93**)

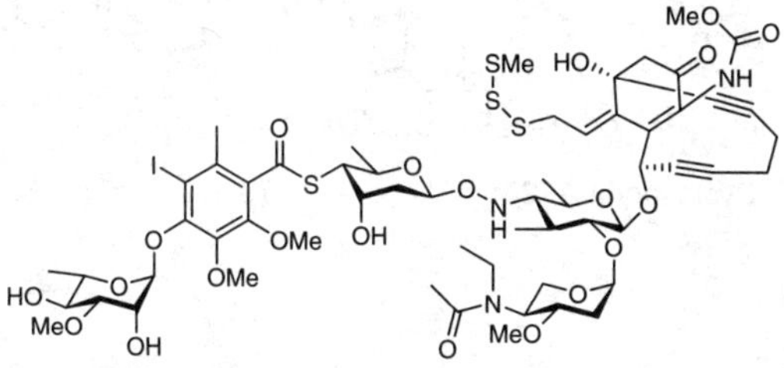

Calicheamicin (**94**)

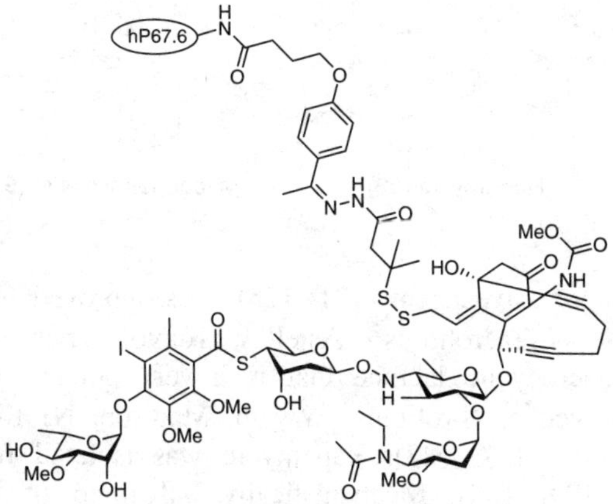

Mylotarg^R (**95**)

12.10 IMMUNOSUPPRESSANT AGENTS

Natural products represent essentially all clinically used immunosuppressant agents. These agents collectively have made organ transplant possible. Cyclosporin (**96**) is an *N*-methyl cyclic peptide and originally was isolated from the fungus *Trichoderma polysporum* as an antifungal agent; almost immediately, the inhibition of T-cell proliferation and *in vivo* immunosuppressive properties were discovered and led to the development and approval of this molecule as a highly effective immunosuppressive agent (116–118). Natural products isolation of related compounds allowed the discovery of new congeners with reduced or no immunosuppressive activity in favor of antifungal and various other biological activities (e.g., antiparasitic activity) and asthma (119, 120). FK506 (**97**), a macrocyclic lactone, discovered from *Streptomyces tsukubaensis*

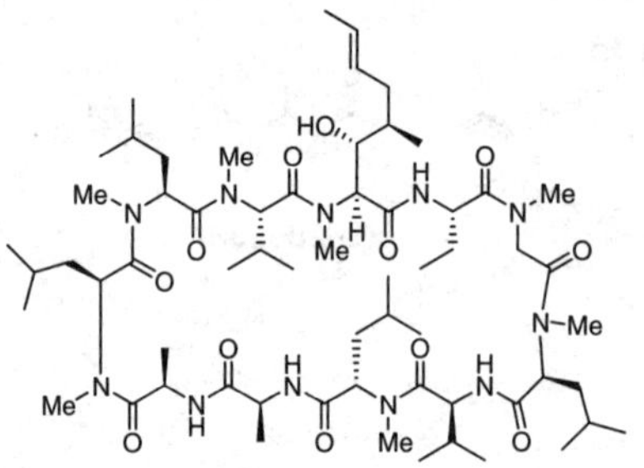

Cyclosporin A (**96**)

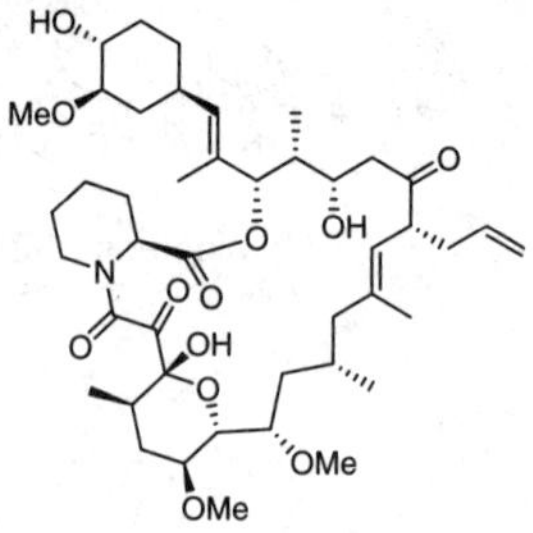

Tacrolimus^R (FK506) (**97**)

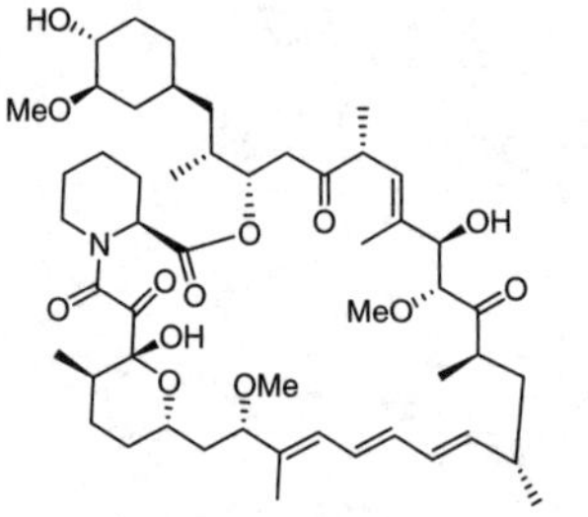

Rapamycin (**98**)

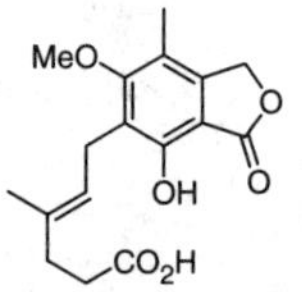

Mycophenolic acid (**99**)

as an immunosuppressive agent (121–125), was approved for clinical use for organ transplant as Tacrolimus® Astellas, Tokyo, Japan (126). Rapamycin (**98**), another macrocyclic lactone that is a very potent immunosuppressive agent, was approved as Sirolimus® Wyeth, Madison, NJ for clinical use for transplant rejection (127, 128). Rapamycin was isolated from *Streptomyces hygroscopicus* (129, 130). Mechanistically, all three of these compounds bind to their specific intracellular receptors, immunophilins, and the resulting complexes target the protein phosphatase, calcineurin (cyclosprin and FK506), and mammalian target of rapamycin (mTOR) to exert their immunosuppressive effects (131–133). Rapamycin and FK506 have played significant roles in studies of signal transduction and identification of various targets for other therapeutic applications, such as mTOR. Mycophenolic acid (**99**) originally was isolated from various species of *Penicillium*, and its antifungal activity has been known since 1932 (134). Mycophenolate was approved for acute rejection of kidney transplant (135, 136).

12.11 CARDIOVASCULAR AGENTS

The biggest impact made by natural products in the treatment of cardiovascular diseases is undoubtedly associated with the discovery of the first of the HMG CoA reductase inhibitors by enzyme-based screening of microbial extracts that led to the isolation (from *Aspergillus terreus*) and characterization of mevino-lin (lovastatin **100**), which is a homolo of compactin (**101**) that was discovered earlier (137, 138). These compounds possess a lipophilic hexahydrodecalin, a 2-methylbutanoate side chain, and a β-hydroxy-δ-lactone connected to the decalin unit with a two-carbon linker. These compounds are potent inhibitors of HMG CoA reductase, the rate-limiting enzyme of cholesterol biosynthesis, and inhibit the synthesis of cholesterol in the liver. Lovastatin (Mevacor®) Merck & Co., Inc., Whitehouse Station, NJ; was the first compound approved for lowering cholesterol in humans and became the cornerstone of all cholesterol-lowering agents generically called "statins." The modification of 2-methylbutanoate to 2,2-dimethylbutanoate led to the semisynthetic derivative, simvastatin (Zocor®, Merck & Co., Inc., Whitehouse Station, NJ; **102**), the second and more effective agent approved for human use (139). Hydroxylation of compactin by biotransfor-mation led to pravastatin (Pravachol®, BMS, NY **103**) (139). The key pharma-cophore of the statins is the β-hydroxy-δ-lactone or open acid. As the importance and value of cholesterol lowering to human pathophysiology became clearer, the search for additional cholesterol-lowering agents became more prominent and led to the discovery and development of several other clinical agents. All these com-pounds retained nature's gift of the pharmacophore, β-hydroxy-δ-lactone (or open acid), with replacement of the decalin unit of the natural products with a variety of aromatic lipophilic groups that resulted in fluvastin (**104**) (140), Atorvastatin (**105**) (141), Cerivastatin (**106**, withdrawn from the clinic) (142), Rosuvastatin (**107**) (143), and Pitavastatin (**108**). The statins have had tremendous impact in improvement of overall human health and quality of life because of the lower-ing of low-density lipoprotein (LDL) particles, which leads to a reduction in the incidence of coronary heart disease; arguably, they are the most successful class of medicines.

Ephedrine (**109**), isolated from the Chinese plant *Ephedra sinaica*, was approved as one of the first bronchodilators and cardiovascular agents. This dis-covery led to a variety of such antihypertensive agents including β-blockers (4).

Angiotensin-converting enzyme (ACE) converts angiotensin I to angiotensin II, and its inhibition has led to several very successful, clinically useful antihy-pertensive agents. Although these inhibitors are of synthetic origin, the original lead was modeled after a nonapeptide, teprotide (**110**). This peptide was isolated from snake (viper, *Bothrops jararaca*) venom by Ondetti et al. It had antihyper-tensive activity in the clinic by parenteral administration (138, 144, 145) but was devoid of oral activity. Ondetti and coworkers worked diligently, and, recogniz-ing that ACE was a metallo-enzyme, they visualized the binding of a smaller snake-venom peptide SQ20475 (**111**) with ACE; they modeled an acyl-proline with a sulfhydryl substitution at the zinc binding site, which led to the design and synthesis of captopril (**112**) as an orally active highly effective antihyper-tensive clinical agent. Additional application of the rational design by Patchett and coworkers led to the synthesis of enalapril (**113**) and other clinically relevant oral ACE inhibitors (138).

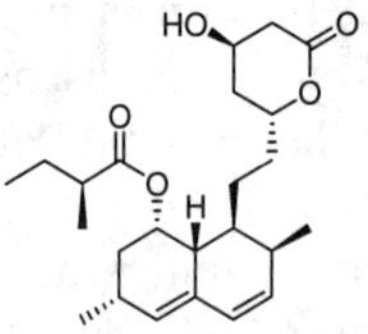

Lovastatin (**100**) 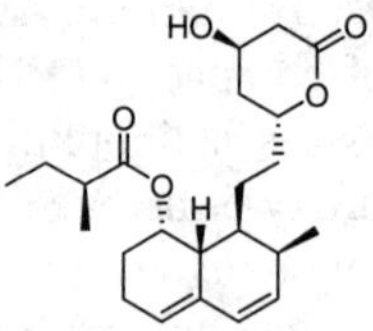   Mevastatin (**101**) 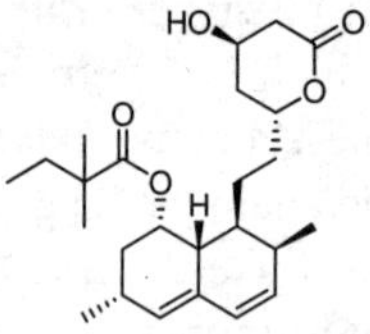   Simvastatin (**102**) 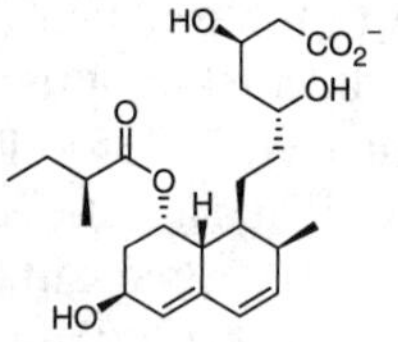   Pravastatin (**103**)

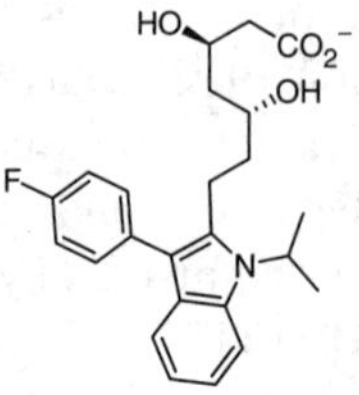

Fluvastin (**104**) 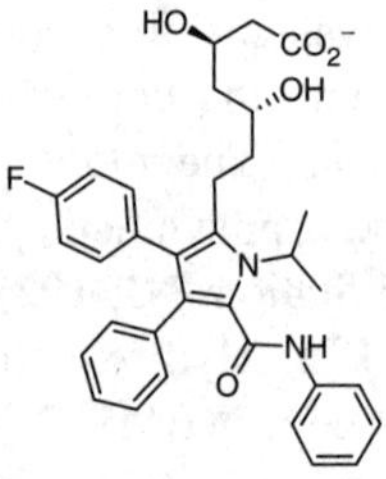   Atorvastatin (**105**) 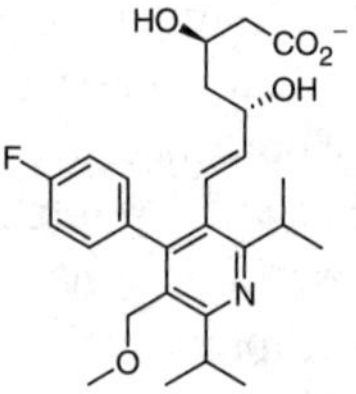   Cerivastatin (**106**)

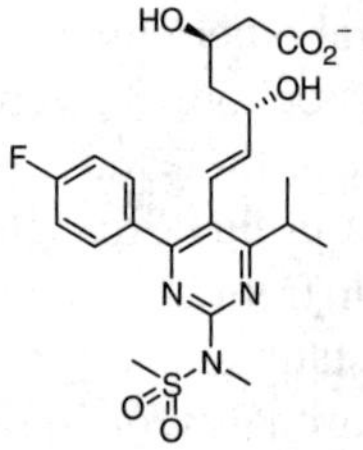

Rosuvastatin (**107**) 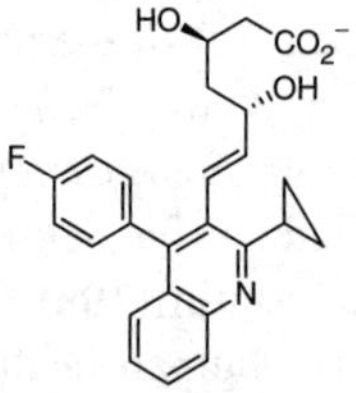   Pitavasatin (**108**)

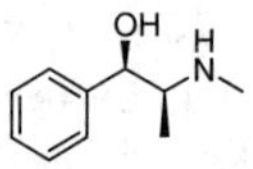

Ephedrine (**109**) 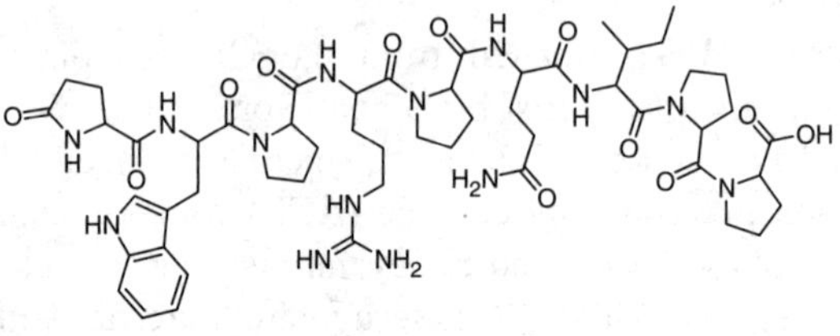   Teprotide (**110**)

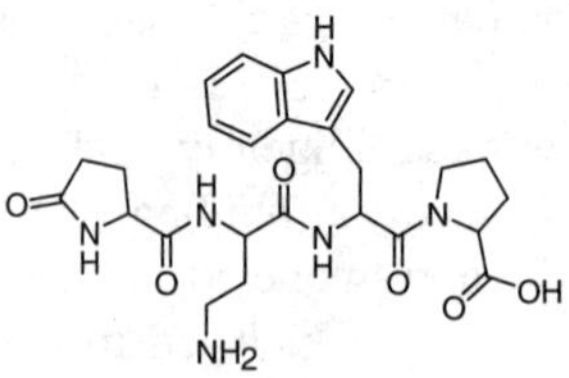

SQ 20575 (**111**) 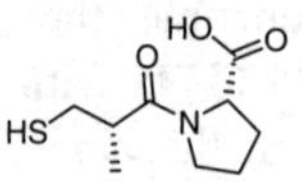   Captropril (**112**) 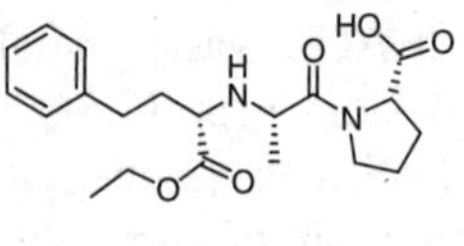   Enalapril (**113**)

12.12 ANTIPARASITIC AGENTS

Avermectins (**114**, **115**) are a series of macrocyclic lactones that are broad-spectrum, highly potent, glutamate-gated, chloride channel-modulator antiparasitic agents produced by *Streptomyces avermitilis* (146, 147). Ivermectin, 23,24-dihydroavermectin B_{1a}/B_{1b} (**116**), was the first product approved in the mid-1980s for treatment of intestinal parasites in domesticated and farm animals, and it remains the standard of care (148, 149). The remarkable activity of ivermectin against *Onchocerca volvulus*, the causative parasitic agent of onchocerciasis (river blindness), led to clinical development and the approval of Mectizan® Merck & Co., Inc., Whitehouse Station, NJ; for the treatment of such diseases. These parasitic diseases have debilitated millions of people in many countries in Africa and South America. Because Mectizan® is a very effective treatment, Merck is providing this drug free of cost to all people in need as a part of the "Mectizan Donation Program," which has had tremendous impact on the health and quality of life of people affected by these diseases (150).

Spinosyns were discovered from the fermentation broth of *Saccharopolyspora spinosa* by screening for mortality of blowfly larvae, and a mixture of spinosyns A (**117**) and D (**118**) was approved and used successfully as a crop protection and an antiparasitic animal health agent. (151) Nodulisporic acids are an indole diterpenoid class discovered from various species of *Nodulisporium* as orally active antiflea and antitick agents for dogs and cats (152, 153). The most active

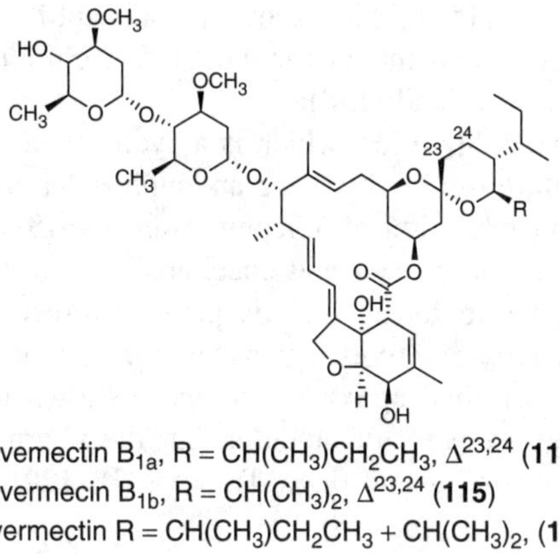

Avemectin B_{1a}, R = $CH(CH_3)CH_2CH_3$, $\Delta^{23,24}$ (**114**)
Avermecin B_{1b}, R = $CH(CH_3)_2$, $\Delta^{23,24}$ (**115**)
Ivermectin R = $CH(CH_3)CH_2CH_3$ + $CH(CH_3)_2$, (**116**)

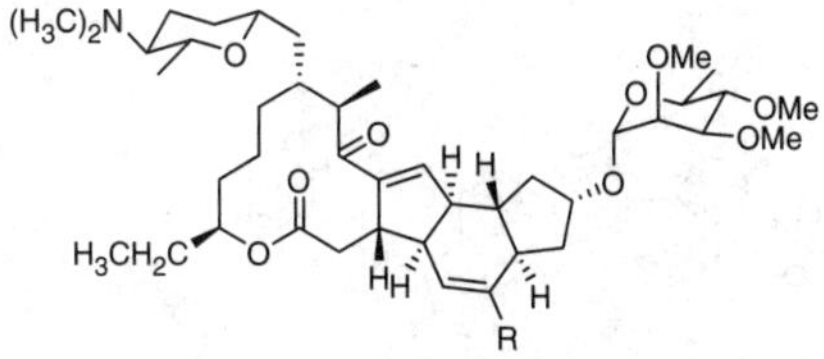

Spinosyn A, R = H (**117**)
Spinosyn D, R = Me (**118**)

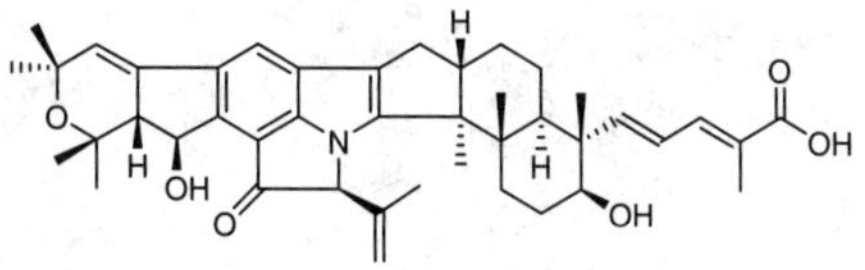

Nodulisporic Acid A (**119**)

of the series is nodulisporic acid A (**119**), which selectively modulates the activity of insect-specific glutamate-gated chloride channels (153).

12.13   PHARMACEUTICAL MODELS

The roles played by natural products as models for design and development of pharmaceutical agents are too many to cover in this overview. A few examples are illustrated during the discussions of specific disease areas above. For example, a marine sponge-derived nucleoside was the precursor for various nucleoside-based antiviral agents, pepstatin for renin and HIV protease inhibitors, snake venom peptide for ACE inhibitors, lovastatin and compactin for all statins, and ephedrine for many painkillers and β-blockers. Below are a few critical examples that have played a big role in defining leads for some therapeutic areas but have not resulted in a drug yet.

Asperlicin (**120**) was isolated from *Aspergillus alliaceus* as a weak cholecystokinin A receptor (CCK-A) antagonist by using CCK receptor binding screening assays (154). It is a competitive antagonist of CCK-A (but not CCK-B) but did not have sufficient potency or oral activity to qualify as a drug candidate. In a remarkable strategy, medicinal chemists simplified the molecule to a benzodiazepine core of asperlicin, which led to the synthesis of potent, safe, and orally active analogs (**121** and **122**) with selectivity for either CCK-A (**121**) or CCK-B (**122**) receptors. They entered human clinical trials but were abandoned because of lack of efficacy (155, 156). The benzodiazepine scaffold was coined as "privileged structures" by Evans et al (155). This discovery is a beautiful demonstration of how a natural product became a model for the CCK program and played a pivotal role in defining the entire field (138).

The second example is apicidin (**123**), which is a cyclic tetrapeptide isolated from a fungus *Fusarium pallidoroseum* by using an empiric antiprotozoal screen (157, 158). It showed potent inhibition of apicomplexan protozoa including the malarial parasite *Plasmodium falciparum* and coccidiosis parasite *Eimeria* spp. It was effective *in vivo* against reducing malaria parasite infection in a mouse model (157) and exhibited strong activity against tumor cell lines (159, 160). Cyclic tetrapeptides with a terminal epoxy-ketone were known to be effective cytotoxic agents before the discovery of apicidin, but the pharmacophore was associated with the epoxy-ketone group (e.g., Trapoxin B, **124**) with covalent

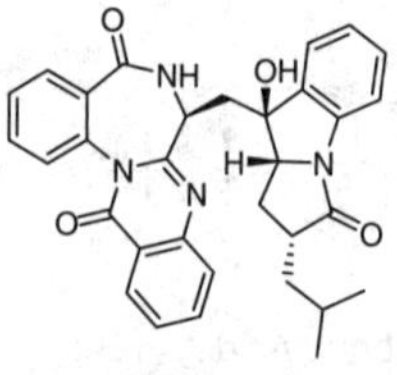

Asperlicin (**120**) 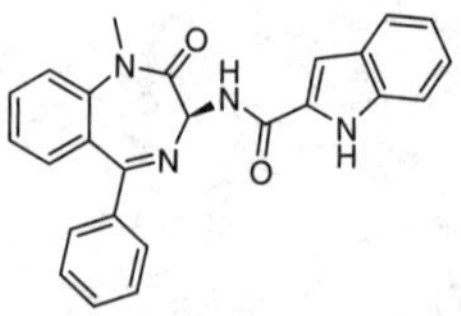   MK-329(**121**) 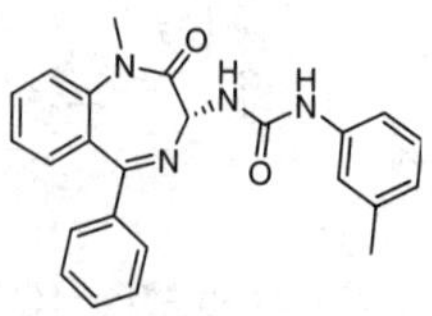   L-365,260 (**122**)

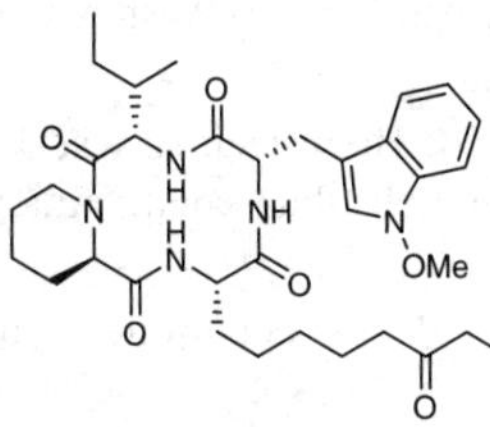

Apicidin (**123**) 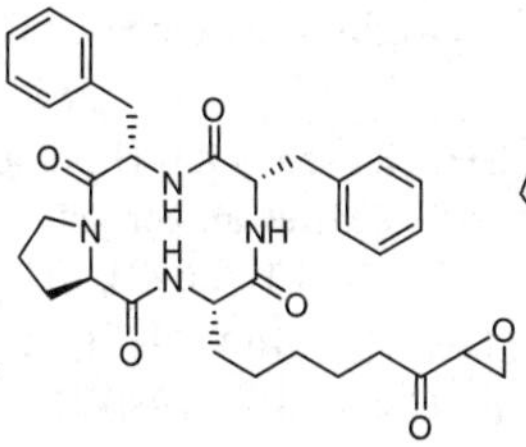 Trapoxin B (**124**) 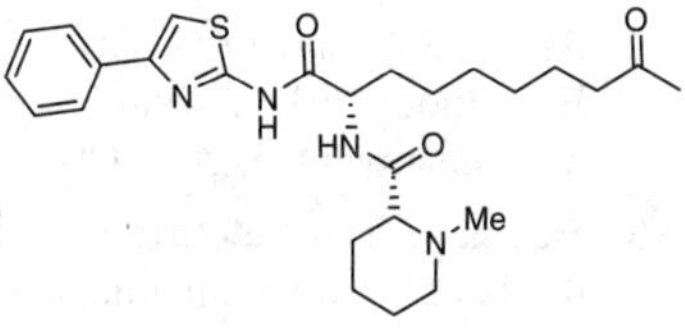 **125**

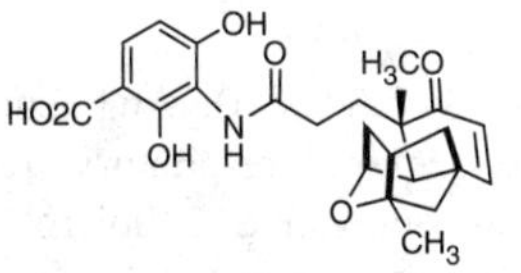

Platensimycin (**126**) 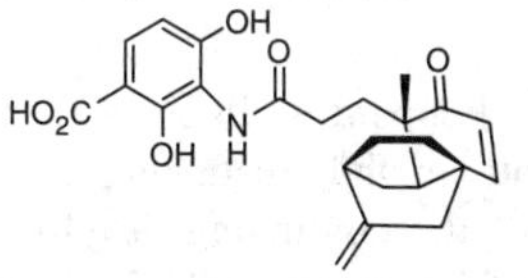 Platencin (**127**)

modification as a mode of action. Apicidin does not contain the epoxy-ketone but showed potent antitumor activity (161). The mode of action of apicidin was shown to be the inhibition of histone deacetylase (HDAC) (157). The amino-*oxo*-decanoic acid (L-Aoda) mimics the acetylated lysine residue and positions itself at the zinc-binding site of HDAC (162). Chemical modification of apicidin with retention of the ethyl or methyl ketone led to the synthesis of small dipeptides (e.g., **125**) that retained the HDAC and tumor cell line inhibitory activities with significant reduction of inhibition of normal cells (161).

In summary, nature has provided a great set of molecules with enormous chemical diversity that has contributed to the treatment of many human diseases. Nature continues to amaze us with novel chemical diversity with unimaginable biological activity and target specificity, as illustrated by the recent discovery of the fatty acid synthesis inhibitor antibiotic platensimycin (**126**) (163, 164) and platencin (**127**) (165, 166). The former shows exquisite selectivity for FabF, whereas the latter compound is a balanced inhibitor of both condensing enzymes, FabF and FabH.

REFERENCES

1. Butler M. The role of natural product chemistry in drug discovery. J. Nat. Prod. 2004;67:2141–2153.

2. Clardy J, Walsh CT. Lessons from natural molecules. Nature 2004;432:829–836.

3. Newman DJ, Cragg GM, Snader KM. Natural products as sources of new drugs over the period 1981–2002. J. Nat. Prod. 2003;66:1022–1037.

4. Newman DJ, Cragg GM, Snader KM. The influence of natural products upon drug discovery. Nat. Prod. Rep. 2000;17:215–234.

5. Singh SB, Barrett JF. Empirical antibacterial drug discovery–foundation in natural products. Biochem. Pharmacol. 2006;71:1006–1015.

6. Walsh CT. Antibiotics: Actions, Origin, Resistance. 2003. ASM Press, Washington, DC.

7. Albers-Schoenberg G, Arison BH, Hensens OD, Hirshfield J, Hoogsteen K, Kaczka EA, Rhodes RE, Kahan JS, Kahan FM, et al. Structure and absolute configuration of thienamycin. J. Am. Chem. Soc. 1978;100:6491–6499.

8. Salzmann TN, Ratcliffe RW, Christensen BG, Bouffard FA. A stereocontrolled synthesis of + -thienamycin. J. Am. Chem. Soc. 1980;102:6161–6163.

9. Zhanel GG, Wiebe R, Dilay L, Thomson K, Rubinstein E, Hoban DJ, Noreddin AM, Karlowsky JA. Comparative review of the carbapenems. Drugs 2007;67: 1027–1052.

10. White AR, Kaye C, Poupard J, Pypstra R, Woodnutt G, Wynne B. Augmentin amoxicillin/clavulanate in the treatment of community-acquired respiratory tract infection: a review of the continuing development of an innovative antimicrobial agent. J. Antimicrob. Chemother. 2004;53:3–20.

11. Guay DR, Koskoletos C. Aztreonam, a new monobactam antimicrobial. Clin. Pharm. 1985;4:516–526.

12. Rubinstein E, Vaughan D. Tigecycline: a novel glycylcycline. Drugs 2005;65: 1317–1336.

13. Yassin HM, Dever LL. Telithromycin: a new ketolide antimicrobial for treatment of respiratory tract infections. Expert. Opin. Investig. Drugs 2001;10:353–367.

14. Woodford N. Novel agents for the treatment of resistant Gram-positive infections. Expert. Opin. Investig. Drugs 2003;12:117–137.

15. Bonfiglio G, Furneri PM. Novel streptogramin antibiotics. Expert. Opin. Investig. Drugs 2001;10:185–198.

16. Georgopapadakou NH. Update on antifungals targeted to the cell wall: focus on beta-1,3-glucan synthase inhibitors. Expert. Opin. Investig. Drugs 2001;10:269–280.

17. Denning DW. Echinocandin antifungal drugs. Lancet 2003;362:1142–1151.

18. Vroman JA, Alvim-Gaston M, Avery MA. Current progress in the chemistry, medicinal chemistry and drug design of artemisinin based antimalarials. Curr. Pharm. Des. 1999;5:101–138.

19. Ro DK, Paradise EM, Ouellet M, Fisher KJ, Newman KL, Ndungu JM, Ho KA, Eachus RA, Ham TS, Kirby J, Chang MC, Withers ST, Shiba Y, Sarpong R, Keasling JD. Production of the antimalarial drug precursor artemisinic acid in engineered yeast. Nature 2006;440:940–943.

20. Withers ST, Keasling JD. Biosynthesis and engineering of isoprenoid small molecules. Appl. Microbiol. Biotechnol. 2007;73:980–990.

21. Chang MC, Eachus RA, Trieu W, Ro DK, Keasling JD. Engineering Escherichia coli for production of functionalized terpenoids using plant P450s. Nat. Chem. Biol. 2007;3:274–277.

22. Bergmann W, Feeney RJ. Contributions to the study of marine products. XXXII. The nucleosides of sponges I. J. Org. Chem. 1951;16:981–987.

23. Bergmann W, Feeney RJ. The isolation of a new thymine pentoside from sponges. J. Am. Chem. Soc. 1950;72:2809–2810.

24. Bergmann W, Burke DC. Contributions to the Study of Marine Products. XL. The nucleosides of sponges. IV. Spongosine. J. Org. Chem. 1956;21:226–228.

25. Wiley RA, Rich DH. Peptidomimetics derived from natural products. Med. Res. Rev. 1993;13:327–384.

26. Dorsey BD, McDonough C, McDaniel SL, Levin RB, Newton CL, Hoffman JM, Darke PL, Zugay-Murphy JA, Emini EA, Schleif WA, Olsen DB, Stahlhut MW, Rutkowski CA, Kuo LC, Lin JH, Chen IW, Michelson SR, Holloway MK, Huff JR, Vacca JP. Identification of MK-944a: a second clinical candidate from the hydroxylaminepentanamide isostere series of HIV protease inhibitors. J. Med. Chem. 2000;43:3386–3399.

27. Terr CE, Pellens M. The Opium Problem. 1928. Bureau of Social Hygeine, New York:

28. Eddy NB, Friebel H, Hahn KJ, Halbach H. Codeine and its alternates for pain and cough relief. 2. Alternates for pain relief. Bull. World Health Organ. 1969;40:1–53.

29. Heel RC, Brogden RN, Speight TM, Avery GS. Buprenorphine: a review of its pharmacological properties and therapeutic efficacy. Drugs 1979;17:81–110.

30. Walker CS, Steel D, Jacobsen RB, Lirazan MB, Cruz LJ, Hooper D, Shetty R, DelaCruz RC, Nielsen JS, Zhou LM, Bandyopadhyay P, Craig AG, Olivera BM. The T-superfamily of conotoxins. J. Biol. Chem. 1999;274:30664–30671.

31. Heading CE. Ziconotide Elan Pharmaceuticals. IDrugs 2001;4:339–350.

32. Miljanich GP. Ziconotide: neuronal calcium channel blocker for treating severe chronic pain. Curr. Med. Chem. 2004;11:3029–3040.

33. Spande TF, Garraffo HM, Edwards MW, Yeh HJC, Pannell L, Daly JW. Epibatidine: a novel chloropyridylazabicycloheptane with potent analgesic activity from an Ecuadoran poison frog. J. Am. Chem. Soc. 1992;114:3475–3478.

34. Daly JW. The chemistry of poisons in amphibian skin. Proc. Natl. Acad. Sci. U.S.A. 1995;92:9–13.

35. Badio B, Daly JW. Epibatidine, a potent analgetic and nicotinic agonist. Mol. Pharmacol. 1994;45:563–9.

36. Holladay MW, Bai H, Li Y, Lin NH, Daanen JF, Ryther KB, Wasicak JT, Kincaid JF, He Y, Hettinger AM, Huang P, Anderson DJ, Bannon AW, Buckley MJ, Campbell JE, Donnelly-Roberts DL, Gunther KL, Kim DJ, Kuntzweiler TA, Sullivan JP, Decker MW, Arneric SP. Structure-activity studies related to ABT-594, a potent nonopioid analgesic agent: effect of pyridine and azetidine ring substitutions on nicotinic acetylcholine receptor binding affinity and analgesic activity in mice. Bioorg. Med. Chem. Lett. 1998;8:2797–2802.

37. Decker MW, Rueter LE, Bitner RS. Nicotinic acetylcholine receptor agonists: a potential new class of analgesics. Curr. Top. Med. Chem. 2004;4:369–384.

38. Decker MW, Meyer MD, Sullivan JP. The therapeutic potential of nicotinic acetylcholine receptor agonists for pain control. Expert. Opin. Investig. Drugs 2001;10:1819–1830.

39. Decker MW, Meyer MD. Therapeutic potential of neuronal nicotinic acetylcholine receptor agonists as novel analgesics. Biochem. Pharmacol. 1999;58:917–923.

40. Hochuli E, Kupfer E, Maurer R, Meister W, Mercadal Y, Schmidt K. Lipstatin, an inhibitor of pancreatic lipase, produced by Streptomyces toxytricini. II. Chemistry and structure elucidation. J Antibiot. (Tokyo) 1987;40:1086–1091.

41. Weibel EK, Hadvary P, Hochuli E, Kupfer E, Lengsfeld H. Lipstatin an inhibitor of pancreatic lipase, produced by Streptomyces toxytricini. I. Producing organism, fermentation, isolation and biological activity. J. Antibiot. (Tokyo) 1987;40: 1081–1085.

42. McNeely W, Benfield P. Orlistat. Drugs 1998;56:241–9;250.

43. Mancino JM. Orlistat: Current issues for patients with type 2 diabetes. Curr. Diab. Rep. 2006;6:389–394.

44. Henness S, Perry CM. Orlistat: a review of its use in the management of obesity. Drugs 2006;66:1625–1656.

45. Miyazaki Y, Godaishi K. Experimental Cultivation of the Plants Containing Galanthamine at Izu. 1 General Growth of Shokiran Lycoris Aurea Herb., Natsuzuisen L. Squamigera Maxim., Snowflake Leucojum Aestivum L., and Snowdrop Galanthus Nivalis L., 1961 to 1962. Eisei. Shikenjo. Hokoku. 1963;81:172–176.

46. Thomsen T, Kewitz H. Selective inhibition of human acetylcholinesterase by galanthamine in vitro and in vivo. Life Sci. 1990;46:1553–1558.

47. Greenblatt HM, Kryger G, Lewis T, Silman I, Sussman JL. Structure of acetylcholinesterase complexed with –galanthamine at 2.3 A resolution. FEBS Lett. 1999;463:321–326.

48. Lee KH. Novel antitumor agents from higher plants. Med. Res. Rev. 1999;19: 569–596.

49. Wani MC, Taylor HL, Wall ME, Coggon P, McPhail AT. Plant antitumor agents. VI. The isolation and structure of taxol, a novel antileukemic and antitumor agent from Taxus brevifolia. J. Am. Chem. Soc. 1971;93:2325–2327.

50. Schiff PB, Horwitz SB. Taxol assembles tubulin in the absence of exogenous guanosine 5′-triphosphate or microtubule-associated proteins. Biochemistry 1981;20:3247–3252.

51. Schiff PB, Horwitz SB. Taxol stabilizes microtubules in mouse fibroblast cells. Proc. Natl. Acad. Sci. U.S.A. 1980;77:1561–1565.

52. Schiff PB, Fant J, Horwitz SB. Promotion of microtubule assembly in vitro by taxol. Nature 1979;277:665–667.

53. Wall ME, Wani MC, Cook CE, Palmer KH, McPhail AT, Sim GA. Plant Antitumor Agents. I. The Isolation and Structure of Camptothecin, a Novel Alkaloidal Leukemia and Tumor Inhibitor from Camptotheca acuminata. J. Am. Chem. Soc. 1966;88:3888–3890.

54. Srivastava V, Negi AS, Kumar JK, Gupta MM, Khanuja SP. Plant-based anticancer molecules: a chemical and biological profile of some important leads. Bioorg. Med. Chem. 2005;13:5892–5908.

55. Sandler A. Irinotecan therapy for small-cell lung cancer. Oncology 2002;16: 419–438.

56. Neuss N, Gorman M, Svoboda GH, Maciak G, Beer CT. Vinca alkaloids. III. Characterization of leurosine and vindaleukoblastine, new alkaloids from Vinca rosea Linn. J. Am. Chem. Soc. 1959;81:4754–4755.

57. Moncrief JW, Lipscomb WN. Structures of leurocristine vincristine and vincaleukoblastine.1 X-Ray analysis of leurocristine methiodide. J. Am. Chem. Soc. 1965;87:4963–4964.

58. Hochster HS, Vogel CL, Burman SL, White R. Activity and safety of vinorelbine combined with doxorubicin or fluorouracil as first-line therapy in advanced breast cancer: a stratified phase II study. Oncologist 2001;6:269–277.

59. Barnett CJ, Cullinan GJ, Gerzon K, Hoying RC, Jones WE, Newlon WM, Poore GA, Robison RL, Sweeney MJ, et al. Structure-activity relationships of dimeric Catharanthus alkaloids. 1. Deacetyl vinblastine amide vindesine sulfate. J. Med. Chem. 1978;21:88–96.

60. Sorensen JB, Hansen HH. Is there a role for vindesine in the treatment of non-small cell lung cancer? Invest. New Drugs 1993;11:103–133.

61. Sorensen JB, Osterlind K, Hansen HH. Vinca alkaloids in the treatment of non-small cell lung cancer. Cancer Treat. Rev. 1987;14:29–51.

62. Ruckdeschel JC. Etoposide in the management of non-small cell lung cancer. Cancer 1991;67:2533.

63. Gordaliza M, Castro MA, del Corral JM, Feliciano AS. Antitumor properties of podophyllotoxin and related compounds. Curr. Pharm. Des. 2000;6:1811–1839.

64. Baldwin EL, Osheroff N. Etoposide, topoisomerase II and cancer. Curr. Med. Chem. Anticancer Agents 2005;5:363–372.

65. You Y. Podophyllotoxin derivatives: current synthetic approaches for new anticancer agents. Curr. Pharm. Des. 2005;11:1695–1717.

66. Pettit GR, Singh SB, Hamel E, Lin CM, Alberts DS, Garcia-Kendall D. Isolation and structure of the strong cell growth and tubulin inhibitor combretastatin A-4. Experientia 1989;45:209–211.

67. Lin CM, Singh SB, Chu PS, Dempcy RO, Schmidt JM, Pettit GR, Hamel E. Interactions of tubulin with potent natural and synthetic analogs of the antimitotic agent combretastatin: a structure-activity study. Mol. Pharmacol. 1988;34:200–208.

68. Kirwan IG, Loadman PM, Swaine DJ, Anthoney DA, Pettit GR, Lippert JW, 3rd, Shnyder SD, Cooper PA, Bibby MC. Comparative preclinical pharmacokinetic and metabolic studies of the combretastatin prodrugs combretastatin A4 phosphate and A1 phosphate. Clin. Cancer Res. 2004;10:1446–1453.

69. Nabha SM, Mohammad RM, Dandashi MH, Coupaye-Gerard B, Aboukameel A, Pettit GR, Al-Katib AM. Combretastatin-A4 prodrug induces mitotic catastrophe in chronic lymphocytic leukemia cell line independent of caspase activation and polyADP-ribose polymerase cleavage. Clin. Cancer Res. 2002;8:2735–2741.

70. Chaplin DJ, Pettit GR, Hill SA. Anti-vascular approaches to solid tumour therapy: evaluation of combretastatin A4 phosphate. Anticancer Res. 1999;19:189–195.

71. Dorr RT, Dvorakova K, Snead K, Alberts DS, Salmon SE, Pettit GR. Antitumor activity of combretastatin-A4 phosphate, a natural product tubulin inhibitor. Invest. New Drugs 1996;14:131–137.

72. Phillips DR, White RJ, Cullinane C. DNA sequence-specific adducts of adriamycin and mitomycin C. FEBS Lett. 1989;246:233–240.

73. Skorobogaty A, White RJ, Phillips DR, Reiss JA. Elucidation of the DNA sequence preferences of daunomycin. Drug. Des. Deliv. 1988;3:125–151.

74. Skorobogaty A, White RJ, Phillips DR, Reiss JA. The 5'-CA DNA-sequence preference of daunomycin. FEBS Lett. 1988;227:103–106.

75. Hecht SM. Bleomycin: new perspectives on the mechanism of action. J. Nat. Prod. 2000;63:158–168.

76. White RJ, Durr FE. Development of mitoxantrone. Invest. New Drugs 1985;3: 85–93.

77. Sasaki Y, Seto M, Komatsu K, Omura S. Staurosporine, a protein kinase inhibitor, attenuates intracellular Ca2 + -dependent contractions of strips of rabbit aorta. Eur. J. Pharmacol. 1991;202:367–372.

78. Omura S. The expanded horizon for microbial metabolites–a review. Gene 1992; 115:141–149.

79. Omura S, Sasaki Y, Iwai Y, Takeshima H. Staurosporine, a potentially important gift from a microorganism. J. Antibiot. (Tokyo) 1995;48:535–548.

80. Thavasu P, Propper D, McDonald A, Dobbs N, Ganesan T, Talbot D, Braybrook J, Caponigro F, Hutchison C, Twelves C, Man A, Fabbro D, Harris A, Balkwill F. The protein kinase C inhibitor CGP41251 suppresses cytokine release and extracellular signal-regulated kinase 2 expression in cancer patients. Cancer Res. 1999;59:3980–3984.

81. Propper DJ, McDonald AC, Man A, Thavasu P, Balkwill F, Braybrooke JP, Caponigro F, Graf P, Dutreix C, Blackie R, Kaye SB, Ganesan TS, Talbot DC, Harris AL, Twelves C. Phase I and pharmacokinetic study of PKC412, an inhibitor of protein kinase C. J. Clin. Oncol. 2001;19:1485–1492.

82. Gerth K, Bedorf N, Hofle G, Irschik H, Reichenbach H. Epothilons A and B: antifungal and cytotoxic compounds from Sorangium cellulosum Myxobacteria. Production, physico-chemical and biological properties. J. Antibiot. (Tokyo) 1996;49:560–563.

83. Bollag DM. Epothilones: novel microtubule-stabilising agents. Expert. Opin. Investig. Drugs 1997;6:867–873.

84. Bollag DM, McQueney PA, Zhu J, Hensens O, Koupal L, Liesch J, Goetz M, Lazarides E, Woods CM. Epothilones, a new class of microtubule-stabilizing agents with a taxol-like mechanism of action. Cancer Res. 1995;55:2325–2333.

85. Altmann KH, Pfeiffer B, Arseniyadis S, Pratt BA, Nicolaou KC. The Chemistry and Biology of Epothilones—The Wheel Keeps Turning. ChemMedChem 2007;2: 396–423.

86. Feling RH, Buchanan GO, Mincer TJ, Kauffman CA, Jensen PR, Fenical W. Salinosporamide A: a highly cytotoxic proteasome inhibitor from a novel microbial source, a marine bacterium of the new genus salinispora. Angew. Chem. Int. Ed. Engl. 2003;42:355–357.

87. Groll M, Huber R, Potts BC. Crystal structures of Salinosporamide A NPI-0052 and B NPI-0047 in complex with the 20S proteasome reveal important consequences of beta-lactone ring opening and a mechanism for irreversible binding. J. Am. Chem. Soc. 2006;128:5136–5141.

88. Macherla VR, Mitchell SS, Manam RR, Reed KA, Chao TH, Nicholson B, Deyanat-Yazdi G, Mai B, Jensen PR, Fenical WF, Neuteboom ST, Lam KS, Palladino MA, Potts BC. Structure-activity relationship studies of salinosporamide A NPI-0052, a novel marine derived proteasome inhibitor. J. Med. Chem. 2005;48:3684–3687.

89. Chauhan D, Catley L, Li G, Podar K, Hideshima T, Velankar M, Mitsiades C, Mitsiades N, Yasui H, Letai A, Ovaa H, Berkers C, Nicholson B, Chao TH, Neuteboom ST, Richardson P, Palladino MA, Anderson KC. A novel orally active proteasome inhibitor induces apoptosis in multiple myeloma cells with mechanisms distinct from Bortezomib. Cancer Cell 2005;8:407–419.

90. Pettit GR, Herald CL, Doubek DL, Herald DL, Arnold E, Clardy J. Isolation and structure of bryostatin 1. J. Am. Chem. Soc. 1982;104:6846–6848.

91. Hennings H, Blumberg PM, Pettit GR, Herald CL, Shores R, Yuspa SH. Bryostatin 1, an activator of protein kinase C, inhibits tumor promotion by phorbol esters in SENCAR mouse skin. Carcinogenesis 1987;8:1343–1346.

92. Baryza JL, Brenner SE, Craske ML, Meyer T, Wender PA. Simplified analogs of bryostatin with anticancer activity display greater potency for translocation of PKCdelta-GFP. Chem. Biol. 2004;11:1261–1267.

93. Wender PA, Hinkle KW, Koehler MF, Lippa B. The rational design of potential chemotherapeutic agents: synthesis of bryostatin analogues. Med. Res. Rev. 1999;19:388–407.

94. Pettit GR. Th dolastatins. Fortschr Chem. Org. Naturst. 1997;70:1–79.

95. Pettit GR, Kamano Y, Herald CL, Tuinman AA, Boettner FE, Kizu H, Schmidt JM, Baczynskyj L, Tomer KB, Bontems RJ. The isolation and structure of a remarkable marine animal antineoplastic constituent: dolastatin 10. J. Am. Chem. Soc. 1987;109:6883–6885.

96. Pettit GR, Singh SB, Hogan F, Lloyd-Williams P, Herald DL, Burkett DD, Clewlow PJ. Antineoplastic agents. Part 189. The absolute configuration and synthesis of natural–dolastatin 10. J. Am. Chem. Soc. 1989;111:5463–5465.

97. Bai R, Pettit GR, Hamel E. Dolastatin 10, a powerful cytostatic peptide derived from a marine animal. Inhibition of tubulin polymerization mediated through the vinca alkaloid binding domain. Biochem. Pharmacol. 1990;39:1941–1949.

98. Bai RL, Pettit GR, Hamel E. Binding of dolastatin 10 to tubulin at a distinct site for peptide antimitotic agents near the exchangeable nucleotide and vinca alkaloid sites. J. Biol. Chem. 1990;265:17141–17149.

99. Saad ED, Kraut EH, Hoff PM, Moore DF, Jr., Jones D, Pazdur R, Abbruzzese JL. Phase II study of dolastatin-10 as first-line treatment for advanced colorectal cancer. Am. J. Clin. Oncol. 2002;25:451–453.

100. Patel S, Keohan ML, Saif MW, Rushing D, Baez L, Feit K, DeJager R, Anderson S. Phase II study of intravenous TZT-1027 in patients with advanced or metastatic soft-tissue sarcomas with prior exposure to anthracycline-based chemotherapy. Cancer 2006;107:2881–2887.

101. Gunasekera SP, Gunasekera M, Longley RE, Schulte GK. Discodermolide: a new bioactive polyhydroxylated lactone from the marine sponge Discodermia dissoluta. J. Org. Chem. 1990;55:4912–4915.

102. ter Haar E, Kowalski RJ, Hamel E, Lin CM, Longley RE, Gunasekera SP, Rosenkranz HS, Day BW. Discodermolide, a cytotoxic marine agent that stabilizes microtubules more potently than taxol. Biochemistry 1996;35:243–250.

103. Mickel SJ. Toward a commercial synthesis of + -discodermolide. Curr. Opin. Drug Discov. Devel. 2004;7:869–881.

104. Mickel SJ, Niederer D, Daeffler R, Osmani A, Kuesters E, Schmid E, Schaer K, Gamboni R, Chen W, Loeser E, Kinder FR, Konigsberger K, Prasad K, Ramsey TM, Repic O, Wang R-M, Florence G, Lyothier I, Paterson I. Large-scale synthesis of the anti-cancer marine natural product + -discodermolide. Part 5: linkage of fragments C and Finale. Org. Process Res. Dev. 2004;8:122–130.

105. Wright AE, Forleo DA, Gunawardana GP, Gunasekera SP, Koehn FE, McConnell OJ. Antitumor tetrahydroisoquinoline alkaloids from the colonial ascidian Ecteinascidia turbinata. J. Org. Chem. 1990;55:4508–4512.

106. Rinehart KL, Holt TG, Fregeau NL, Stroh JG, Keifer PA, Sun F, Li LH, Martin DG. Ecteinascidins 729, 743, 745, 759A, 759B, and 770: potent antitumor agents from the Caribbean tunicate Ecteinascidia turbinata. J. Org. Chem. 1990;55:4512–4515.

107. Corey EJ, Gin DY, Kania RS. Enantioselective total synthesis of ecteinascidin 743. J. Am. Chem. Soc. 1996;118:9202–9203.

108. Martinez EJ, Corey EJ. A new, more efficient, and effective process for the synthesis of a key pentacyclic intermediate for production of ecteinascidin and phthalascidin antitumor agents. Org. Lett. 2000;2:993–996.

109. Martinez EJ, Owa T, Schreiber SL, Corey EJ. Phthalascidin, a synthetic antitumor agent with potency and mode of action comparable to ecteinascidin 743. Proc. Natl. Acad. Sci. U.S.A. 1999;96:3496–3501.

110. Anderson HJ, Coleman JE, Andersen RJ, Roberge M. Cytotoxic peptides hemiasterlin, hemiasterlin A and hemiasterlin B induce mitotic arrest and abnormal spindle formation. Cancer Chemother. Pharmacol. 1997;39:223–226.

111. Loganzo F, Discafani CM, Annable T, Beyer C, Musto S, Hari M, Tan X, Hardy C, Hernandez R, Baxter M, Singanallore T, Khafizova G, Poruchynsky MS, Fojo T, Nieman JA, Ayral-Kaloustian S, Zask A, Andersen RJ, Greenberger LM. HTI-286, a synthetic analogue of the tripeptide hemiasterlin, Is a potent antimicrotubule agent that circumvents P-Glycoprotein-mediated resistance in vitro and in Vivo. Cancer Res. 2003;63:1838–1845.

112. Maiese WM, Lechevalier MP, Lechevalier HA, Korshalla J, Kuck N, Fantini A, Wildey MJ, Thomas J, Greenstein M. Calicheamicins, a novel family of antitumor antibiotics: taxonomy, fermentation and biological properties. J. Antibiot. (Tokyo) 1989;42:558–563.

113. Giles F, Estey E, O'Brien S. Gemtuzumab ozogamicin in the treatment of acute myeloid leukemia. Cancer 2003;98:2095–2104.

114. Lee MD, Dunne TS, Siegel MM, Chang CC, Morton GO, Borders DB. Calichemicins, a novel family of antitumor antibiotics. 1. Chemistry and partial structure of calichemicin.gamma.1I. J. Am. Chem. Soc. 1987;109:3464–3466.

115. Lee MD, Dunne TS, Chang CC, Ellestad GA, Siegel MM, Morton GO, McGahren WJ, Borders DB. Calichemicins, a novel family of antitumor antibiotics. 2. Chemistry and structure of calichemicin.gamma.1I. J. Am. Chem. Soc. 1987;109: 3466–3468.

116. Ruegger A, Kuhn M, Lichti H, Loosli HR, Huguenin R, Quiquerez C, von Wartburg A. Cyclosporin A, a peptide metabolite from trichoderma polysporum link ex pers. Rifai, with a remarkable immunosuppressive activity. Helv. Chim. Acta 1976;59: 1075–1092.

117. Wenger RM. Pharmacology of cyclosporin sandimmune. II. Chemistry. Pharmacol. Rev. 1990;41:243–247.

118. Zenke G, Baumann G, Wenger R, Hiestand P, Quesniaux V, Andersen E, Schreier MH. Molecular mechanisms of immunosuppression by cyclosporins. Ann. N. Y. Acad. Sci. 1993;685:330–335.

119. Bua J, Ruiz AM, Potenza M, Fichera LE. In vitro anti-parasitic activity of Cyclosporin A analogs on Trypanosoma cruzi. Bioorg. Med. Chem. Lett. 2004;14: 4633–4637.

120. Eckstein JW, Fung J. A new class of cyclosporin analogues for the treatment of asthma. Expert. Opin. Investig. Drugs 2003;12:647–653.

121. Kino T, Hatanaka H, Miyata S, Inamura N, Nishiyama M, Yajima T, Goto T, Okuhara M, Kohsaka M, Aoki H, et al. FK-506, a novel immunosuppressant isolated from a Streptomyces. II. Immunosuppressive effect of FK-506 in vitro. J. Antibiot. (Tokyo) 1987;40:1256–1265.

122. Kino T, Inamura N, Sakai F, Nakahara K, Goto T, Okuhara M, Kohsaka M, Aoki H, Ochiai T. Effect of FK-506 on human mixed lymphocyte reaction in vitro. Transplant. Proc. 1987;19:36–39.

123. Ochiai T, Nagata M, Nakajima K, Suzuki T, Sakamoto K, Enomoto K, Gunji Y, Uematsu T, Goto T, Hori S, et al. Studies of the effects of FK506 on renal allografting in the beagle dog. Transplantation 1987;44:729–733.

124. Kino T, Goto T. Discovery of FK-506 and update. Ann. N. Y. Acad. Sci. 1993;685: 13–21.

125. Tanaka H, Kuroda A, Marusawa H, Hatanaka H, Kino T, Goto T, Hashimoto M, Taga T. Structure of FK506, a novel immunosuppressant isolated from Streptomyces. J. Am. Chem. Soc. 1987;109:5031–5033.

126. Spencer CM, Goa KL, Gillis JC. Tacrolimus. An update of its pharmacology and clinical efficacy in the management of organ transplantation. Drugs 1997;54: 925–975.

127. Kahan BD, Camardo JS. Rapamycin: clinical results and future opportunities. Transplantation 2001;72:1181–1193.

128. Camardo J. The Rapamune era of immunosuppression 2003: the journey from the laboratory to clinical transplantation. Transplant. Proc. 2003;35:18–24.

129. Baker H, Sidorowicz A, Sehgal SN, Vezina C. Rapamycin AY-22,989, a new antifungal antibiotic. III. In vitro and in vivo evaluation. J. Antibiot. (Tokyo) 1978;31: 539–545.

130. Sehgal SN, Baker H, Vezina C. Rapamycin AY-22,989, a new antifungal antibiotic. II. Fermentation, isolation and characterization. J. Antibiot. (Tokyo) 1975;28: 727–732.

131. Schreiber SL. Chemistry and biology of the immunophilins and their immunosuppressive ligands. Science 1991;251:283–287.

132. Ho S, Clipstone N, Timmermann L, Northrop J, Graef I, Fiorentino D, Nourse J, Crabtree GR. The mechanism of action of cyclosporin A and FK506. Clin. Immunol. Immunopathol. 1996;80:40–45.

133. Schreiber SL, Crabtree GR. The mechanism of action of cyclosporin A and FK506. Immunol. Today 1992;13:136–142.

134. Campbell IM, Calzadilla CH, McCorkindale NJ. Some new metabolites related to mycophenolic acid. Tetrahedron Lett.1966:5107–5111.

135. Behrend M. Mycophenolate mofetil Cellcept. Expert Opin. Investig. Drugs 1998;7: 1509–1519.

136. Sollinger HW. Mycophenolate mofetil for the prevention of acute rejection in primary cadaveric renal allograft recipients. U.S. Renal Transplant Mycophenolate Mofetil Study Group. Transplantation 1995;60:225–232.

137. Alberts AW, Chen J, Kuron G, Hunt V, Huff J, Hoffman C, Rothrock J, Lopez M, Joshua H, Harris E, Patchett A, Monaghan R, Currie S, Stapley E, Albers-Schonberg G, Hensens O, Hirshfield J, Hoogsteen K, Liesch J, Springer J. Mevinolin: a highly potent competitive inhibitor of hydroxymethylglutaryl-coenzyme A reductase and a cholesterol-lowering agent. Proc. Natl. Acad. Sci. U.S.A. 1980;77:3957–3961.

138. Patchett AA. 2002 Alfred Burger Award Address in Medicinal Chemistry. Natural products and design: interrelated approaches in drug discovery. J. Med. Chem. 2002;45:5609–5616.

139. Coukell AJ, Wilde MI. Pravastatin. A pharmacoeconomic review of its use in primary and secondary prevention of coronary heart disease. Pharmacoeconomics 1998;14:217–236.

140. Plosker GL, Wagstaff AJ. Fluvastatin: a review of its pharmacology and use in the management of hypercholesterolaemia. Drugs 1996;51:433–459.

141. Roth BD. The discovery and development of atorvastatin, a potent novel hypolipidemic agent. Prog. Med. Chem. 2002;40:1–22.

142. Plosker GL, Dunn CI, Figgitt DP. Cerivastatin: a review of its pharmacological properties and therapeutic efficacy in the management of hypercholesterolaemia. Drugs 2000;60:1179–1206.

143. Chong PH, Yim BT. Rosuvastatin for the treatment of patients with hypercholesterolemia. Ann. Pharmacother. 2002;36:93–101.

144. Ondetti MA, Williams NJ, Sabo EF, Pluscec J, Weaver ER, Kocy O. Angiotensin-converting enzyme inhibitors from the venom of Bothrops jararaca. Isolation, elucidation of structure, and synthesis. Biochemistry 1971;10:4033–4039.

145. Ondetti MA, Rubin B, Cushman DW. Design of specific inhibitors of angiotensin-converting enzyme: new class of orally active antihypertensive agents. Science 1977;196:441–444.

146. Burg RW, Miller BM, Baker EE, Birnbaum J, Currie SA, Hartman R, Kong YL, Monaghan RL, Olson G, Putter I, Tunac JB, Wallick H, Stapley EO, Oiwa R, Omura S. Avermectins, new family of potent anthelmintic agents: producing organism and fermentation. Antimicrob, Agents Chemother. 1979;15:361–367.

147. Cully DF, Vassilatis DK, Liu KK, Paress PS, Van der Ploeg LH, Schaeffer JM, Arena JP. Cloning of an avermectin-sensitive glutamate-gated chloride channel from Caenorhabditis elegans. Nature 1994;371:707–711.

148. Davies HG, Green RH. Avermectins and milbemycins. Nat. Prod. Rep. 1986;3:87–121.

149. Chabala JC, Mrozik H, Tolman RL, Eskola P, Lusi A, Peterson LH, Woods MF, Fisher MH, Campbell WC, Egerton JR, Ostlind DA. Ivermectin, a new broad-spectrum antiparasitic agent. J. Med. Chem. 1980;23:1134–1136.

150. Goa KL, McTavish D, Clissold SP. Ivermectin. A review of its antifilarial activity, pharmacokinetic properties and clinical efficacy in onchocerciasis. Drugs 1991;42:640–658.

151. Kirst HA, Creemer LC, Naylor SA, Pugh PT, Snyder DE, Winkle JR, Lowe LB, Rothwell JT, Sparks TC, Worden TV. Evaluation and development of spinosyns to control ectoparasites on cattle and sheep. Curr. Top. Med. Chem. 2002;2:675–699.

152. Ondeyka JG, Helms GL, Hensens OD, Goetz MA, Zink DL, Tsipouras A, Shoop WL, Slayton L, Dombrowski AW, Polishook JD, Ostlind DA, Tsou NN, Ball RG,

Singh SB. Nodulisporic Acid A, a novel and potent insecticide from a nodulisporium sp. Isolation, structure determination, and chemical transformations. J. Am. Chem. Soc. 1997;119:8809–8816.

153. Meinke PT, Smith MM, Shoop WL. Nodulisporic acid: its chemistry and biology. Curr. Top. Med. Chem. 2002;2:655–674.

154. Chang RS, Lotti VJ, Monaghan RL, Birnbaum J, Stapley EO, Goetz MA, Albers-Schonberg G, Patchett AA, Liesch JM, Hensens OD, et al. A potent nonpeptide cholecystokinin antagonist selective for peripheral tissues isolated from Aspergillus alliaceus. Science 1985;230:177–179.

155. Evans BE, Bock MG, Rittle KE, DiPardo RM, Whitter WL, Veber DF, Anderson PS, Freidinger RM. Design of potent, orally effective, nonpeptidal antagonists of the peptide hormone cholecystokinin. Proc. Natl. Acad. Sci. U.S.A. 1986;83:4918–4922.

156. Bock MG, DiPardo RM, Evans BE, Rittle KE, Whitter WL, Veber DE, Anderson PS, Freidinger RM. Benzodiazepine gastrin and brain cholecystokinin receptor ligands: L-365,260. J. Med. Chem. 1989;32:13–16.

157. Darkin-Rattray SJ, Gurnett AM, Myers RW, Dulski PM, Crumley TM, Allocco JJ, Cannova C, Meinke PT, Colletti SL, Bednarek MA, Singh SB, Goetz MA, Dombrowski AW, Polishook JD, Schmatz DM. Apicidin: a novel antiprotozoal agent that inhibits parasite histone deacetylase. Proc. Natl. Acad. Sci. U.S.A. 1996;93: 13143–13147.

158. Singh SB, Zink DL, Liesch JM, Mosley RT, Dombrowski AW, Bills GF, Darkin-Rattray SJ, Schmatz DM, Goetz MA. Structure and chemistry of apicidins, a class of novel cyclic tetrapeptides without a terminal alpha-keto epoxide as inhibitors of histone deacetylase with potent antiprotozoal activities. J. Org. Chem. 2002;67: 815–825.

159. Han JW, Ahn SH, Park SH, Wang SY, Bae GU, Seo DW, Kwon HK, Hong S, Lee HY, Lee YW, Lee HW. Apicidin, a histone deacetylase inhibitor, inhibits proliferation of tumor cells via induction of p21WAF1/Cip1 and gelsolin. Cancer Res. 2000;60:6068–6074.

160. Kim MS, Son MW, Kim WB, In Park Y, Moon A. Apicidin, an inhibitor of histone deacetylase, prevents H-ras-induced invasive phenotype. Cancer Lett. 2000;157:23–30.

161. Jones P, Altamura S, Chakravarty PK, Cecchetti O, De Francesco R, Gallinari P, Ingenito R, Meinke PT, Petrocchi A, Rowley M, Scarpelli R, Serafini S, Steinkuhler C. A series of novel, potent, and selective histone deacetylase inhibitors. Bioorg. Med. Chem. Lett. 2006;16:5948–5952.

162. Remiszewski SW. Recent advances in the discovery of small molecule histone deacetylase inhibitors. Curr. Opin. Drug Discov. Devel. 2002;5:487–499.

163. Singh SB, Jayasuriya H, Ondeyka JG, Herath KB, Zhang C, Zink DL, Tsou NN, Ball RG, Basilio A, Genilloud O, Diez MT, Vicente F, Pelaez F, Young K, Wang J. Isolation, structure, and absolute stereochemistry of platensimycin, a broad spectrum antibiotic discovered using an antisense differential sensitivity strategy. J. Am. Chem. Soc. 2006;128:11916–11920.

164. Wang J, Soisson SM, Young K, Shoop W, Kodali S, Galgoci A, Painter R, Parthasarathy G, Tang YS, Cummings R, Ha S, Dorso K, Motyl M, Jayasuriya H, Ondeyka J, Herath K, Zhang C, Hernandez L, Allocco J, Basilio A, Tormo JR, Genilloud O, Vicente F, Pelaez F, Colwell L, Lee SH, Michael B, Felcetto T, Gill C,

Silver LL, Hermes JD, Bartizal K, Barrett J, Schmatz D, Becker JW, Cully D, Singh SB. Platensimycin is a selective FabF inhibitor with potent antibiotic properties. Nature 2006;441:358–361.

165. Wang J, Kodali S, Lee SH, Galgoci A, Painter R, Dorso K, Racine F, Motyl M, Hernandez L, Tinney E, Colletti SL, Herath K, Cummings R, Salazar O, Gonzalez I, Basilio A, Vicente F, Genilloud O, Pelaez F, Jayasuriya H, Young K, Cully DF, Singh SB. Discovery of platencin, a dual FabF and FabH inhibitor with in vivo antibiotic properties. Proc. Natl. Acad. Sci. U.S.A 2007;104:7612–7616.

166. Jayasuriya H, Herath KB, Zhang C, Zink DL, Basilio A, Genilloud O, Diez MT, Vicente F, Gonzalez I, Salazar O, Pelaez F, Cummings R, Ha S, Wang J, Singh SB. Isolation and structure of platencin: a FabH and FabF dual inhibitor with potent broad-spectrum antibiotic activity. Angew. Chem. Int. Ed. Engl. 2007;46: 4684–4688.

13

NATURAL PRODUCTS
AS ANTICANCER AGENTS

DAVID G. I. KINGSTON

Department of Chemistry, Virginia Polytechnic Institute and State University, Blacksburg, Virginia

DAVID J. NEWMAN

Natural Products Branch, Developmental Therapeutics Program, Division of Cancer Treatment and Diagnosis, National Cancer Institute, Frederick, Maryland

Natural products have provided some of the most effective drugs for the treatment of cancer, including such well-known drugs as paclitaxel (Taxol™; Bristol-Myers Squibb) adriamycin, vinblastine, and vincristine. Natural products have also provided many compounds that have led to the discovery of new biochemical mechanisms. This review summarizes the major natural products in clinical use today and introduces several new ones on the cusp of entering clinical practice. The review is organized by mechanism of action, with compounds that interact with proteins discussed first, followed by compounds that interact with RNA or DNA.

Natural products were the original source of almost all the drugs used by mankind before 1900, and they continue to be a major source of new drugs and drug leads (1, 2). The reasons for the continued importance of natural products are not hard to discover. In the first place, a high correlation exists between the properties of drugs and those of natural products (3, 4). In addition, natural products usually have built-in chirality, and they are thus uniquely suited to bind to complex proteins and other biologic receptors. Finally, natural products have been enormously successful as drugs and drug leads, not only in the anticancer area but also in many other pharmaceutical areas (5). It is thus unsurprising

Natural Products in Chemical Biology, First Edition. Edited by Natanya Civjan.
© 2012 John Wiley & Sons, Inc. Published 2012 by John Wiley & Sons, Inc.

that several authors have gone on record as advocating an increase in the drug discovery effort assigned to natural products (4, 6, 7).

This review summarizes the contributions of natural products to the discovery and development of anticancer agents. It includes information on many natural products and natural product analogs that are in clinical use as anticancer drugs, and it describes some natural product drugs in late-stage clinical trials. Because of space limitations, it is does not provide a comprehensive listing of all natural product and natural product-derived anticancer agents; readers interested in such a listing should consult a recent review (8).

13.1 OVERVIEW OF NATURAL PRODUCTS AS ANTICANCER AGENTS

As of the time of writing, 178 drugs are approved world-wide for the treatment of cancer in all of its manifestations, and 175 of these are listed together with their classifications as to source in a recent review (8). Of the 178 approved antitumor agents, 25 (14%) are natural products, 48 (27%) are modified natural products, and 20 (11%) are synthetic compounds derived from a natural product pharmacophore. Natural products have thus led to 52% of the approved drugs against the collection of diseases that go by the collective name of cancer. Another 20 drugs (11%) are biologics, with the remainder being synthetic compounds. The highly significant contribution of natural products to anticancer drug discovery is clear from these figures.

The term "natural product" describes a broad class of anticancer agents, which range from complex compounds like paclitaxel and vinblastine to relatively simple compounds such as combretastatin-A4. Associated with these different structures are several different mechanisms of action, some of which were only discovered when the corresponding natural product was investigated. The following sections are divided on the basis of the mechanism of action of the drugs rather than their source, and consequently, any given section may include plant, microbial, or marine-derived agents. The broadest division is between those compounds that act by targeting proteins in some way and those compounds that act by direct interactions with DNA or RNA. Thus, these two broad areas provide the two major sections of this review.

13.2 COMPOUNDS THAT TARGET PROTEINS

The mammalian cell cycle is a complex and carefully regulated biologic process that leads to cell division, and faulty regulation of this cycle is one feature of most cancers. The cell cycle thus offers several targets for therapeutic intervention, and several of the proteins involved either directly or indirectly in controlling this cycle are the targets of some important anticancer agents. The most important targets, in terms of the number of drugs that target them, are the proteins tubulin, topoisomerase I, and topoisomerase II, but other proteins such as the checkpoint kinases chk1 and ckh2 and the heat shock protein Hsp90 are also the targets of some drugs.

The protein tubulin is an interesting and important target. It exists in both α and β forms, and during a normal cell cycle, these two monomeric proteins polymerize into microtubules by noncovalent interaction; these microtubules are then involved in the reorganization of the chromosomes into the nuclei of the mother and daughter cells. After mitosis, the microtubules dissociate to α and β tubulin monomers. Therapeutic intervention, which controls inhibition of mitosis and eventual apoptotic cell death, can occur either by inhibition of the assembly of tubulin monomers into microtubules or by inhibition of the dissociation of microtubules to α and β tubulin monomers. This process is dynamic, and even very small perturbations in assembly and/or disassembly of the monomers/dimers may lead to cell death, frequently via the apoptotic cascade(s), but at agent concentrations that are sometime orders of magnitude lower than those quoted in the literature to give formal inhibition of tubulin *in vitro*.

The topoisomerases are enzymes that change the topology of DNA. Normal supercoiled DNA needs to unwind to undergo replication, and this unwinding requires that the DNA be cleaved and religated. This cleavage can be brought about by a single-strand break, which is mediated by topoisomerase I, or a double-strand break, which is brought about by topoisomerase II.

Other mechanisms of action involve inhibitors of the heat shock protein Hsp-90, of checkpoint kinases, and inhibitors of protein degradation by interaction with the proteasome.

13.2.1 Compounds that Inhibit Tubulin Assembly

13.2.1.1 The vinca alkaloids The antitumor alkaloids vinblastine (**1**) and vincristine (**2**) were the first natural products to be used on a large scale as anticancer agents, and they thus blazed the trail for others that came afterward. Vinblastine (as vincaleukoblastine) was isolated from *Catharanthus roseus* (L.) G. Don, which was formerly known as *Vinca rosea* L., by two independent teams during the 1950s (9, 10), and vincristine (as leurocristine) was isolated and structurally characterized by Svoboda (11) and Neuss et al. (10) in 1961 and early 1962, respectively. These alkaloids inhibit the polymerization of tubulin to microtubules. Vinblastine is used in combination with other agents for treatment of Hodgkin's disease and bladder and breast cancers, whereas vincristine is used for treatment of acute lymphocytic leukemias and lymphomas. The semisynthetic analogs vindesine (**3**) and vinorelbine (**4**) have been developed more recently. Vindesine, which was first developed in the 1970s, is in clinical use; it seems to be more active than vincristine against non-small-cell lung cancer, but it also has a higher hematological toxicity than vincristine, so its utility is still being evaluated (12). Vinorelbine has been approved for treatment of non-small-cell lung cancer, and the fluorinated analog vinflunine (structure not shown) has entered clinical trials (13). For a recent general review of the vinca alkaloids and their analogs, see Reference 14.

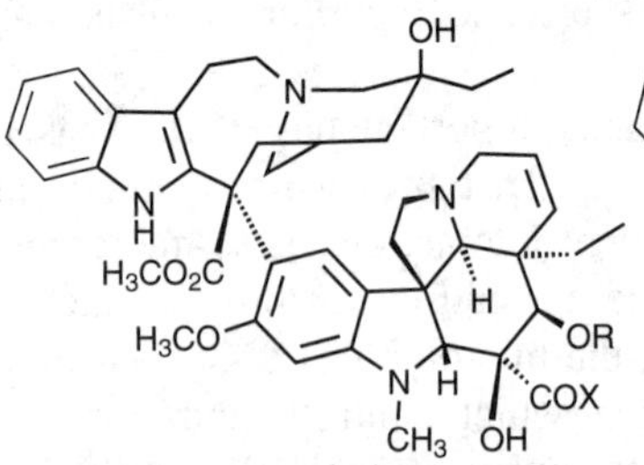

1 Vinblastine X = OCH₃, R = COCH₃
3 Vindesine X = NH₂, R = H

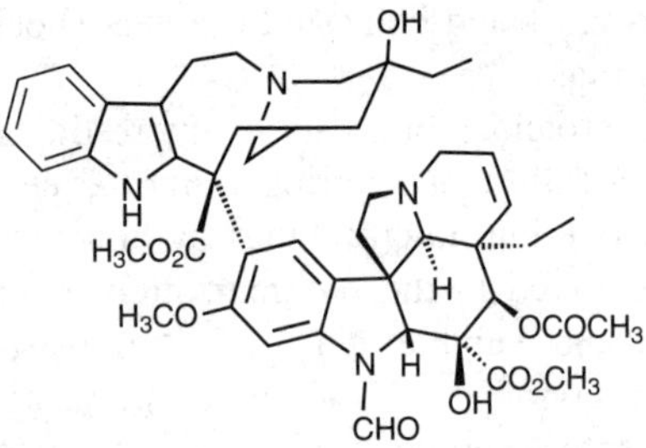

2 Vincristine

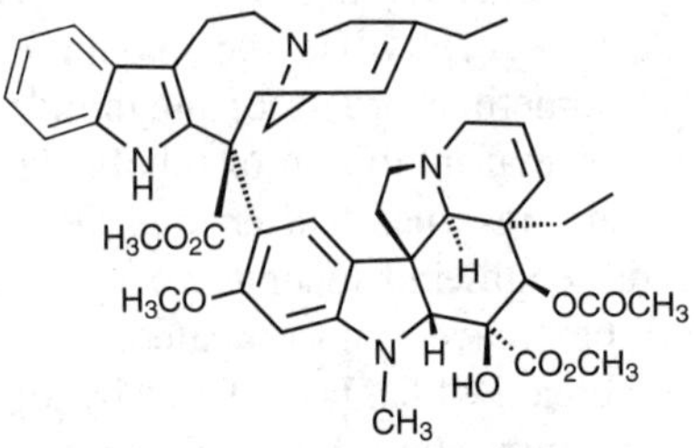

4 Vinorelbine

Figure 13.1 Structures of the vinca alkaloids vinblastine, vincristine, videsine, and vinorelbine.

13.2.1.2 Combretastatin The first member of this class of compounds, (−)-combretastatin (**5**), was isolated from *Combretum caffrum* in 1982 (15). Subsequent studies led to the isolation of many additional combretastatins, including combretastatins A1 (**6**) and A2 (**7**), which are two of the most active members of this class, with potent activity as inhibitors of tubulin assembly. Additional development by the Pettit group led to the design of combretastatin A4 phosphate (CA4-P) (**8**) as a promising drug candidate. This compound is actually a prodrug, which only functions as an inhibitor of tubulin assembly after hydrolysis of the phosphate was subsequently found to be an important member of a new class of compounds known as tumor-specific vascular targeting agents. CA4-P operates by binding to endothelial cell tubulin and causing changes in the morphology of the endothelial cells lining the microvessels feeding the tumors. This process causes disruption of the blood flow, which makes the microvessels unable to deliver oxygen to the tumor and leads ultimately to tumor necrosis. It received orphan drug approval by the Food and Drug Administration (FDA) in 2003 for thyroid cancer, and the FDA has approved it for a "fast track" Phase II clinical trial against anaplastic thyroid cancer (16).

13.2.1.3 Eribulin (E7389) The natural product analog E7389 is not yet in clinical use, but it is in Phase III clinical trials and hopefully will enter clinical use

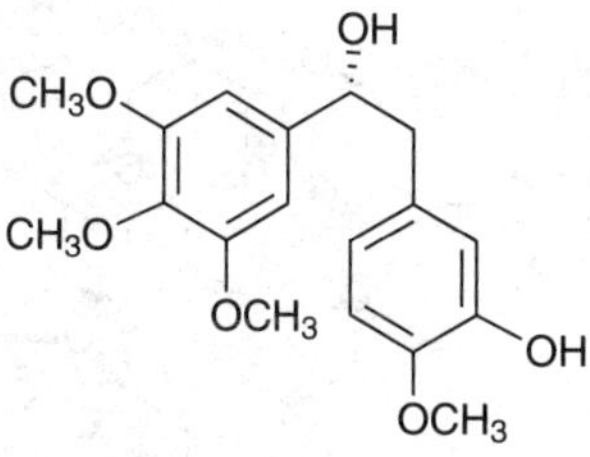

5 (-)-Combretastatin

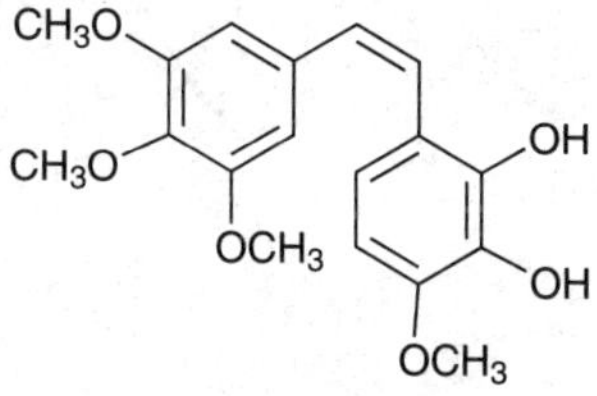

6 Combretastatin A1

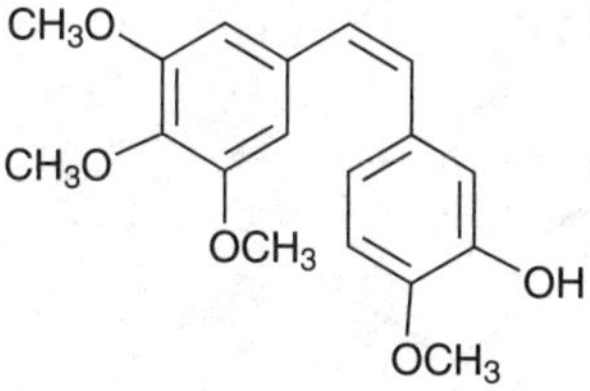

7 Combretastatin A2

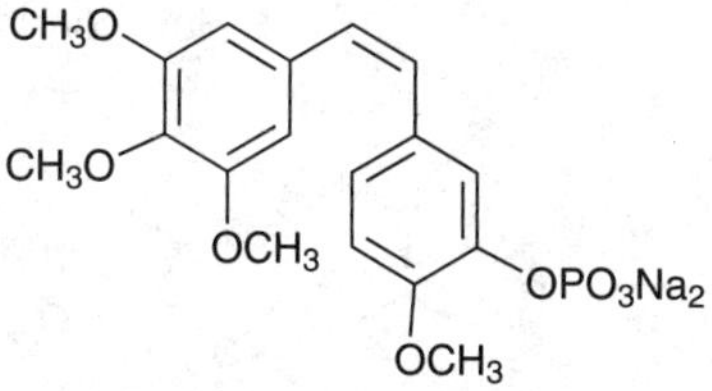

8 Combretastatin A4 phosphate

Figure 13.2 Structures of combretastatin and it analogs.

within the next few years. The natural product on which it is based, halichondrin B (**9**), is a member of a relatively large family of congeners with a polycyclic macrolide structure, which is reminiscent of ionophores. Many years of synthetic work by Kishi's group at Harvard and a group at the Eisai Research Institute in the United States led to the selection of the two lead compounds E7389 (**10**) and E7390 (**11**) (17). These compounds were compared in both *in vitro* and *in vivo* assays at the National Cancer Institute (NCI) with the natural product obtained by deep-water dredging of a producing sponge. In an example of the skill of synthetic chemists when given what initially seemed to be an almost impossible task, E7389 was prepared in large quantity and entered clinical trials in 2001 as an inhibitor of tubulin assembly. It has a mechanism of action different from that of other tubulin interactive agents; it inhibits microtubule growth, but not shortening, and sequesters tubulin into aggregates (18). It showed good activity against refractory breast carcinoma in Phase II studies and entered Phase III studies for the same indication in late August 2006. Very recently, Hamel's group at NCI have reported additional investigations on the mechanism of binding of both halichondrin B and E7389 to tubulin, which indicates that these agents may well form small, highly unstable aberrant tubulin polymers rather than the conventional massive stable structures found with vinca alkaloids and the antimitotic peptides (19).

13.2.1.4 Dolastatin The dolastatins are a class of bioactive peptides isolated by the Pettit group from the Indian Ocean nudibranch, *Dolabella auricularia*. A

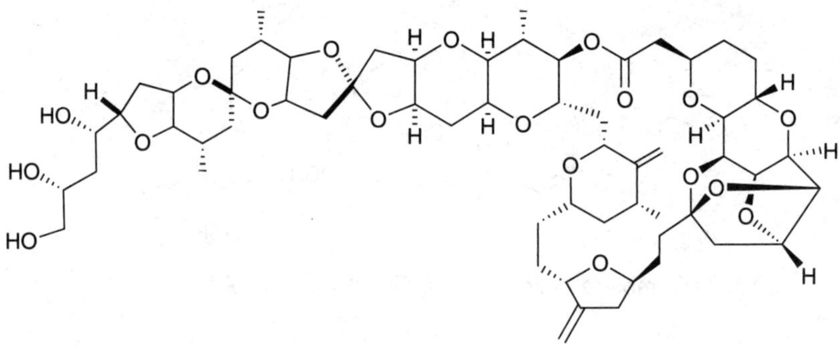

9 Halichondrin B

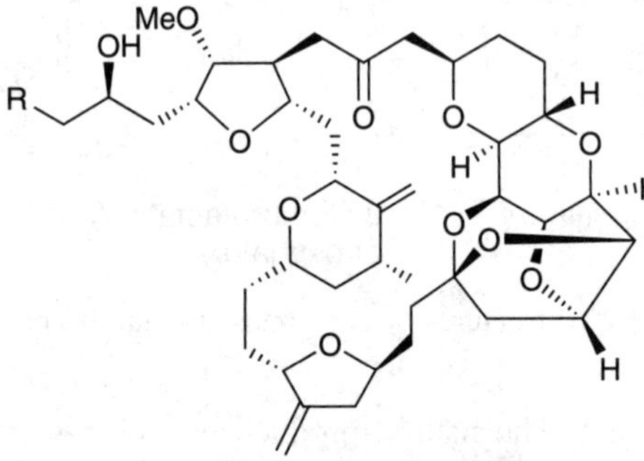

10 Eribulin (E7389); R = NH$_2$
11 E7390; R = OH

Figure 13.3   Structures of halichondrin B and its synthetic analogs E7389 and E7390.

total of 18 compounds were isolated over a 20-year period, with structures varying from relatively simple linear peptides to cyclic peptidolactones with nonpeptide components. Dolastatin 10 (**12**) was the most potent compound, with cytotoxicity in the subnanomolar range. It was shown to be a potent antimitotic agent, binding strongly to the β-subunit of tubulin (20), but it could not be isolated in sufficient quantity because of the scarcity of the source and of the low levels of secondary metabolites in the nudibranch. Pettit and colleagues thus devised many synthetic schemes, which led to the production of enough dolastatin 10 to go into human clinical trials as a tubulin interactive agent. Although it progressed to Phase II, it did not continue further because of a lack of activity and toxicity. However, the base structure led to the synthesis and biologic evaluation by various groups of a large number of related compounds, and synthadotin (also known as tasidotin, **13**) emerged as a lead compound (21, 22). Synthadotin is an orally available synthetic derivative of dolastatin 15, and it is in Phase II trials as a tubulin interactive agent. Its discovery provides another example of the skill of synthetic chemists and the potential of novel natural products to be developed into drugs.

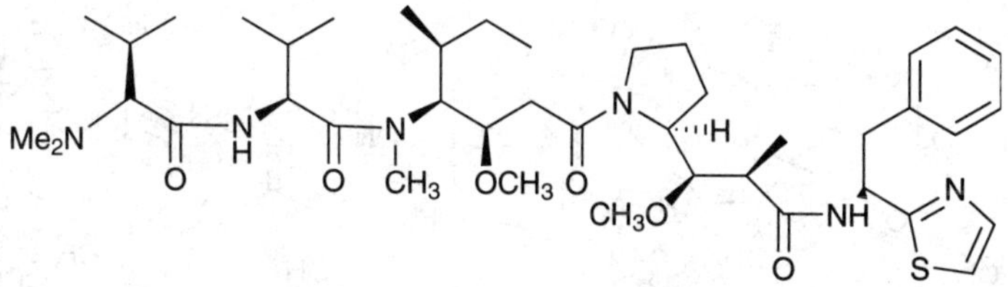

12 Dolastatin 10

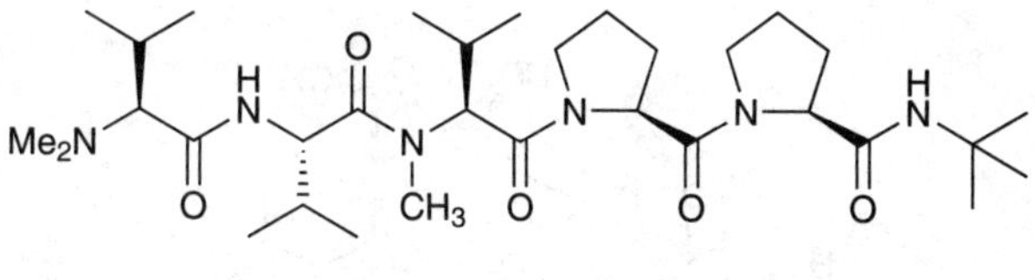

13 ILX-651 (Synthatodin or Tasidotin)

Figure 13.4   Structures of dolastatin 10 and ILX-651 (synthadotin or tasidotin).

13.2.2   Compounds that Promote Tubulin Assembly

13.2.2.1 Paclitaxel (Taxol)   Paclitaxel (Taxol™; Bristol Myers Squibb **14**) and its semisynthetic analog docetaxel (**15**) are two of the most important anticancer agents of the last 25 years. Paclitaxel was isolated originally by Wall and Wani from *Taxus brevifolia* (23) and named taxol; this name was later trademarked by Bristol-Myers Squibb (New York, NY), and the name paclitaxel was substituted. The semisynthetic paclitaxel analog docetaxel was prepared by Potier and his collaborators (24). Paclitaxel was selected as a development candidate in 1977 based on its good activity against human tumor xenografts in nude mice. Its drug development was challenging because of problems with solubility and supply, but the supply problem was overcome by a variety of methods (25, 26), and a formulation in ethanol and Cremophor EL enabled clinical work to proceed (27). The discovery of its mechanism of action as a promoter of tubulin assembly (28) was important in maintaining interest in its development, which was a particularly challenging one. As noted, the normal function of a cell requires that microtubules be in dynamic equilibrium with monomeric tubulins. Paclitaxel was the first compound found to disrupt this equilibrium by promoting microtubule assembly.

Paclitaxel and docetaxel (**15**) are now used, either as single agents or in combination with other drugs such as cisplatin, for the treatment of ovarian cancer (29), breast cancer (30), and non-small-cell lung cancer (31). Paclitaxel has also found an important application as a coating in stents to prevent restenosis (32). It and docetaxel are major drugs, with combined sales of taxane anticancer drugs being over $3 billion in 2004.

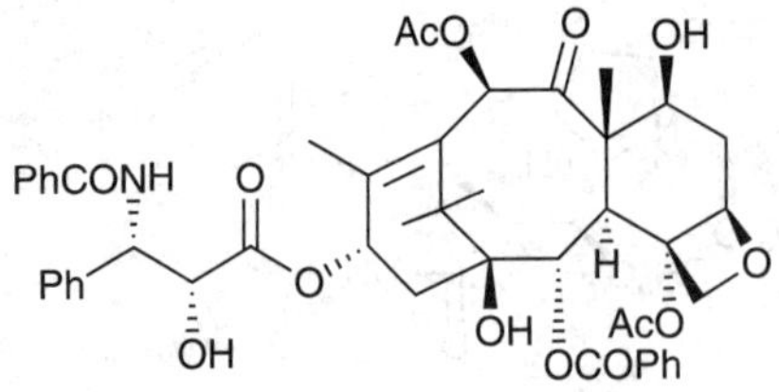

Figure 13.5 Structures of paclitaxel (Taxol), docetaxel, and britaxel-5.

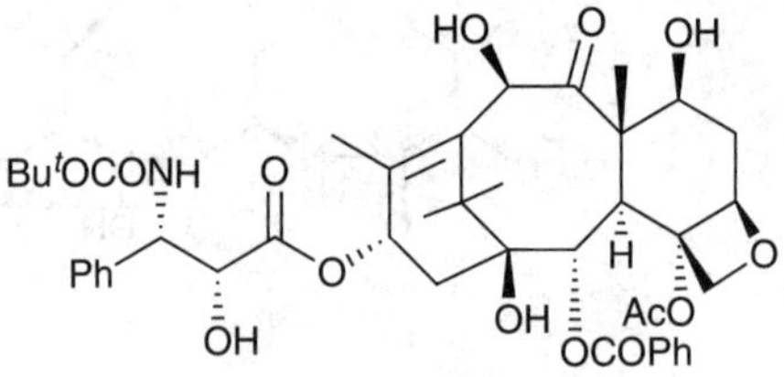

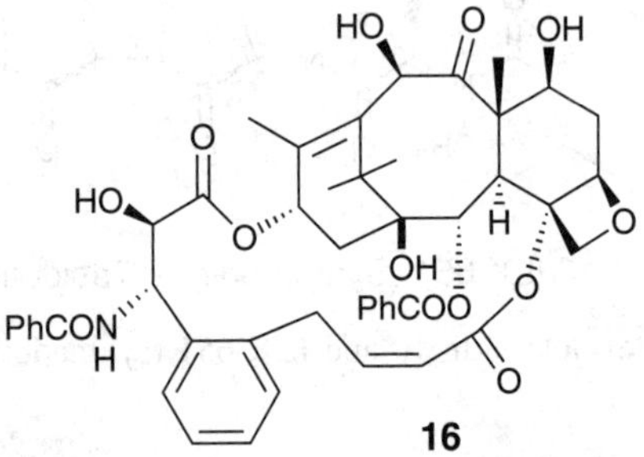

Numerous analogs and prodrugs of paclitaxel have been developed, and several of these are in clinical trials. Full coverage of these compounds is beyond the scope of this chapter, but additional details are provided in a recent review (33). As of early 2007, the only new form of paclitaxel in clinical use is the albumin-bound paclitaxel known as Abraxane (Abraxis BioScience, Schaumberg, IL) (34).

The binding of paclitaxel to microtubules has been studied extensively. The polymeric and noncrystalline nature of the tubulin–taxol complex prevents a direct approach by X-ray crystallography, but Lowe et al. (35) could determine the structure of tubulin at 3.5 Å resolution by electron diffraction. Using this structure, various possible binding orientations of paclitaxel on tubulin have been proposed, but recent REDOR NMR studies have established T-Taxol as the most probable conformation (36). The synthesis of the highly active bridged analog britaxel-5 (**16**), which is constrained to a T-Taxol conformation, confirmed this hypothesis (37).

13.2.2.2 Epothilones The epothilones A-D (**17–20**) were isolated from the myxobacterium *Sorangium cellulosum* as antifungal agents (38), but they were found subsequently to have the same mechanism of action as paclitaxel, which promotes the assembly of tubulin into microtubules (39). The epothilones are thus of great interest as potential antitumor agents because of their mechanism of action and because they are also active against some paclitaxel-resistant cell lines. At first glance, they would seem to have a very different shape

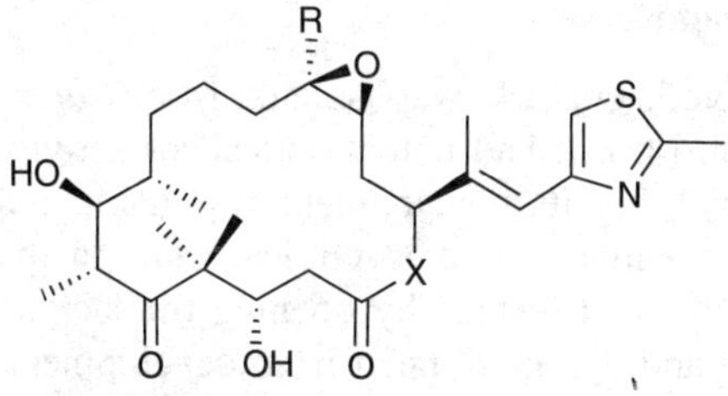

17 Epothilone A R = H, X = O
18 Epothilone B R = Me, X = O
21 Ixabepilone R = Me, X = NH

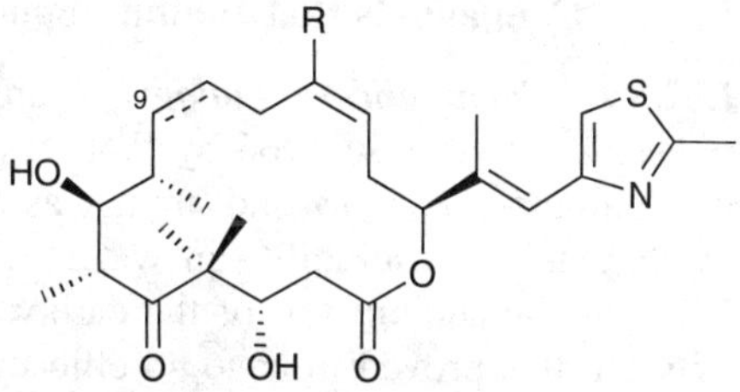

19 Epothilone C R =H, 9,10-dihydro
20 Epothilone D R = Me, 9,10-dihydro
22 R = CF$_3$, 9,10-dehydro

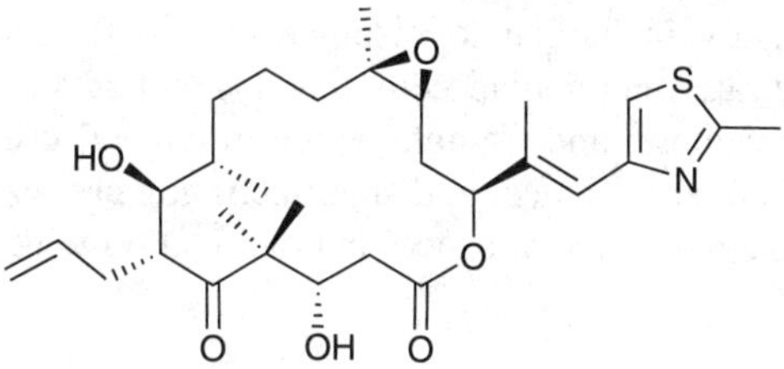

23 ZK-EPO

Figure 13.6 Structures of the epothilones.

from paclitaxel, but molecular modeling has shown that some significant common structural features exist in the two base molecules (40). Originally, the epothilones were difficult to obtain in large quantity, and a significant amount of work was performed in academia and industry to synthesize both epothilones A and B and their more active precursors, epothilones C (**19**) and D (**20**). However, by using genetic manipulation, Frykman et al. (41) cloned and expressed the polyketide gene cluster that produces epothilones A and B. Subsequent removal of the terminal gene for the P$_{450}$ enzyme and transfer to a different host enabled them to produce crystalline epothilone D from a large-scale fermentation.

The aza-analog of epothilone B (ixabepilone, **21**), which was synthesized by Bristol-Myers Squibb, and epothilone B (patupilone) are in Phase III trials. Epothilone D (KOS-862, **20**) and ZK-EPO (**23**) (42) are in Phase II clinical trials, and the synthetic analog fludelone (**22**) looks very promising in animal trials (43). However, an evaluation of the epothilones as a class is less optimistic, concluding that "[d]isappointingly, however, clinical activity has been limited to taxane-sensitive tumor types (prostate cancer and breast cancer) and does not seem to be distinctly different to the activity of taxanes .... Epothilones should certainly not be considered as alternative taxanes, but whether epothilones are here to stay or will fade away has yet to be determined." (44).

13.2.3   Compounds that Inhibit Topoisomerase I

13.2.3.1 Camptothecin analogs   Camptothecin (**24**) was isolated from *Camptotheca acuminata* in 1966 by Wall et al. (45). It had potent anticancer activity in preliminary *in vitro* and animal assays, but its development was hampered by its extreme insolubility in water. It eventually entered clinical trials in the 1970s as the sodium salt of the carboxylic acid formed by opening the lactone ring, but this proved to have no efficacy and it was dropped from development. Interest in camptothecin was rekindled by the discovery that its primary cellular target was inhibition of topoisomerase I (46). Extensive medicinal chemical studies then led to the development of the two water-soluble derivatives topotecan (Hycamtin; GlaxoSmithKline, Brentford, Middlesex, United Kingdom, **25**) and irinotecan (Camptosar; Pfizer, New York, NY, **26**). The camptothecins are unique pharmacologically in having topoisomerase I as their only target, and in being able to penetrate mammalian cells readily, and several analogs are in clinical trials (47). Hycamptin and Camptosar are in clinical use for second-line treatment of metastatic ovarian cancer and small-cell lung cancer (topotecan) and for treatment of metastatic colorectal cancer in combination with 5FU/leucovorin (irinotecan) (47, 48).

13.2.3.2 Rebeccamycin   Rebeccamycin is an indolocarbazole; a comprehensive review of this class has appeared recently (49). Compounds related to rebeccamycin (**27**) are extremely interesting from a mechanistic standpoint,

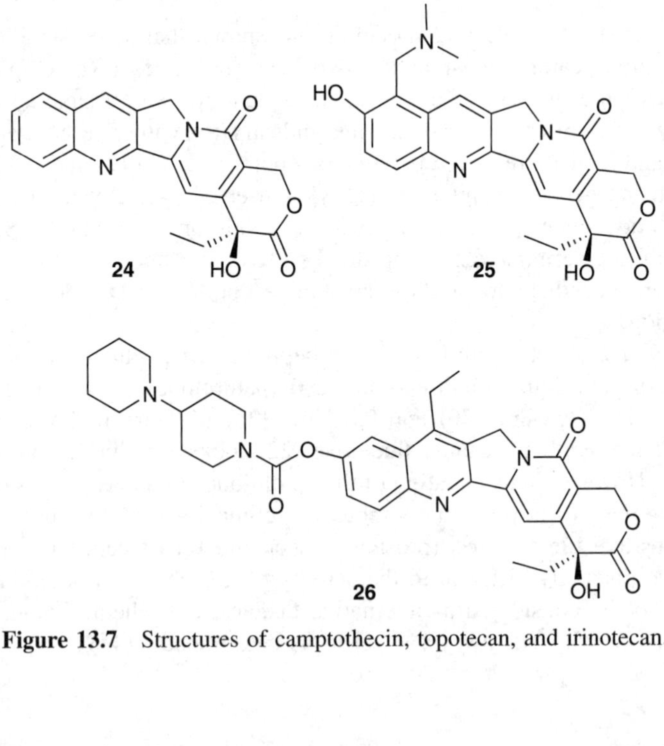

Figure 13.7   Structures of camptothecin, topotecan, and irinotecan.

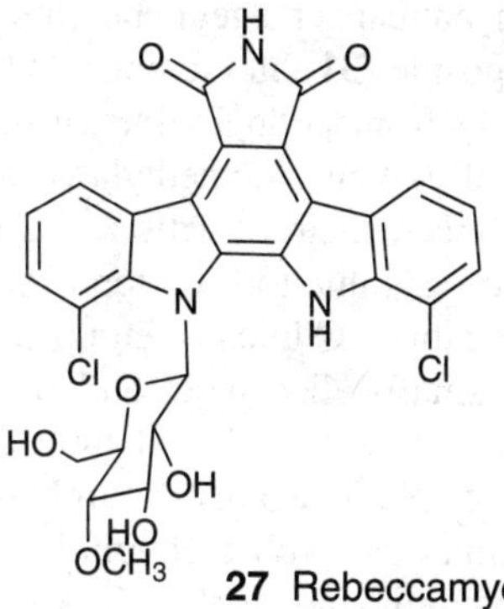

27 Rebeccamycin

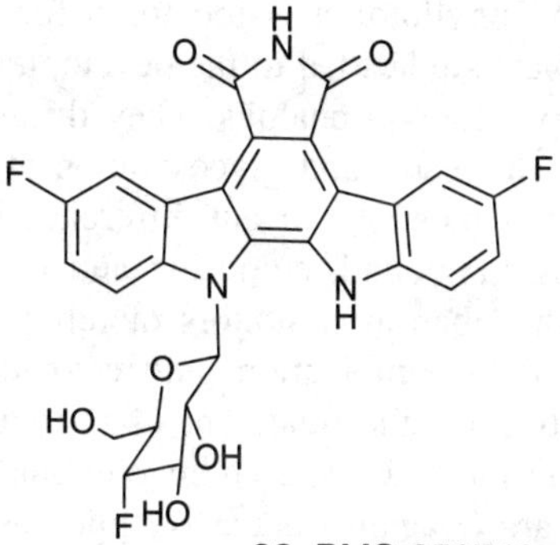

28 BMS-250749

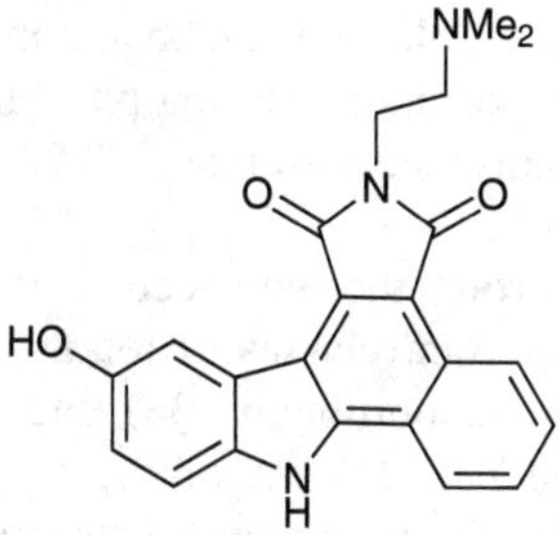

29

Figure 13.8 Structures of rebeccamycin and BMS-250749.

because relatively simple modifications of the indolocarbazole skeleton generate molecules with enhanced topoisomerase I activity. Thus, active agents can be made by modification of the rebeccamycin skeleton using fluorine substitution, which gives rise to BMS-250749 (**28**); this compound is headed for Phase I trials as a topoisomerase I inhibitor (47, 50). Second, modification of the base skeleton to include other heterocyclic and carbocyclic rings extends the compounds into previously unexplored chemical space. An example is demonstrated by asymmetric phenyl substitution, which produces compounds such as (**29**) with significant cytotoxic activity in cell lines, blocking at G_2/M or S phase in the cell cycle (51).

13.2.4 Compounds that Inhibit Topoisomerase II

13.2.4.1 Podophyllotoxins Podophyllotoxin is a major constituent of the rhizome of the American May apple, *Podophyllum peltatum*. It was shown as early as 1947 to inhibit formation of the mitotic spindle, and its structure (**30**) was elucidated in 1951 (52).

Podophyllotoxin is too toxic for use as an anticancer agent, but medicinal chemical studies led to the development of etoposide (**31**) and teniposide (**32**) as podophyllotoxin analogs. They differ chemically from podophyllotoxin in their stereochemistry and glycosylation at C4 as well as being demethylated at C4', but their most significant difference is in their mechanism of action. Unlike the parent compound, etoposide and teniposide act as inhibitors of topoisomerase II rather than as inhibitors of tubulin polymerization. Clinically etoposide (**31**) is used in combination with cisplatin against small-cell cancer, and it is also effective for the treatment of testicular cancer and non-small-cell lung cancer. Teniposide (**32**) is used in combination with cisplatin against neuroblastoma, with ara-C against acute lymphoblastic leukemia, and with carboplatin against small-cell lung cancer. One problem with these compounds is their lack of water solubility, and the soluble compound Etopophos (Bristol Myers Squibb **33**) was developed to circumvent this problem. It can be administered intravenously and is then rapidly converted to etoposide by plasma phosphatase. For a recent general review of the podophyllotoxins, see Reference 53.

13.2.4.2 Anthracyclines   From the perspective of the number of patients treated, one of the most important classes of topoisomerase II inhibitors is that of the anthracyclines, with daunorubicin (**34**) and its derivative doxorubicin (adriamycin) (**35**) being the best known of these agents currently in clinical use. Adriamycin is still a major component of the treatment regimen for breast cancer, despite its known cardiotoxicity (54). The mechanism of action of these molecules is now known to be inhibition of topoisomerase II (55), although they are also effective DNA binders (56). Both drugs are used for the treatment of acute non-lymphocytic leukemia, Hodgkin and non-Hodgkin lymphomas, and sarcomas, in addition to breast cancer. Derivatives of doxorubicin, such as epirubicin, idarubicin, pirirubicin, and valrubicin, have also been approved for clinical use, and the expansion of the efficacy of doxorubicin is being explored through targeted delivery techniques, including both liposomally encapsulated and monoclonal-linked derivatives.

13.2.5   Compounds that Interact with Other Proteins

13.2.5.1 Geldanamycin   The first signal transduction modulator to enter clinical trials, other than a formal cyclin-dependent kinase or protein kinase C inhibitor, was the microbial product 17-allylamino-geldanamycin (17-AAG, **36**). This modulator entered Phase I trials in 2001 and is currently in Phase II trials in a variety of cancers. Geldanamycin and its derivatives bind at the major ATP-binding site of the protein chaperone Hsp-90. The protein chaperones are emerging as attractive targets for cancer chemotherapy, and the reader is referred to three recent reviews for additional information (57–59).

13.2.5.2 Staurosporine   The indolocarboxazoles first came into prominence with the identification of staurosporine (**37**) and its simple derivative UCN-01

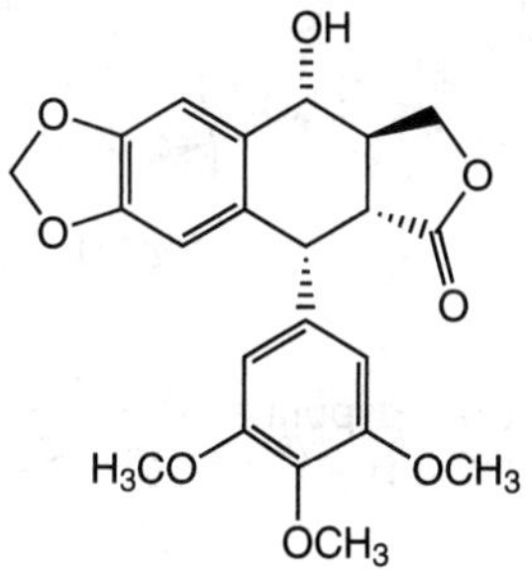

30

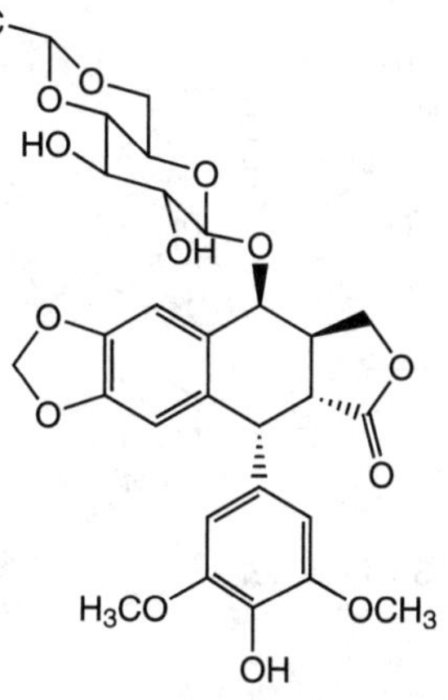

31 R = H
33 R = PO_3H_2

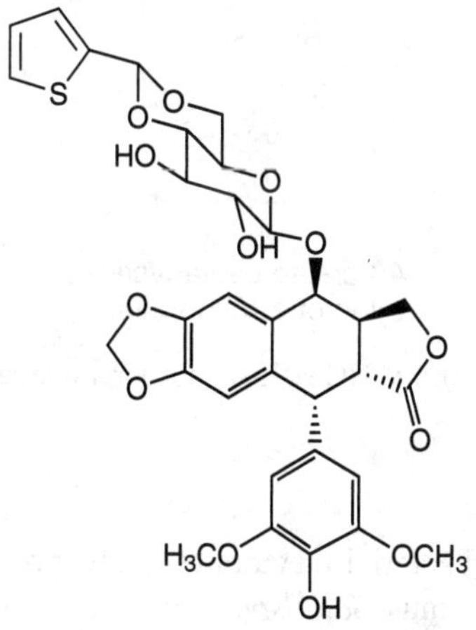

32

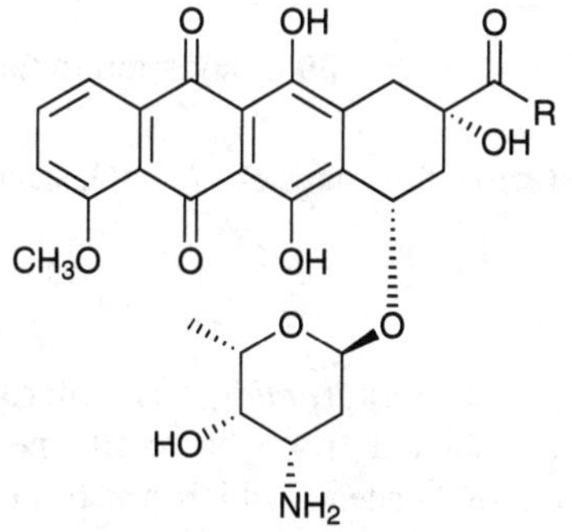

34 Daunorubicin; R = CH_3
35 Doxorubicin; R = CH_2OH

Figure 13.9 Structures of podophyllotoxin, etoposide, teniposide, etopophos, daunorubicin, and doxorubicin.

(**38**) as inhibitors of components of the eukaryotic cell cycle and of protein kinase C. Although these compounds are related structurally to rebeccamycin, they have very different mechanisms of action, in that they are highly potent but entirely nonselective inhibitors of protein kinases. UCN-01 has entered Phase I/II clinical trials against a variety of cancers, including leukemias, lymphomas, various solid tumors, melanoma, and small-cell lung cancer. Its clinical development has been hampered by its high binding to human plasma proteins (60, 61).

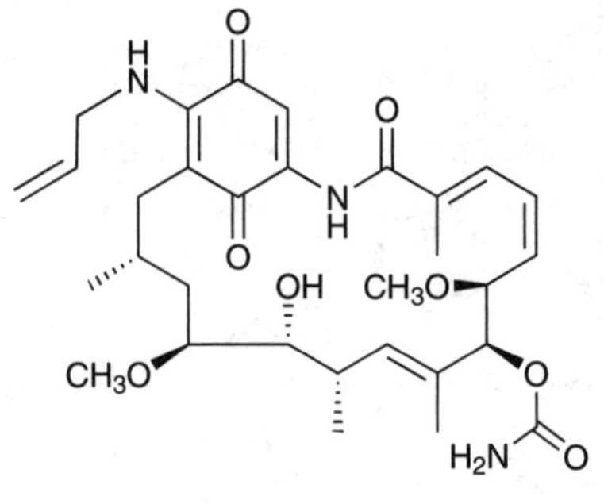

36 17-AAG

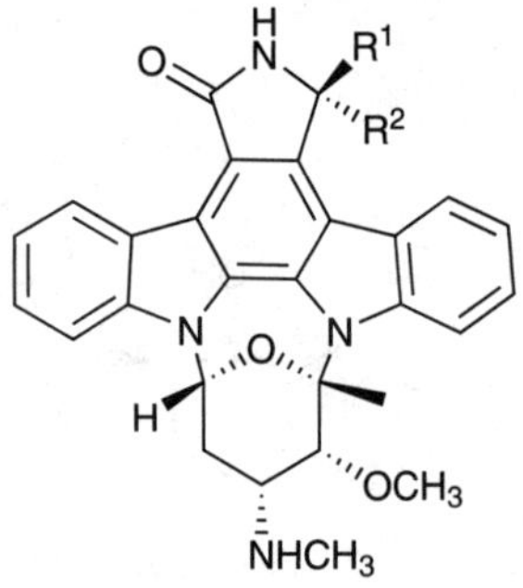

37 Staurosporine R^1 = R^2 = H
38 UCN01 R^1 = H; R^2 = OH

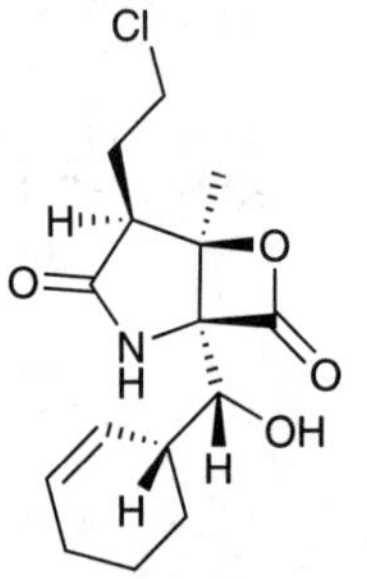

39 Salinosporamide A

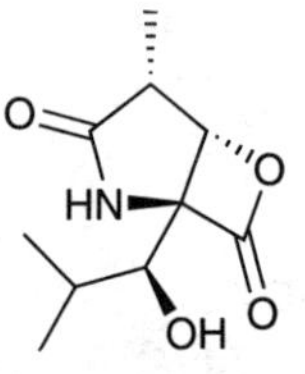

40 *clasto*-Lactacystin-
β-lactone (Omuralide)

Figure 13.10 Structures of 17-AAG, staurosporine, UCN01, salinosporamide, and omuralide.

13.2.5.3 Salinosporamide The marine bacterial metabolite salinosporamide A (**39**) was isolated from the totally new genus *Salinispora* that mapped to the Micromonosporaceae, which are found in marine sediments across the tropics. It demonstrated activity as a cytotoxic proteasome inhibitor (62) similar to that observed for the structurally related compound omuralide (**40**) (63, 64), which resulted from a spontaneous rearrangement of the microbial metabolite lactacystin in neutral aqueous media. Salinosporamide has been synthesized (65, 66) and has been fermented in saline media on a large scale under cGMP conditions. It entered Phase I clinical trials in May 2006.

13.3 COMPOUNDS THAT TARGET DNA OR RNA

The second major class of anticancer agents consists of compounds that act directly on DNA or RNA, either by intercalation, by alkylation, or by cleavage.

13.3.1 Actinomycin D

The first microbial-derived agent in clinical use for cancer was actinomycin D (**41**) (which was systematically named as D-actinomycin C_1 and generically named dactinomycin) that was introduced in the early 1960s. Despite extensive research into the preparation of analogs, no derivatives have progressed beyond clinical trials (67). Its mechanism of action is inhibition of DNA-dependent RNA synthesis, which in turn depends on the strong intercalation of actinomycin into double-helical DNA (68, 69). It is used clinically in the treatment of trophoblastic tumors in females, in metastatic carcinoma of the testis, and in Wilms's tumor in children (67). In recent years there have been reports that actinomycin D may also act on the signal transduction cascade(s) at the level of transcription factor(s), and it will be interesting to see whether these activities rejuvenate interest in this class of molecules (70).

13.3.2 Bleomycins

Another important class is the family of glycopeptolide antibiotics known as bleomycins (e.g., bleomycin A_2 and Blenoxane; Nippon Kayaku Co., Ltd., Tokyo, Japan) (**42**); the bleomycins are structurally related to the phleomycins (71, 72). Bleomycin was originally thought to act through DNA cleavage, because it cleaves both DNA and RNA in an oxidative, sequence-selective, metal-dependent fashion in the presence of oxygen. Recent studies, however, suggest that an alternative mechanism of action may be inhibition of t-RNA from experiments reported recently by the Hecht group (73). Bleomycins are used clinically in combination therapy for the treatment of squamous cell carcinomas and malignant lymphomas.

13.3.3 Mitomycins

The mitomycins (mitosanes) were discovered in the late 1950s, and mitomycin C (**43**) was approved for clinical use in Japan in the 1960s and in the United States in 1974. Its serious bone marrow toxicity has led to extensive synthetic studies aimed at developing a less toxic analog but without significant success; it remains the only clinically used member of this class. It alkylates DNA only after undergoing a one-electron reduction. The current model postulates that mitomycin C alkylates and cross-links DNA by three competing pathways (74). Clinically it is used primarily in combination with other drugs for the treatment of gastric and pancreatic carcinomas (75).

13.3.4 Calicheamicin

Calicheamicin (**44**) is a member of a large group of antitumor enediyne antibiotics. It was isolated from *Micromonospora echinospora* ssp. *calichensis* by workers at Lederle Laboratories (Pearl River, NY, now Wyeth) (76, 77); the

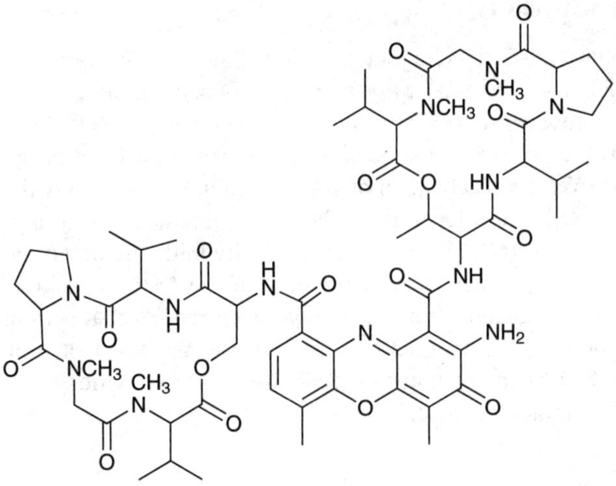

41 Actinomycin D

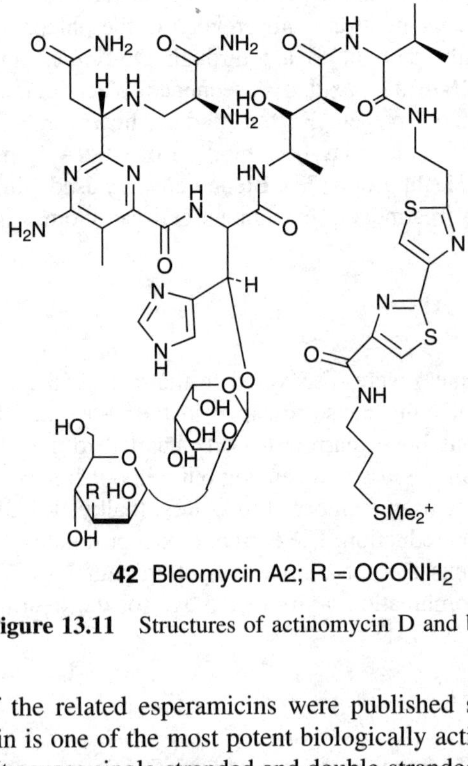

42 Bleomycin A2; R = OCONH$_2$

Figure 13.11 Structures of actinomycin D and bleomycin A2.

structures of the related esperamicins were published simultaneously (78, 79). Calicheamicin is one of the most potent biologically active natural products ever discovered. It causes single-stranded and double-stranded DNA cleavage through a unique mechanism that involves reductive cleavage of the trisulfide "trigger" followed by Bergman cyclization to a diradical. It proved to be too potent and

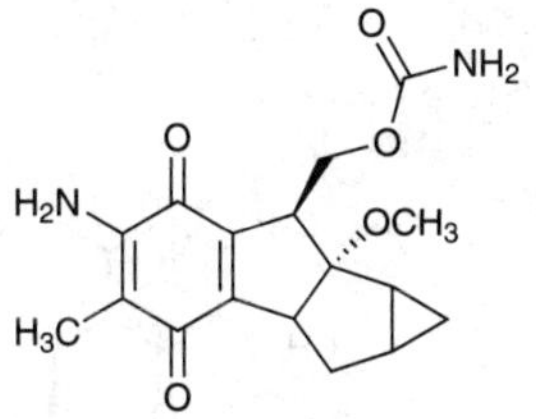

43 Mitomycin C

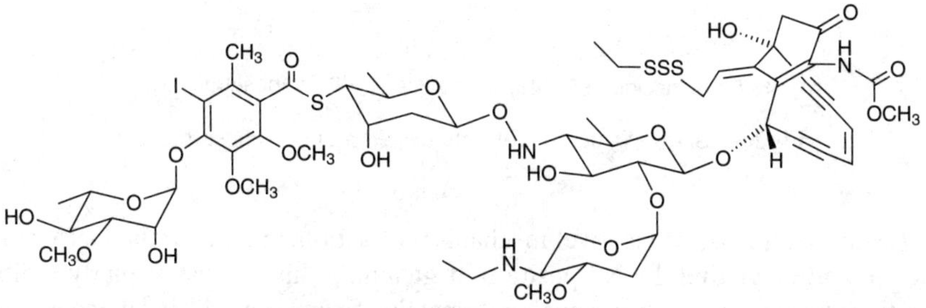

44 Calicheamicin

Figure 13.12   Structures of mitomycin C and calicheamicin.

too toxic for direct clinical use, but it has been used as the "warhead" in the antibody-targeted chemotherapeutic agent Mylotarg (Wyeth Laboratories, Collegeville, PA), which was approved by the FDA in 2000 for clinical use for the treatment of acute myelogenous leukemia (80). Mylotarg is the first such antibody-targeted agent to be approved for use on humans.

13.3.5   Ecteinascidin

Currently, no approved antitumor drugs directly are derived from marine sources, but ecteinascidin (Yondelis; PharmaMar, Madrid, Spain) was submitted to the EMEA in early August 2006 for approval as an antisarcoma agent and was recommended for approval by the EMEA advisory committee in July 2007 (entered in proof) Approved in September 2007 for Sarcoma. This compound (**45**) was isolated originally from the Caribbean tunicate, *Ecteinascidia turbinata* (81, 82). The original supplies for preclinical studies came from a combination of wild harvesting and aquaculture both in sea and on land. The supply problem was finally overcome by the Spanish company, PharmaMar, which developed a 21-step semisynthetic route from the bacterial product cyanosafracin B (**46**) that could be carried out under cGMP conditions (83). This route provided an adequate source for advanced clinical trials and an assured supply if the drug is approved for clinical use.

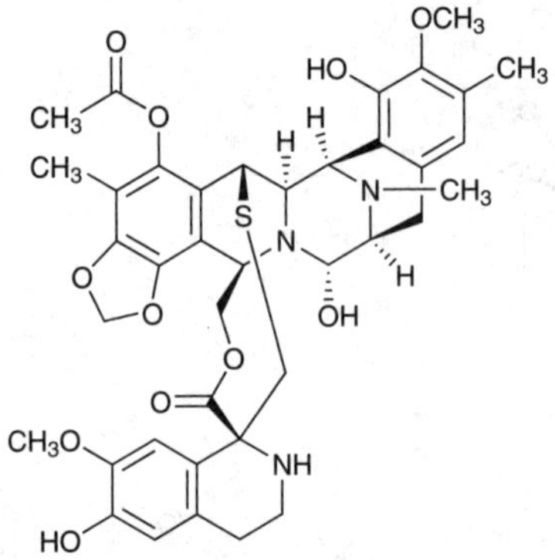

45 Ecteinascidin (ET-743) 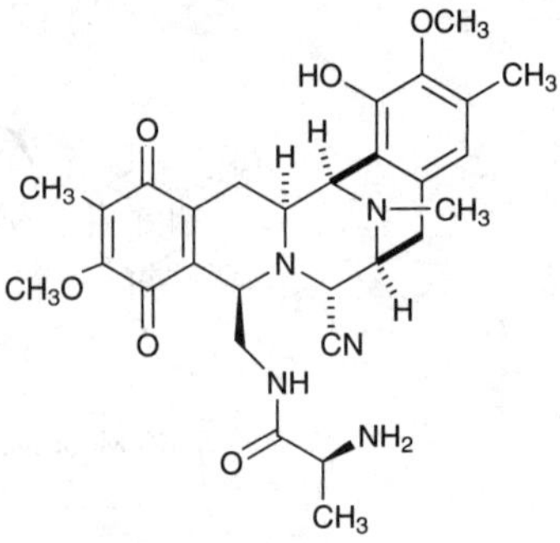   **46** Cyanosafracin B

Figure 13.13   Structures of ecteinascidin and cyanosafracin B.

Ecteinascidin 743 has a novel mechanism of action, binding to the DNA minor groove and alkylating the N2 position of guanine. This process strongly inhibits the transcription of specific genes. Ultimately it causes a p53-independent cell-cycle block, which leads to apoptosis. It has shown a clinical benefit rate close to 40% in Phase II studies on sarcomas (84).

13.3.6   Other Agents

Mithramycin (**47**) is an antitumor antibiotic isolated from *Streptomyces plicatus*. It currently is used to a limited extent for the treatment of embryonal cell carcinoma of the testes and of cancer-related hypercalcemia (85). It is reported to be a specific inhibitor of the Sp1 transcription factor in hematopoietic cells (86).

Streptozotocin (**48**) is an N-nitroso urea isolated from *Streptomyces achromogenes*. It acts as a DNA-alkylating agent (87), and it is recommended for use in combination with doxorubicin (**35**) as the drug of choice for the chemotherapy for patients with malignant neuroendocrine pancreatic tumors (88).

13.4   CONCLUSIONS

As noted in the Introduction, natural products have served historically as the major source of drugs and lead compounds for the treatment of cancer, and the examples provided in this short review indicate that important discoveries in this area are still being made. Despite this impressive track record, many pharmaceutical companies have deemphasized natural product-based drug discovery efforts in favor of approaches such as combinatorial chemistry. Sadly de novo combinatorial chemistry, which was expected to be a panacea for the discovery of small-molecule drug leads over the last 15 or so years, has so far yielded only one drug for antitumor therapy. This drug is the orally active multikinase inhibitor,

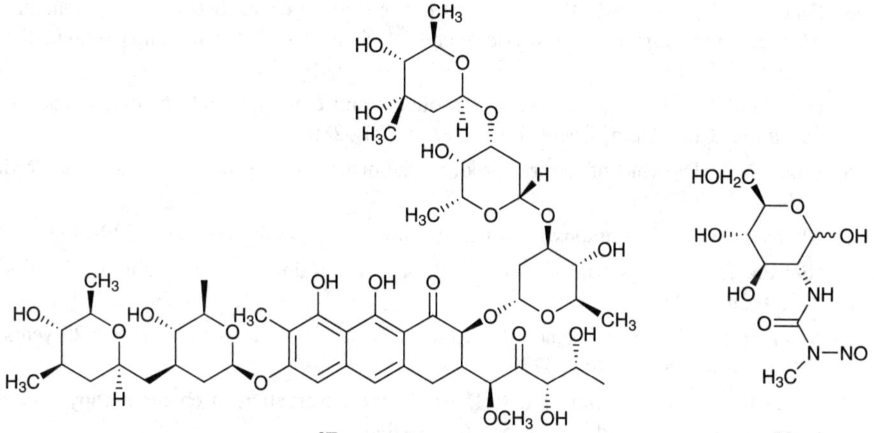

Figure 13.14   Structures of mithramycin and streptozotocin.

Sorafenib from Bayer AG (Leverkusen, Germany), which was approved in 2005 (89). The comments of Ortholand and Ganesan (4) are appropriate here

> The early years of combinatorial chemistry suffered from an excess of hype, and a major victim was natural-product screening. Many organizations went through an irreversible shift in policy, and prematurely discontinued their efforts in this area. We are now seeing the backlash from this knee-jerk reaction. The early combinatorial strategies were flawed and unproven, and have yet to deliver any blockbuster drugs. Meanwhile, we have lost the uniqueness of screening natural-product space as a complement to synthetic compounds. If past indicators are any guide, there are undoubtedly many more unique and potent biologically active natural products waiting to be discovered.

The data in this review support this statement and show clearly that natural products continue to provide both tools to probe biologic mechanisms and skeletons upon which to "improve" on the properties of the natural product. Scientists are still discovering or rediscovering the truth that natural products are the best lead structures from which to begin a search for novel mechanisms and novel treatments for a multitude of diseases, not just cancer.

REFERENCES

1. Kingston DGI, Newman DJ. The search for novel drug leads for predominately antitumor therapies by utilizing mother nature's pharmacophoric libraries. Curr. Opin. Drug Disc. Develop. 2005;8:207–227.

2. Newman DJ, Cragg GM, Snader KM. Natural products as sources of new drugs over the period 1981-2002. J. Nat. Prod. 2003;66:1022–1037.

3. Feher M, Schmidt JM. Property distributions: differences between drugs, natural products, and molecules from combinatorial chemistry. J. Chem. Inf. Comput. Sci. 2003;43:218–227.

4. Ortholand J-Y, Ganesan A. Natural products and combinatorial chemistry: back to the future. Curr. Opin. Chem. Biol. 2004;8:271–280.

5. Butler MS. The role of natural product chemistry in drug discovery. J. Nat. Prod. 2004;67:2141–2153.

6. Clardy J, Walsh C. Lessons from natural molecules. Nature 2004;432:829–837.

7. Nielsen J. Combinatorial synthesis of natural products. Curr. Opin. Cell. Biol. 2002;6:297–305.

8. Newman DJ, Cragg GM. Natural products as sources of drugs over the last 25 years. J. Nat. Prod. 2007;70:461–477.

9. Noble RL, Beer CT, Cutts JH. Role of chance observation in chemotherapy: *Vinca rosea*. Ann. N.Y. Acad. Sci. 1958;76:882–894.

10. Neuss N, Cone NJ, Gorman M, Boaz HE. Vinca alkaloids. II. Structures of leurocristine (LCR) and vincaleukoblastine (VLB). J. Am. Chem. Soc. 1962;84:1509–1510.

11. Svoboda GH. Alkaloids of *Vinca rosea* (*Catharanthus rosea*).9. Extraction and characterization of leurosidine and leurocristine. Lloydia 1961;24:173–178.

12. Joel S. The comparative clinical pharmacology of vincristine and vindesine: Does vindesine offer any advantage in clinical use? Cancer Treat. Rev. 1995;21:513–525.

13. Duflos A, Kruczynski A, Barret J-M. Novel aspects of natural and modified vinca alkaloids. Curr. Med. Chem.: Anti-Cancer Agents 2002;2:55–70.

14. Gueritte F, Fahy J. The vinca alkaloids. In: Anticancer Agents from Natural Products. Cragg GM, Kingston DGI, Newman DJ, eds. 2005. Taylor and Francis, Boca Raton, FL. pp. 123–135.

15. Pettit GR, Cragg GM, Herald DL, Schmidt JM, Lohavanijaya P. Isolation and structure of combretastatin. Can. J. Chem. 1982;60:1374–1376.

16. Pinney KG, Jelinek C, Edvardsen K, Chaplin DJ, Pettit GR. The discovery and development of the combretastatins. In: Anticancer Agents from Natural Products. Cragg GM, Kingston DGI, Newman DJ, eds. 2005. Taylor and Francis, Boca Raton, FL. pp. 23–46.

17. Zheng W, Seletsky BM, Palme MH, Lydon PJ, Singer LA, Chase CE, Lemelin CA, Shen Y, Davis H, Tremblay L, et al. Macrocyclic ketone analogues of halichondrin B. Bioorg. Med. Chem. Lett. 2004;14:5551–5554.

18. Jordan MA, Kamath K, Manna T, Okouneva T, Miller HP, Davis C, Littlefield BA, Wilson L. The primary antimitotic mechanism of action of the synthetic halichondrin E7389 is suppression of microtubule growth. Mol. Cancer Ther. 2005;4:1086–1095.

19. Dabydeen DA, Burnett JC, Bai R, Verdier-Pinard P, Hickford SJ, Pettit GR, Blunt JW, Munro MHG, Gussio R, Hamel E. Comparison of the activities of the truncated halichondrin B analog NSC 707389 (E7389) with those of the parent compound and a proposed binding site on tubulin. Mol. Pharmacol. 2006;70:1866–1875.

20. Bai R, Pettit GR, Hamel E. Binding of dolastatin-10 to tubulin at a distinct site for peptide antimitotic agents near the exchangeable nucleotide and vinca alkaloid sites. J. Biol. Chem. 1990;265:17141–17149.

21. Cunningham C, Appleman LJ, Kirvan-Visovatti M, Ryan DP, Regan E, Vukelja S, Bonate PL, Ruvuna F, Fram RJ, Jekunen A, et al. Phase I and pharmacokinetic study of the dolastatin-15 analogue tasidotin (ILX651) administered intravenously on days 1, 3, and 5 every 3 weeks in patients with advanced solid tumors. Clin. Cancer Res. 2005;11:7825–7833.

22. Ebbinghaus S, Rubin E, Hersh E, Cranmer LD, Bonate PL, Fram RJ, Jekunen A, Weitman S, Hammond LA. A phase I study of the dolastatin-15 analogue tasidotin (ILX651) administered intravenously daily for 5 consecutive days every 3 weeks in patients with advanced solid tumors. Clin. Cancer Res. 2005;11:7807–7816.

23. Wani MC, Taylor HL, Wall ME, Coggon P, McPhail AT. Plant antitumor agents. VI. The isolation and structure of taxol, a novel antileukemic and antitumor agent from *Taxus brevifolia*. J. Am. Chem. Soc. 1971;93:2325–2327.

24. Gueritte-Voegelein F, Senilh V, David B, Guenard D, Potier P. Chemical studies of 10-deacetyl baccatin III. Hemisynthesis of taxol derivatives. Tetrahedron 1986;42:4451–4460.

25. Denis J-N, Greene AE, Guenard D, Gueritte-Voegelein F, Mangatal L, Potier P. A highly efficient, practical approach to natural taxol. J. Am. Chem. Soc. 1988;110:5917–5919.

26. Holton RA, Biediger RJ, Boatman PD. Semisynthesis of taxol and taxotere. In: Taxol: Science and Applications. Suffness M, ed. 1995. CRC Press, Inc., Boca Raton, FL. pp. 97–121.

27. Suffness M, Wall ME. Discovery and development of taxol. In: Taxol: Science and Applications. Suffness M, eds. 1995. CRC Press, Inc., Boca Raton, FL. pp. 3–25.

28. Schiff PB, Fant J, Horwitz SB. Promotion of microtubule assembly *in vitro* by taxol. Nature 1979;277:665–667.

29. Piccart MJ, Cardoso F. Progress in systemic therapy for breast cancer: an overview and perspectives. Eur. J. Can. 2003;1 (supp): 56–69.

30. Ozols RF. Progress in ovarian cancer: an overview and perspective. Eur. J. Can. 2003;1 (supp): 43–55.

31. Davies AM, Lara PN, Mack PC, Gandara DR. Docetaxel in non-small cell lung cancer: a review. Exp. Opin. Pharmacother. 2003;4:553–565.

32. Kamath KR, Barry JJ, Miller KM. The taxus drug-eluting stent: a new paradigm in controlled drug delivery. Adv. Drug Deliv. Rev. 2006;58:412–436.

33. Kingston DGI, Newman DJ. Taxoids: cancer-fighting compounds from nature. Curr. Opin. Drug Disc. Dev. 2007;10:130–144.

34. Gradishar WJ. Albumin-bound paclitaxel: a next-generation taxane. Exp. Opin. Pharmacother. 2006;7:1041–1053.

35. Lowe J, Li H, Downing KH, Nogales E. Refined structure of α β-tubulin at 3.5 Å resolution. J. Mol. Biol. 2001;313:1045–1057.

36. Paik Y, Yang C, Metaferia B, Tang S, Bane S, Ravindra R, Shanker N, Alcaraz AA, Johnson SA, Schaefer J, et al. REDOR NMR distance measurements for the tubulin-bound paclitaxel conformation. J. Am. Chem. Soc. 2007;129:361–370.

37. Ganesh T, Guza RC, Bane S, Ravindra R, Shanker N, Lakdawala AS, Snyder JP, Kingston DGI. The bioactive taxol conformation of β-tubulin: Experimental evidence from highly active constrained analogs. Proc. Natl. Acad. Sci. U.S.A. 2004;101:10006–10011.

38. Hofle G, Bedorf N, Steinmetz H, Schomburg D, Gerth K, Reichenbach H. Epothilone A and B - novel 16-membered macrolides with cytotoxic activity: Isolation, crystal structure, and conformation in solution. Angew Chem Int. Ed. 1996;35:1567–1569.

39. Bollag DM, McQueney PA, Zhu J, Hensens O, Koupal L, Liesch J, Goetz M, Lazarides E, Woods CM. Epothilones, a new class of microtubule-stabilizing agents with a taxol-like mechanism of action. Cancer Res. 1995;55:2325–2333.

40. Giannakakou P, Gussio R, Nogales E, Downing KH, Zaharevitz D, Bollbuck B, Poy G, Sackett D, Nicolaou KC, Fojo T. A common pharmacophore for epothilone and taxanes: Molecular basis for drug resistance conferred by tubulin mutations in human cancer cells. Proc. Natl. Acad. Sci. U.S.A. 2000;97:2904–2909.

41. Frykman S, Tsuruta H, Lau J, Regentin R, Ou S, Reeves C, Carney J, Santi D, Licari P. Modulation of epothilone analog production through media design. J. Ind. Microbiol. Biotech. 2002;28:17–20.

42. Klar U, Buchmann B, Schwede W, Skuballa W, Hoffmann J, Lichtner RB. Total synthesis and antitumor activity of ZK-EPO: the first fully synthetic epothilone in clinical development. Angew Chem Int. Ed. 2006;45:7942–7948.

43. Wu K-D, Cho YS, Katz J, Ponomarev V, Chen-Kiang S, Danishefsky SJ, Moore MAS. Investigation of antitumor effects of synthetic epothilone analogs in human myeloma models *in vitro* and *in vivo*. Proc. Natl. Acad. Sci. U.S.A. 2005;102:10640–10645.

44. de Jonge M, Verweij J. The epothilone dilemma. J. Clin. Oncol. 2005;23:9048–9050.

45. Wall ME, Wani MC, Cook CE, Palmer KH, McPhail AT, Sim GA. Plant anti-tumor agents. 1. The isolation and structure of camptothecin, a novel alkaloidal leukemia and tumor inhibitor from *Camptotheca acuminata*. J. Am. Chem. Soc. 1966;88:3888–3890.

46. Hsiang YH, Hertzberg R, Hecht S, Liu LF. Camptothecin induces protein-linked DNA breaks via mammalian DNA topoisomerase I. J. Biol. Chem. 1985;260:14873–14878.

47. Pommier Y. Topoisomerase I inhibitors: camptothecins and beyond. Nat. Rev. Cancer 2006;6:789–802.

48. Rahier NJ, Thomas CJ, Hecht SM. Camptothecin and its analogs. Anticancer Agents from Natural Products. Cragg GM, Kingston DGI, Newman DJ, eds. 2005. Taylor and Francis, Boca Raton, FL. pp. 5–21.

49. Sanchez C, Mendez C, Salas JA. Indolocarbazole natural products: occurrence, biosynthesis, and biological activity. Nat. Prod. Rep. 2006;23:1007–1045.

50. Saulnier MG, Balasubramanian BN, Long BH, Frennesson DB, Ruediger E, Zimmerman K, Eummer JT, St. Laurent DR, Stoffan KM, Naidu BN, et al. Discovery of a fluoroindolo[2,3-a]carbazole clinical candidate with broad spectrum antitumor activity in preclinical tumor models superior to the marketed oncology drug, CPT-11. J. Med. Chem. 2005;48:2258–2261.

51. Routier S, Merour J-Y, Dias N, Lansiaux A, Bailly, C, Lozach, O, Meijer, L. Synthesis and biological evaluation of novel phenylcarbazoles as potential anticancer agents. J. Med. Chem. 2006;49:789–799.

52. Hartwell JL, Schrecker AW. Components of podophyllin. V. The constitution of podophyllotoxin. J. Am. Chem. Soc. 1951;73:2909–2916.

53. Lee K-H, Xiao Z. Podophyllotoxin and analogs. 832: In: Anticancer Agents from Natural Products. Cragg GM, Kingston DGI, Newman DJ, eds. 2005. Taylor and Francis, Boca Raton, FL. pp. 71–87.

54. Arcamone F. Anthracyclines. In: Anticancer agents from natural products. Cragg GM, Kingston DGI, Newman DJ, eds. 2005. Taylor and Francis, Boca Raton, FL. pp. 299–320.

55. Zunino F, Capranico G. DNA topoisomerase II as the primary target of anti-cancer anthracyclines. Anticancer Drug Des. 1990;5:307–317.

56. Frederick CA, Williams LD, Ughetto G, Van der Marel GA, Van Boom JH, Rich A, Wang AHJ. Structural comparison of anticancer drug-DNA complexes: adriamycin and daunomycin. Biochemistry 1990;29:2538–2549.

57. Isaacs JS. Heat-shock protein 90 inhibitors in antineoplastic therapy: is it all wrapped up? Exp. Opin. Investig. Drugs 2005;14:569–589.

58. Janin YL. Heat shock protein 90 inhibitors. A text book example of medicinal chemistry J. Med. Chem. 2005;48:7503–7512.

59. Chiosis G. Targeting chaperones in transformed systems - a focus on hsp90 and cancer. Exp. Opin. Ther. Targ. 2006;10:37–50.

60. Prudhomme M. Staurosporines and structurally related indolocarbazoles as antitumor agents. In: Anticancer Agents from Natural Products. Cragg GM, Kingston DGI, Newman DJ, eds. 2005. Taylor and Francis, Boca Raton, FL. pp. 499–517.

61. Sanchez C, Mendez C, Salas JA. Indolocarbazole natural products: Occurrence, biosynthesis, and biological activity. Nat. Prod. Rep. 2006;23:1007–1045.

62. Feling RH, Buchanan GO, Mincer TJ, Kauffman CA, Jensen PR, Fenical W. Salinosporamide A: a highly cytotoxic proteasome inhibitor from a novel microbial source, a marine bacterium of the new genus *Salinospora*. Angew Chem. Int. Ed. 2003;42:355–357.

63. Corey EJ, Li WD. Total synthesis and biological activity of lactacystin, omuralide and analogs. Chem. Pharm. Bull. 1999;47:1–10.

64. Corey EJ, Li WD, Nagamitsu T, Fenteany G. The structural requirements for inhibition of proteasome function by the lactacystin-derived β-lactone and synthetic analogs. Tetrahedron 1999;55:3305–3316.

65. Reddy LR, Saravanan P, Corey EJ. A simple stereocontrolled synthesis of salinosporamide A. J. Am. Chem. Soc. 2004;126:6230–6231.

66. Reddy LR, Fournier J-F, Reddy BVS, Corey EJ. New synthetic route for the enantioselective total synthesis of salinosporamide A and biologically active analogues. Org. Lett. 2005;7:2699–2701.

67. Mauger AM, Lackner H. The actinomycins. In: Anticancer Agents from Natural Products. Cragg GM, Kingston DGI, Newman DJ, eds. 2005. Taylor and Francis, Boca Raton, FL. pp. 281–297.

68. Kawamata J, Imanishi M. Interactions of actinomycin with deoxyribonucleic acid. Nature 1960;187:1112–1113.

69. Kersten W, Kersten H, Rauen HM. Action of nucleic acids on the inhibition of growth by actinomycin of *Neurospora crassa*. Nature 1960;187:60–61.

70. Gniazdowski M, Denny WA, Nelson SM, Czyz M. Transcription factors as targets for DNA-interacting drugs. Curr. Med. Chem.: Anti-Cancer Agents 2003;10:909–924.

71. Galm U, Hager MH, Van Lanen SG, Ju J, Thorson JS, Shen B. Antitumor antibiotics: bleomycin, enediynes, and mitomycin. Chem. Rev. 2005;105:739–758.

72. Hecht SM. Bleomycin group antitumor agents. In: Anticancer Agents from Natural Products. Cragg GM, Kingston DGI, Newman DJ, eds. 2005. Taylor and Francis, Boca Raton, FL. pp. 357–381.

73. Tao Z-F, Konishi K, Keith G, Hecht SM. An efficient mammalian transfer RNA target for bleomycin. J. Am. Chem. Soc. 2006;128:14806–14807.

74. Kumar SG, Lipman R, Cummings J, Tomasz M. Mitomycin c-DNA adducts generated by dt-diaphorase. Revised mechanism of the enzymatic reductive activation of mitomycin C. Biochemistry 1997;36:14128–14136.

75. Remers WA. The mitomycins. In: Anticancer Agents from Natural Products. Cragg GM, Kingston DGI, Newman DJ, eds. 2005. Taylor and Francis, Boca Raton, FL. pp. 475–497.

76. Lee MD, Dunne TS, Siegel MM, Chang CC, Morton GO, Borders DB. Calichemicins, a novel family of antitumor antibiotics. 1. Chemistry and partial structure of calichemicin γ J. Am. Chem. Soc. 1987;109:3464–3466.

77. Lee MD, Dunne TS, Chang CC, Ellestad GA, Siegel MM, Morton GO, McGahren WJ, Borders DB. Calichemicins, a novel family of antitumor antibiotics. 2. Chemistry and structure of calichemicin γ J. Am. Chem. Soc. 1987;109:3466–3468.

78. Golik J, Clardy J, Dubay G, Groenewold G, Kawaguchi H, Konishi M, Krishnan B, Ohkuma H, Saitoh K, Doyle TW. Esperamicins, a novel class of potent antitumor antibiotics. 2. Structure of esperamicin X. J. Am. Chem. Soc. 1987;109:3461–3462.

79. Golik J, Dubay G, Groenewold G, Kawaguchi H, Konishi M, Krishnan B, Ohkuma H, Saitoh K, Doyle TW. Esperamicins, a novel class of potent antitumor antibiotics. 3. Structures of esperamicins A1, A2, and A1b. J. Am. Chem. Soc. 1987;109:3462–3464.

80. Hamann PR, Upeslacis J, Borders DB. Enediynes. In: Anticancer Agents from Natural Products. Cragg GM, Kingston DGI, Newman DJ, eds. 2005. Taylor and Francis, Boca Raton, FL. pp. 451–474.

81. Rinehart KL, Holt TG, Fregeau NL, Stroh JG, Keifer PA, Sun F, Li LH, Martin DG. Ecteinascidins 729, 743, 745, 759a, 759b, and 770: potent antitumor agents from the caribbean tunicate *Ecteinascidia turbinata*. J. Org. Chem. 1990;55:4512–4515.

82. Wright AE, Forleo DA, Gunawardana GP, Gunasekera SP, Koehn FE, McConnell OJ. Antitumor tetrahydroisoquinoline alkaloids from the colonial ascidian *Ecteinascidia turbinata*. J. Org. Chem. 1990;55:4508–4512.

83. Cuevas C, Perez M, Martin M, Chicharro JL, Fernandez-Rivas C, Flores M, Francesch A, Gallefo P, Zarzuelo M, de la Calle F, et al. Synthesis of ecteinascidin ET-743 and phthalascidin PT-650 from cyanosafracin B. Org. Lett. 2000;2:2545–2548.

84. Fayette J, Coquard IR, Alberti L, Ranchere D, Boyle H, Blay J-Y. ET-743: a novel agent with activity in soft tissue sarcomas. Oncologist 2005;10:827–832.

85. Lynch GR, Lane M. Other antitumor antibiotics. In: Cancer Management in Man. Woolley PV, ed. 1989. Kluwer Academic Publishers, Dordrecht. pp. 134–146.

86. Koschmieder S, Agrawal S, Radomska HS, Huettner CS, Tenen DG, Ottmann OG, Berdel WE, Serve HL, Muller-Tidow C. Decitabine and vitamin D3 differentially affect hematopoietic transcription factors to induce monocytic differentiation. Int. Oncol. 2007;30:349–355.

87. Dolan ME. Inhibition of DNA repair as a means of increasing the antitumor activity of DNA reactive agents. Adv. Drug Deliv. Rev. 1997;26:105–118.

88. Arnold R, Rinke A, Schmidt C, Hofbauer L. Chemotherapy. Best Pract. Res. Clin. Gastroent. 2005;19:649–656.

89. Wilhelm S, Carter C, Lynch M, Lowinger T, Dumas J, Smith RA, Schwartz B, Simantov R, Kelley S. Discovery and development of sorafenib: a multikinase inhibitor for treating cancer. Nature Rev. Drug Discov. 2006;5:835–844.

FURTHER READING

Cragg GM, Kingston DGI, Newman DJ, eds. Anticancer Agents from Natural Products. 2005. CRC Press; Boca Raton, FL.

14

PLANT-DERIVED NATURAL PRODUCTS RESEARCH IN DRUG DISCOVERY

KUO-HSIUNG LEE, HIDEJI ITOKAWA, TOSHIYUKI AKIYAMA, AND SUSAN L. MORRIS-NATSCHKE

Natural Products Research Laboratories, School of Pharmacy, University of North Carolina, Chapel Hill, North Carolina

Numerous important bioactive compounds have been, and continue to be, isolated worldwide from natural sources. These compounds include both primary and secondary metabolites isolated mainly from plants, as well as from the animal and mineral kingdoms. The recent development of new bioassay methods has facilitated progress in the BDFI (bioactivity-directed fractionation and isolation) of many useful bioactive compounds from natural sources (1). These active principles could be developed or additionally modified to enhance the biologic profiles as clinical trials candidates. Many natural pure compounds have become medicines, dietary supplements, and other useful commercial products. This chapter, current as of 2008, summarizes research on many different useful compounds isolated or developed from plants with an emphasis on those discovered recently by the laboratories of the authors as antitumor and anti-HIV clinical trial candidates.

From ancient times, many crude herbs have been used as remedies for various diseases. In Asia, these therapies include traditional Chinese medicine (TCM), Japanese–Chinese medicine (Kampo), Korean–Chinese medicine, Jamu (Indonesia), and Ayurvetic medicine (India). In Europe, phytotherapy and homeopathy have found medicinal use. These herbal therapies generally are classified

Natural Products in Chemical Biology, First Edition. Edited by Natanya Civjan.
© 2012 John Wiley & Sons, Inc. Published 2012 by John Wiley & Sons, Inc.

as "alternative medicines" in America. Alternative medicine, which comes mainly from the aforementioned traditional and folk medicines used worldwide, also is being combined with conventional medicine (Western medicine), which results in integrative medicine. In TCM, crude herbal drugs formerly were divided into three categories: upper-, middle-, and lower-class medicines. Upper-class medicines usually are not toxic, have moderate physiologic effects, and often are used to maintain good health. Thus, they sometimes are called supplementary drugs. Both upper- and middle-class medicines are used as therapeutic drugs, but the former medicines are not as toxic as the latter. Lower-class medicines can contain very toxic substances, which can be used judiciously as medicines. TCM prescriptions usually mix herbs that belong to these three categories, according to a unique principle (2). The experience gained by using TCM for centuries provides a rich source of information for modern research in drug discovery.

14.1 ANTICANCER AND ANTITUMOR COMPOUNDS

Suffness and Douros (3) suggested the following definitions to avoid confusion of terminology. *Cytotoxicity* is used when compounds or extracts show *in vitro* (in cells) activity against tumor cell lines. *Antitumor* or *antineoplastic* indicates that the materials are effective *in vivo* (in animals) in experimental systems. *Anticancer* refers to compounds that are active clinically against human cancer.

The development of novel, clinically useful anticancer agents is highly dependent on the bioassay screening systems, as well as the sample sources. Two bioassay types have been used: cell-based and mechanism of action-(MOA)-based. Initially, cell-based assays used primarily L1210, P388, and nasopharyngeal (KB) cells in preliminary screening for antitumor activity. Screening against a panel of human cancer cell lines was implemented to discover agents active against different types of cancer. For active agents, the *in vitro* studies are followed by *in vivo* xenograft studies to ensure efficacy. New MOA-based bioassay systems aimed at particular molecular targets also have revolutionized the discovery of potential drug candidates. Important anticancer drug targets include tubulin, DNA topoisomerases I and II (topo I and topo II), cyclin-dependent kinases (CDKs), growth and transcription factors, and so forth.

Higher plants have yielded many effective, clinically useful anticancer drugs, including those derived from *Catharanthus* alkaloids, *Taxus* diterpenes, *Camptotheca* alkaloids, and *Podophyllum* lignans. Research in this area has been reviewed extensively (3–13).

The seminal discoveries of taxol (tubulin-interactive) and camptothecin (topo I-interactive) by Wall and Wani (14–16) represent how natural products have influenced the additional development of natural product-derived and synthetic entities. The following discussion of the discovery and development

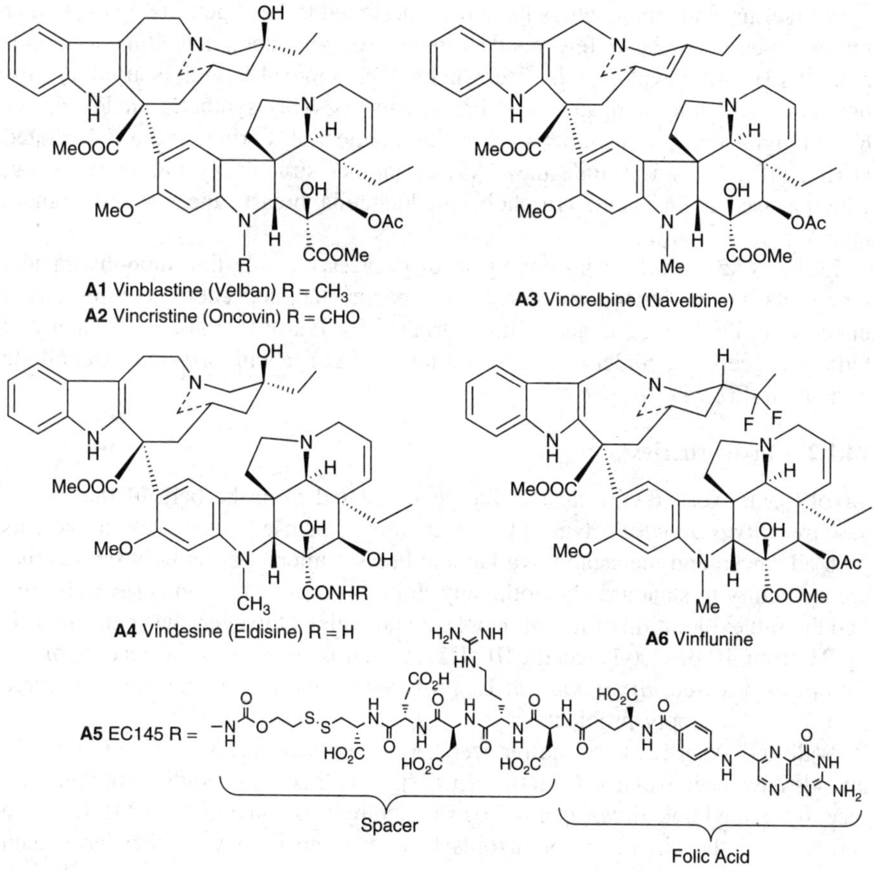

Figure 14.1 Structures of natural and synthetic *Catharanthus* alkaloids.

of current important antineoplastic compounds will be organized by plant species.

14.1.1 Catharanthus species

Catharanthus alkaloids, particularly vinblastine (**A1**) and vincristine (**A2**), are well known anticancer drugs, which are used clinically to treat Hodgkin's lymphoma and acute childhood lymphoblastic leukemia, respectively. These alkaloids interact with tubulin, a protein necessary for cell division, and are inhibitors of mitosis (the process of cell division).

Originally, these compounds were isolated from *Catharanthus roseus* (L.) G. Don [formerly known as *Vinca rosea* (Apocynaceae family)], which is used as folk medicine in Madagascar to inhibit milk secretion and as a hypotensive agent, astringent, and emetic. Moreover, native people in England and the West Indies have used this species to lower blood sugar.

Numerous synthetic analogs have been designed to have activity against other tumor types or to have fewer side effects. Among them, navelbine (vinorelbine) (**A3**) was developed by Burroughs Wellcome (17) and is used against non-small cell lung and advanced breast cancers. This synthetic analog of **A1** has an eight-membered, rather than a nine-membered, C ring, and a dehydrated D ring (18). Eldisine (vindesine) (**A4**) is another structurally modified analog, which is used against acute lymphoblastic leukemia, breast cancer, and malignant melanoma.

EC145 (**A5**), a folic acid conjugate of desacetyl vinblastine monohydrazide, represents a new generation of receptor-specific targeted chemotherapy and is undergoing Phase I anticancer clinical trials (19). Phase II trials for bladder and kidney cancers are underway with vinflunine (**A6**), a bifluorinated vinorelbine derivative (20, 21)

14.1.2 Taxus species

Taxol (paclitaxel) (**B1**), a taxane diterpene isolated from the bark of the Pacific yew tree *Taxus brevifolia* Nutt. (Taxaceae family), is used extensively in patients with advanced and metastatic ovarian and breast tumors, particularly tumors that are refractory to standard chemotherapy. Initially, supply problems severely limited the full exploration of its antineoplastic potential. However, the semisynthesis of **B1** from 10-deacetylbaccatin III (**B2**), which is isolated from needles of the European yew tree, *Taxus baccata* L., provided an alternative renewable resource to resolve the supply problems.

Wall and Wani (22) are the pioneers in taxol discovery. To date, around 400 taxoids have been isolated from the *Taxus* species. *Taxus* alkaloids were reviewed recently in the book *Taxus, genus Taxus* edited by Itokawa and Lee (23). Biologic activity and the chemistry of taxoids from the Japanese yew also have been reviewed (24).

B1 interacts with cellular tubulin via promotion of microtubule assembly and inhibits mitosis. It is active against breast, brain, tongue, endometrial, and ovarian, as well as other, cancers (25, 26). The clinically used analog docetaxel (taxotere) (**B3**) is synthesized from the more readily available **B2**. Taxotere is used particularly against non-small cell lung cancers.

Extensive structure–activity relationship (SAR) studies have led to many related antineoplastic taxane analogs, including ortataxel (**B4**), an orally administered taxoid in Phase II clinical trials (27). SAR studies of ring C-secotaxoids were published recently (28).

In addition, conjugates between taxol and various other compounds, such as 3,17β-estradiol (29), various fatty acids (30), or a biodegradable polymer (poly-L-glutamic acid, paclitaxel polyglumex) (31, 32), seek to improve drug targeting or tissue distribution. The laboratories of the authors have conjugated taxoids with other anticancer agents, including epipodophyllotoxins (**B5**) (33) and camptothecin (**B6**) (34). A recent review provides an overview of novel formulations of taxanes, including many in clinical trials, developed to overcome the solubility issues with **B1** and **B3** (35).

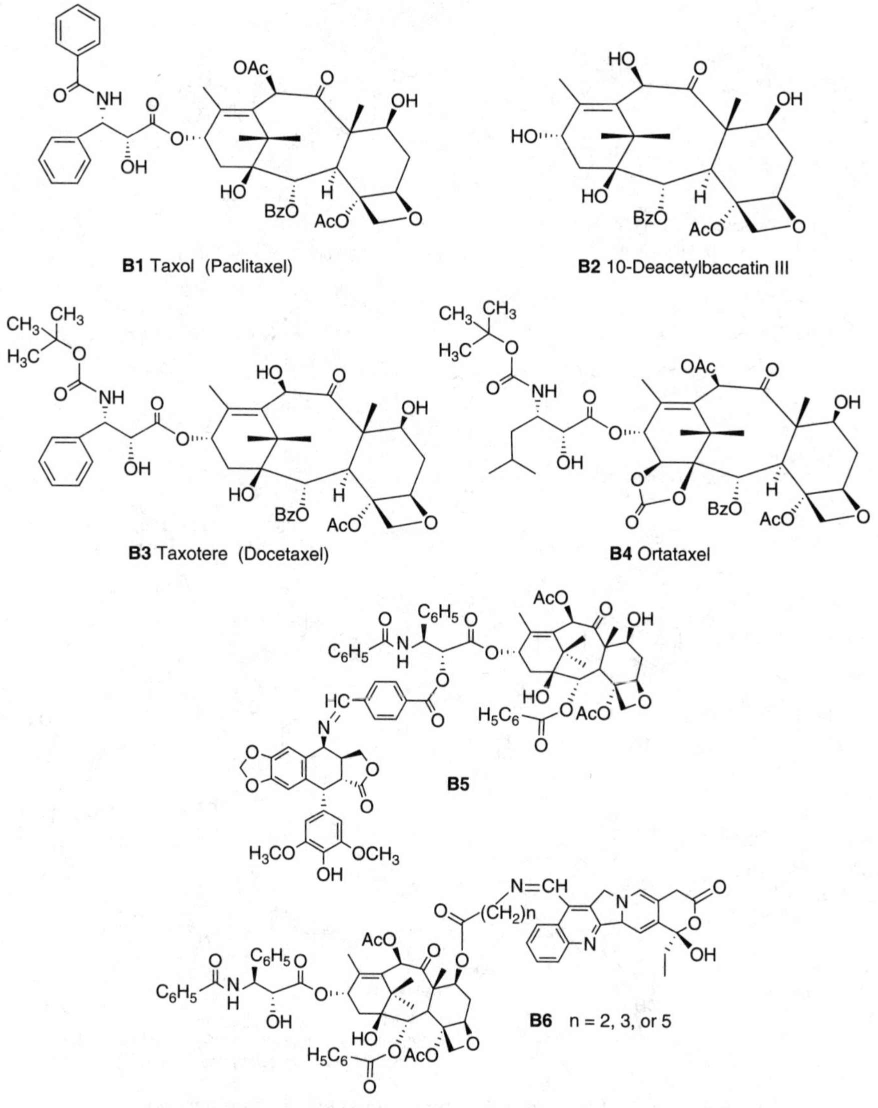

Figure 14.2 Structures of natural and synthetic taxoids.

14.1.3 Camptotheca species

Camptothecin (CPT, **C1**) is a potent antitumor pentacyclic alkaloid isolated from *Camptotheca acuminata* Decne. (Nyssaceae family) and originating in China (36, 37). Interest in CPT was sparked by the discovery that its primary cellular target is DNA topo I (38). 10-Hydroxycamptothecin (**C2**), which also occurs naturally, has a better therapeutic index and is used in China for treating many cancers.

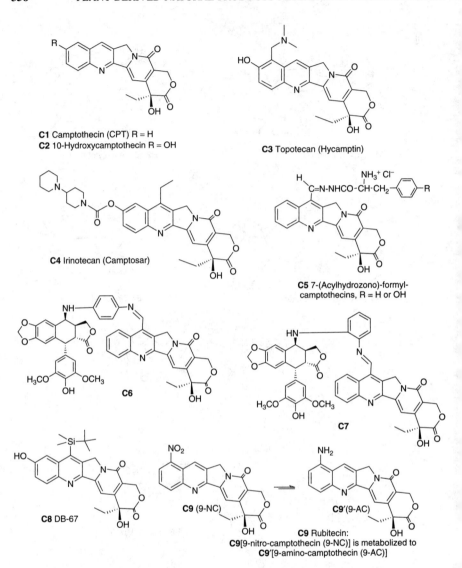

C1 Camptothecin (CPT) R = H
C2 10-Hydroxycamptothecin R = OH

C3 Topotecan (Hycamptin)

C4 Irinotecan (Camptosar)

C5 7-(Acylhydrozono)-formyl-
camptothecins, R = H or OH

C6

C7

C8 DB-67

C9 (9-NC)

C9′(9-AC)

C9 Rubitecin:
C9[9-nitro-camptothecin (9-NC)] is metabolized to
C9′[9-amino-camptothecin (9-AC)]

Figure 14.3   Structures of natural and synthetic camptothecins.

Poor water solubility of the natural products led to the development of the semisynthetic, more water-soluble analogs topotecan (Hycamptin, **C3**) and irinotecan (Camptosar, **C4**), which are used primarily for the treatment of advanced ovarian and metastatic colorectal cancers, respectively (39). Some CPT analogs, such as **C5**, synthesized in the laboratories of the authors showed greater topo I inhibition than CPT (40). Additional CPT analogs also are of interest in combination regimens as radiation sensitizers (41). Two epipodophyllotoxin-campothecin conjugates, **C6** and **C7**, from the laboratories of the authors exhibit dual mechanisms of action, being both topoisomerase I/II-inhibitory (42). They have improved *in vitro* anticancer profiles (42) and are active against etoposide- and camptothecin- resistant KB cells (KB7D and KB/CPT100, respectively).

Clinical application and perspectives on the CPTs have been discussed in several excellent reviews (43–45). New CPT analogs in anticancer clinical trials include DB-67 (a silatecan or 7-silylcamptothecan, **C8**) (46) and rubitecan (9-nitrocamptothecin, **C9**) (47).

14.1.4 Podophyllum species

The genus *Podophyllum* (Berberidaceae family), including the American *P. peltatum* L. (American mayapple) and Indian or Tibetan *P. emodi* Wall (syn. *P. hexandrum* Royle), has been used for centuries for its medicinal properties. Podophyllin, a resin obtained from an alcoholic extract of *Podophyllum* rhizome, has been used for a long time as a remedy for warts and was listed in the U.S. Pharmacopoeia from 1820 to 1942, when it was removed because of undesirable toxicity (48).

Podophyllotoxin (**D1**) is an aryltetralinlactone cyclolignan with a flat, rigid five-ring skeleton; it was isolated in 1880 from rhizomes of *P. peltatum*. It was found to show antineoplastic activity, but it is highly toxic and failed the U.S. National Cancer Institute (NCI) Phase I clinical trials as an antitumor drug in the 1970s. Chemical modification of **D1** led to successful development of the clinically useful anticancer drugs etoposide (**D2**) and teniposide (**D3**). These compounds inhibit cellular DNA topo II and are used to treat small cell lung and testicular cancers and lymphomas/leukemias. Etopophos (etoposide phosphate, **D4**) is a clinically used, water-soluable phosphate ester of etoposide. It lessens the chance of precipitation of the drug during intravenous administration.

Limitations of **D2**, including myelosuppression, drug resistance development, and poor water solubility, prompted extensive SAR studies. Using drug improvement principles, several series of 4-alkylamino and 4-arylamino epipodophyllotoxin analogs were synthesized and showed increased inhibition of DNA topo II activity and increased cytotoxicity in **D2**-resistant cell lines (49–51). From the preclinical development in the laboratories of the authors, GL-331 (**D5**) (52), which contains a *p*-nitroanilino moiety at the 4β position of **D2**, emerged as a clinical trials candidate. Compared with **D2**, GL-331 has advantages of better water solubility, easier manufacturability, and fewer side effects. It also shows cytotoxic activity against **D2**-resistant cancer cell lines. GL-331 progressed to Phase IIa clinical trials as an anticancer drug (53).

The rational design of improved **D2**-analogs has made use of several new computational strategies (52)(54–56). In 2004, Gordaliza et al. (57) reviewed the distribution, sources, application, and new cytotoxic derivatives of **D1**. More recently, Lee and Xiao (58) also have reviewed podophyllotoxins and related analogs, including GL-331, to demonstrate how plant natural products can lead to successful preclinical drug candidates.

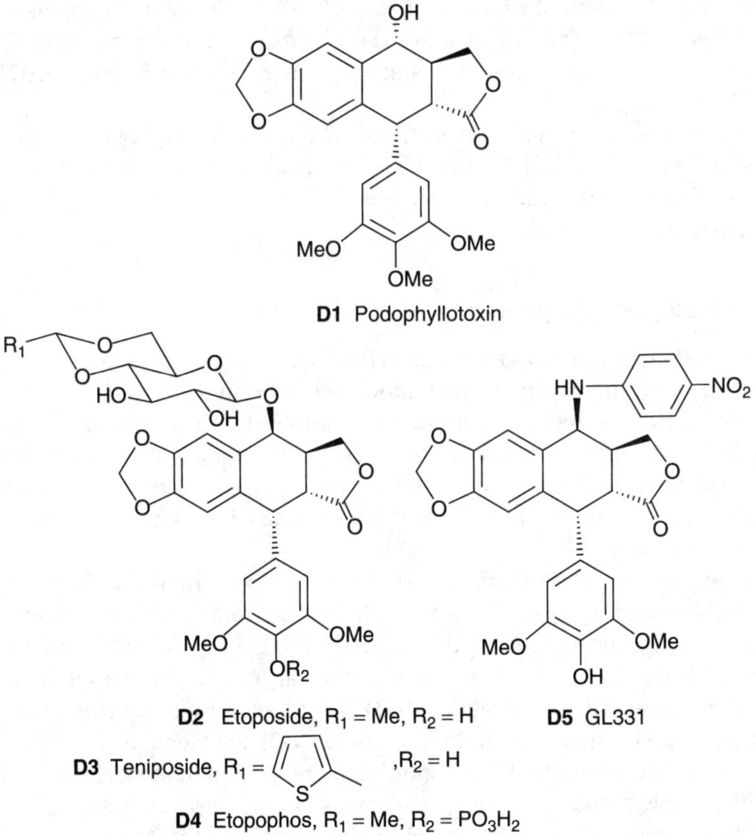

D1 Podophyllotoxin

D2 Etoposide, R_1 = Me, R_2 = H   **D5** GL331

D3 Teniposide, R_1 = [thiophene structure] , R_2 = H

D4 Etopophos, R_1 = Me, R_2 = PO$_3$H$_2$

Figure 14.4   Structures of natural and synthetic podophyllotoxins.

14.1.5   Cephalotaxus species

The genus *Cephalotaxus* (Cephalotaxaceae family) contains coniferous evergreen trees and shrubs that are indigenous to Asia. Historically, the bark has long been used in TCM for various indications.

Powell et al. (59–61) originally isolated the antitumor alkaloids homoharringtonine (**E1**) and harringtonine (**E2**). Subsequently, Chinese investigators discovered that alkaloids from *C. fortunei* Hook. F possess antitumor activity (62). Consequently, **E1** was obtained from the Chinese evergreen tree *C. harringtonia* K. Koch var. *harringtonia* (63), and other active alkaloids were isolated from various *Cephalotaxus* species (64, 65). Interestingly, the parent compound, cephalotaxine (**E3**), is devoid of antitumor activity.

E1 has been investigated in anticancer clinical trials at the U.S. National Cancer Institute (NCI), particularly for the treatment of myeloid leukemia (66, 67). However, its side effects still remain an issue. Thus, the authors have continued to study new natural constituents of *Cephalotaxus* species and to develop

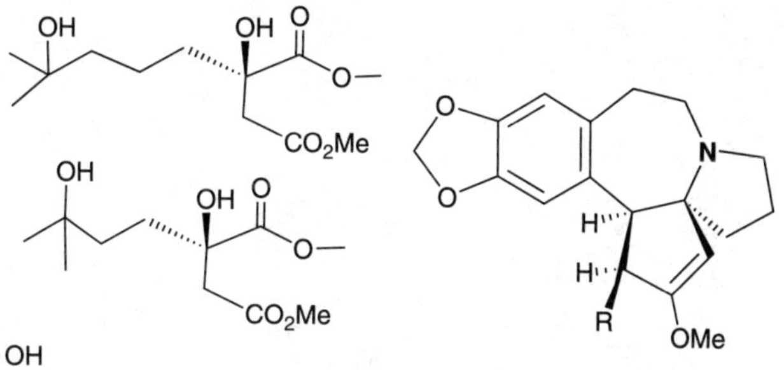

E1 Homoharringtonine R =

E2 Harringtonine R =

E3 Cephalotaxine, R = OH

Figure 14.5 Structures of *Cephalotaxus* alkaloids.

new analogs on the basis of SAR studies, as presented in a review by Itokawa et al. (68).

14.1.6 Colchicum species

The alkaloid colchicine (**F1**) isolated from the medicinal plant *Colchicum autumnale* L. (Liliaceae family) still is used to treat gout and familial Mediterranean fever. **F1** and thiocolchicine (**F2**) (SCH_3 rather than OCH_3 at C-10), which is more stable and more potent but slightly more toxic, are mitotic inhibitors that inhibit polymerization of tubulin (69). Although they show antileukemic activity, they are too toxic to use as anticancer agents, which prompts the synthesis of new, less toxic analogs.

Replacing the C-7 acetoamide group on the B ring with various oxygen-containing groups [ketone (**F3**, thiocolchicone), hydroxyl (**F4**), and ester (**F5**, **F6**)] (70) led to compounds that were equally or more active *in vitro* than **F2**. The C-ring contracted colchinol-7-one thiomethyl ether or allo-ketone (**F7**) was equipotent with the seven-membered ring natural product **F1**. Three-related, ring-contracted colchicinoids (**F8–F10**) showed excellent activity in drug-sensitive and drug-resistant KB cell lines (71).

The above synthetic colchinoids have three methylated phenolic groups in the A ring, which are needed for full potency as tubulin/mitotic inhibitors. Removing one or two of the methyl groups reduces potency, and tridemethylated colchicines and thiocolchicines (**F11–F14**) no longer interact with tubulin but rather are a new class of DNA topo II inhibitory agents (72). They show *in vitro* activity against bone and breast cancers (73).

14.1.7 Salvia species

Salvia officinalis L. (Labiatae family) is native to Europe and America, but the roots and rhizome of *S. miltiorrhiza* Bunge (called Tanshen) have been used widely in China for treating various cardiac (heart) and vascular (blood vessel) disorders, such as atherosclerosis or blood-clotting abnormalities. This plant

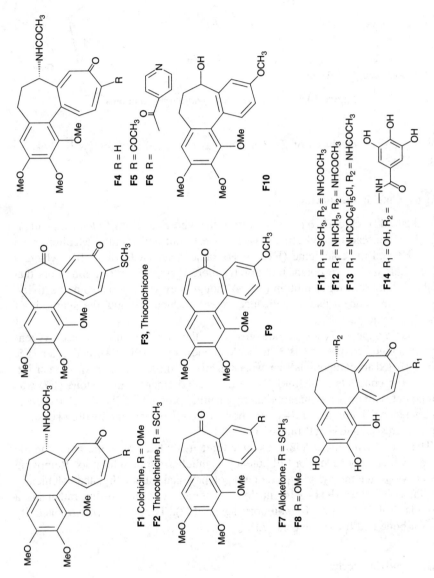

Figure 14.6 Structures of colchicine and related compounds.

F4 R = H
F5 R = COCH₃
F6 R =

F3, Thiocolchicone

F1 Colchicine, R = OMe
F2 Thiocolchicine, R = SCH₃

F7 Alloketone, R = SCH₃
F8 R = OMe

F9

F10

F11 R₁ = SCH₃, R₂ = NHCOCH₃
F12 R₁ = NHCH₃, R₂ = NHCOCH₃
F13 R₁ = NHCOC₆H₅Cl, R₂ = NHCOCH₃

F14 R₁ = OH, R₂ =

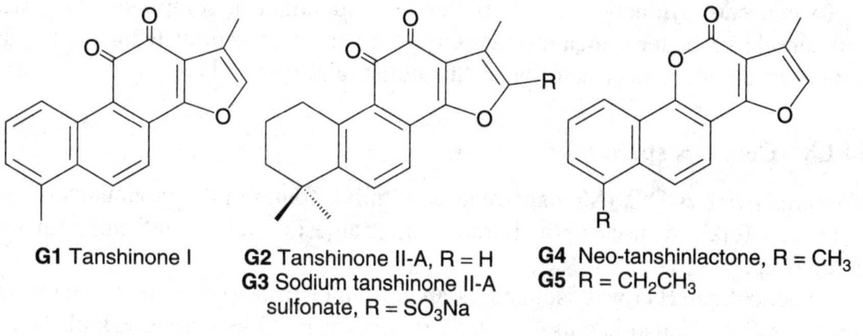

G1 Tanshinone I

G2 Tanshinone II-A, R = H
G3 Sodium tanshinone II-A
sulfonate, R = SO₃Na

G4 Neo-tanshinlactone, R = CH₃
G5 R = CH₂CH₃

Figure 14.7 Structures of tanshinlactones and neo-tanshinlactone.

exhibits hypotensive effects, causes coronary artery vasodilation, and inhibits platelet aggregation. Accordingly, it should not be used in combination with warfarin. Tanshen also has been applied for hemorrhage, dysmenorrhea, miscarriage, swelling, inflammation, chronic hepatitis, and insomnia (74, 75). Clinically available preparations of a *S. miltiorrhiza*/*Dalbergia odorifera* mixture may have potential as an anti-anginal drug (76).

Tanshinone diterpenoids, including tanshinone I (**G1**) and tanshinone IIA (**G2**), are bioactive compounds from *S. miltiorrhiza* (77). Sodium tanshinone sulfate (**G3**) is a water-soluble derivative of **G2**, is used clinically to treat angina pectoris and myocardial infarction, and exhibits strong membrane-stabilizing effects on red blood corpuscles. In addition, novel *seco*-abietane rearranged diterpenoids were isolated recently from this species (78).

Other studies have shown that *S. miltiorrhiza* exerts clear cytotoxic effects and strongly inhibits the proliferation of liver cancer cells (79). Various tanshinones were tested in SAR studies against several human tumor cells, namely nasopharyngeal (KB), cervical (Hela), colon (Colon 205), and laryngeal (Hep-2) cell systems (74, 75).

Neo-tanshinlactone (**G4**), also originally isolated from Tanshen, has a unique and different structure compared with other compounds from *S. miltiorrhiza* Bunge. Preliminary studies of this compound showed unique specific activity against the MCF-7 breast cancer cell line but insignificant activity against other cell lines in the tested panel. Extended bioassay studies showed that **G4** is quite active against estrogen receptor positive (ER+) human breast cancer cell lines (MCF-7 and ZR-75-1) but inactive against ER negative (ER−) human breast cancer lines (MDA MB-231 and HS 587-T) (80). This finding is significant because more than 60% of breast cancer cases in postmenopausal women are ER+. Compared with the breast cancer drug tamoxifen citrate, **G4** was 10-fold more potent and 20-fold more selective against two ER+ cell lines (80). It also was potent against an ER− cell line that overexpresses the protein HER2+, which plays a key role in regulating cell growth and affects 25–30% of breast cancer patients (80). These results indicate that **G4** is an excellent candidate for additional development toward anti-breast cancer clinical trials.

To that end, synthetic analog studies have identified a compound (**G5**) with comparable or better anticancer activity and certain structural features that are critical to the anticancer activity of this compound type (81).

14.1.8 Cocculus species

Cocculus trilobus DC. (Menispermaceae family), found in the mountains of east Asia, has folkloric uses as a diuretic, an analgesic, and an anti-inflammatory crude drug.

Sinococuline (**H1**) was isolated as an antitumor principle from the stems and rhizomes (82). It also has *in vivo* activity against P388 leukemia. **H1** likely is a general cytotoxic rather than a cell-specific agent (83).

14.1.9 Maytenus species

Maytenus illicifolia Mart. ex Reiss. (Celastraceae family), more commonly known as "Cangorosa," is found in South America where it is used for its analgesic, antipyretic, antiseptic, and anticancer properties and for birth control, particularly in Paraguay.

Bioactivity-directed fractionation and isolation by various research groups has led to the isolation of various active principles. Kupchan et al. (84, 85) first identified antileukemic maytansinoids, for example, maytansine (**I1**), from the African plant *M. ovatus* [later renamed *M. serrata* (Hochst. ex A. Rich.) R. Wilczek]. Although it advanced to Phase II clinical trials, testing then was suspended because of neurotoxicity. The related maytanprine (**I2**), isolated from *M. diversifolia* (Maxim.) Ding Hou, has been investigated for growth-inhibiting and apoptosis-inducing activities in K562 leukemia cells (86). Cytotoxic monotriterpenes include pristimerin (**I3**) and isotingenone III (**I4**) from *M. illicifolia* (87), as well as the triterpene dimers dihydroisocangorosin A (**I5**) and cangorosin B (**I6**) (88, 89). The authors also have identified cytotoxic sesquiterpene pyridine alkaloids, including emarginatines B (**I7**) and F (**I8**), from *M. emarginata* (Willd.) Ding Hou (90, 91).

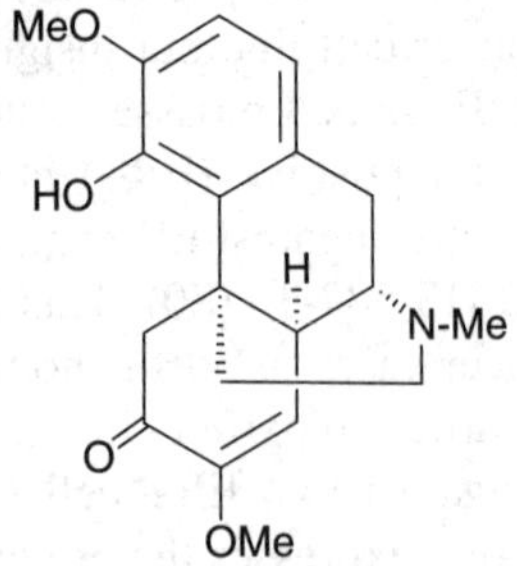

H1 Sinococuline

Figure 14.8 Structure of sinococuline.

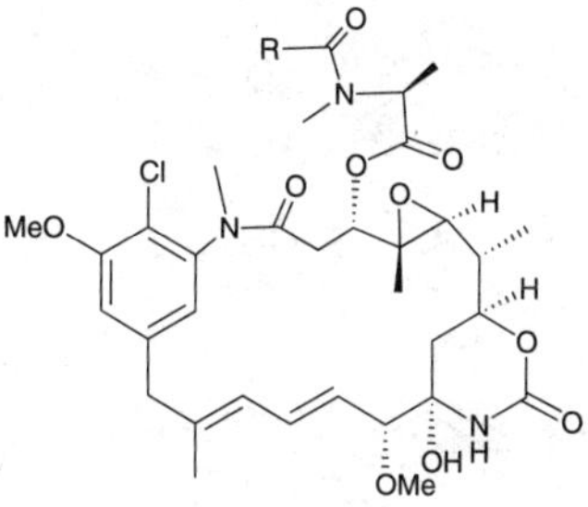

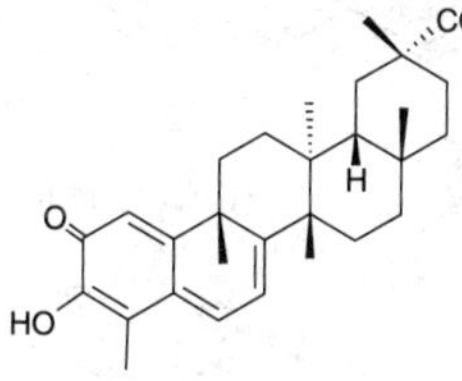

I3 Pristimerin

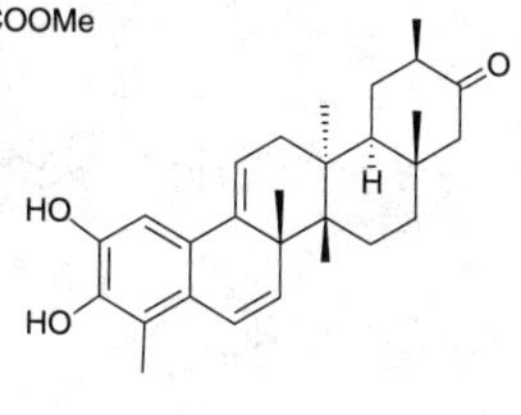

I4 Isotingenone III

I1 Maytansine, R = CH_3
I2 Maytanprine, R = CH_2CH_3

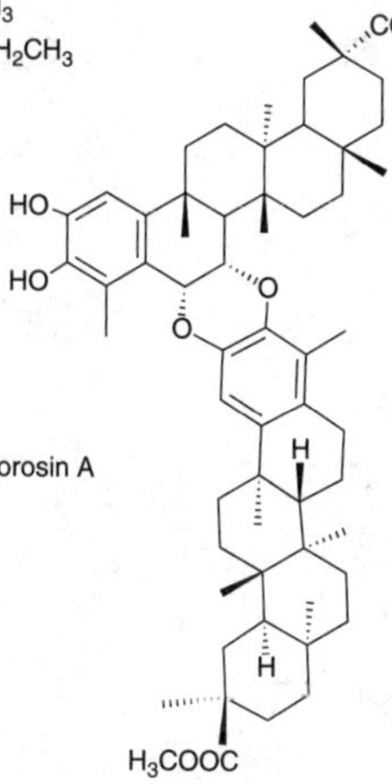

I5 Dihydroisocangorosin A

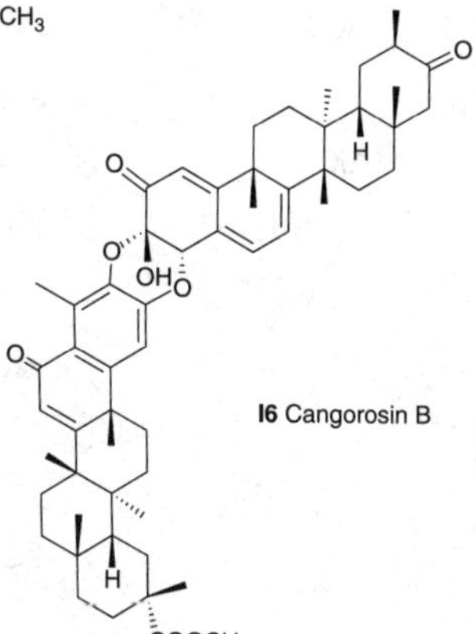

I6 Cangorosin B

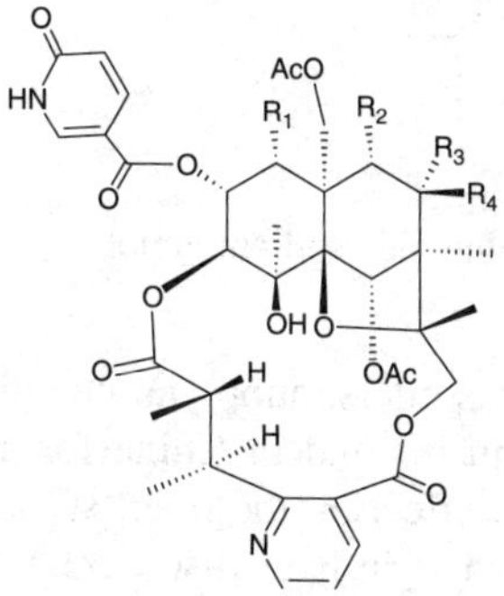

I7 Emarginatine B, $R_1 = R_3 = OAc$, $R_2 = OBz$, $R_4 = H$
I8 Emarginatine F, $R_1 = OBz$, $R_2 = OAc$, $R_3 = H$, $R_4 = OH$

Figure 14.9 Structures of *Maytenus* natural products.

Recently, new triterpenes were isolated from *M. chuchuhuasca* R. Hamet and Colas (92), along with sesquiterpenes from the same plant (93). New compounds also were reported from *M. ilicifolia* (94). In addition, antitumor promoting sesquiterpenes were isolated from *M. cuzooina* Loes. (95).

14.1.10 Curcuma species (turmeric)

The turmeric rhizome is a main ingredient of curry powder. It gives color and flavor to food, and it has aromatic, stimulant, and carminative properties. This herb

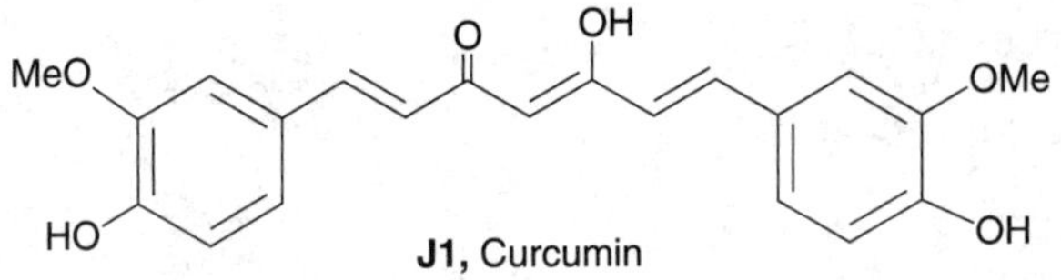

Figure 14.10　Structures of curcuminoids and curcumol.

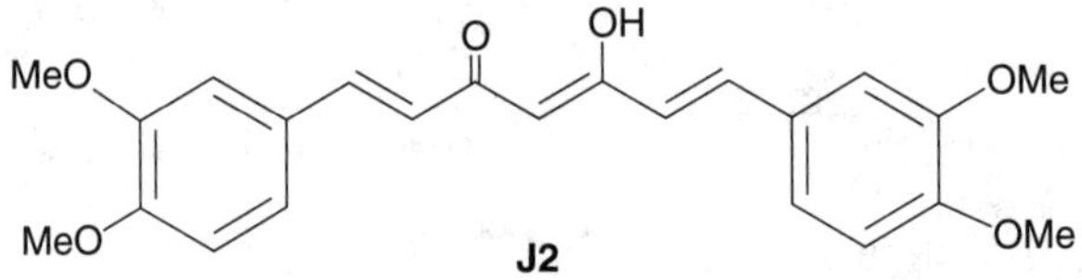

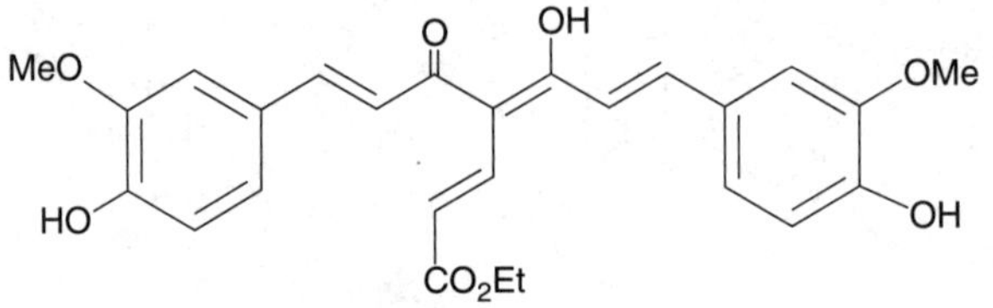

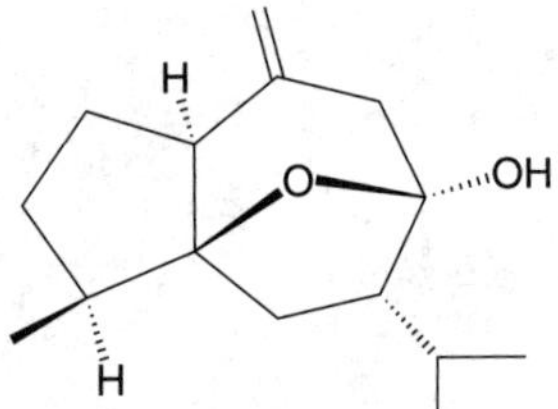

is used traditionally in India to treat biliary disorders, anorexia, cough, diabetic wounds, liver disorders, rheumatism, and sinusitis and in China for abdominal pains and jaundice. Turmeric has a protective effect on the liver, stimulates bile secretion in animals, and is recommended for use in liver disorders.

The major pigment in *Curcuma* species (Zingiberaceae family) is the yellow phenolic diarylheptanoid curcumin (**J1**). Curcumin and its analogs have potent antioxidant and anti- inflammatory effects, cytotoxicity against tumor cells, and antitumor-promoting activity (96). The biologic effects and targets of curcumin, as well as its possible roles in cancer prevention and therapy, have been reviewed recently (97, 98).

Several synthetic curcumin analogs, including **J2**, showed potent antian-drogenic activities against two human prostate cancer cell lines, PC-3 and DU-145 (99). In expanded *in vitro* testing, these synthetic curcumin derivatives showed antiprostate cancer activity superior to that of hydroxyflutamide, the currently available and preferred anti-androgen for treating prostate cancer

(100). In continuing work in the laboratories of the authors, dimethoxy-4-ethoxycarbonylethylenyl-curcumin (**J3**) showed potent anti-androgenic activity and therefore is a promising prostate cancer drug candidate (101, 102). A recent review from the laboratories of the authors discusses the design and development of curcumin analogs as promising candidates for chemotherapy of prostate cancer (103). In addition, the sesquiterpene curcumol (**J4**), obtained from *Curcuma aromatica* Salisb., was effective against cervical cancer (104).

14.1.11 Euphorbia species

Euphorbia kansui Liou (Euphorbiaceae family) is distributed widely in northwest China. The dried roots of the plant are known as "kansui" and classified as a lower-class medicine. It is used as an herbal remedy for ascites (abdominal fluid accumulation) and cancer in China.

Ingenol diterpenoids are among the bioactive chemical constituents. Kansuiphorins A–D (**K1–K4**) were isolated as cytotoxic principles of *E. kansui* by the laboratories of the authors (105, 106). In particular, **K1** and **K2** demonstrated potent antileukemic activity against P-388 leukemia in mice (105). A related ingenol-type diterpene (DBDI, **K5**) showed unique suppression of mast cell activation, a process that occurs during inflammation, and thus, might have the potential to treat allergic diseases (107). In other pharmacological studies, two isolated ingenols from an immuno-enhancing *E. kansui* extract increased immune activity in a dose-dependent manner (108). Three new cytotoxic diterpenoids, yuexiandajisus D (**K6**), E (**K7**), and F (**K8**), were isolated from the species *E. ebrateolata* Hayata. (109). In additional studies, **K6** showed moderate cytotoxicity against additional (HCT-8 and Bel-7402) cell lines (110).

14.2 ANTI-HIV COMPOUNDS

Acquired immunodeficiency syndrome (AIDS), a degenerative disease of the immune system, is caused by the human immunodeficiency virus (HIV) and results in life-threatening opportunistic infections and malignancies. Antiviral and immunomodulating natural products have been investigated as treatments for AIDS (111).

14.2.1 Lomatium suksdorfii (coumarin derivatives)

Suksdorfin (**L1**), a dihydroseselin-type angular pyranocoumarin isolated from *Lomatium suksdorfii* Coult. Rose (Apiaceae family) was identified as a lead anti-HIV natural product through BDFI (112). Substitution of two camphanoyl esters for the acetate and isovaleroyl groups in the natural product lead to the extremely potent lead compound 3′R,4′R-di-O-(−)-camphanoyl-(+)-*cis*-khellactone (DCK) (**L2**) (113).

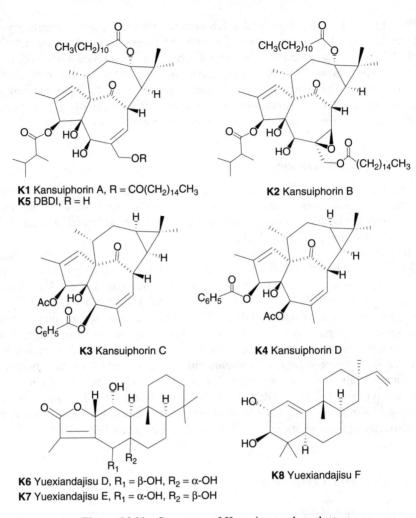

K1 Kansuiphorin A, R = CO(CH₂)₁₄CH₃
K5 DBDI, R = H

K2 Kansuiphorin B

K3 Kansuiphorin C

K4 Kansuiphorin D

K6 Yuexiandajisu D, R₁ = β-OH, R₂ = α-OH
K7 Yuexiandajisu E, R₁ = α-OH, R₂ = β-OH

K8 Yuexiandajisu F

Figure 14.11   Structures of Kansui natural products.

After additional synthetic modification to improve potency, 4-methyl DCK (**L3**) and then 3-hydroxymethyl-4-methyl DCK (**L4**) were found. The latter compound was selected as a clinical trial candidate (114).

Furthermore, a positional isomer of DCK, 3'R,4'R-di-O-(−)-camphanoyl-2',2'dimethyldihydropyrano(2,3-f)chromone (DCP) (**L5**) is even more promising because most DCP analogs are active against drug-resistant HIV strains, although DCK analogs are not. Adding an ethyl group at the 2-position of DCP decreased toxicity to cells compared with DCP, so, to date, the most likely clinical trials candidate in the DCP series is 2-ethyl DCP (**L6**) (114, 115).

The DCK and DCP compounds exert their antiviral activity by blocking the HIV reverse transcriptase (RT), however, at a later step than that affected by

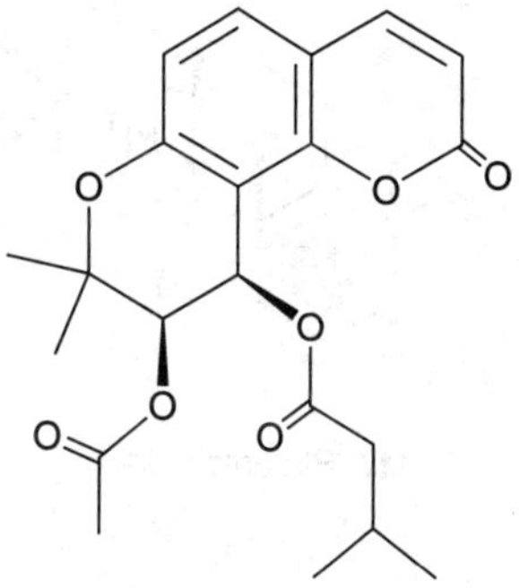

L1 Suksdorfin

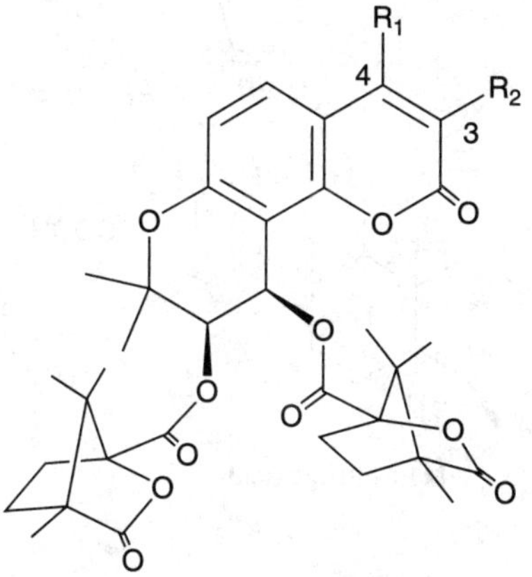

L2 DCK, $R_1 = R_2 = H$
L3 4-MethylDCK, $R_1 = Me$, $R_2 = H$
L4 3-Hydroxymethyl-4-methylDCK,
    $R_1 = Me$, $R_2 = CH_2OH$

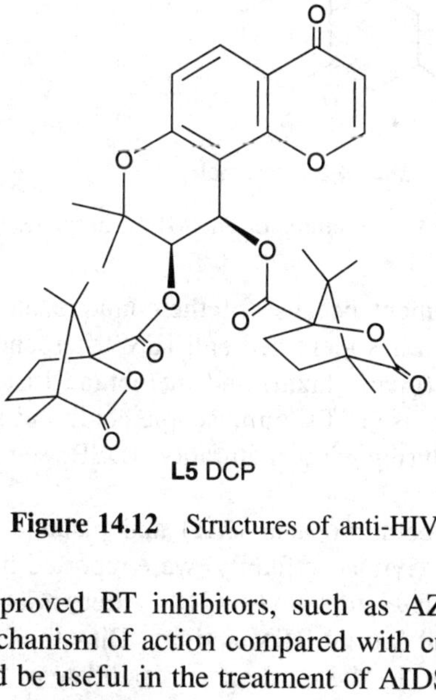

L5 DCP

Figure 14.12 Structures of anti-HIV coumarins.

the clinically approved RT inhibitors, such as AZT. Thus, these compounds have a novel mechanism of action compared with current drugs. DCK and DCP compounds could be useful in the treatment of AIDS, although additional investigation is merited and needed (116).

14.2.2 Syzigium claviflorum (triterpene betulinic acid derivatives)

Triterpenes represent a structurally varied class of natural products existing in many plant species. Thousands of triterpenes have been reported with hundreds of new derivatives described each year. Some naturally occurring triterpenes exhibit moderate anti-HIV-1 activity and, therefore, provide good leads

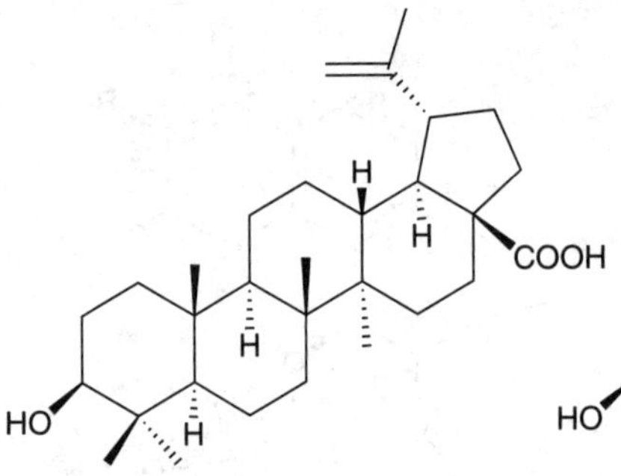

M1 Betulinic acid

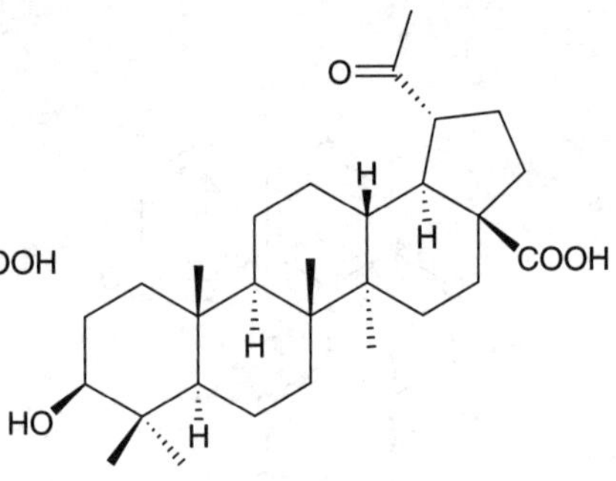

M2 Platanic acid

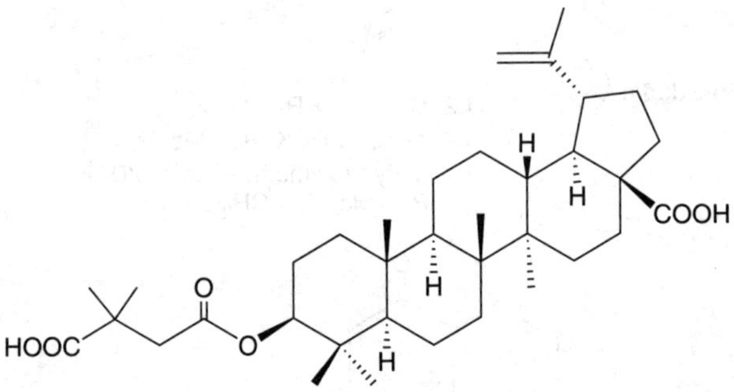

M3 DSB (Bevirimat)

Figure 14.13 Structures of anti-HIV triterpenes.

for additional drug development because of their unique mode of action and chemical structures. Research has identified anti-HIV triterpenes that block HIV entry, including absorption (glycyrrhizin) and membrane fusion (RPR103611); inhibit viral reverse transcriptase (RT) (mimuscopic acid) and protease (ganoderiol, geumonoid); and act during viral maturation (DSB, see more description below) (117).

Two lupane triterpenes, betulinic acid (**M1**) and platanic acid (**M2**), from *Syzigium claviflorum* Wall. (Myrtaceae family) were reported first to reduce HIV IIIB reproduction by 50% in H9 lymphocytes (118). Afterwards, many derivatives were synthesized and studied for anti-HIV activity. Dimethyl succinyl betulinic acid (DSB, **M3**) was found to be the most useful candidate as an anti-HIV agent (119, 120).

DSB is the first in a new class of drugs to treat HIV infection. Its novel viral target—maturation—is unlike that of any currently approved anti-AIDS drug. DSB disrupts the late-stage viral maturation processes of HIV and causes the viral core structure to be defective and noninfectious (121).

DSB was discovered originally by the Natural Products Research Laboratories (NPRL), University of North Carolina, directed by the author (K.H.L.) (119), and then licensed to Panacos Pharmaceuticals, Inc. (Watertown, MA) for development. Panacos has named the compound Bevirimat and lists it as the lead antiviral product of the company.

During 2004, two Phase I studies and a Phase I/II study of **M3** were completed. In the Phase I studies, the drug was well tolerated and showed good anti-HIV levels in the body. In the Phase I/II study, **M3** showed activity in HIV-infected patients and significantly reduced viral blood levels (known as viral load) (122, 123). Another 2004 milestone was that the U.S. Food and Drug Administration (FDA) granted Fast Track Status for **M3**.

The Phase IIa study demonstrated the antiviral potency of **M3**, following once-daily oral dosing for 10 days in HIV-infected subjects not on other antiretroviral therapy. Viral load was reduced significantly compared with placebo. On day 11, following complete dosing, the median reduction at the 200-mg dose was a 91% decrease. In the Phase IIa trial, **M3** was well tolerated, all adverse experiences were mild or moderate, and no dose-limiting toxicity was identified (122, 123).

Subsequently, studies have shown that **M3** can be administered successfully in a tablet form rather than by an oral solution. Also, two drug interaction clinical trials of **M3**, in combination with the approved HIV drugs ritonavir and atazanavir, have been completed and showed little likelihood of significant adverse drug–drug interactions when used in combination therapy (124).

In summary, **M3** shows potent viral load reduction, a strong safety profile (with no evidence of organ toxicity or clinical intolerance), no evidence of clinically significant drug interactions, and, quite importantly, no evidence of rapid resistance development, which is a primary cause for antiretroviral treatment failure (125–127)

Phase IIb clinical trials began in 2006 and still are ongoing. One of the trial goals will be to determine an optimal dose of **M3**. These trials will involve HIV-infected patients who are failing current therapy and will be randomized, blinded, and placebo-controlled (124).

In Phase III clinical trials targeted for 2007/2008, combination therapy studies will be performed in a total of 300 to 500 patients at a commercial dose. The target for New Drug Application (NDA) is 2008/2009 (124). The efficient clinical trials progress of **M3** continues to mark it as a leading new treatment for AIDS.

14.3   ACTIVE COMPOUNDS ISOLATED FROM WELL-KNOWN FOLKLORIC MEDICINE

In addition to anticancer and anti-HIV agents, various types of active compounds that are active against other diseases and disorders (e.g., malaria, inflammation, and so forth.) also have been isolated from natural sources, especially well-known folkloric medicine. These compounds and their plant sources are described below.

14.3.1   Artemisia annua (qinghao, artemisinin derivatives)

Qinghao (Sweet Wormwood) is the dried aerial parts of the herb *Artemisia annua* L. (Asteraceae family), which has been used in China for centuries to treat fever

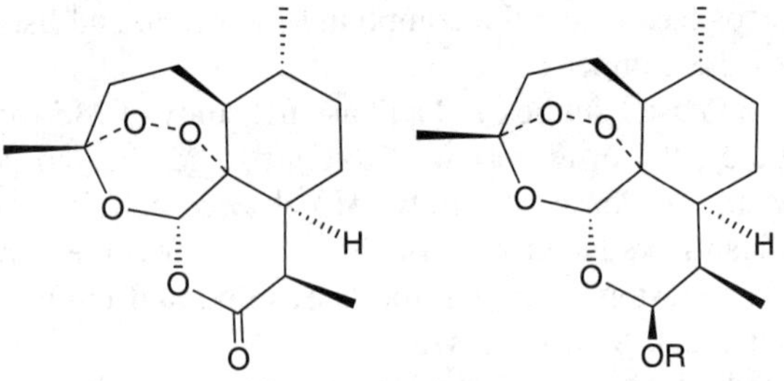

N1 Artemisinin

N2 Artemether, R = CH$_3$
N3 Arteether, R = CH$_2$CH$_3$
N4 Sodium artesunate,
     R = COCH$_2$CH$_2$COONa

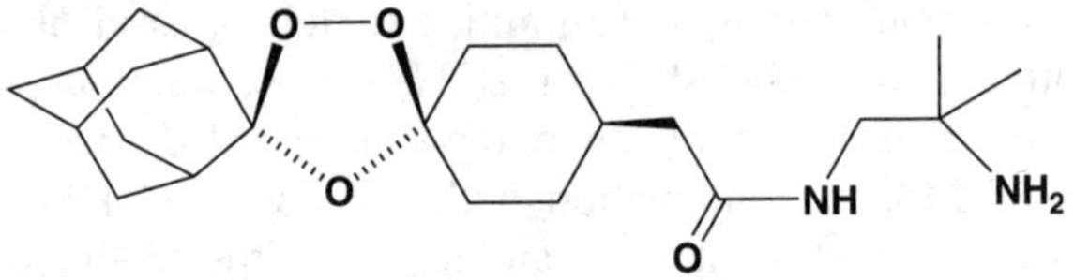

N5 OZ-277 (RBx11160)

Figure 14.14   Structures of artemisinin and related compounds.

and malaria. Artemisinin (**N1**) (Qing Hao Su) (128), the active principle, directly kills *Plasmodium falciparum* (malaria parasites) with little toxicity to animals and humans. Thus, it is a clinically effective, safe, and rapid antimalarial agent (129, 130). The novel endo-peroxide link is essential for the antimalarial activity.

Artemether (**N2**) and arteether (**N3**) are the most well-studied analogs among many synthetic derivatives and are used in malaria-prone regions, particularly India (131). Artemether and sodium artesunate (a hemisuccinate derivative of dihydroartemisinin) (**N4**) have been added by the World Health Organization to its Model List of Essential Medicines (132).

The laboratory of the authors has synthesized analogs related to artemisinin (133). Recently, an antimalarial synthetic trioxolane drug development candidate called **OZ-277** (**N5**, also known as RBx11160) (134) has sparked great interest and has progressed to Phase II clinical trials in India, Thailand, and Africa. Modification and pharmacological studies are ongoing (135–137). A recent review discusses artemisinin and related antimalarials (138).

14.3.2   Ginsengs: Asian, American, Sanchi, and Siberian

Ginseng is the root of *Panax ginseng* C.A. Meyer (Asian ginseng) (Araliaceae family). In Oriental medicine, it has enjoyed a strong reputation since ancient times for being tonic, regenerating, and rejuvenating. The genus name *Panax* is formed from the Greek pan (all) and akos (remedy). This panacea (panakeia) was believed to be the universal remedy. Wild ginseng is scarce and has been replaced

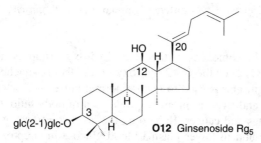

Ginsenoside		R_1	R_2
O1	Rb$_1$	-glc(2-1)glc	-glc(6-1)glc
O2	Rb$_2$	-glc(2-1)glc	-glc(6-1)ara$_p$
O3	Rc	-glc(2-1)glc	-glc(6-1)ara$_f$
O4	Rd	-glc(2-1)glc	-glc
O9	Rg$_3$	-glc(2-1)glc	-H
O11	Rh$_2$	-glc	-H

Ginsenoside		R_1	R_2
O5	Re	-glc(2-1)rha	-glc
O6	Rf	-glc(2-1)glc	-H
O7	Rg$_1$	-glc	-glc
O8	Rg$_2$	-glc(2-1)rha	-H
O10	Rh$_1$	-glc	-H

O12 Ginsenoside Rg$_5$

Figure 14.15   Structures of *Panax* ginsenosides.

by cultivated ginseng or "true" ginseng. Species include American ginseng (*P. quinquefolium* L.), cultivated in North America; Japanese ginseng (*P. japonicus* (Nees.) C.A. Mey., widely distributed in Japan; San-chi ginseng (*P. notoginseng* (Burk.) F.H. Chen), reputed as a tonic and hemostatic in China; and Siberian ginseng (*Eleutherococcus senticosus* Maxim.).

14.3.2.1 Asian ginseng (Panax ginseng C.A. Meyer)   Many compounds have been isolated from the root of Asian ginseng: polysaccharides, glycopeptides (panaxanes), vitamins, sterols, amino acids and peptides, essential oil, and polyalkynes (139–141). About 30 saponins (called ginsenosides) isolated from the root are dammarane triterpenoids, which generally contain three or four hydroxyl (OH) groups [a 3β,12β,20(*S*) trihydroxylated-type (protopanaxadiol-type) and a 3β,6α,12β,20(*S*) tetrahydroxylated-type (protopanaxatriol-type)], which are attached to various sugars. The individual saponins (e.g., ginsenosides Rb$_{1-2}$, Rc-f, Rg$_{1-3}$, and Rh$_{1-2}$, **O1–O11**) differ in the mono-, di-, or tri-saccharide nature of the two sugars attached at the C-3 or C-6 and C-20

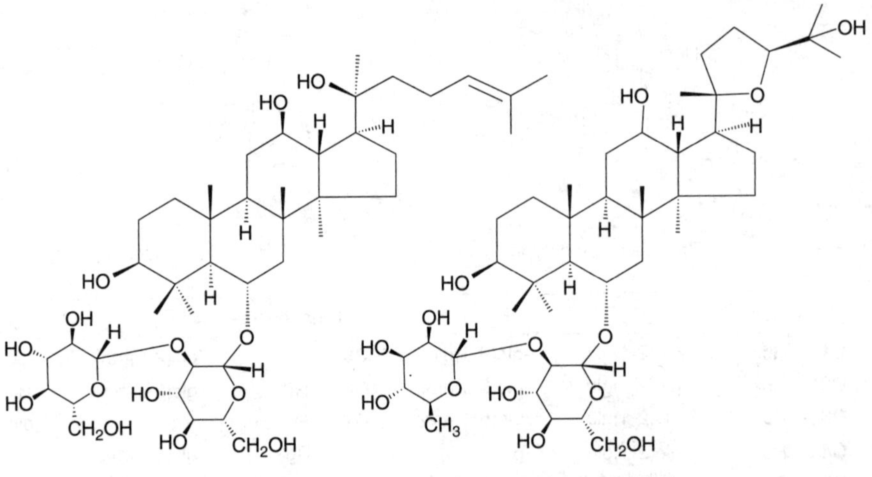

O6 Ginsenoside Rf
(found in *P. ginseng*)

O13 24(R)-Pseudoginsenoside F$_{11}$
(found in *P. quinquefolium*)

Figure 14.16 Structures of two different ginsenosides in *P. ginseng* and *P. quinquefolium*.

hydroxy groups. In some cases, the C-20 hydroxy group is absent (e.g., in ginsenoside Rg$_5$, **O12**) or the C-12 hydroxy group also is attached to a sugar.

Traditionally, ginseng is used to restore normal pulse and remedy collapse, to benefit the spleen and liver, to promote production of body fluid, to calm nerves, and to treat diabetes and cancer. Regarding the last two effects, ginsenoside Rh$_2$ (**O11**) has been found to lower plasma blood glucose in streptozotocin-induced diabetic rats (142), and recently, ginseng and its constituents have been studied for cancer prevention and anticarcinogenic effects against chemical carcinogens. Ginsenoside Rg$_3$ (**O9**) and Rg$_5$ (**O12**) were found to reduce significantly lung tumor incidence, and Rg$_3$ (**O9**), Rg$_5$ (**O12**), and Rh$_2$ (**O11**) showed active anti-carcinogenic activity (143). In addition, effects of ginseng on quality of life have been discussed (144).

14.3.2.2 *American ginseng (*Panax quinquefolium L.*)*   American ginseng contains almost the same components as *Panax ginseng* (139). Thus, it could be used for the same medical conditions as Asian ginseng. However, in Chinese theory, some differences exist: American ginseng is "cool" and is used mainly to reduce the internal heat and promote the secretion of body fluids, whereas Asian ginseng is "warm." Correspondingly, differences in biologic activity also exist (140): American ginseng stimulates the production of human lymphocytes, whereas Asian ginseng does not have a significant effect, and Siberian ginseng enhances production.

The main chemical difference between Asian and American ginseng is in the presence or absence of ginsenoside Rf (**O6**) and 24(R)-pseudoginsenoside F$_{11}$ (**O13**). Ginsenoside Rf is found in Asian ginseng but not in American ginseng,

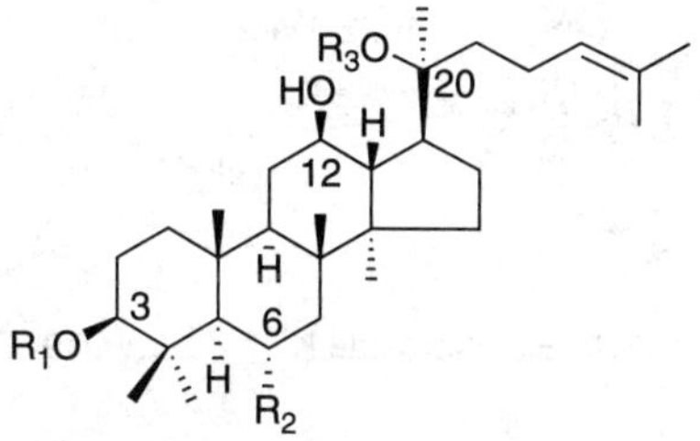

Notoginsenoside	R_1		R_2	R_3
O14	K	-glc(6-1)glc	-H	-glc
O15	R_1	-H	-glc(2-1)xyl	-glc
O16	U	-H	-OH	-glc(6-1)glc

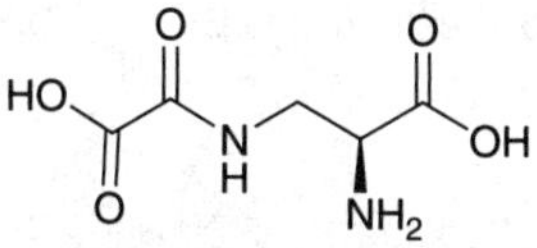

O17 Dencichine

Figure 14.17   Structures of *P. notoginseng* natural products.

and although 24(R)-pseudoginsenoside F_{11} is abundant in American ginseng, only trace amounts are present in Asian ginseng (141)(145–147).

14.3.2.3 Sanchi ginseng [Panax notoginseng (Burk.) F.H. Chen]   This ginseng exerts a major effect on the cardiovascular system. It dilates the coronary vessels and reduces vascular resistance, which results in increased coronary flow and decreased blood pressure. Chinese traditional medicine used this ginseng to arrest bleeding, remove blood stasis, and relieve pain. Recent studies have shown that, in the treatment of angina pectoris, this herb can produce a 95.5% improvement in symptoms. The herb usually can stop bleeding in cases of respiratory bleeding or vomiting of blood.

P. notoginseng contains saponins similar to those found in *P. ginseng*, both ginsenosides and notoginsenosides. Certain saponins of both types, including notoginsenosides K, R_1, and U (**O14–O16**), showed immunological adjuvant activity (stimulated immune response against antigens) (148, 149). Another important bioactive constituent of *P. notoginseng* is the non-protein amino acid dencichine (**O17**), which can increase platelets and stop bleeding (150).

14.3.2.4 Siberian ginseng (Eleutherococcus senticosus maxim.)   Siberian ginseng is harvested from its natural habitat in Russia and northeast China and has been used in China for over 2000 years. It is not a true ginseng like *Panax ginseng* or *P. quinquefolia*, but it does belong to the same Araliaceous family. Siberian ginseng has its own bioactive ingredients with unique and proven medicinal values. It possesses significant adaptogenic action (antistress and antifatigue)

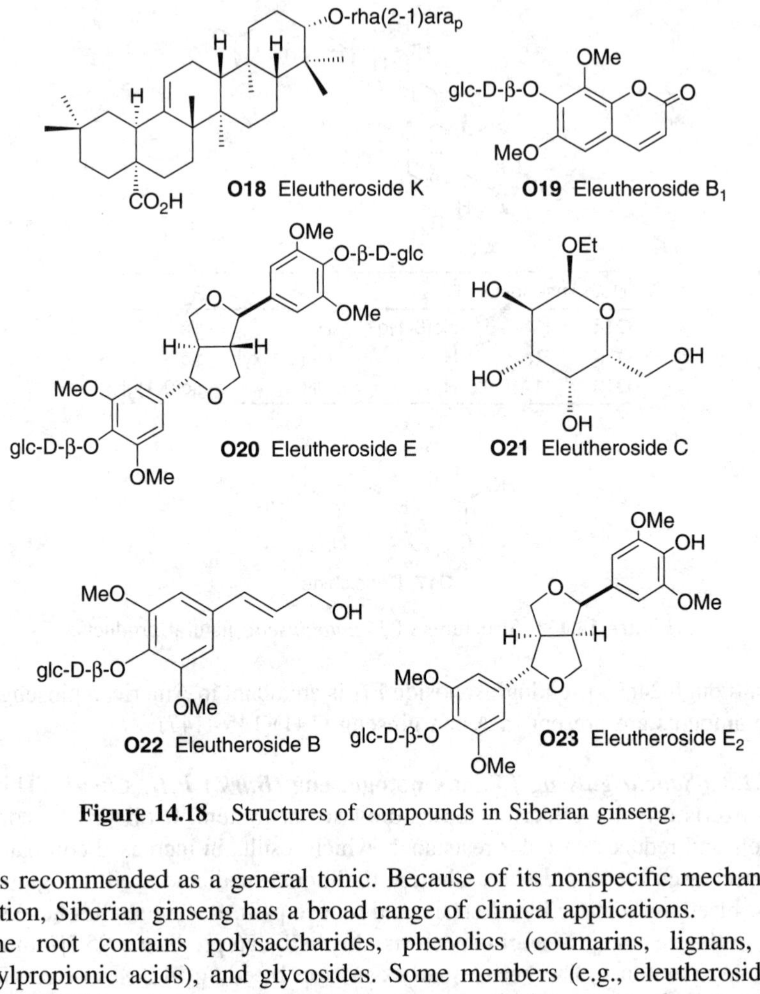

Figure 14.18 Structures of compounds in Siberian ginseng.

and is recommended as a general tonic. Because of its nonspecific mechanisms of action, Siberian ginseng has a broad range of clinical applications.

The root contains polysaccharides, phenolics (coumarins, lignans, and phenylpropionic acids), and glycosides. Some members (e.g., eleutheroside K, **O18**) of the latter group are specifically triterpenoid in nature, whereas others, including isofraxoside (eleutheroside B_1, **O19**), glycosides of syringaresinol (e.g., eleutheroside E, **O20**), and the ethyl ether of galactose (eleutheroside C, **O21**), belong in a miscellaneous series. Eleutherosides E (**O20**) and B (**O22**, also called syringin) are two major glycosides and typically are used as marker compounds associated with bioactivity, particularly antifatigue action (151). A new lignan glycoside eleutheroside E_2 (**O23**) was isolated recently from *E. senticosus* (152).

14.3.3 Ganoderma lucidum (fungus)

Chinese people consider *Ganoderma lucidum* (Polyporaceae family) as the "Miraculous King of Herbs." It is highly regarded for its medicinal properties, which include promoting the healing ability of the human body, strengthening the immune system, and increasing longevity. Accordingly, *Ganoderma* works in the treatment of cancer because it helps cleanse the body from toxins and strengthen the immune system. It also enhances liver detoxification, thus improving liver function and stimulating the regeneration of liver cells.

The chemical composition includes polysaccharides and lanosteroid triterpenes. In the former class, many polysaccharides have been linked to immune-stimulating effects (153–155). The latter class contains over 130 different triterpenoids with diverse pharmacological activities (156). Examples of the polyoxygenated triterpene structures are ganoderic acids B (**P1**), C_2 (**P2**), and D (**P3**), as well as ganoderiol F (**P4**), ganodermanondiol (**P5**), and ganodermanontriol (**P6**).

Because of its widespread use as a health food as well as for medicinal purposes, the quality control of *G. lucidum* is quite important, and modern analytical

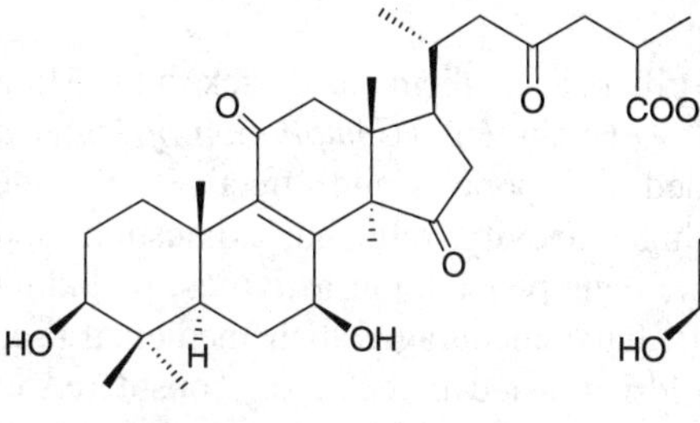

P1 Ganoderic acid B

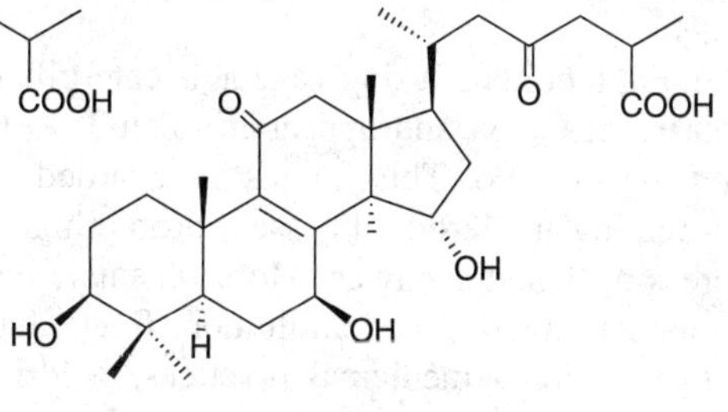

P2 Ganoderic acid C_2

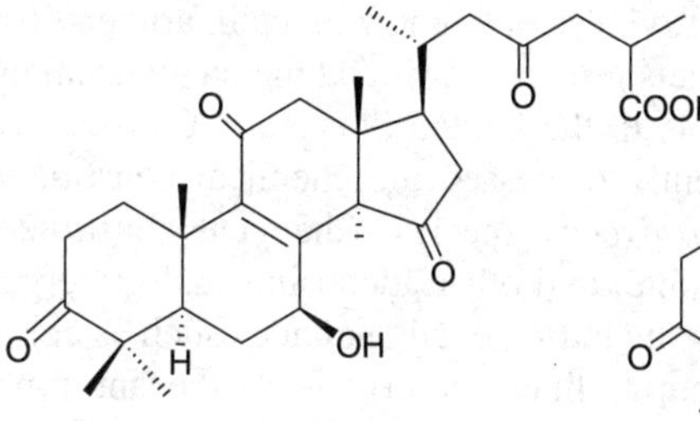

P3 Ganoderic acid D

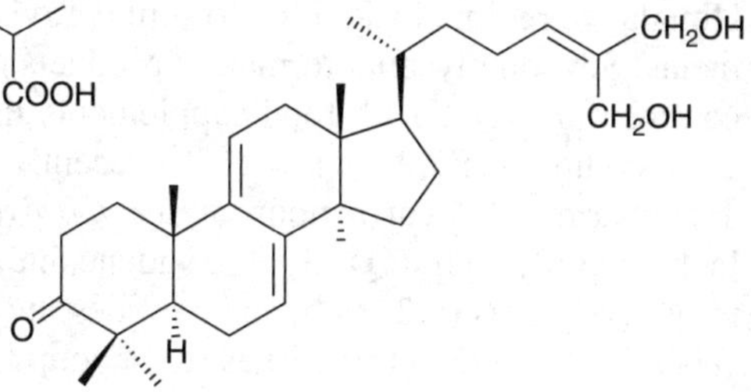

P4 Ganoderiol F

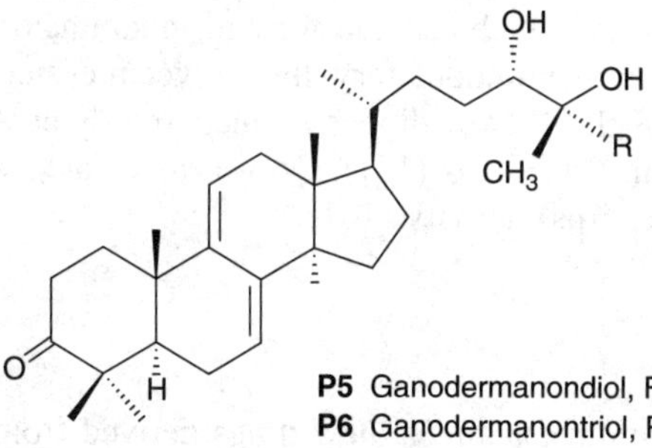

P5 Ganodermanondiol, R = CH_3
P6 Ganodermanontriol, R = CH_2OH

Figure 14.19 Structures of *Ganoderma* triterpenes.

methods are being established for the quantitative determination of major triter-penoids, including **P1–P3**, as marker compounds (157).

14.3.4 Cordyceps sinensis (Tung Chung Hsia Tsao)

Vegetable caterpillars are called "Tung Chung Hsia Tsao" in Chinese, which trans-lates as "winter worm and summer grass." They result from the infection of large underground caterpillars by a fungus. An entomopathogenic fungus grows in the larva of the sphinx moth in autumn. Although the larva hibernates underground through the winter, the fungus kills the infected host and grows throughout the cadaver. In summer, a rod-like stroma of the fungus grows out from the mummi-fied shell of the dead host, looking like a sprouting, dark blue to black grass. The mycelia of *Cordyceps sinensis* (Berk.) Saccar. (Clavicipitaceae family) that col-onize the larvae of *Hepialus armoricans* (Hepialidae family) are representative vegetable caterpillars and are highly valued in the Chinese traditional medicinal system.

Initial records of using vegetable caterpillars as medicine date back to the Ming dynasty in China and appear in "Pen-Tsao-Kang-Mu" (*Compilation of Materia Medica*) in 1596. This fungus is regarded as a popular and effective medicine for treating numerous illnesses, promoting longevity, relieving exhaustion, and increasing athletic power. More recently, their physiologic activities, including immunostimulating and antitumor effects, have encouraged their medicinal use.

In traditional medicinal practices, wild-harvested mycelia are considered to have higher therapeutic benefits and, therefore, command higher prices than cul-tivated fungus. Native occurrence of the fungus is confined to the highlands of the Himalayan region, in addition to some provinces of China with cold, arid environ-ments. Several mycelial formation products grown in artificial media are available commercially as health food supplements in the United States and Canada.

A comprehensive review (158) recently discussed the chemical constituents and pharmacological actions of the *Cordyceps* species. Chemical constituents include cordycepin (**Q1**, 3′-deoxyadenosine) and other adenosine analogs, ergos-terol derivatives (**Q2–Q4**), and peptides, including cyclic peptides such as cordy-cepeptide A (**Q5**) (159). These compounds likely contribute to the antitumor, antibacterial, antifungal, and anti-inflammatory activities. Regarding the latter activity, **Q1** has been found to inhibit platelet aggregation (160). *C. sinensis* also contains polysaccharides, which account for anti-inflammatory, antioxidant, antitumor, antimetastatic, immunomodulatory, hyperglycemic, steroidogenic, and hypolipidemic effects (158, 159). New diphenyl ethers (cordyols A–C, **Q6–Q8**) were discovered recently in *Cordyceps* (161). **Q6** was associated with significant anti-HSV (herpes simplex virus) activity (161).

14.4 CONCLUSION

Based on the above successful examples, new drugs derived from natural prod-uct leads will be discovered continuously, and modern medicinal chemistry-based

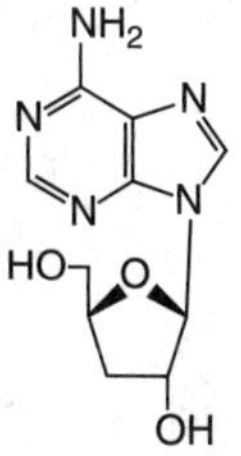

Q1 Cordycepin

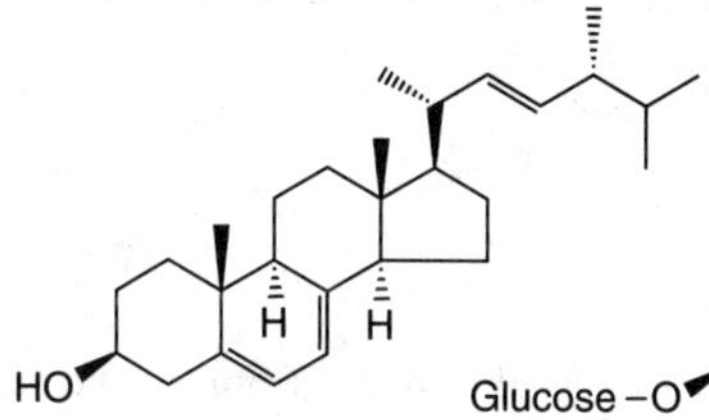

Q2 Ergosterol

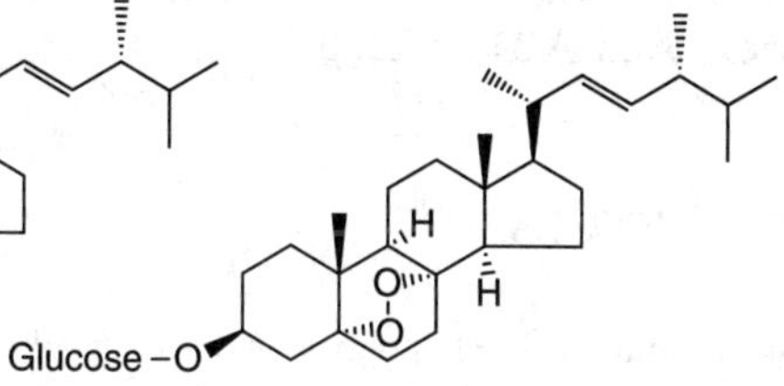

Q3

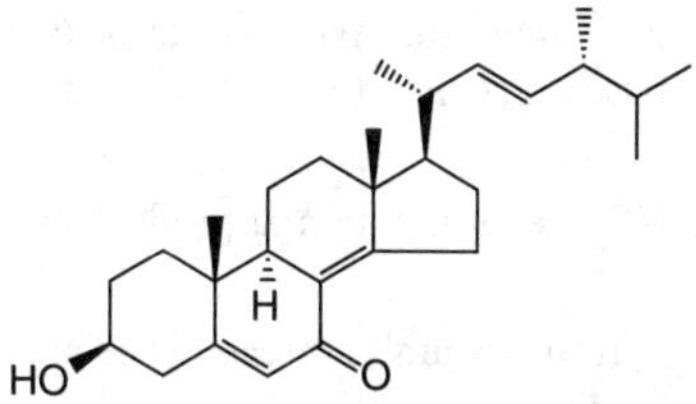

Q4

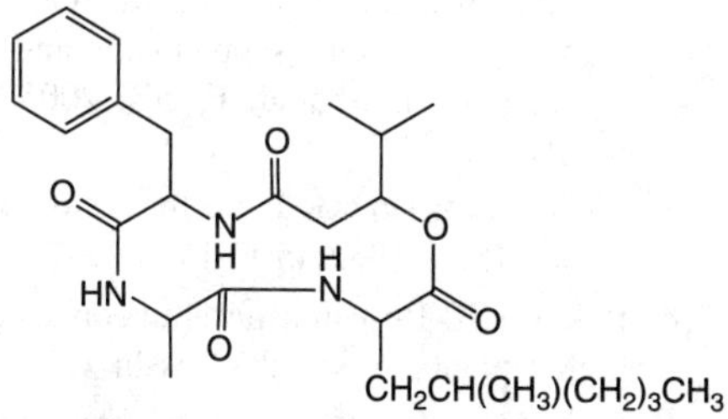

Q5 Cordycepeptide A

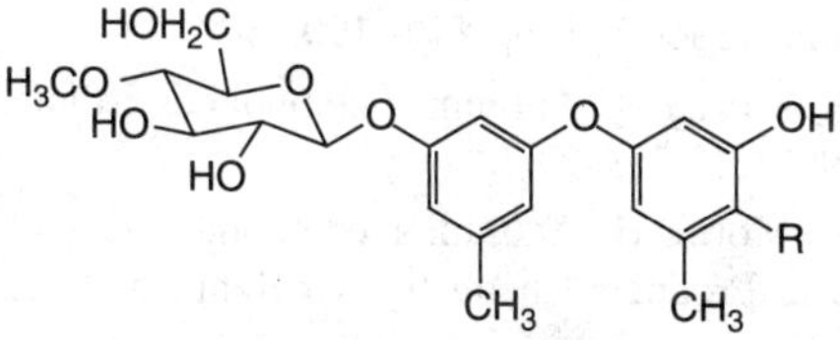

Q6 Cordyol A, R = H
Q7 Cordyol B, R = CO_2CH_3

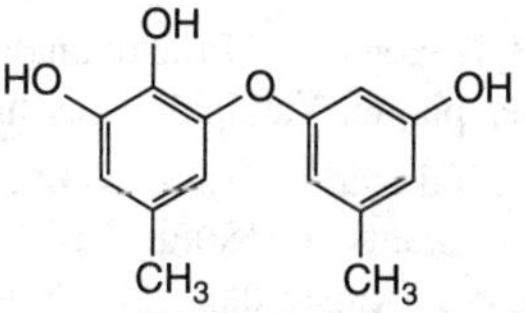

Q8 Cordyol C

Figure 14.20 Structures of *Cordyceps* natural products.

molecular modification will play an important role in developing the new leads into useful drug candidates. Highly efficient bioactivity-directed fractionation and isolation, characterization, analog synthesis, and mechanistic studies are prerequisites for the development of new plant-derived compounds as clinical candidates for world-class new medicines. Drug discovery also will benefit from the discovery of new biologic targets and the continual improvement of bioassay technology (162–169). The development of new, effective, and safe world-class drugs from medicinal plants, which have been appreciated for centuries for treating illness, will be long lasting, as they are the best and most effective source for generating new medicines by use of modern scientific technology.

14.5 ACKNOWLEDGMENT

This investigation was supported by grants from the National Cancer Institute, the National Institutes of Health (NIH) (CA-17625), and the National Institute of Allergy and Infectious Diseases, NIH (AI-33066) that were awarded to K.H.

Lee. This article is No. 257 in the series "Antitumor Agents" and No. 71 in the series "Anti-AIDS Agents."

REFERENCES

1. Balunas MJ, Kinghorn AD. Drug discovery from medicinal plants. Life Sci. 2005;78:431–441.

2. Lee KH, Itokawa H, Kozuka M. Asian herbal products: the basis for development of high-quality dietary supplements and new medicines. In: Asian Functional Foods. Shi J, Ho CT, Shahidi F, eds. 2005.Marcel Dekker/CRC Press, Boca Raton, FL. pp. 21–72.

3. Suffness M, Douros J. Current status of the NCI plant and animal product program. J. Nat. Prod. 1982;45:1–14.

4. Itokawa H. Research on antineoplastic drugs from natural sources especially from higher plants. Yakugaku Zasshi 1988;108:824–841.

5. Lee KH. Antineoplastic agents and their analogues from Chinese traditional medicine. In: Human Medicinal Agents from Plants. Kinghorn AD, Balandrin M, eds. 1993. Amer. Chem. Soc. Symposium Series 534. pp 170–190.

6. Itokawa H, Takeya K, Hitotsuyanagi Y, Morita H. Antitumor compounds isolated from higher plants. Yakugaku Zasshi 1999;119:529–583.

7. Itokawa H, Takeya K, Hitotsuyanagi Y, Morita H. Antitumor compounds isolated from higher plants. In: Studies in Natural Products Chemistry. Ur-Rahman A, ed. 2000. Elsevier, Amsterdam, pp. 269–350.

8. Itokawa H, Takeya K, Lee KH. Anticancer compounds from higher plants. In: Biomaterials from Aquatic and Terrestrial Organisms. Fingerman M, Nagabhushanam R, eds. 2006. Science Publishers, Enfield, NH, pp. 255–283.

9. Tang W, Hemm I, Bertram B. Recent development of antitumor agents from Chinese herbal medicines; Part I. Low molecular compounds. Planta Med. 2003;69:97–108.

10. Tang W, Hemm I, Bertram B. Recent development of antitumor agents from Chinese herbals medicines; Part II. Low molecular compounds. Planta Med. 2003;69:193–201.

11. Lee KH. Current developments in the discovery and design of new drug candidates from plant natural product leads. J. Nat. Prod. 2004;67:273–283.

12. Mukherjee AK, Basu S, Sarkar N, Ghosh AC. Advances in cancer therapy with plant based natural products. Curr. Med. Chem. 2001;8:1467–1486.

13. Cragg GM, Newman DJ. Discovery and development of antineoplastic agents from natural sources. Cancer Invest. 1999;17:153–163.

14. Cragg GM, Newman DJ. A tale of two tumor targets: topoisomerase I and tubulin. The Wall and Wani contribution to cancer chemotherapy. J. Nat. Prod. 2004;67:233–244.

15. Wall ME, Wani MC. Camptothecin and taxol: from discovery to clinic. J. Ethnopharmacol. 1996;51:239–253.

16. Oberlies NH, Kroll DJ. Camptothecin and taxol: historic achievements in natural products research. J. Nat. Prod. 2004;67:129–135.

17. Potier P. The synthesis of navelbine: prototype of a new series of vinblastine derivatives. Sem. Oncol. 1989;16:2–4.

18. Jenks S, Smigel K. Updates: cancer drug approved; new leukemia treatment. J. Nat. Cancer Inst. 1995;87:167–170.

19. Vlahov IR, Santhapuram HKR, Kleindl PJ, Howard SJ, Stanford KM, Leamon CP. Design and regioselective synthesis of a new generation of targeted chemotherapeutics. Part I: ED145, a folic acid conjugate of desacetylvinblastine monohydrazide. Bioorg. Med. Chem. Lett. 2006;16:5093–5096.

20. Okouneva T, Hill BT, Wilson L, Jordan MA. The effects of vinflunine, vinorelbine, and vinblastine on centromere dynamics. Mol. Cancer Ther. 2003;2:427–436.

21. Kruczynski A, Barret JM, Erievant C, Colpaert F, Fahy J, Hill BT. Antimitotic and tubulin-interacting properties of vinflunine, a novel fluorinated Vinca alkaloid. Biochem. Pharmacol. 1998;55:635–648.

22. Wani MC, Tayler HI, Wall ME, Coggon P, McPhail AT. Plant antitumor agents. VI. The isolation and structure of taxol, a novel antileukemia and antitumor agent from Taxus brevifolia. J. Am. Chem. Soc. 1971;93:2325–2327.

23. Itokawa H, Lee KH, eds. Taxus: Genus Taxus. 2003. Taylor & Francis, London.

24. Shigemori H, Kobayashi J. Biological activity and chemistry of taxoids from the Japanese yew, Taxus cuspidate. J. Nat. Prod. 2004;67:245–256.

25. Cragg GM, Suffness M. Metabolism of plant-derived anticancer agents. Pharmcol. Ther. 1988;37:425–461.

26. Kingston DG. The chemistry of taxol. Pharmcol. Ther. 1991;52:1–34.

27. Geney R, Chen J, Ojima I. Recent advances in the new generation taxane anticancer agents. Med. Chem. 2005;1:125–139.

28. Appendino G, Bettoni P, Noncovich A, Sterner O, Fontana G, Bombardelli E, Pera P, Bernack RJ. Structure-activity relationship of ring C-secotaxoids. 1. Acylative modifications. J. Nat. Prod. 2004;67:184–188.

29. Liu C, Strobl JS, Bane S, Schilling JK, McCracken M, Chatterjee SK, Rahim-Bata R, Kingston DGI. Design, synthesis, and bioactivities of steroid-linked taxol analogues as potential targeted drugs for prostate and breast cancer. J. Nat. Prod. 2004;67:152–159.

30. Kuznetsova L, Chen J, Sun L, Wu X, Pepe A, Veith JM, Pera P, Bernacki RJ, Ojima I. Syntheses and evaluation of novel fatty acid-second generation taxoid conjugates as promising anticancer agents. Bioorg. Med. Chem. Lett. 2006;16:974–977.

31. Raez LE, Lilenbaum R. New developments in chemotherapy for advanced non-small cell lung cancer. Curr. Opin. Oncol. 2006;18:156–161.

32. Singer JW, Shaffer S, Baker B, Gbernareggi A, Stromatt S, Nienstedt D, Besman M. Paclitaxel poliglumex (XYOTAX; CT-2103): an intracellularly targeted taxane. Anticancer Drugs 2005;16:243–254.

33. Shi Q, Wang HK, Bastow KF, Tachibana Y, Chen K, Lee FY, Lee KH. Antitumor agents 210. Synthesis and evaluation of taxoid-epipodophyllotoxin conjugates as novel cytotoxic agents. Bioorg. Med. Chem. 2001;9:2999–3004.

34. Ohtsu H, Nakanishi Y, Bastow KF, Lee FY, Lee KH. Antitumor agents 216. Synthesis and evaluation of paclitaxel-camptothecin conjugates as novel cytotoxic agents. Bioorg. Med. Chem. 2003;11:1851–1857.

35. Hennenfent KL, Govindan R. Novel formulations of taxanes: a review. Old wine in a new bottle? Ann. Oncol. 2006;17:735–749.

36. Wall ME, Wani MC, Cook CE, Palmer KH, McPhail AT, Sim GA. Plant antitumor agents. 1. The isolation and structure of camptothecin, a novel alkaloidal leukemia and tumor inhibitor from Camptotheca acuminate. J. Am. Chem. Soc. 1966;88: 3888–3890.

37. Wall ME. In: Chronicles of Drug Discovery, Volume 3. Lednicer D, ed. 1993. American Chemical Society, Washington D.C., pp. 327–348.

38. Covey JM, Jaxel C, Kohn KW, Pommier Y. Protein-linked DNA strand breaks induced in mammalian cells by camptothecin, an inhibitor of topoisomerase I. Cancer Res. 1989;49:5016–5022.

39. Vanhoefer U, Harstrick A, Achterrath W, Cao S, Seeber S, Rustum YM. Irinotecan in the treatment of colorectal cancer: clinical overview. J. Clin. Oncol. 2001;19: 1501–1518.

40. Wang HK, Liu SY, Hwang KM, Taylor G, Lee KH. Synthesis of novel water-soluble 7-(aminoacylhydrazone)-formyl camptothecins with potent inhibition of DNA topoisomerase I. Bioorg. Med. Chem. 1994;2:1397–1402.

41. Dallavalle S, Merlini L, Morini G, Muso L, Penco S, Beretta GL, Tinelli S, Zunino F. Synthesis and cytotoxic activity of substituted 7-aryliminomethyl derivatives of camptothecin. Eur. J. Med. Chem. 2004;39:507–513.

42. Bastow KF, Wang HK, Cheng YC, Lee KH. Antitumor agents 173. Synthesis and evaluation of camptothecin-4β-amino-4′-O-demethyl epipodophyllotoxin conjugates as inhibitors of mammalian DNA topoisomerases and as cytotoxic agents. Bioorg. Med. Chem. 1997;5:1481–1488.

43. O'Leary J, Muggia FM. Camptothecins: a review of their development and schedules of administration. Eur. J. Cancer 1998;34:1500–1508.

44. Garcia-Carbonero R, Supko JG. Current perspectives on the clinical experience, pharmacology, and continued development of the camptothecins. Clin. Cancer Res. 2002;8:641–661.

45. Thomas CJ, Rahier NJ, Hecht SM. Camptothecin: current perspectives. Bioorg. Med. Chem. 2004;12:1585–1604.

46. Du W, Kaskar B, Blumbergs P, Subramanian PK, Curran DP. Semisynthesis of DB-67 and other silatecans from camptothecin by thiol-promoted addition of silyl radicals. Bioorg. Med. Chem. 2003;11:451–458.

47. Clark JW. Rubitecan. Expert Opin. Investig. Drugs 2006;15:71–79.

48. Imbert TF. Discovery of podophyllotoxins. Biochimie 1998;80:207–222.

49. Lee KH, Imakura Y, Haruna M, Beers SA, Thurston LS, Dai HJ, Chen CH, Liu SY, Chan YC. New cytotoxic 4-alkylamino analogues of 4′-demethyl-epipodophyllotoxin as inhibitors of human DNA topoisomerase II. J. Nat. Prod. 1989;52:606–613.

50. Chang JY, Han FS, Liu SY, Wang HK, Lee KH, Cheng YC. Effect of 4-β-arylamino derivatives of 4′-O-demethylpodophyllotoxin on human DNA topoisomerase II, tubulin polymerization. KB cells, and their resistant variants. Cancer Res. 1991;51:1755–1759.

51. Wang ZW, Kuo YH, Schnur D, Bowen JP, Liu SY, Han FS, Chan JY, Chen YC, Lee KH. New 4 β-arylamino derivatives of 4′-O-demethylepipodophyllotoxin and related compounds as potent inhibitors of human DNA topoisomerase II. J. Med. Chem. 1990;33:2660–2666.

52. Cho SJ, Tropsha A, Suffness M, Cheng YC, Lee KH. Three dimensional quantitative structure-activity relationship study of 4'-O-demethylepipodophyllotoxin analogs using the modified CoMFA/q2-GRS approach. J. Med. Chem. 1996;39:1383–1385.

53. Communication between Genelabs Technologies, Inc. and Fossella FV. MD. University of Texas MD Anderson Cancer Center, July 18, 1994.

54. Zheng J, Wang HK, Bastow KF, Zhu XK, Cho SJ, Cheng YC, Lee KH. Antitumor agents. 177. Design, syntheses, and biological evaluation of novel etoposide analogs bearing pyrrolecarboxamidino group as DNA topoisomerase II inhibitors. Bioorg. Med. Chem. Lett. 1997;7:607–612.

55. Zhu XK, Guan J, Tachibana Y, Bastow KF, Cho SJ, Cheng HH, Cheng YC, Gurwith M, Lee KH. Antitumor agents. 194. Synthesis and biological evaluations of 4β-mono-, -di-, and –trisubstituted aniline-4'-O-demethylpodophyllotoxin and related compounds with improved pharmacological profiles. J. Med. Chem. 1999;42: 2441–2446.

56. Xiao Z, Xiao YD, Feng J, Golbraikh A, Tropsha A, Lee KH. Modeling of epipodophyllotoxin derivatives using variable selection k nearest neighbor QSAR method. J. Med. Chem. 2002;45:2294–2309.

57. Gordaliza M, Garcia PA, del Corral JMM, Castro MA, Gomez-Zurita MA. Podophyllotoxin: distribution, sources, application, and new cytotoxic derivatives. Toxicon 2004;44:441–459.

58. Lee KH, Xiao Z. The podophyllotoxins and analogs. In: Antitumor Agents from Natural Sources. Kingston DGI, Cragg GM, Newman DJ, eds. 2005. CRC Press, Boca Raton, FL.

59. Powell RG, Madrigal RV, Smith CR, Mikolajczak KL. Alkaloids of Cephalotaxus harringtonia var. drupacea. 11-Hydroxycephalotaxine and drupacine. J. Org. Chem. 1974;39:676–680.

60. Powell RG, Weisleder D, Smith CR. Antitumor alkaloids for Cephalotaxus harringtonia: structure and activity. J. Pharm. Sci. 1972;61:1227–1230.

61. Powell RG, Weisleder D, Smith CR, Rohwedder WK. Structures of harringtonine, isoharringtonine, and homoharringtonine. Tetrahedron Lett. 1970;11:815–818.

62. Huang CC, Han CS, Yue XF, Shen CM, Wang SW, Wu FG, Xu B. Cytotoxity and sister chromatid exchanges induced in vitro by six anticancer drugs developed in the People's Republic of China. J. Natl. Cancer Inst. 1983;71:841–847.

63. Spencer GF, Plattner RD, Powell RG. Quantitative gas chromatography and gas chromatography-mass spectrometry of Cephalotaxus alkaloids. J. Chromatogr. 1976; 120:335–341.

64. Grem JL, Cheson BD, King SA, Leyland-Jones B, Suffness M. Cephalotaxine esters: antileukemic advance or therapeutic failure? J. Natl. Cancer Inst. 1988;80: 1095–1103.

65. Paudler WW, Kerley GI, McKay J. The alkaloids of Cephalotaxus drupacea and Cephalotaxus fortunei. J. Org. Chem. 1963;28:2194–2197.

66. Luo CY, Tang JY, Wang YP. Homoharringtone: a new treatment option for myeloid leukemia. Hematology 2004;9:259–270.

67. Kantarjian HM, Cortes J. New strategies in chronic myeloid leukemia. Intl. J. Hematol. 2006;83:289–293.

68. Itokawa H, Wang X, Lee KH. Homoharringtonine and related compounds. In: Antitumor Agents from Natural Sources. Kingston DGI, Cragg GM, Newman DJ, eds. 2005. CRC Press, Boca Raton, FL.

69. Caprard HG, Brossi A. Tropolonic colchicum alkaloids. In: The Alkaloids. Brossi A, ed. 1984. Academic Press, New York. pp. 48–54.

70. Shi Q, Verdier-Pinard P, Brossi A, Hamel E, McPhail AT, Lee KH. Antitumor agents 172. Synthesis and biological evaluation of novel deacetamidothiocolchicin-7-ols and ester analogs as antitubulin agents. J. Med. Chem. 1997;40:961–966.

71. Shi Q, Chen K, Brossi A, Verdier-Pinard P, Hamel E, McPhail AT, Lee KH. Antitumor agents 184. Syntheses and antitubulin activity of compounds derived from reaction of thiocolchicone with amines: lactams, alcohols, and esters analogs of allothiocolchicinoids. Helv. Chim. Acta 1998;81:1023–1037.

72. Guan J, Zhu XK, Tachibana Y, Bastow KF, Brossi A, Hamel E, Lee KH. Antitumor agents. 185. Synthesis and biological evaluation of tridemethylthiocolchicine analogues as novel topoisomerase II inhibitors. J. Med. Chem. 1998;41:1956–1961.

73. Bastow KF, Tatematsu H, Bori ID, Fukushima Y, Sun L, Goz G, Lee KH. Induction of reversible protein-linked DNA breaks in human osteogenic sarcoma cells by novel cytocidal colchicine derivatives which inhibit DNA topoisomerase II in vitro: absence of cross-resistance in a colchicine-resistant sub-clone. Bioorg. Med. Chem. Lett. 1993;3:1045–1050.

74. Wu WL, Chang WL, Chen CF. Cytotoxicity activities of tanshinones against human crcinoma cell lines. Am. J. Chinese Med. 1991;14:207–216.

75. Ryu SY, Lee CO, Choi SU. In vitro cytotoxicity of tanshinones from Salvia miltiorrhiza. Planta Med. 1997;63:339–342.

76. Sugiyama A, Zhu BM, Takahara A, Satoh Y, Hashimoto K. Cardiac effects of Salvia miltoirrhiza/Dalbergia odorifera mixture, an intravenously applicable Chinese medicine widely used for patients with ischemic heart disease in China. Circulat. J. 2002;66:182–184.

77. Li HB, Chen F. Preparative isolation and purification of six diterpenoids from the Chinese medicinal plant Salvia miltiorrhiza by high-speed counter-current chromatography. J. Chromatogr. A 2001;925:109–114.

78. Chang J, Xu J, Li M, Zhao M, Ding J, Zhang JS. Novel cytotoxic seco-abietane rearranged diterpenoids from Salvia prionitis. Planta Med. 2005;71:361–366.

79. Liu J, Shen HM, Ong CN. Salvia militiorrhiza inhibits cell growth and induces apoptosis in human hepatoma HepG(2) cells. Cancer Lett. 2000;153:85–93.

80. Wang X, Bastow KF, Sun CM, Lin YL, Yu HJ, Don MJ, Wu TS, Nakamura S, Lee KH. Antitumor Agents. 239. Isolation, structure elucidation, total synthesis, and anti-breast cancer activity of neo-tanshinlactone from Salvia miltiorrhiza. J. Med. Chem. 2004;47:5816–5819.

81. Wang X, Nakagawa-Goto K, Bastow KF, Don MJ, Lin YL, Wu TS, Lee KH. Antitumor agents. 254. Synthesis and biological evaluation of novel neo-tanshinlactone analogues as potent anti-breast cancer agents. J. Med. Chem. 2006;49:5631–5634.

82. Itokawa H, Tsuruoka S, Takeya K, Mori N, Sonobe T, Kosemura S, Hamanaka T. An antitumor morphinane alkaloid, sinococuline, from Cocculus trilobus. Chem. Pharm. Bull. 1987;35:1660–1662.

83. Liu WK, Wang XK, Che CT. Cytotoxic effects of sinococuline. Cancer Lett. 1996;99:217–224.

84. Kupchan SM, Komoda Y, Court WA, Thomas GJ, Smith RM, Karim A, Gilmore CJ, Haltiwanger RC, Bryan RF. Maytansine, a novel antileukemic ansa macrolide from Maytenus ovatus. J. Am. Chem. Soc. 1972;94:1354–1356.

85. Kupchan SM, Sneden AT, Branfman AR, Howie GA, Rebhun LI, McIvor WE, Wang RW, Schanitman TC. Structural requirements for antileukemic activity among the naturally occurring and semisynthetic maytansinoids. J. Med. Chem. 1978;21:31–37.

86. Nakao H, Senokuchi K, Umebayashi C, Kanemaru K, Masuda T, Oyama Y, Tonemori S. Cytotoxic activity of maytanprine isolated from M. diversifolia in human leukemia K562 cells. Biol. Pharm. Bull. 2004;27:1236–1240.

87. Itokawa H, Shirota O, Ikuta H, Morita H, Takeya K, Iitaka Y. Triterpenes from rhizomes of Maytenus ilicifolia. Phytochemistry 1991;30:3713–3716.

88. Itokawa H, Shirota O, Morita H, Takeya K, Tomioka N, Iitaka Y. Triterpene dimers from Maytenus ilicifolia. Tetrahedron Lett. 1990;31:6881–6882.

89. Shirota O, Morita H, Takeya K, Itokawa H. Revised structures of cangorosins, triterpene dimers from Maytenus ilicifolia. J. Nat. Prod. 1997;60:111–115.

90. Kuo YH, Chen CH, Kuo LM, King ML, Wu TS, Haruna M, Lee HK. Antitumor agents 112. Emarginatine B, a novel potent cytotoxic sesquiterpene pyridine alkaloid from Maytenus emarginata. J. Nat. Prod. 1990;53:422–428.

91. Kuo YH, King ML, Chen CF, Chen HY, Chen CH, Chen K, Lee HK. Two new macrolide sesquiterpene pyridine alkaloids from Maytenus emarginata: emarginatine G and the cytotoxic emarginatine F. J. Nat. Prod. 1994;57:262–269.

92. Shirota O, Sekita S, Satake M, Morita H, Takeya K, Itokawa H. Nine regioisomeric and stereoisomeric triterpene dimers from Maytenus chuchuhuasca. Chem. Pharm. Bull. 2004;52:739–746.

93. Shirota O, Sekita S, Satake M, Morita H, Takeya K, Itokawa H. Two new sesquiterpene pyridine alkaloids from Maytenus chuchuhuasca. Heterocycle 2004;63:1891–1896.

94. Ohsaki A, Imai Y, Naruse M, Ayabe S, Komiyama K, Takashima J. Four new triterpenoids from Maytenus ilicifolia. J. Nat. Prod. 2004;67:469–471.

95. Gonzalez AG, Tincusi BM, Bazzocchi IL, Tokuda H, Nishino H, Konoshima Y, Jimenez IA, Ravelo AG. Anti-tumor promoting effects of sesquiterpenes from Maytenus cuzcoina (Celastraceae). Bioorg. Med. Chem. 2000;8:1773–1778.

96. Aggarwal BB, Kumar A, Bharti AC. Anticancer potential of curcumin: preclinical and clinical studies. Anticancer Res. 2003;23:363–398.

97. Shishodia S, Sethi G, Aggarwal BB. Curcumin: getting back to the roots. Ann. NY Acad. Sci. 2005;1056:206–217.

98. Singh S, Khar A. Biological effects of curcumin and its role in cancer chemoprevention and therapy. Anticancer Agents Med. Chem. 2006;6:259–270.

99. Ohtsu H, Xiao Z, Ishida J, Nagai M, Wang HK, Itokawa H, Su CY, Shih C, Lee Y, Tsai MY, Chang C, Lee KH. Antitumor agents 217. Curcumin analogues as novel androgen receptor antagonists with potential as anti-prostate cancer agents. J. Med. Chem. 2002;45:5037–5042.

100. Ohtsu H, Itokawa H, Su CY, Shih C, Chiang T, Chang E, Lee YF, Chiu SY, Chang C, Lee KH. Antitumor agents 222. Synthesis and anti-androgen activity of new diarylheptanoids. Bioorg. Med. Chem. 2003;11:5083–5090.

101. Lin L, Shi Q, Su CY, Shih CCY, Lee KH. Antitumor agents 247. New 4-ethoxy-carbonylethylcurcumin analogs as potential antiandrogenic agents. Bioorg. Med. Chem. 2006;14:2527–2534.

102. Lin L, Shi Q, Nyarko AK, Bastow KF, Wu CC, Su CY, Shih CCY, Lee KH. Antitumor agents 250. Design and synthesis of new curcumin analogues as potential anti-prostate cancer agents. J. Med. Chem. 2006;49:3963–3972.

103. Lin L, Lee KH. Structure-activity relationships of curcumin and its analogs with different biological activities. In: Studies in Natural Products Chemistry, Volume 33. Ur-Rahman A, ed. Elsevier, New York, pp. 785–812.

104. Lee KH. Antineoplastic agents and their analogues from Chinese traditional medicine. In: Human Medicinal Agents from Plants. Kinghorn AD, Balandrin M, eds. 1993. Amer. Chem. Soc. Symposium Series 534, pp. 170–190.

105. Wu TS, Lin YM, Haruna M, Pan DJ, Shingu T, Chen YP, Hsu HY, Nakano T, Lee KH. Antitumor agents, 119. Kansuiphorins A and B, two novel antileukemic diterpene esters from Euphorbia kansui. J. Nat. Prod. 1991;54:823–829.

106. Pan DJ, Hu CQ, Chang JJ, Lee TTY, Chen YP, Hsu HY, McPhail DR, McPhail AT, Lee KH. Kansuiphorin-C and -D, cytotoxic diterpenes from Euphorbia kansui. Phytochemistry 1991;30:1020–1023.

107. Nunomura S, Kitanaka S, Ra C. 3-O-(2,3-Dimethylbutanoyl)-13-O-decanoylingenol from Euphorbia kansui suppresses IGE-mediated mast cell activation. Biol. Pharm. Bull. 2006;29:286–290.

108. Matsumoto T, Cyong JC, Yamada H. Stimulatory effects of ingenols from Euphorbia kansui on the expression of macrophage Fc receptor. Planta Med. 1992;58:255–258.

109. Shi HM, Williams ID, Sung HHY, Zhu HX, Ip NY, Min ZD. Cytotoxic diterpenoids from the roots of Euphorbia ebracteolata. Planta Med. 2005;71:349–354.

110. Fu GM, Qin HL, Yu SS, Yu BY. Yuexiandajisu D, a novel 18-nor-rosane-type dimeric diterpenoid from Euphorbia ebracteolata Hayata. J. Asian Nat. Prod. Res. 2006;8:29–34.

111. Cos P, Maes L, Berghe DV, Hermans N, Pieters L, Vlietinck A. Plant substances as anti-HIV agents selected according to their putative mechanism of action. J. Nat. Prod. 2004;67:284–293.

112. Lee TT, Kashiwada Y, Huang L, Sneider J, Cosentino M, Lee KH. Suksdorfin: an anti-HIV principle from Lomatium suksdorfii, its structure-activity correlation with related coumarins, and synergistic effects with anti-AIDS nucleosides. Bioorg. Med. Chem. 1994;2:1051–1056.

113. Huang L, Kashiwada Y, Cosentino LM, Fan S, Chen CH, McPhail AT, Fujioka T, Mihashi K, Lee KH. Anti-AIDS agents 15. Synthesis and anti-HIV activity of dihydroseselins and related analogs. J. Med. Chem. 1994;37:3947–3955.

114. Yu D, Suzuki M, Xie L, Morris-Natschke SL, Lee KH. Recent progress in the development of coumarin derivatives as potent anti-HIV agents. Med. Res. Rev. 2003;23:322–345.

115. Yu D, Chen CH, Brossi A, Lee KH. Anti-AIDS agents 60. Substituted 3′R,4′R-di-O-(−)-camphanoyl-2′2′-dimethyldihydropyrano[2,3-f.chromone (DCP) analogs as potent anti-HIV agents. J. Med. Chem. 2004;47:4072–4082.

116. Yu D, Lee KH. Anti-AIDS agents 63. Recent progress and prospects on plant-derived anti-HIV agents and analogs. In: Medicinal Chemistry of Bioactive Natural Products. Liang XT, Fang WS, eds. 2006 Wiley, Inc., Hoboken, NJ, pp. 357–398.

117. Kashiwada Y, Hashimoto F, Cosentino LM, Chen CH, Garrren PE, Lee KH. Beturinic acid and dihydrobetulinic acid derivatives as potent anti-HIV agents. J. Med. Chem. 1996;39:1016–1017.

118. Fujioka T, Kashiwada Y, Kilkuskie RE, Cosentino LM, Ballas LM, Jiang JB, Janzen WP, Chen IS, Lee KH. Anti-AIDS agents 11. Betulinic acid and platonic acid as anti-HIV principles from Syzigium claviflorum, and the anti-HIV activity of structurally related triterpenoids. J. Nat. Prod. 1994;57:243–247.

119. Lee KH, Kashiwada Y, Hashimoto F, Cosentino LM, Manak M. Betulinic acid derivatives and antiviral use. University of North Carolina, at Chapel Hill and Biotech, Research Laboratories, PCT Int. Appl. WO 9639033. December 12, 1996.

120. Sun IC, Kashiwada Y, Morris-Natschke SL, Lee KH. Plant-derived terpenoids and analogues as anti-HIV agents. Curr. Top. Med. Chem. 2002;3:155–169.

121. Li F, Goila-Gaur R, Salzwedel K, Kilgore NR, Reddick M, Matallana C, Castillo A, Zoumplis D, Martin DE, Orenstein JM, Allaway FP, Freed EO, Wild CT. PA-457: a potent HIV inhibitor that disrupts core condensation by targeting a late step in Gag processing. Proc. Natl. Acad. Sci. U.S.A. 2003;100:13555–13560.

122. Results of the Phase IIa study were presented as an oral late breaker presentation at the 45th Interscience Conference on Antimicrobial Agents and Chemotherapy, 2005.

123. Triangle Business Journal, September 9, 2005.

124. Data provided by Panacos, Inc.

125. Yu D, Wild CT, Martin DE, Morris-Natschke SL, Chen CH, Allaway G, Lee KH. Anti-AIDS agents 64. The discovery of maturation inhibitors and their potential in the therapy of HIV. Expert Opin. Investig. Drugs 2005;14:681–693.

126. Yu D, Lee KH. Anti-AIDS agents 63. Recent progress and prospects on plant-derived anti-HIV agents and analogs. In: Medicinal Chemistry of Bioactive Natural Products. Liang XT, Fang WS, eds. 2006. John Wiley & Sons, Hoboken, NJ, pp. 357–398.

127. Yu D, Morris-Natschke SL, Lee KH. Anti-AIDS agents 67. New developments in natural products-based anti-AIDS research. Med. Res. Rev. 2007;27:133–148.

128. Li Y, Huang H, Wu YL. Qinghaosu (Artemisinin) - A fantastic antimalarial drug from a traditional Chinese medicine. In: Medicinal Chemistry of Bioactive Natural Products. Liang XT, Fang WS, eds. 2006. John Wiley & Sons, Hoboken, NJ, pp 183–256.

129. Meshnick SR. In: Antimalarial Chemotherapy. Rosenthal PJ, ed. 2001. Humana Press, Totowa, NJ, pp 191–201.

130. Avery MA, McLean G, Edwards G, Ager A. Structure–activity relationships of peroxide-based artemisinin antimalarials. In: Biologically Active Natural Products: Pharmaceuticals. Cutler SJ, Cutler HG, eds. 2000. CRC Press, Boca Raton, FL, pp. 121–132.

131. Pareek A, Nandy A, Kochar D, Patel KH, Mishra SK, Mathur PC. Efficacy and safety of β-arteether and α/β-arteether for treatment of acute Plasmodium falciparum malaria. Am. J. Trop. Med. 2006;75:139–142.

132. http://mednet3.who.int/EMLib/ .

133. Imakura Y, Yokoi T, Yamagishi T, Koyama J, Hu H, McPhail DR, McPhail AT, Lee KH. Synthesis of deethanoqinghaosu, a novel analog of the antimalarial qinghaosu. J. Chem. Soc. Commun. 1988;372.

134. Vennerstrom JL, Arbe-Barnes S, Brun R, Charman SA, Chiu FC, Chollet J, Dong Y, Dorn A, Hunziker D, Matile H, McIntosh K, Padmanilayam M, Santo Tomas J, Scheurer C, Scorneaux B, Tang Y, Urwyler H, Wittlin S, Charman WN. Identification of an antimalarial synthetic trioxolane drug development candidate. Nature 2004;430:900–904.

135. Dong Y, Chollet J, Matile H, Charman SA, Chiu FC, Charman WN, Scorneaux B, Urwyler H, Santo Tomas J, Scheurer C, Snyder C, Dorn A, Wang X, Karle JM, Tang Y, Wittlin S, Brun R, Vennerstrom JL. Spiro and dispiro-1,2,4-trioxolanes as antimalarial peroxides: charting a workable structure-activity relationship using simple prototypes. J. Med. Chem. 2005;48:4953–4961.

136. Perry CS, Charman SA, Prankerd RJ, Chiu FC, Dong Y, Vennerstrom JL, Charman WN Chemical kinetics and aqueous degradation pathways of a new class of synthetic ozonide antimalarials. J. Pharm. Sci. 2006;95:737–747.

137. Dong Y, Tang Y, Chollet J, Matile H, Wittlin S, Charman SA, Charman WN, Tomas JS, Scheurer C, Snyder C, Scorneaux B, Bajpai S, Alexander SA, Wang X, Padmanilayam M, Cheruku SR, Brun R, Vennerstrom JL. Effect of functional group polarity on the antimalarial activity of spiro and dispiro-1,2,4-trioxolanes. Bioorg. Med. Chem. 2006;14:6368–6382.

138. Haynes RK. From artemisinin to new artemisinin antimalarials: Biosynthesis, extraction, old and new derivatives, stereochemistry and medicinal chemistry requirements. Curr. Top. Med. Chem. 2006;6:509–537.

139. Shibata S, Fujita M, Itokawa H, Tanaka O. Studies on the constituents of Japanese and Chinese crude drugs. XI. Panaxadiol, a sapogenin of Ginseng roots. Chem. Pharm. Bull. 1963;11:759.

140. Shibata S. Chemistry and cancer preventing activities of ginseng saponins and some related triterpenoid compounds. J. Korean Med. Sci. 2001;16(suppl S):S28–S37.

141. Tanaka O. Ginseng and its congeners - Traditional oriental food drugs. In: Food Phytochemicals for Cancer Prevention II, ACS Symposium Series 1994;547:335–341.

142. Lai DM, Tu YK, Liu IM, Chen PF, Cheng JT. Mediation of β-endorphin by ginsenoside Rh2 to lower plasma glucose in streptozotocin-induced diabetic rats. Planta Med. 2006;72:9–13.

143. Yun TK, Choi SY. The relationship between cancer and ginseng intake. Intl. J. Epidemiol. 1990;19:871–876.

144. Ellis JM, Reddy P. Effects of Panax ginseng on quality of life. Ann. Pharmacother. 2002;36:375–379.

145. Yun TK, Lee YS, Lee YH, Kim SI, Yun HY. Anticarcinogenic effect of Panax ginseng CA Meyer and identification of active compounds. J. Korean Med. Sci. 2001;16(suppl S)S6–S18.

146. Attele AS, Wu JA, Yuan CS. Ginseng pharmacology. Biochem. Pharmacol. 1999;58: 1685–1693.

147. Surh YJ, Na HK, Lee JY, Keum YS. Molecular mechanisms underlying anti-tumor promoting activities of heat-processed Panax ginseng CA Meyer. J. Korean Med. Sci. 2001;16(suppl S):S38–S41.

148. Sun H, Ye Y, Pan Y. Immunological-adjuvant saponins from the roots of Panax notoginseng. Chem. Biodivers. 2005;2:510–515.

149. Sun H, Chen Y, Ye Y. Ginsenoside Re and notoginsenoside R$_1$: immunological adjuvants with low haemolytic effect. Chem. Biodivers. 2006;3:718–726.

150. Xie GX, Qiu YP, Qiu MF, Gao XF, Liu YM, Jia W. Analysis of dencichine in Panax notoginseng by gas chromatography-mass spectrometry with ethyl chloroformate derivatization. J. Pharm. Biomed. Anal. 2007;43:920–925.

151. Kimura Y, Sumiyoshi M. Effects of various Eleutherococcus senticosus cortex on swimming time, natural killer activity and corticosterone level in forced swimming stressed mice. J. Ethnopharmacol. 2004;95:447–453.

152. Li XC, Barnes DL, Khan IA. A new lignan glycoside from Eleutherococcus senticosus. Planta Med. 2001;67:776–778.

153. Bao XF, Zhen Y, Ruan L, Fang JM. Purification, characterization, and modification of T lymphocyte-stimulating polysaccharides from spores of Ganoderma lucidum. Chem. Pharm. Bull. 2002;50:623–629.

154. Bao XF, Wang XS, Dong Q, Fang JN, Li XY. Structural features of immunologically active polysaccharides from Ganoderma lucidum. Phytochemistry 2002;29:175–181.

155. Lin YL, Lee SS, Lou SM, Chiang BL. Polysaccharide purified from Ganoderma lucidum induces gene expression changes in human dendritic cells and promotes T helper 1 immune response in BALB/c mice. Mol. Pharmacol. 2006;70:637–644.

156. Shiao MS. Natural products of the medicinal fungus Ganoderma lucidum: occurrence, biological activities, and pharmacological functions. Chem. Rec. 2003;3:172–180.

157. Wang XM, Yang M, Guan SH, Liu RX, Xia JM, Bi KS, Guo DA. Quantitative determination of six major triterpenoids in Ganoderma lucidum and related species by high performance liquid chromatography. J. Pharm. Biomed. Anal. 2006;41:838–844.

158. Ng TB, Wang HX. Pharmaceutical actions of Cordyceps, a prized folk medicine. J. Pharm. Pharmacol. 2005;57:1509–1519.

159. Li SP, Yang FQ, Tsim KWK. Quality control of Cordyceps sinensis, a valued traditional Chinese medicine. J. Pharm. Biomed. Anal. 2006;41:1571–1584.

160. Ho HJ, Cho JY, Rhee MH, Park HJ. Cordycepin (3'-deoxyadenosine) inhibits human platelet aggregation in a cyclic AMP- and cyclic GMP-dependent manner. Eur. J. Pharmacol. 2007;558:43–51.

161. Bunyapaiboonsri AT, Yoiprommarat S, Intereya K, Kocharin K. New diphenyl ethers from the insect pathogenic fungus Cordyceps sp. BCC 1861. Chem. Pharm. Bull. 2007;55:304–307.

162. Kaelin WG Jr. Taking aim at novel molecular targets in cancer therapy. J. Clin. Invest. 1999104:1495.

163. Keshet E, Ben-Sasson SA. Anticancer drug targets: approaching angiogenesis. J. Clin. Invest. 1999;104:1497–1501.

164. Kaelin WG Jr. Choosing anticancer drug targets in the postgenomic era. J. Clin. Invest. 1999;104:1503–1506.

165. Shapiro GI, Harper JW. Anticancer drug targets: cell cycle and checkpoint control. J. Clin. Invest. 1999;104:1645–1653.

166. Sellers WR, Fisher DE. Apoptosis and cancer drug targeting. J. Clin. Invest. 1999; 104:1655–1661.

167. Gibbs JB. Anticancer drug targets: growth factors and growth factor signaling. J. Clin. Invest. 2000;105:9–13.

168. Reddy A, Kaelin WG Jr. Using cancer genetics to guide the selection of anticancer drug targets. Curr. Opin. Pharmacol. 2002;2:366–373.

169. Cummings J, Ward TH, Ranson M, Dive C. Apoptosis pathway-targeted drugs–from the bench to the clinic. Biochim. Biophys. Acta 2004;1705:53–66.

INDEX

Natural Products in Chemical Biology, First Edition. Edited by Natanya Civjan.
© 2012 John Wiley & Sons, Inc. Published 2012 by John Wiley & Sons, Inc.